Humidification and Ventilation Management in Textile Industry

Humidification and Ventilation Management in Textile Industry

B. Purushothama

CRC Press
Taylor & Francis Group
Boca Raton London New York

CRC Press is an imprint of the
Taylor & Francis Group, an **informa** business

WOODHEAD PUBLISHING INDIA PVT LTD

New Delhi

First published 2024
by CRC Press
4 Park Square, Milton Park, Abingdon, Oxon, OX14 4RN

and by CRC Press
2385 NW Executive Center Drive, Suite 320, Boca Raton FL 33431

Routledge is an imprint of the Taylor & Francis Group, an informa business

British Library Cataloguing-in-Publication Data
A catalogue record for this book is available from the British Library

Print edition not for sale in South Asia (India, Sri Lanka, Nepal, Bangladesh, Pakistan or Bhutan).

ISBN13: 9781032630168 (hbk)
ISBN13: 9781032630182 (pbk)
ISBN13: 9781032630199 (ebk)

DOI: 10.4324/9781032630199

Typeset in Times New Roman
by Bhumi Graphics, New Delhi

WPI

Contents

Preface

The importance of maintaining and managing humidity and temperature in a textile mill is not a new concept, but understanding the requirements, the equipment capabilities and utilising them to get the best results is a challenge the technicians face all the time. Now the systems have changed from manually operated to fully automatic; however, unless one knows how to monitor, he shall still have the problem. Some guidelines are needed for the shop floor technicians relating to maintenance of humidity and monitoring the air.

The idea of writing a book on Managing of Humidity came from my friend Mr. Ananth Harnahalli. I hesitated first, as I was only a Textile Technologist and not a Humidification Engineer. Mr. Ananth reminded me that I had faced lot of problems due to improper maintenance of humidity while working on the shop floor and had struggled a lot to get correct conditions. The problems faced were always unique as the systems were getting changed, materials were getting changed and also the working conditions. Mr. Ananth told that the purpose of this book should be to guide the shop floor technicians and engineers in maintaining required conditions and to act in advance. They need the basic concepts and the choice available in the market to update their humidification plants, hence this book.

An attempt was made to collect and provide information starting from the basic concepts, developments, varying needs of the industry, the problems associated with maintenance of plants to get the required conditions, designing of plant capacity, modification or designing of building to get the best results, various issues of health and hygiene, the pollution control issues, various models available in the market, etc. Although this book was written by keeping textile technicians in mind, the use of air conditioning and humidification examples of other industries are also given as a comparison and for easy understanding of the concepts.

Woodhead Publication India came forward and published the book in 2009. There was a good response from the industry for this book, and hence a decision was taken to revise the book and bring out the second edition considering the developments taken place in last one decade. I hope this second edition shall be more useful to the industry and expecting the same support by all readers.

I am thankful for all the information providers, without which this book would not have come out. Also I am thankful to all my friends who encouraged me to write this book, and my family members for their complete cooperation. My special thanks are to the management and staff of Woodhead Publishing India, who not only published the first edition of this book but also published number of my books relating to textiles and management since 2009.

B. Purushothama

1

Need for maintaining humidity in textile industry

1.1 Introduction

Spinning of yarn from cotton fibres and weaving or knitting fabric from the yarn is known to mankind from time immemorial. It is probably one of the first crafts developed as man thought of civilization. As this primitive craft flourished in tune with civilization, the ancients learnt that if water is sprayed on the floor on the hot days or if wet cloth is kept over the warp, the working is easier due to less yarn breakage. This was the state of affairs until two hundred years back, but as the industrial revolution took place and mass production was aimed, different methods were found to give moisture to the material in process. As more people were concentrated under one roof along with number of machines working with a speed, the generation of temperature also added to the problem. Though frequent use of water cans for humidity and wide windows for fresh air were provided, it did not help in improving the working. The increase in speed of machinery also liberated fibrous dust.

Air is an important element for every human being. The capacity for work and general health might be seriously impaired by defective ventilation. The purity of air, the temperature and the movement of air are few of the many factors to be considered. The comfort of an occupied space depends as much on its condition, as on its freshness. Although humidity is invisible to our eyes, we can easily observe its effects. Human are more comfortable and efficient with proper humidification. In business and industrial environments, the performance of equipment and materials is enhanced by effective humidity control. Maintaining indoor air quality through humidity management can lower energy costs, increase productivity, save labour and maintenance costs and ensure product quality. Controlled humidification helps protect humidity-sensitive materials, personnel, delicate machinery and equipment. Beyond the important issues of comfort and process control, humidity control can help safeguard against explosive atmospheres. In short, humidification can provide a better environment and improve the quality of life and work.

Air, free from dirt debris and fibres, that is closely maintained within fixed limits of temperature and humidity, is a vital necessity to the textile industry. It is not only because of the changes in dielectric properties and tensile properties of fibres due to varying humidity and temperature but also for maintaining clean working environment. The occurrence of static can be a major problem when processing textiles, and it is directly related to levels of relative humidity. The electrical sensitivity that determines whether static electrification will occur is dependent on the moisture content of the air and fibres. As the fibres lose moisture, they increase their electrical resistance. This means they can no longer easily dissipate the electrical charge which is generated by the frictional contact with the machinery. The generation of static electricity while processing in spinning and weaving creates dust and fibre fly (fluff). Higher moisture content lowers the insulation resistance and helps to carry off the electrostatic charge. Hence relative humidity needs to be maintained above the lower limit of RH range, specified for various textile processes so as to avoid the problems of yarn breakage in dry and brittle condition and also minimise the buildup of static charge so as to reduce dust and fibre fly. By maintaining humidity at around 50%RH, static build-up is eliminated, and all these associated problems are avoided.

Another advantage of maintaining the correct humidity in processing facilities is that it reduces airborne particles. A higher humidity encourages airborne lint, dust and fly to precipitate out of the atmosphere. Also, if a cold water humidification system is used, the evaporation of the water into the air causes an adiabatic cooling effect that can reduce ambient temperatures by between 2 and 6°C. These additional benefits of using humidifiers create a healthier, less polluted, more pleasant atmosphere for workers and a more productive workforce.

1.2. Relation of humidity to working in the textile mills

Correct ambient conditions are essential to prevent degradation of textile materials upon which a series of operations right from beating in blow room to weaving fabric at loom shed or knitting the fabric or producing nonwoven sheets. Fibres should have requisite properties so that the final product retains its basic shape, size and strength. Above certain moisture limit, i.e., above the upper limit of relative humidity for the fibre and the process, fibres tend to stick and lead to formation of laps on the rolls, which disrupt the production process. Removal of laps is not only a manual and time-consuming process but also results in the damage of machine parts, especially the rubber coatings. Fibres become brittle and store electric charges generated because of friction between the fibres during their individualization process when atmospheric relative humidity is very low. In case of weaving, as the warp yarns are coated with size film, the environment should be suitable for the size film on the yarn. Too low a humidity makes size film brittle resulting in cracking of the film, whereas a too high humidity makes the beam soft.

Modern spinning equipment are designed to operate at high speeds; however, the increase in ambient temperature curtails the speed limits of operation. Moreover, the sophisticated electronic controls in modern textile machinery also require controlled temperature, which should not exceed 33°C or so. It is also necessary to limit the range of temperature to which the textile machinery is exposed, since the structure of the machinery containing many steel and aluminum parts, which expand at different rates with temperature rise (due to difference in coefficient of thermal expansion) will be subjected to mechanical stress. Hence, along with maintenance of stable relative humidity conditions recommended for different textile processes, it is also desirable to maintain the temperature level within a range, without fluctuation.

Mechanical properties of fibres and yarns also depend on the surrounding temperature conditions to which these are exposed during the textile process. Apart from the dust levels, the stickiness in some of the cottons also demands controlled weather. When cotton is sticky, higher humidity creates sticking of fibres to rollers and other parts of the machine. The general reasons for controlling temperature and humidity in a textile mill are as follows.

- Dry air causes lower moisture regain that contributes to lower yarn strength, poor quality and lower productivity.
- Yarns with low moisture content are weaker, more brittle and less elastic, create more friction and are more prone to static electrification.
- Materials at optimum regain are less prone to breakage, heating and friction effects, have fewer imperfections, are more uniform and feel better.
- Higher humidity reduces static problems. Reduced static makes materials more manageable and increases machine speeds.
- Textile weights are standardized at 65% RH and 20°C. Low humidity causes lower material weight and lowered profits.
- Low humidity causes fabric shrinkage. Maintained humidity permits greater reliability in cutting and fitting during garment creation and contributes to the maintenance of specification where dimensions are important, such as in the carpet industry.
- Humidification reduces fly and micro-dust, giving a healthier and more comfortable working environment.

Adequate yarn humidity (moisture in yarn) is needed to enhance the strength and the elasticity and to have smooth yarn surface. Both tensile strength and elasticity depend on fibre and spinning characteristics, on warp pretreatment (slashing) and increase with moisture content of the yarn being fed into the weaving process. Hairiness depends on the spinning system, speed, humidity and the fibre quality. Higher spindle speed, lower humidity and abrasion while spinning are the major reasons for increased hairiness. It is reduced by slashing, as fibres protruding from the yarn are glued to it. Moisture content smoothens the hairs and lubricates the yarn surface. Abrasion between yarns, mainly in the shed area, removes short fibres (lint) and size dust from the warp yarn. Adequate yarn moisture reduces the fall out.

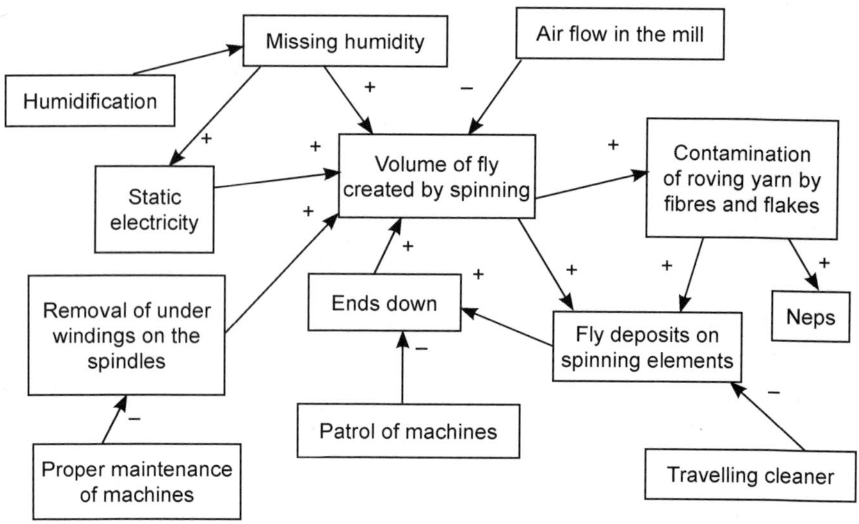

Figure 1.1 Influence of fibre fly on the spinning process

Fig. 1.1 is a relation diagram showing the influence of fibre fly on spinning process as developed by Luwa. The + and – marks near the arrows explain whether the effect will increase or decrease.

While weaving, the yarn adsorbs water from the air. Lint and dust falling out from the yarn are incorporated into the room air. Power consumed by the loom and other devices in the room is converted into heat and incorporated into the room air. This heat evaporates the moisture from yarn. Experience shows that yarns perform best in weaving machines when their moisture content is 7–9% (parts of water in 100 parts of dry yarn). Less moisture reduces strength, elasticity and smoothness. Higher moisture may make the size glue the warp yarns together. Therefore, there is a need to humidify the area with suitable controls.

Maintaining sufficiently high RH levels provides the most effective and economical means of preventing the build-up of static charges. With high RH, an invisible film of moisture forms on surfaces in the room. The presence of normal impurities makes this film a conductor that carries static electricity harmlessly to the ground before it can harm. RH of at least 45% is needed to reduce or prevent the accumulation of static charges, although some materials such as wool and certain synthetic fabrics may require higher RH levels. Similarly, heat-generating machines may require higher RH to provide sufficient moisture in proximity to the machine to dissipate static charges.

Due to high heat dissipation from spinning as well as weaving and knitting equipment, there is a significant increase in temperature conditions particularly in the vicinity of the machinery and their driving motors. The natural wax covering cotton fibres softens at these raised temperature conditions, thereby adversely affecting the lubricating property of wax for controlling static and dynamic friction.

Increase in temperature beyond the design limit also reduces the RH condition near the processing elements of the machinery. Hence textile air-engineering design has to take care of controlled air flow within the textile machinery for dissipating heat generated at the source, and it is customary to carry the waste heat along with the return air to the return air trench. The quantity of return air going to exhaust or recirculation is regulated for controlling the inside design conditions.

The requirement of RH is lower at blow room at around 45 to 50%, moderate at spinning processes from carding to ring spinning at around 55%, around 65% in winding and warping, whereas weaving rooms need high RH of 80% to 85% at the warp sheet level, i.e., at 'loom sphere', whereas it would suffice to maintain general humidity condition in the room at around 65% R.H. Knitting operation also requires a stable relative humidity condition at 55% ± 5% for precise control of yarn tension. Hence, it is important to maintain stable RH conditions within the prescribed tolerance limits at all steps of textile processing.

Workers are part of manufacturing process. Therefore, the conditions maintained in the shed should not only be comfortable for the process and the product, but they should also be comfortable to the people. Table 1.1 gives generally recommended humidity levels in a Textile Mill.

Table 1.1 General recommendations of RH% in a textile mill.

Department	Cotton %	Man-Made Fibres %	Department	Wool %
Opening and Picking	45–60	50–55	Raw Wool Storage	50–55
Carding	50–55	50–60	Mixing and Blending	65–70
Sliver Lapping	55–60	55–65	Carding – Worsted	60–70
Ribbon Lapping	55–60	55–65	Carding – Woolen	60–75
Combing	55–65	55–65	Combing – Worsted	65–75
Drawing	50–60	50–60	Drawing – Worsted	
Roving	50–60	50–60	Bradford System	50–60
Spinning	45–60	50–65	French System	65–70
Winding and Spooling	60–65	60–65	Spinning-Bradford Worsted	50–55
Twisting	60–65	50–65	French (Mule)	75–85
Warping	55–70	50–65	Woolen (Mule)	65–75
Knitting	60–65	50–60	Winding and Spooling	55–60
Weaving	70–85	60–70	Warping – Worsted	50–55

Similar to the requirement of humidity, the temperature also plays a very important role in the textile processes. Table 1.2 gives the normal temperature levels followed in textile mills.

Table 1.2 Normal temperature levels followed in textile mills.

Department	Min Temperature		Max Temperature	
	°C	°F	°C	°F
Cotton Mixing	27	80	33	92
Blow Room	27	80	35	95
Cards and Draw Frames	27	80	35	95
Comber	27	80	33	92
Ring Frame	30	85	35	95
Winding	27	80	33	92
Warping	27	80	33	92
Weaving	21	70	31	88

1.3 Dust control

Normally the name 'Textile Mills' reminds of cotton dust-laden environment. Major problem of dust exists in the blow room and carding section of spinning mill, whereas exposure level in other areas is comparatively not much. In spinning mill, the extent of cotton dust contamination varies from section to section, as it is worst in the blow room and minimum at the cone-winding section.

Cotton dust is defined as dust present in the air during the handling or processing of cotton, which may contain a mixture of many substances including plant matter, fibre, bacteria, fungi, soil, pesticides, non-cotton plant matter and other contaminants which may have accumulated with the cotton during the growing, harvesting and subsequent processing or storage periods. Any dust present during the handling and processing of cotton through the weaving or knitting of fabrics and dust present in other operations or manufacturing processes using raw or waste cotton fibres and cotton fibre byproducts from textile mills are considered as cotton dust within this definition. The workers exposed to such working environment inhale fibrous particles and dust whole day. Generally, air suction system exists nearly in all departments to maintain certain humidity and to remove air contaminants; however, at some places, it works effectively but at certain areas air exchange is not proper resulting into suffocation and inconvenience for the workers. In weaving mill, fibrous particles are present in the working environment though not much are generally inhaled by most of the workers. These small fibrous particles generated during weaving activities disperse in occupational air. It is therefore essential to have sufficient circulation of filtered air. Air washers and ventilation systems are very essential for this.

All textile-manufacturing processes, except garment making, generate environmental pollution. Dust and fly during cotton spinning and fly during weaving gets released into the air streams of the production departments. Most of the modern textile mills are equipped with automatic waste removal, dust filtration and humidification plants. The dust and fly released by the machines are sucked away by suction nozzles and ducts. The dust-laden air is filtered, humidified and recirculated. The number of air changes per hour (ACH) is optimized in each department to keep the air streams clean, hygienic to prevent any risk to the health of the workers. The normal ACH are given in Table 1.3.

Table 1.3 Normal air changes per hour.

Department	Number of Air Changes Per Hour
Blow room, Drawing, Combing and Roving	15
Carding	20
Spinning	45
Winding	30
Twisting, Warping, Sizing and Weaving	20

In order to minimize risk of industrial diseases such as byssinosis (lung disease) among the workers, the Occupational Safety and Health Authority (OSHA) of USA has specified concentration limits of dust in the air streams of production rooms for compliance by the concerned industries as follows:

- 0.5 mg per cubic metre, from blowing to roving preparation and for manufacture of nonwovens.
- 0.2 mg per cubic meter, for spinning, twisting, winding and warping.
- 0.75 mg per cubic metre, for sizing and weaving.

With the industrial growth, the quality of air has been considerably deteriorated. Atmospheric air contains lot of aerosol. This has created the need for air conditioning, which implies fresh air supply, removal of aerosols and heat and air motion for cooling and refreshing. Compliance with the above listed limits for air cleanliness brings in economic benefits for the textile mills in the form of improved worker attendance, product quality, process efficiency, reduced end-breakage rate in spinning and weaving mills and improved yield of yarn.

Because of this need, the textile industry is one of the largest industrial users of air washing equipment. Similar equipment are found in other industries such as automotive industry spray paint booths, tobacco industry, hospital surgery and nursery rooms, photographic film manufacturing plants and aircraft industry clean rooms. Each of these industries normally uses water in a gas-scrubbing device to clean and process air so that it meets their particular clean air standards. The use of air-washing equipment in the textile industry is more difficult to understand than in other industries since there are many different processes and different combinations of processes in one plant. Air washers are utilized throughout the various processes in cotton mills where raw cotton is processed into woven cotton fabric. They are also used in blending plants where raw cotton is processed and blended with synthetic staple into yarn and then woven into blended fabric. Manmade fiber plants producing nylon or polyester yarn also use air washers. Fiberglass plants producing fiberglass yarn for tyre-cord and other industrial uses also utilize air washers extensively. Some of these plants have gas scrubber systems which double as plant air washers and which present some of the most difficult water treatment problems. Air washers are found extensively in knitting plants, including women's hosiery plants, and in carpet mills where carpet yarn is processed and dyed, as well as in the carpet weaving plants. Dyeing, finishing and bleaching processes generally do not require air washer systems, but these processes are often located in the same plant as one or more of the previously described processes. For example, a plant blending cotton with synthetic staple to produce drapery material may have another section of the plant where they dye this material, finish it and possibly run it through a printing process. Although dyeing and finishing operations do not require air washers, they are starting to use various types of smoke abatement equipment to control vapours emanating from the plant from some of these finishing processes.

1.4 Control of air pollution

The textile industry is plagued by air pollution problems, which must be resolved. In particular, smoke and odour arising in the process require abatement. The major air pollution problem in the textile industry occurs during the finishing stages, where various processes are employed for coating the fabrics. After the coatings are applied, the coated fabrics are cured by heating in ovens, dryers, tenter frames, etc. A frequent result is the vaporization of the organic compounds into high-molecular-weight volatile organic (usually hydrocarbon) compounds (VOCs). In terms of actual emissions, the industry must also deal with larger particles, principally lint. The other problem is the creation VOCs, which take the form of visible smoke and invisible but objectionable odour. Smoke is made up of tiny solid or liquid particles of VOCs less than 1 μm in size that are suspended in the gaseous discharge. When the smoke/gas mix goes up the stack, the Environmental Protection Agency (EPA) measures the density or opacity of the emission using an official transparent template and labels the opacity as ranging from 0% to 100%. The more opaque the smoke, the more visible it is. Opacity of smoke is related to the quantity, rather than the weight of particles resent in the gas.

The problem with odour is that it is not practically measurable. It is a sensation originated by interaction between molecules of hydrocarbons and millions of nerve fibres in the human olfactory membranes. The human nose can differentiate between 4,000 different odours. The odours associated with textile plant emissions are usually caused by hydrocarbons with molecular weight less than 200 and fewer than 15 carbon molecules (C_1 to C_{14}). These odourous molecules attach themselves to the particles of smoke and can be carried great distances from their point of origin and cause complaints. Proper system for collecting the polluted air and directing them for suitable treatment is an important aspect of air controlling.

1.5 Air changes

The number of air changes is an important factor to ensure that air is clean and safe. The number of air changes required depends on the type of activity being taking place. Table 1.4 gives the specified air changes for different levels of air controls.

Table 1.4 Specified air changes at different levels of air control.

ISO Class	Controls	Air Velocity at Table Level in FPM	Air Changes Rate per Hour	HEPA Coverage as % of Ceiling
1	Stringent	70–130	>750	100
2	Stringent	70–130	>750	100
3	Stringent	70–130	>750	100
4	Stringent	70–110	500–600	100
5	Stringent	70–90	150 - 400	100
6	Intermediate	25–40	60–100	33–40
7	Intermediate	10–15	25–40	10–15
8	Less Stringent	3–5	10–15	05–10

The method used to calculate CFM requirements for a given fan or fans is based on complete changes of air in a structure or room in a given time period. To determine the CFM required, the room volume (in cubic ft.) is divided by the appropriate 'Minutes per Air Change'. In textile industry, the minutes per air change ranges from 5 to 15. Additional considerations are local code requirements on air changes, specific use of the space and the type of climate in the area. Air changes are required more when more people are working in a section, the generation of heat is more or the generation of dust is more comparing the maximum dust level allowed in the air.

When considering air purification systems, it is typical to evaluate the filter media, product efficiency claims and the size and portability of the unit. The most important factor in the success of any system is the frequency of ACH that the system can create. The rate of ACH determines the rate at which the total volume of air in the room is cleaned by an air purification system, which is a major factor in the degree of air cleaning that can be achieved.

1.6 Humidity and the health

Providing ideal indoor air quality includes air purity, temperature and moisture level. A low humidity level can be as unhealthy and uncomfortable as excessive humidity. Dry air causes dry, rough and flaky skin because the skin's outer layers lose moisture to the surrounding air. Respiratory passages such as the nose and throat also lose moisture from their membranes causing dryness and irritation. Low levels of humidity can also contribute to respiratory infections, allergic and asthmatic symptoms and an increase in airborne dust and allergens. If the humidity level is too low, bacteria, viruses, respiratory infections and allergic asthma will increase. If the humidity level is too high, dust mites and fungi/mold will proliferate and allergic asthma will increase.

Numerous studies done by ASHRAE and other indoor air quality experts suggest an optimum RH range of 40 to 60%. Dryer or wetter causes different kinds of problems, at each extreme. See the chart below, based on ASHRAE sponsored research. The shaded portions indicate problems. For example, bacteria problems shall be less at 26 to 60%RH and virus's problem from 43 to 70%. Most of the problems are not there between 45 and 55%, as can be seen from the table.

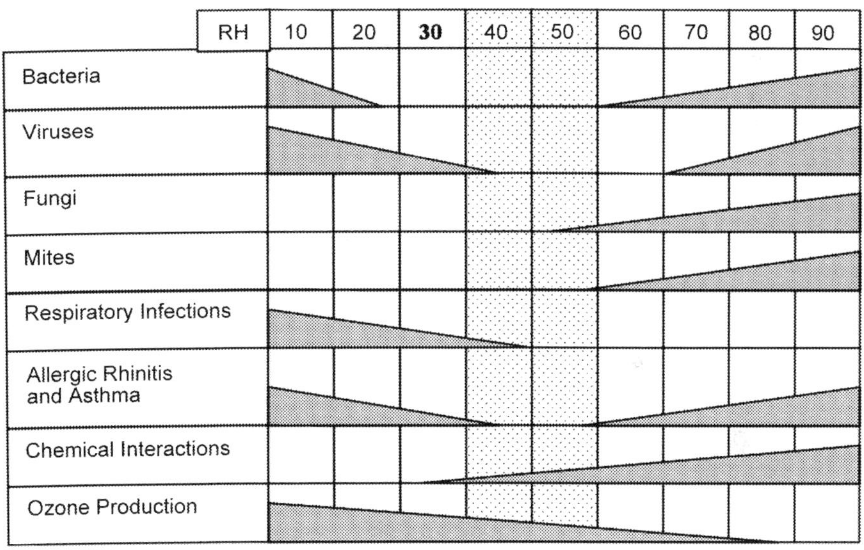

Figure 1.2 Comfortable levels of humidification [ASHRAE findings]

1.7 Protection of electronics and equipment

The last 3 decades have seen a major shift in the technology adopted in textile industry. The manual operations are being replaced by automations controlled by programmed electronic logic systems. It is not only the testing laboratories, but also monitoring of speeds, settings, temperature, humidity and various other factors are controlled by electronic gadgets. Central to all electronic circuits today is the IC (integrated circuit) or 'chip'. The heart of the IC is a wafer-thin miniature circuit engraved in semiconductor material. Electronic components and chips in particular can be overstressed by electrical transients (voltage spikes). This may cause cratering and melting of minute areas of the semiconductor, leading to operational upsets, loss of memory or permanent failure. The damage may be immediate or the component may fail sooner than an identical part not exposed to an electrical transient. A major cause of voltage spikes is electrostatic discharge (ESD). Although of extremely short duration, transients can be lethal to the wafer-thin surfaces of semiconductors. ESD may deliver voltage as high as lightning, and it strikes faster. In addition to the risk of damage to electronic devices from static electricity charges, there are grave risks associated with sparks from static charges in many process applications. Static electricity is extremely dangerous in the presence of gases, volatile liquids or explosive dusts such as is found in munitions plants, paint spray booths, printing plants, pharmaceutical plants and other places. Many static control products (special mats, carpeting, sprays, straps, etc.) are available, but cannot replace the work done by humidification, which is a passive static-control means working to control static all the time. However, care should be taken to ensure that there are no condensations on electronic parts. The RH preferred is from 40 to 70%.

1.8 Ideal conditions

Normal question in front of a technician in shop floor of a textile mills is "What is the ideal humidity?" This depends on the type of textile and the process undertaken. Natural fibres are far more susceptible to moisture than manmade, in terms of performance. However, manmade textiles suffer more with static charge build-up.

Cotton and linen need to be processed at very high RH levels, 70 to 85% in weaving and 50 to 65% in spinning, because they are very brittle. By humidifying each process, from the combing of the raw material, through carding, twisting, spinning and weaving, the manufacturer can ensure that the product remains flexible and is prevented from breaking. This is important since the longer the fibre, the finer the thread that can be spun from it.

Wool is similarly susceptible to dry air, although a little more forgiving, requiring humidity levels of around 65%RH. Manmade fibres also require the correct, albeit lower, level of humidity since below 45%RH, they are prone to a build-up of static electricity. Silk should be processed at between 65 and-70% RH, although artificial silk spinning requires a higher level of 85%RH.

It is clear that the need for providing conditioned and controlled air, monitoring the air movement in textile industry is very essential not only from the point of view of product quality and productivit, but also by considering the health of employees and the community around. Extensive works have been done to design the best possible combination considering the effectiveness and the cost implications. It is not possible to discuss all the available systems. Some of the widely used systems are discussed in the further chapters.

2
A glance at the developments

2.1 General

While moving heat via machinery to provide air conditioning is a relatively a modern invention, but not the cooling of buildings. The concept of air conditioning is known to have been applied in ancient Rome, where aqueduct water was circulated through the walls of certain houses to cool them. Medieval Persia had buildings that used cisterns and wind towers to cool buildings during the hot season. Cisterns (large open pools in a central courtyard, not underground tanks) collected rainwater; wind towers had windows that could catch wind and internal vanes to direct the airflow down into the building, usually over the cistern and out through a downwind cooling tower. Cistern water evaporated, cooling the air in the building.

The concept of modern air conditioning was first established by an American Mr. Stuart W. Cramer in the beginning of20th century. The words 'air conditioning'was coined by him and defined as 'A process of treating the air so that its temperature, humidity, cleanliness and distribution within the room are controlled simultaneously'. He was exploring ways to add moisture to the air in his textile mill. Cramer coined the term 'air conditioning', as an analogue to 'water conditioning', then a well-known process for making textiles easier to process. He combined moisture with ventilation to 'condition' and change the air in the factories.Mr. Willis Carrier further developed the system and in 1906 the first Buffalo humidifier and air washerplant was designed. He adopted the term and incorporated it into the name of his company. This evaporation of water in air, to provide a cooling effectis now known as evaporative cooling.

Early commercial applications of air conditioning were manufactured to cool air for industrial processing rather than personal comfort. Modern air conditioning emerged from advances in chemistry during the 19th century, and the first large-scale electrical air conditioning was invented and used in 1902 by Willis Havilland Carrier. Designed to improve manufacturing process control in a printing plant, his invention controlled not only temperature but also humidity. The low heat and humidity were to help maintain consistent paper dimensions and ink alignment. Later carrier's technology was applied to increase productivity in the workplace, and 'The Carrier Air Conditioning Company of America' was formed to meet rising demand. Over time, air conditioning came to be used to improve comfort in homes and automobiles. Residential sales expanded dramatically in the 1950s.

Early textile mills had north light roofing to get a uniform sunlight. Further, there used to be windows and number of openings so that the air could move freely. However, these buildings were not able to protect the temperature and humidity inside the working area. Any change in outside temperature or humidity was affecting the working. Normally, there used to be problems in the evenings when temperature was changing and also in the afternoons when outside temperature was high. The supervisors had the main task of getting water sprinkled on the floor in the afternoons and getting all windows closed in the nights. There used to be closed steam pipes to heat the section when the temperature was low. Heating lamps on draw frames, speed frames and combers were a common feature. The experience of the technician and the speed in which he could monitor the facilities were very important for getting smooth working. After the false ceilings were introduced, the effect of outside humidity and temperature reduced, and also the amount of air to be controlled became less. Then the windows were permanently closed which helped further to control the temperature and humidity.

As the speed of machinery increases, the precision in controlling temperature and humidity became prime necessity. The central air washer plants with automatic controls became an inseparable part of textile mills.

Just a few years ago, textile air washers were designed primarily as low-air velocity, non-chilled water systems. They were so large that air-washer rooms were an integral part of the building housing the textile operation. Lint and fibre screens ahead of the washers were nonexistent in many cases, and the washer rooms were typically filled with cotton lint. The warm water in the washer was a perfect environment for bacteria. Lint, dirt and other suspended matter in the water continually plugged air washer spray nozzles, reducing the efficiency. As the industry began to install air conditioning equipment, these units were upgraded and modified to perform properly with chilled water.

2.2 Unit humidifiers

Unit humidifiers came into picture in the middle of last century, which almost became an integral part of textile mills.

Unit humidifier or semi-central duct unit is a combination of axial flow fan, one or more centrifugal atomizers, air mixing chamber with interconnected fresh air and return air dampers to take fresh air as well as to recirculate departmental air, a set of large area V-shaped air filters and fabricated distribution duct with eliminator type grills. If required, steam heating coils can also be mounted on return air damper of the mixing chambers

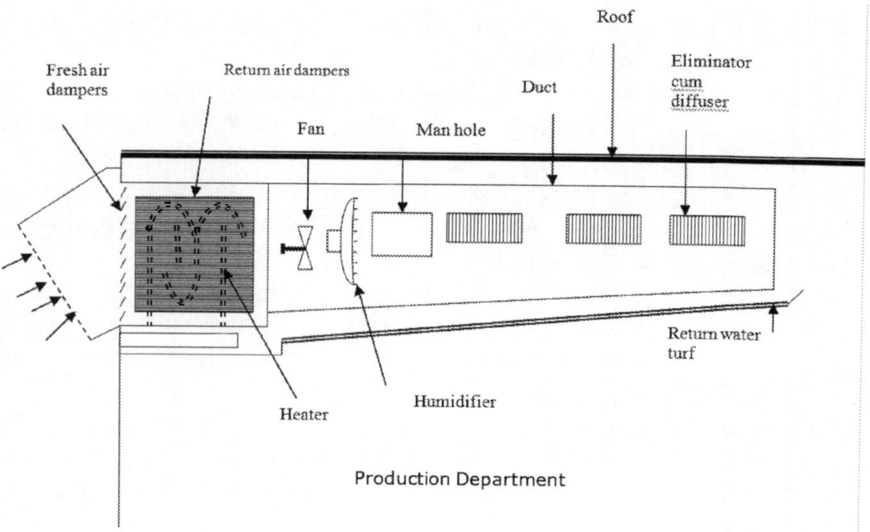

Figure 2.1 Unit-type plant

Different models with various fan capacities from 12,000 C.F.M to 25,000 C.F.M are available in the market. Unit humidifiers are normally fitted in the top near the ceiling. The ducts normally shall be extended up to 50 feet.

Figure 2.2 Semi central humidification unit

2.3 Central station-type plant

Early system of air cooling consisted of a cooling coil arrangement in which supply air was cooled and dehumidified by indirect heat exchange with cold water pumped through tubing. Willies Carrier, the developer of air washers surmised that an alternative design using direct contact between the air and chilled water might improve performance. In his early air conditioning systems, he observed that, although the air came in contact with water on the cooling coil surface, dehumidification occurred. In a typical air washer, water droplets are brought into direct contact with the air resulting in the exchange of heat and mass (moisture) from the air to the water droplets. If the supplied water temperature is below the entering air dew-point temperature, moisture from the air will condense onto the water droplets, thereby dehumidifying the supply air stream. If the incoming air is hot and dry, it takes up the moisture and increases the humidity.

Figure 2.3 Air washer plant developed by willies carrier

Air washer plants are big centralized units, which use water spray from a bank of nozzles on air moving with a force. The water spray not only adds humidity but also cleans the air from dust by making them wet and heavy, which fall down. A set of eliminators prevent the movement of water droplets and the dust particles along with the humidified air. Hence, the production area gets humidified clean air.

The Central Station-type plants have two units. The main plant consisting of a fan, air washer and other accessories is located in a plant room, which is outside the conditioned space. Only the air distribution system is in conditioned area. The distribution duct is connected with the plant, but is not an integral part of it. The fan, air washer and circulating pumps are normally at floor level and have an easy access, where the ducts are at roof level.

The circulating fan may be either centrifugal type or an axial flow type. The fan may be installed before the air washer or after the air washer. The first type is called as blow-through, whereas the second one is called as suck-through.

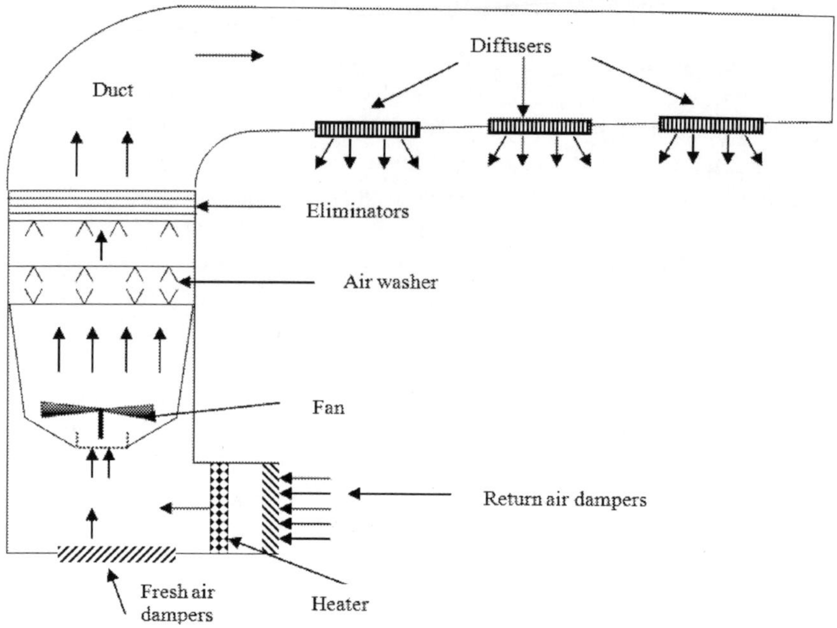

Figure 2.4 Central station – air washer

Central station plants are more costly than the unit type, but are more reliable and easily accessible for inspection and maintenance. The life of these plants is also higher. Air coming out of air washer plants is normally at 95%RH, whereas in unit type, it is above 100%, and some moisture is carried with the air in the form of fine mist. Therefore, for the same quantity of air handled, the unit humidifiers give more evaporative cooling. However, dry bulb temperature shall normally be higher inside the department because of less air circulation in case of unit humidifiers. The central station type is more suitable for refrigerative cooling as chilled water can be added as required. Also, dehumidification is possible by supplying water with a temperature less than the dew point temperature of the air.

As the requirement of air with predetermined temperature and humidity became the need of the day, air heaters and air coolers were developed and installed along with the humidification plants. Normally, closed steam radiators or steam coils were installed in the path of airflow after the humidification unit. Later electrical heating coils with thermostat controls were introduced. In case of steam coils, the steam pressure was manually adjusted and monitored, whereas with thermostat the temperature is controlled automatically. A combination of controls of humidity, temperature and circulation of air led to the concept of air conditioning. Most air conditioning systems perform the functions of providing the cooling and heating energy required, conditioning the supply air, that is, heat or cool, humidify or dehumidify, clean and purify and attenuate any objectionable noise produced by the HVAC & R equipment, distributing the conditioned air containing sufficient outdoor air to the conditioned space and controlling and maintaining the indoor environmental parameters such as temperature, humidity, cleanliness, air movement, sound level and pressure differential between the conditioned space and surroundings within predetermined limits

In central hydronic air conditioning systems, also called as central air conditioning systems, air is cooled or heated by coils filled with chilled or hot water distributed from a central cooling or heating plant. It is mostly applied to large-area buildings with many zones of conditioned space. An air system, which is called as air-handling system is used to condition, transport and to distribute the conditioned or recirculated air, get outdoor air or exhaust air to control the indoor environment according to requirements. The major components of an air system are the air-handling units, supply/return ductwork, fan-powered boxes, space diffusion devices and exhaust systems.

Different types of air handling units are now available which are tailor made for the industry or the purpose for which it is applied. The humidification and ventilation systems used in textile industry can be grouped in different categories. However, the control of desired conditions inside the work area requires a scientific knowledge of the properties of air and the competency of the equipment underuse. If we are just running a plant without understanding, the basic properties of air and its relations with water and water vapour, we shall not be able to achieve the required results, but might end up with more losses. The ventilation requirement varies widely depending upon the type of industries, nature of work. In certain industries beside normal ventilation, air pressurization is also required.

2.4 Supplementary humidification

As the seasons are changing as autumn, summer, rainy season and winter, the outside temperature and humidity condition shall be continuously varying. If the air washer capacity is designed to handle the driest weather, the plant has to be underutilized in other seasons. Therefore, the plants are designed considering the weather normally prevailing for maximum number of days in a year. The deficiency in capacity can be supplemented by direct water atomization units especially for use in dry summer days. Bahnson L-type centrifugal fans were widely used earlier, but because of its problem of letting coarse drops of water, its usage is reducing. The pneumatic atomizers are becoming popular because of good atomization and better distribution.

2.5 All-air system

Where high relative humidity levels are to be maintained as in the loom shed, the common practice was to provide supplementary systems. The main plant can provide around 60% RH, and the remaining 20% is provided by Bahnson L type fans. It is possible to achieve the required 80% humidity without using supplementary humidification, but by increasing the number of air changes. Use of two systems like central unit and supplementary units are called as split system as the responsibility of providing required conditions is split into two units. The system eliminating supplementary unit and managing with only central units are called as all-air system. Air circulation capacity of the main plant in the case of all-air system should be quite high, around three times, as compared to split system. This results in very high capital cost as well as operational cost. However, the all-air system provides uniform humidity throughout the department compared to supplementary humidification. Further, the water droplets that are condensed in split humidification can corrode machine parts. The studies have shown that with all-air system, the loom efficiency is higher compared to supplementary humidification. This increase in efficiency compensates the increased capital expenses and the running expenses.

Although the air washer plants are supposed to provide 95% humidity at the diffusers, it fails in a number of cases to reach even 80%. With All-Air system, the provision for supplementing the humidity by additional humidifiers is not possible. Hence while designing a plant; this point needs to be considered.

2.6 Energy management

The recent developments in humidification are mainly concentrating on energy savings and pollution control. Helmut Stubble's invention (US Patent No 283026 dated 5th March 1996) relates to a process for the cooling and conditioning of a waste air generated in a textile room from machines carrying out a textile process. The process comprises of drawing the heated waste air from the textile room and passing the air through a liquid medium heat exchanger; drawing a liquid cooling medium at a desired temperature from a constant source and passing the liquid cooling medium through the heat exchanger to cool the heated waste air, the liquid cooling medium thereby being heated by the heated waste air; recycling at least a portion of the cooled air exiting the heat exchanger back to the textile room and directing the heated liquid cooling medium from the heat exchanger to an operative location in the textile process requiring a separate medium of approximately the temperature of the heated liquid cooling medium, and using the heated liquid cooling medium at the location as a source of the separate medium to reduce the overall energy consumption of the textile process. Open waters, well and/or tap water of the water supply network are used for the heat exchange, whereby this water is used after utilisation in the heat exchange at other locations for the cooling of the process and/or as hot water for further utilisation in the process.

SITRA–PCRA Climo Control developed in 2005, which was sponsored by the Petroleum Conservation Research Association (PCRA), helps save energy to the tune of 25–60% in the existing condition with a payback period of 18 months. Climo Control consists of three different modules to vary the speed of supplier fan, exhaust fan and water pump in the humidification plant based on the outside climate. SITRA also developed energy-saving overhead cleaners in the same year and named as SITRA-PCRA Ener Optimisers. SITRA has developed five different methods and based on this, the investment would vary from Rs 4,000 to Rs 50,000 with a payback period ranging from four months to five years.

Mr. Arun Shourie, chairman of PackPlast India, a pioneer in energy-saving projects in textile mill humidification developed Auto Return Air Cleanser which cleans the return air of textile mill and reuses, reducing the air pollution as well as conditioning cost.

An apparatus and a method for humidifying a continuous textile material are described in European Patent EP1428923 by Bertoldo, Franco 06/16/2004. The apparatus comprises a rotatable drum having an internal cavity, the sidewall of which is formed externally by a side surface and a tank facing the side surface. A conveying means for conveying the textile material through the tank and means for supplying into the tank substantially saturated steam at a pressure Pv and at a temperature Tv and for supplying into the internal cavity substantially saturated steam at a pressure Pc and at a temperature Tc. The conveying means comprise a permeable belt for pushing the textile material into direct contact with the side surface at least in the zone of tank, the steam supplied into the internal cavity being kept at a pressure Pc less than steam pressure Pv inside the tank by a certain value so as to cause partial and controlled condensation of the steam supplied into the tank onto the side surface and onto the textile material.

In a development (Patent Application No. 28/MUM/2007A dated 26th Jan 2007) by Rakesh Pramodbhai Shah, Nilesh Prafulchandra Varia and Devand Prafulchandra Varia, steam generated in the steam generator is injected into the moist air generating chamber through the insulated pipeline and water is sprayed in the chamber through nozzles. Humidifier and conditioner chamber areconnected with moist air generating chamber through the insulated pipeline. An electronic control panelcontrols the velocity of moist air and temperature in the humidifier andconditioner chamber.

United States Patent 4183224 relates to a method and apparatus for precise temperature and humidity control within a confined or determined chamber, which involves determining a desired condition in a controlled space. The essence of the concept is to provide humidity control by the addition of moisture at all operating conditions and, in effect, never having to control dehumidification. In addition, the method for heating or cooling to insure that the heating or cooling does not affect the humidity, but that humidity is added after achieving all the desired temperature relationships, which enables the system to eliminate control into the dehumidification mode. Further, the process includes a higher airflow rate than normally utilised, all of which improves the precision of the humidification and temperature control process.

Geiger Stephan, Leu Karl and Subuis Robert (Ref International Patent WO/2007/090313 dated 16th Aug 2007 Application No PCT/CH2007/000064) developed a device for evaporation of water.

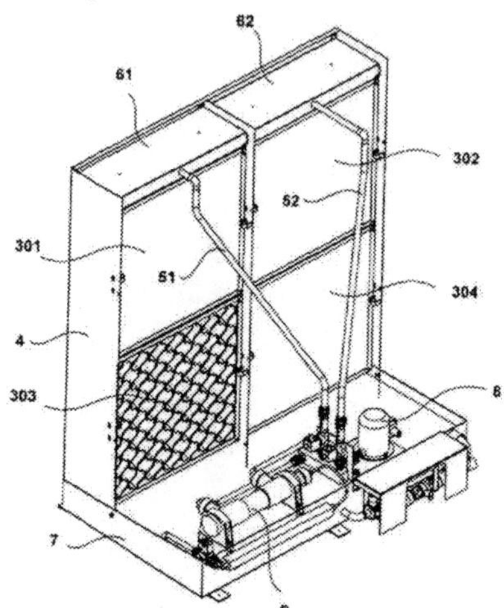

Figure 2.5 Device for evaporation of water

In this device for the evaporation of liquid, in particular water, an evaporation mat is used which is wetted with the liquid on which the liquid evaporates. The evaporation mat comprises a textile fabric having fibres, wherein the surface of the fibres is coated with a covering,(which comprises a cured reaction product of a polyamine and a polyalkylene glycol etherified with end groups of the structure X-CH2[CH(OR)]WCH2–), in which structure w is an integer from 0 to 1 and, when w is 0, X is a halogen, and, when w is 1, X is halogen and R is hydrogen, or X and R together are –O–. Preferably, the evaporation mat is a consolidated nonwoven thatcontains fibres made of a synthetic thermoplastic which are bonded to one another by means of thermoplastic hot-melt glue at their intersection points. The devices are used for air humidification, for concentrating solutions, or for evaporative cooling.

3

Different humidification systems

There are several ways to add moisture to air. Adding thermal energy to vapourise the water and simply spraying water through misting nozzles are the two popular categories. The best method depends on how much water needs to be vapourised, how quickly and what equipment are already available to help distribute the moisture. In case of very cold and dry weather, direct injection of steam is used, which gives immediate relief. However, it is not suggested if the temperature is above normal human body temperature, as it can cause uneasiness among the operators. The systems adopted for humidifying differ significantly. Humidifying a hospital operating room is different from those for a textile mill, an office building or even a laboratory. Different types of operations have substantially different requirements for the achievement of proper relative humidity. These requirements determine what means of humidification one should use.

3.1 Steam humidification

Steam, evaporative pan and water spray are the popular humidifying systems used conventionally. Each has particular advantages and limitations which determine its suitability for a particular application. Steam is ready-made water vapour that needs only to be mixed with the air. With evaporative pan humidification, air flows across the surface of heated water in the pan and absorbs the water vapour. Both steam and evaporative pan humidification do not affect the temperature of the humidified air. Water spray humidification disperses water as a fine mist into the air stream where it evaporates. As it evaporates, it draws heat from the air and cools it.

3.1.1 Direct steam injection

In this system, water is heated to the boiling point where it turns to steam vapour. The steam is allowed to rise from the unit without the aid of a motorized fan. This can be used virtually at all commercial, institutional and industrial applications. Where steam is not readily available, self-contained steam generating units or central system steam humidifiers are used. Steam humidifiers use natural fuels such ascoal, wood, husk, bagasse, furnace oil, diesel or developed fuel such aselectricity to generate steam droplets. It produces the noisy sound of boiling water. Although the product may be cheaper, it costs more to operate than the impeller and ultrasonic type humidifiers due to the high consumption of electricity used to heat the water.

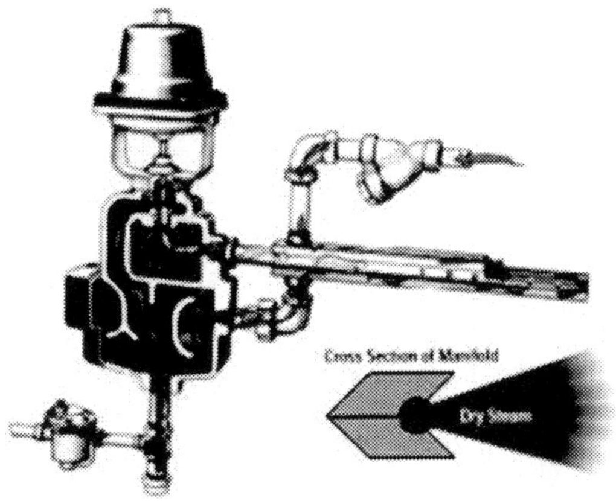

Figure 3.1 Direct Steam Humidification

The early industrial systems used space-heating steam boilers for humidification. If there were not enough steam, system leaks to meet humidification needs, additional steam would be vented into the space. When air handlers came along, various steam injection systems were developed for large-scale controlled humidification. Generally, the same steam was used that provided space heat and perhaps ran industrial processes. However, this steam contained traces of chemicals used for boiler water treatment. During the initial days of indoor air quality concerns, people were scared of using boiler chemicals. In some cases, special more environmental friendly chemicals were used, but in more cases, dedicated boilers were needed. This became expensive to install, especially for small facilities, and facilities that would otherwise not even have a boiler. This gave rise to the electric humidification technology. A recently developed technology is the direct-fired gas humidification unit. Steam systems are the most economical, when there is an existing boiler and the application requires a lot of volume. Since boiler chemicals are still a concern in most commercial applications, steam-to-steam heat exchangers have been developed. This way, the low operating cost of the central boiler is obtained, and there is a lower first cost than buying a dedicated boiler system. There are also chemicals that are certified safe for humidification applications.

Direct-fired natural gas humidifiers produce steam directly. They are designed to operate with no extra steam pressure build-up and deal with hard water fouling without the use of treatment chemicals. Direct-fired units are the most economical systems for large applications where there is no existing boiler.

The first direct-fired unit was introduced by DriSteem, and is called the GTS humidifier (gas to steam). It resembles a large fat-fryer with water instead of oil in the tank. Units are available from 75 up to 600 lbs. per hour steam output, 100,000 to 800,000 BTUs per hour gas input

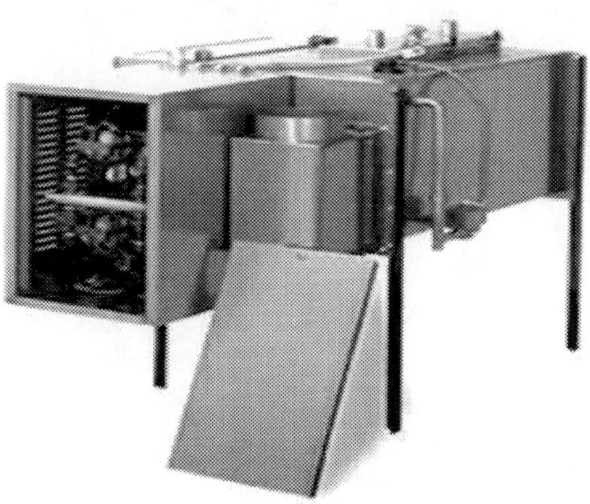

Figure 3.2 Direct-fired natural gas humidifier

Direct steam injection was normally practiced in dry cold winters, as it give both temperature and humidity, almost instantly. However, there were problems of the steam getting condensed when it came in contact with cold roof, and water drops were falling down. This was very dangerous when water drops fell on running material or on the machine parts, resulting in jamming, lapping and breakdown of machines. Sometimes this led to stains on the fabrics. Before refilling the water tank, the remaining water including mineral deposits collected should be tipped out. Depending on the water hardness, the insert must be de-scaled once or twice a month by using de-scaling agent. The deposits can also be scraped off with a tool, e.g., a screw driver.

The main advantages of steam humidification are favourable price, a high humidification performance and little maintenance. The vapourisation principle guarantees absolutely germ, bacteria and mineral-free humidification as residues remain in the tank.

3.1.2 Immersion humidifiers

Immersion humidifiers incorporate electric heating elements in a reservoir of water to provide humidification. Immersed heating elements raise the temperature in a tank, boiling the water and generating steam that is sent to a dispersal unit for absorption into the air. This type of humidifier is slow to react to humidity changes if it has been off for a while and typically does not control the amount of minerals from the water that get introduced into the air. This method is generally not recommended for data center applications. This system is normally not followed in textile industry.

3.1.3 Steam canister

Steam canister humidifiers use electrodes inserted into the water reservoir to pass current through the water, causing it to boil and to discharge pure steam at 212°F (100°C) at atmospheric pressure through a steam distributor. This approach provides greater application flexibility than the infrared since the canister does not need to be mounted in the conditioned air stream, but it cannot react to changes in humidity as quickly if it has been off for a while. The canister portion of the humidifier can be installed inside the air conditioner or outside the air conditioner on ducts, on walls or in other locations. The distributor can be mounted in the air conditioner, in duct work or with a special blower box for free discharge in the space. The piping for the distributor is specifically sized to match the humidifier's capacity and must be located to avoid condensate blockage during operation. A steam canister humidifier must bring the water in the canister up to boiling temperature before it is at full capacity, a process that can take several minutes. It does this by passing an electrical current through the water between electrodes. The mineral content in the water provides the conductivity for this to occur. The unit is consuming energy during this time, and the relative significance of this energy consumption varies based on how many times the unit turns on and off. Consequently, if operation is intermittent, steam canister systems will operate less efficiently than anticipated, and the efficiency of the system may be reduced over time as the system ages and the electrodes are consumed.

As moisture is conveyed into the air unit's air stream, efficiency can be affected by losses in the hose and distributor system. The amount of loss depends primarily on the length of the hose and number of bends. For example, if the canister is generating 20 lbs. /hr. of moisture at the canister and 4 lbs. /hr. is lost due to condensation, the net output is reduced to 16 lbs. /hr. emission. This loss is evidenced by water running back down the hose to the drain or back into the bottle. Efficiency can also be negatively impacted if the distributor is located close to the cold metal surfaces on the leaving side of the evaporator coil as some of the hot steam condenses on the metal rather than being absorbed into the air. Another consideration for steam canister systems is water quality. They require water conductivity levels of between 200MM (micromho) and 500MM for optimal performance. Higher levels may result in excessive arcing, and at levels below 060MM, there is insufficient conductivity for current to flow between the electrodes. The water quality and number of hours of operation determine the life of the electrodes, which are part of the replaceable canister assembly. The minerals in the water cause the consumption of the electrodes, and accordingly, canisters can last for several weeks or several months. This is typically the only maintenance item in the system.

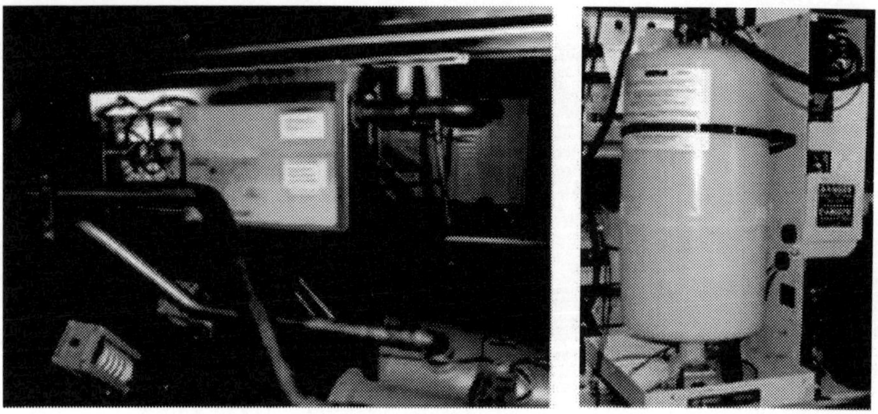

Figure 3.3 Steam canister

3.1.4 Infrared humidification

Infrared humidifiers are typically installed within precision air conditioning units. They use high-intensity quartz infrared lamps over a stainless steel humidifier pan. The infrared radiation from the lamps breaks the surface tension of the water, allowing the air flowing across it to evaporate and carry the moisture away as a particle-free vapour. This provides very precise and fast humidification.The number and wattage of the lamps depends on the size of humidifier and voltage utilised. When humidification is called for, heat energy from the lamps is reflected onto the water in the pan. The infrared radiation breaks the surface tension of the water, allowing the air flowing across it to evaporate and carry the moisture away as a particle-free vapour. The process takes less than six seconds from turning on the bulbs to achieving full capacity, making infrared humidifiers very responsive. The only time capacity is reduced is if a bulb fails, but it can be easilyand quickly replaced.

Because they do not depend on boiling water for evaporation to occur, and as there is no loss due to condensation, infrared units typically provide full rated capacity when operating. Butas airflow for an infrared humidifier comes from the precision air unit through a bypass, the precision air unit's airflow must be at the recommended level for the humidifier to operate at full capacity. Water quality as measured in micromho (MM) and total dissolved solids (TDS) has a negligible effect on the performance and effectiveness of infrared humidifiers. Infrared humidifiers can be used at any MM level, even where water conductivity levels are less than 060MM. The minerals in the water are not evaporated into the air stream in the infrared process. Instead, they are flushed through the system and either go down the drain or settle out in the bottom of the pan. Serviceability of infrared humidifiers is simple, with all components easily accessible for periodic cleaning. The quartz lamps are located above and away from the water in the pan, minimizing the possibility of corrosion and burnout. The water pan must be cleaned or changed out periodically because sediment drops out of the water and scales on the pan. Most users simply keep two or more pans on hand that can be changed out as necessary and cleaned as maintenance resources permit.

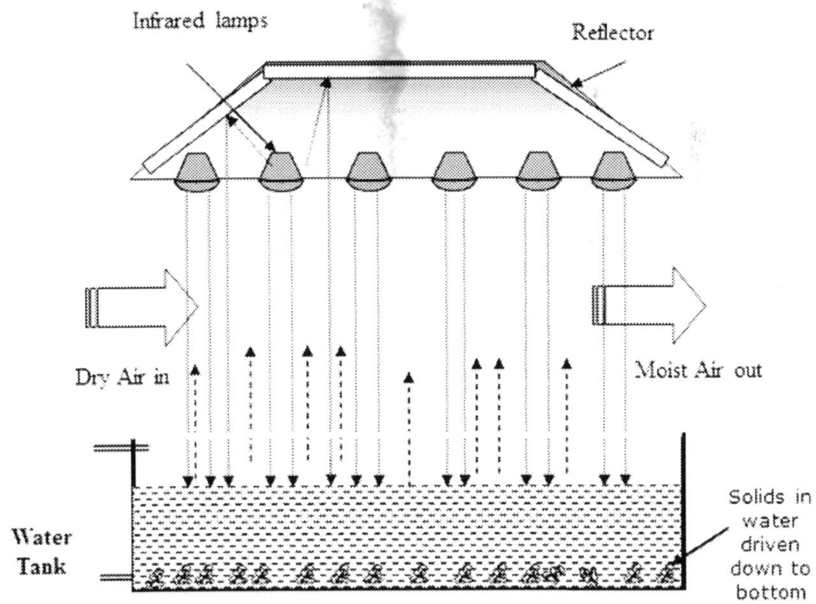

Figure 3.4 Infrared humidifier

3.1.5 Steam-to-steam humidifiers

Steam-to-steam humidifiers use a heat exchanger and the heat of treated steam to create a secondary steam for humidification from untreated water. The secondary steam is typically at atmospheric pressure, placing increased importance on equipment location. Maintenance of steam-to-steam humidifiers depends on water quality. Impurities such as calcium, magnesium and iron deposit as scale, requiring frequent cleaning. Response to control is slower than with direct steam because of the time required to boil the water.

3.1.6 Electronic steam humidifiers (electrode)

Electronic steam humidifiers are used when a source of steam is not available. Electricity and water create steam at atmospheric pressure. Electrode-type units pass electrical current through water to provide proportional output. Use with pure demineralized, de-or distilled water alone will generally not provide sufficient conductivity for electrode units. Water quality affects the operation and maintenance of electrode-type humidifiers. Use with hard water requires more frequent cleaning, and pure softened water can shorten electrode life. Microprocessor-based diagnostics assist with troubleshooting. Electrode units are easily adaptable to different control signals and offer full-modulated output. However, the need to boil the water means control will not compare with direct-injection units.

3.1.7 Electronic steam humidifiers (ionic bed)

Ionic bed electronic humidifiers typically use immersed resistance heating elements to boil water. Since current does not pass through water, conductivity is not a concern. Therefore, ionic bed technology makes the humidifier versatile enough to accommodate various water qualities. These units work by using ionic bed cartridges containing a fibrous media to attract solids from water as its temperature rises, minimizingthe build-up of solids inside the humidifier. Water quality does not affect operation, and maintenance typically consists of simply replacing the cartridges. Ionic bed humidifiers are adaptable to different control signals and offer full-modulated output. Control is however exercised by the need to boil the water.

3.2 Evaporative pan

The evaporative approach to humidification uses an evaporative pad with air blowing across it. Hot water is run over the pad and evaporated as the air travels through it. Water in a glass pan absorbed by a wick evaporates providing humidity.The warmer the water, the more effective the process is, using the energy in the water to aid the process. The evaporator pad may be pressed out (not wrung out) in lukewarm water. Depending on the level of scaling, the pad must be replaced once or twice per season. This approach is commonly seen in residential applications and is an inexpensive method of humidification. However, it generally does not provide the precision, cleanliness or speedof response required by data center operations. This is recommended only as an alternative to self-contained steam generating unit humidifiers for small load commercial or institutional applications. This type of humidification is used in small laboratories.

Evaporative pan humidification can increase dry-bulb temperature as measured on the psychrometric chart. This unwanted temperature change occurs as air is forced across the warmed water in the pan. The increase in DB can cause damaging results in process applications and increase the need for humidity control. The psychrometric chart helps illustrate that evaporative pan humidification is not a constant DB process.

Maintenance of evaporative pan humidification systems demands regular cleaning of the heating coils and pan, which are subject to 'liming up'. The use of chemical additives added either automatically or manually to the water in the pan can reduce this problem. Response to control with the evaporative pan method is slow due to the time required for evaporation to take place before humidified air can be circulated. Output is determined by water temperature and surface area. Evaporative pan humidifiers can sustain bacteria colonies in the reservoir and distribute them throughout the humidified space. High water temperatures, water treatment and regular cleaning and flushing of the humidifier help to minimise the problem.

I remember the days when we were pouring water on floor during dry and hot summer days. The water used to evaporate, and we were getting good working. This system had a problem of lint accumulation, and keeping the working area clean was very difficult. Sometimes, the people used to slip on the wet floor. In case any material fell on floor, everything was to be thrown as waste because of staining. There are different styles in evaporative pan humidifiers.

3.2.1 Drum style

A pipe brings water directly to a reservoir (a pan) attached to the furnace. The level of water in the pan is controlled by a float valve. The wick is typically a foam pad mounted on a drum and attached to a small motor. Hot air enters the drum at one end and is forced to leave through the sides of the drum. When the hygrostat calls for humidity, the motor is turned on causing the drum to rotate slowly through the pan of water and preventing the foam pad from drying out. Advantages include low cost and inexpensive maintenance. The disadvantages are requirement for frequent inspections of cleanliness and pad condition, water evaporation even when humidification is not required and mold growth in the pan full of water.

3.2.2 Disc Wheel Style

Very similar in design to the drum style humidifiers, this type of furnace humidifier replaces the foam drum with a number of plastic discs. The discs look slightly like traditional gramophone records with small grooves on both sides. This allows for a very large evaporative surface area, without requiring a great deal of space. Unlike the drum style humidifiers, the disc wheel does not need replacing. Advantages include low maintenance, no replacement parts necessary, high output due to large evaporative surface area, can be installed in hard water situations and maintains efficiency throughout lifespan. The disadvantages are higher price and water evaporation even when humidification is not required.

3.2.3 Flow-through style (also known as 'biscuit style' or many other, similar variant names)

A pipe brings water directly to an electricallycontrolled valve at the top of the humidifier. Air passes through an aluminum 'biscuit' (often called a pad; using the term 'biscuit' to emphasise the solid rather than foamy form), which is conceptually similar to a piece of extremely coarse steel wool. The biscuit has a coating of a matte ceramic, resulting in an extremely large surface area within a small space. When the hygrostat calls for humidity, the valve is opened, causing a spray of water onto the biscuit. Hot air is passed through the biscuit, causing the water to evaporate from the pad and be carried into the building.

Advantages include reduced maintenance (new biscuit only when clogged with dust or mineral deposits), lack of a pan of potentially stagnant water to serve as a breeding ground for mould as with a drum-style humidifier, no incidental humidification caused by a constantlyreplenished pan of water in a high velocity air stream, reduced requirement for expensive air filters and replacement of the biscuit. Disadvantages are higher purchase price, manufacturer and model-specific replacement biscuits versus the relatively generic drum-style pads and the requirement for a drain provision.

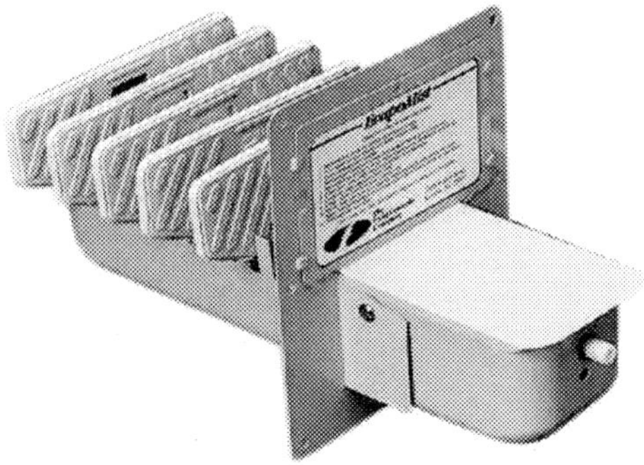

Figure 3.5 Flow through type humidifier

3.2.4 Fan-forced humidifiers

The most common humidifier, an 'evaporative' or 'wick humidifier', consists of just a few basic parts: a reservoir, wick and fan. These units usually have a wicking device that absorbs water from a reservoir and a motorised fan blows air through the water soaked wick to distribute moisture into the air through evaporation. Reservoirs are common to all humidifiers and come in different shapes and sizes. The reservoir is a tank of water filled prior to operation and provides the water for the moisture output. The wick is a filter that absorbs water from the reservoir. Evaporation of water from the wick is dependent on relative humidity. A room with low humidity will have a higher evaporation rate compared to a room with high humidity. Therefore, this type of humidifier is self-regulating: As the humidity of the room increases, the water vapour output naturally decreases. These wicks regularly need cleaning and replacement to get effective results. The fan is adjacent to the wick and blows air onto the wick, thus aiding in the evaporation of the water within the wick.

3.2.5 Atmospheric forced draft evaporator

Where the quantity of wastewater is small and difficult to treat, atmospheric forced draft evaporator is the ideal choice. Various models from 100 to 5000 Liters per day capacities are standard. **Advantages and features** are low capital cost, ease of operation and maintenance, no addition of chemicals, very small quantity of sludge and can handle difficult to treat wastewaters.

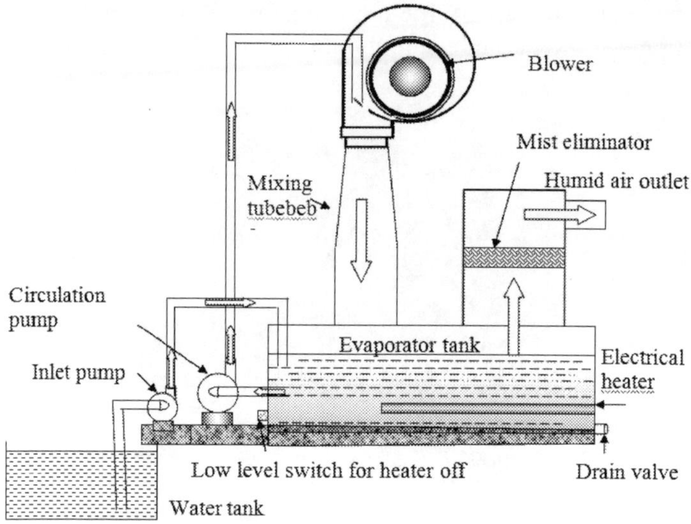

Figure 3.6 Atmospheric forced draft evaporator

3.2.5.1 Working principle

Atmospheric evaporator consists of feed pump, evaporator tank with electrical or diesel heater, circulation pump, blower, mixing tube, mist eliminator and humid air outlet. The wastewater from the wastewater tank is pumped by the feed pump to the evaporator tank. Electric or diesel heater heats the effluent in the evaporator tank. The hot effluent is pumped into the mixing tube, through which air is blown from atmosphere by the blower. The air and wastewater are thoroughly mixed in the mixing tube and passes to the evaporator tank, before getting out through the mist eliminator and humid air outlet. Electrical controls limits the operation of feed pump, circulation pump and blower depends on the levels in the evaporator tank and wastewater tank. Control system with automatic ON/OFF, safety shut off and switchgears with control panel are provided with the system.

3.2.6 Low temperature vacuum evaporator

The low temperature vacuum evaporator vaporizeswater at lower temperature of around 40°C than normal 100°C due to vacuum. Hence,they are useful for recovery of heat-sensitive chemicals and chemicals sensitive to air oxidation such as cyanide plating bath and Stannous tin bath and recovery of solutions containing volatile components.

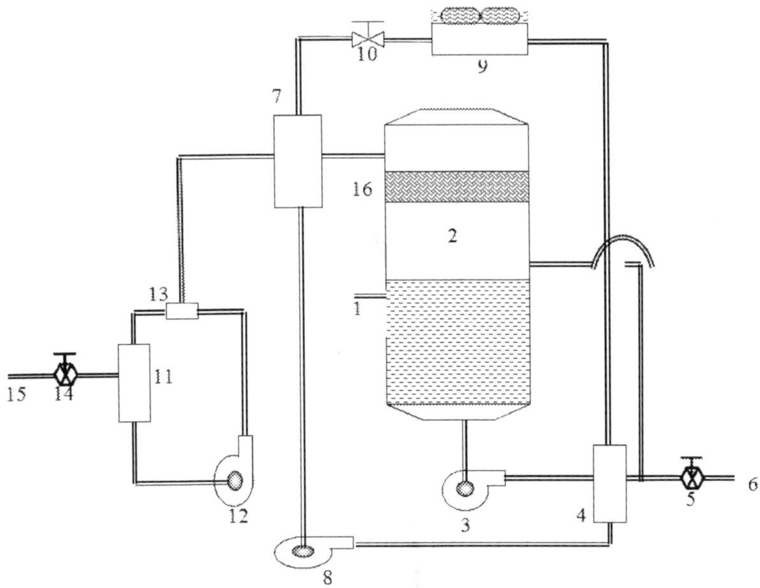

Figure 3.7 Low temperature vacuum evaporator

The low temperature vacuum evaporator (Fig. 3.7) consists of an evaporator tank 2, evaporator heat exchanger 4, process water circulation pump 3, refrigerant compressor 8, condenser heat exchanger 7, distillate tank 11, vacuum producing venture 13 and distillate circulating pump 12. The feed water enters the evaporator tank 2, and is circulated by the pump 3 to evaporator heat exchanger 4, where the water is heated by high pressure refrigerant from the refrigerant compressor 8. The hot wateris flashed as water vapour into the evaporator tank at the top. This water vapour passes through the mist eliminator 16 and the condenser heat exchanger 7. The refrigerant from the evaporator heat exchanger passes through air cooler and expansion valve and becomes cool liquid. The cool refrigerant cools the hot water vapour to distillate. The distillate is collected in the distillate tank. The distillate is circulated by the distillate circulation pump through a venture, which produces vacuum in the evaporator tank. The vacuum produced by the venturi in the evaporation tank causes process water to boil between 40°Cand50°C. The concentrated process water from the evaporator tank and the distillate from the distillate tank are automatically drained by the electrical controls. All necessary controls and switch gears are part of the system. Advantages and features of this system areindependent of solution heating requirements, less or no air pollution, can handle heat sensitive compounds and carbonates do not build up scales as with other evaporators.

3.2.7 Mechanical vapour recompression and multiple effect evaporator

3R Technology India offers MVR (Mechanical Vapour Recompression) and Multiple Effect Evaporators (MEE) for large volumes and energy economy, which is energy efficient, since the latent heat of vaporisation is fully utilised through vapour recompression and condensation. Typical layout of a MVR Evaporator is shown in Fig.3.8.

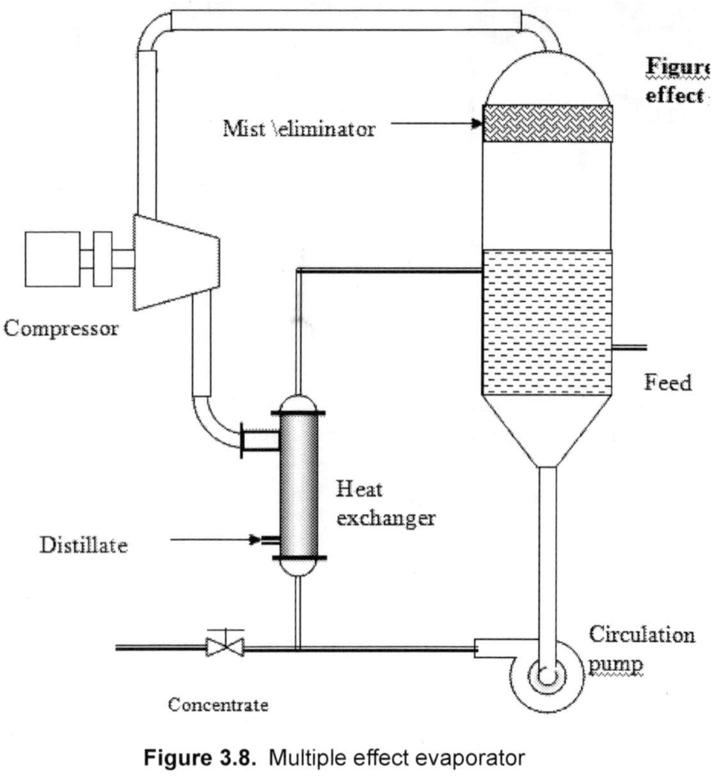

Figure 3.8. Multiple effect evaporator

3.2.7.1 Working Principle

The evaporated vapour flows through the mist eliminator to the suction of the compressor. The vapour is pressurised in the compressor and becomes superheated. The higher temperature compressed vapour flows to the shell side of the heat exchanger. The compressed vapour condenses by leaving its latent heat to the liquid film circulated by the circulation pump on the tube side. MEE are designed to improve the energy economy of the evaporation process. The basic principle is to use the heat given up by condensation in one effect to provide heat for another effect. The vapour from the separator of the first effect enters the falling film heat exchanger of the second effect, while live steam heats in the falling film heat exchanger of the first effect. The steam and cooling water rates for the double effect unit are approximately 50% of those required for a single effect unit.Unit up to nine effects can be built depending on flow rate and recovery economics. Typical flow rates for multiple effect range from 3000 LPH to 50,000 LPH. The major advantage of MVR and MEE is the energy economy. Typical MVR energy requirement is 0.05 to 0.15 KWH per kg of water evaporated. The energy economy of multiple effect evaporators depends on the number of effects, and it vary from 220 kcal of thermal energy per 1 kg water evaporation for a triple effect evaporator to 120 kcal for a six effect evaporator. As the operating cost of MVR and MEE is low, large flow system favours its application in all sectors of industry and also desalination of sea and brackish water.

3.2.8 Ultrasonic humidifiers

3.2.8.1 Principle of operation

A piezoelectric transducer immersed in a water bed converts a high-frequency electronic signal into a high-frequency mechanical oscillation. As the oscillation speed is increased to a level where the water particles can no longer follow the oscillating surface, a momentary vacuum and strong compression occur, leading to the explosive formation of air bubbles (cavitation). At cavitation, broken capillary waves are generated, and tiny (1µm diameter) droplets break the surface tension of the water and are quickly dissipated into the air, taking vapour form and absorbed into the air stream.

The principle of ultrasonic humidification is based on the superposition of two effects, Capital bubble implosion and capillary wave theory which are as follows:

- **Cavitation bubble implosion:** The change in amplitude of the oscillator gives rise to a powerful water hammer effect that releases tiny cavitation bubbles. The implosion of the bubbles on the surface emits tiny water aerosols into the ambient air.
- **Capillary wave theory:** The ultrasonic oscillators generate regularly formed Rayleigh surface waves in the water tank. Minute water aerosols are also emitted into the ambient air on the crests of these waves.

By superimposing these two effects, the use of ultrasonic humidifiers enables a homogeneous aerosol mist to be produced with minimal energy consumption!

Ultrasonic humidifiers use a multiple of piezoelectric transducers (physically similar in size to a nickel) that are mounted under a pan of water. A metal diaphragm vibrating at an ultrasonic frequency creates water droplets that exit the humidifier in the form of a cool fog. The transducers vibrate 1.65 million times per second, causing a mushroom-shaped water finger to form off producing micro droplets 0.52 cubic micron similar to steam. The droplet size remains constant regardless of the life of the transducer. Heat used to change water to vapour is absorbed from the surrounding air. High-frequency oscillations are created by a membrane contacting the water. These oscillations produce a micro-fine water mist which is blown into the room by a small fan where it evaporates immediately. The ultrasonic system, with its high-frequency vibrations, destroys most bacteria and viruses. Although silent, ultrasonic humidifiers should be cleaned regularly to avoid bacterial contamination which may be projected into the air. In case of visible lime scale deposits the nebuliser must be cleaned with a brush which is normally included with the unit. The water tank can easily be cleaned of mineral deposits by means of regular household decalcifying agents.

Figure 3.9 Ultrasonic humidifier

Ultrasonic humidifiers get high marks for energy efficiency, but they require mineral-free water for trouble-free operation. The cost and maintenance of the water treatment system plus the cost of the humidifier itself can often negate potential energy savings. Additionally, humidifier placement is very critical to proper operation.

3.3 Water spray humidifiers

Circulating discs pull water up from a water-filled pan and a motorised fan blows air over these discs to distribute moisture into the air through evaporation. They are referred to as air washers because the water in the discs traps some airborne particles in the water reservoir. This type is recommended for industrial applications where evaporative cooling is required. It is typically employed in textile mills in the hot regions during summer time. When properly designed and installed, humidification systems offer great benefits for the safe operation of machinery and equipment, the optimisation of employee comfort and large potential savings in heating and cooling costs.

Drossopher system has a jet of water with atmospheric pressure of about 7 to 8 striking against a steel plunger breaking into fine water droplets. This cools air around the water jet. This is installed overhead, and water is allowed to flow down gradually over the loom. The requirement of high-pressure piping system, collection of return water, overhead mounting of the humidifier making it difficult to clean and maintain are the main drawbacks of this system. As new and simple systems are available, this system has almost become obsolete.

The water spray process can create potential temperature control problems. In order to become water vapour or humidity, water requires approximately 1,000 Btu per pound to vaporise. This heat must be drawn from the air, where it will hopefully vaporise. If not enough heat is available quickly enough, the water remains a liquid. This unvapourised water can result in over humidification, and the water can 'plate out' on surfaces, creating a sanitation hazard. Water spray contains virtually none of the heat of vaporisation required to increase the RH of the air to desired conditions. For this reason, water spray humidification is a virtually constant enthalpy process.Response of water spray humidifiers to control is slow due to the need for evaporation to take place before humidified air can be circulated. On/off control of output means imprecise response to system demand and continual danger of saturation. Water spray systems can distribute large amounts of bacteria, and unevaporated water discharge can collect in ducts, around drains and drip pans and on eliminator plates, encouraging the growth of algae and bacteria. Corrosion is another ongoing problem with water spray humidification. Scale and sediment can collect on nozzles, ductwork, eliminator plates, etc., leading to corrosion and high maintenance costs.

3.3.1 Impeller humidifier (cool mist humidifier)

A rotating disc flings water at a diffuser, which breaks the water into fine droplets that float into the air. This is used as a localised humidification unit, especially above the mixing bins where stack mixing is followed. This type of humidifiers hasa problem that dust in the air gets choked at the edges and the water gets accumulated and big drops start falling. Hence, periodic cleaning with a hard brush fitted to a long bamboo stick is practiced.

Figure 3.10 Cool Mist Humidifier

3.3.2 Water atomization with compressed air

Compressed air attains high velocity as it expands and this energy is utilised to atomise water. There are two systems in pneumatic atomisers, viz.,high-pressure system and low pressuresystem. The high-pressure system employs air at 2 to 3 Kg/cm2.

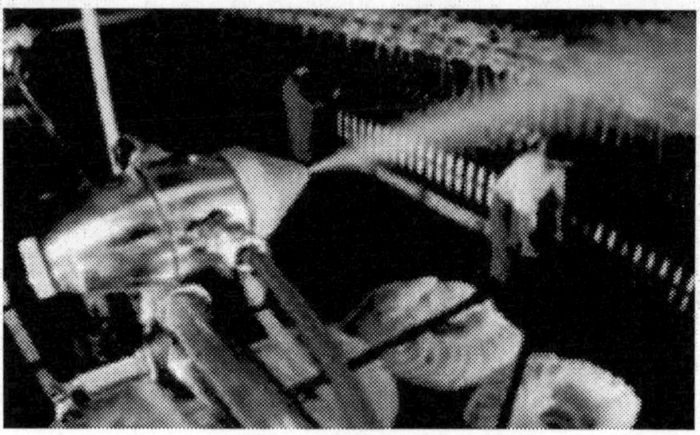

Figure 3.11 Water atomisation with compressed air

Water may be introduced into the atomiser by vacuum created by using compressed air as an ejector or it may also be supplied to the atomiser at the same or at a slightly higher pressure than air pressure. The capacity varies from 2.5 Litres/Hour to 7 Litres/Hour. The low pressure system employs air at about 0.5 kg/cm2, and water is also supplied at low pressure.

The water atomisation with compressed air is one of the cheapest methods adopted. Although it has the disadvantage of absence of ventilation or air movement, it is being used in a number of small installations and old textile units. Also, it is used for increasing humidity in a localised area. Its low cost, ease in installation and absence of any structure to house it are the factors in its favour. Now mobile units are also available.

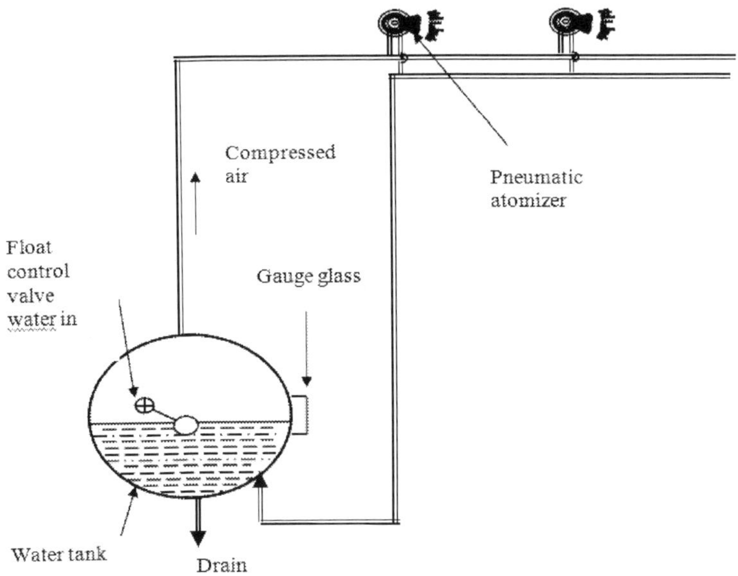

Figure 3.12 Compressed air humidification

3.3.3 The nozzle-free, horizontal airflow fogger

Nozzle-free horizontal airflow foggers have unique design that simplifies principal moving parts to one motor and its integral fan. Water or other fluids are fed into the specialised hub of fan blade assembly where it is subdivided and channeled into passage ways running the length of each blade.

Figure 3.13 Nozzle-free horizontal airflow fogger

As liquid exits, it is atomised into ultra-fine fog particles. Aquafog is an example of this type of units. Unlike other fogging systems, which may require permanent plumbing installation, high pressure pumps, complicated filtration and constant maintenance of clogged nozzles, Aquafog's streamlined one motor / one blade design produces and disperses fog using a single nozzle-free blade assembly.

3.3.4 Centrifugal disc humidifier

In this type of system, water is spinned around a disc and thrown away by centrifugal force and meets a circular comb that breaks the water into fine particles, which are carried away by airflow by the fan behind the rotating disc. Because of the air circulation, there is a good atomisation and humidification. The commercially available humidifiers atomise 20 to 25 litres per hour water with about 1700 to 3500 M3/hour air circulation. This type, also called as L type centrifugal humidifier, throws lot of water in the form of coarse water particles and hence its utility is getting reduced. It is not used where sophisticated machines are working.

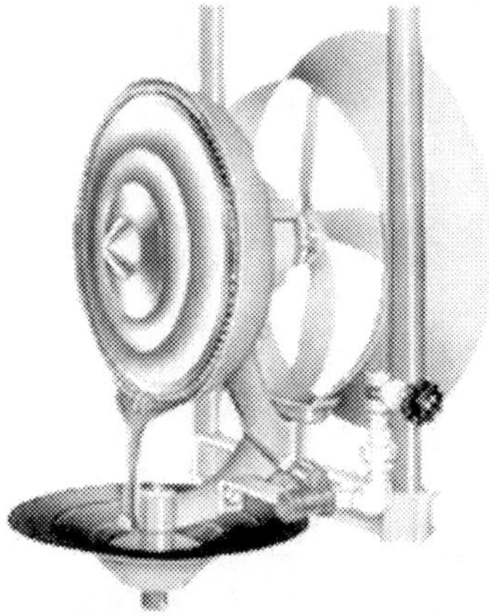

Figure 3.14 Disc humidifier

New methods are being developed world over with an intention to get smallest particles of moisture to have effective humidification and to have lower maintenance cost. In this chapter, we just glanced some of the systems. In the next chapter, let us study in detail the systems normally used in textile mills.

4

Air handling units in textile industry

4.1 Introduction

Willies Carrier designed spray-type air conditioner (air washer) for his textile mill that became the basis for the modern air handling units used in textile industry. He used air washers extensively for cooling and dehumidification in many industrial applications. Over a time, this air conditioning approach became less popular in offices and commercial places, but is still very popular in textile industry as in addition to conditioning a supply air stream, air washers are capable of cleaning the supply air stream with low air-side pressure drop and minimum maintenance.

Figure 4.1 Air washer designed by Willies Carrier

Air washer plants are big centralized units, which use water spray from a bank of nozzles on air moving with a force. The water spray not only adds humidity but also cleans the air from dust by making them wet and heavy, which fall down. A set of eliminators prevent the movement of water droplets and the dust particles along with the humidified air. Hence, the production area gets humidified clean air. Functions of an air washer are cooling the air, humidifying the air, dehumidifying the air and cleaning the air of particles and hyper ionizing the air.

Fig. 4.2 illustrates the major components in an air washer unit. Chilled water is sprayed directly into a pre filtered supply air stream. The chilled water absorbs heat from the air and condenses moisture from the air when the supply water temperature is below the entering air dew-point temperature. The water spray nozzles can be arranged in co-flow or counter-flow with the air stream. Also, an extended media fill material can be used to increase the surface area of water in contact with the supply air stream. A downstream mist eliminator may be needed depending on the face velocity of air through the unit as well as the characteristics of the water spray droplets.

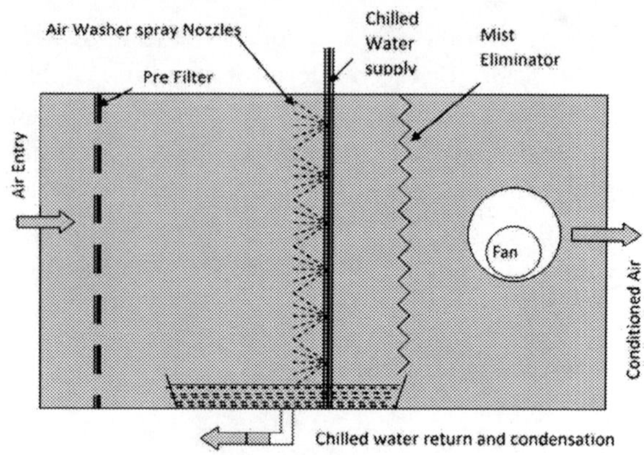

Figure 4.2 Components of an air washer plant

Water sprays in an air washer configuration have the ability to filter air with an efficiency of 95% for particles larger than 5 μm. Although many microorganisms are much smaller (around 1 μm), most microorganisms tend to agglomerate to form 'macro particles' with aggregate sizes greater than their mono-disperse sizes. These microorganisms can be effectively scrubbed from a contaminant-laden air stream by an air washer.

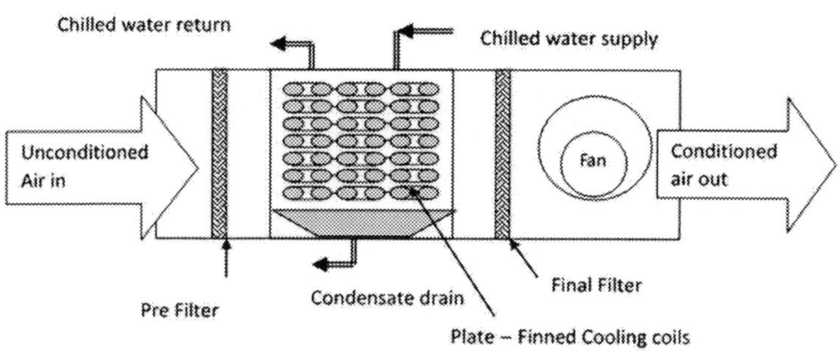

Figure 4.3 Central station air handling unit

In contrast to the air washer, Fig. 4.3 shows the major components in a traditional central station air-handling unit using a plate-finned chilled water cooling coil. In this arrangement, the water is supplied to a multi-row cooling coil and does not come into direct contact with the supply air stream. The chilled water cools the coil surface which, in turn, cools and dehumidifies the supply air stream. Moisture from the supply air condenses onto the coil surface and drains by gravity to a pan located beneath the coil. Pre and final filters are needed for particulate control. Although pre filters are generally located upstream of coils, designers may locate the final filters either upstream or downstream of the cooling coil.

Air washers, as they are big in size are normally a centralised system. Central humidification plants are of masonry type, the components of the plant housed inside a civil room with brick wall and RCC slabs and usually located adjacent to the production department at strategic locations based on the nature of the truss for the feasibility of layout of overhead ducting with minimum length of travel and based on the ease of connecting the underground trenches with minimum bends. Normally modern air washer plant components have unique features contributing towards energy saving, ease of maintenance and long life, such as multiple fans with direct drive, PVC diffusers, PVC air washer components with galvanised MS / SS hardware, filters of GI /SS and control dampers of powder coated sheet metal type / aero foil aluminum extruded type. Spray systems are designed and fitted with nozzles to produce a maximum of 90% to 95% saturation efficiency. The mills need to provide information on size of the department (volume to be controlled, the installed heat generating units such as motors, power load, machines, men employed), type of roof, false ceiling, machinery load, light load, RH required, maximum and minimum outside conditions (dry-bulb and wet-bulb temperatures) in summer and winter, layout of the machinery showing the proposed location of the plants, so as to design a plant appropriate to the production facility.

Processes used in conditioning air from one state to another state vary widely depending on factors such as the volume and qualities of the air to be conditioned, the temperature and humidity to which the air must be conditioned and whether a portion of the conditioned air will be re-circulated or 100% outside air will be used. Additionally, the qualities desired in the conditioned air vary. In some instances, only the temperature of the air is important whereas in other instances the humidity of the air is more important.

The various components of air handling units in textile industry include centrifugal fans or axial flow fans, high-pressure blowers, heavy duty exhaust fans, air volume dampers, input air filters, mist eliminators, water sump, spray nozzle bank, condensers, chillers, heat exchangers, steam heaters, cooling coils, ducting, air distribution baffles or louvers, supply air diffusers and grilles, return air grilles, split grilles, etc.

Air washers using spray water as the medium for adiabatic cooling of air (by direct evaporation of water into the air stream thereby reducing the air's dry-bulb temperature and raising its humidity) are extensively used in humidification systems for textile mills, due to the following advantages.

- Significant saving in initial capital cost and energy cost with evaporative cooling systems, compared to air conditioning systems.
- Reduced power demand with reduction in electrical maximum demand charges.
- Improved air quality due to air cleaning properties of spray water.
- Flexibility to use chilled water in air washer instead of normal temperature water, when outside air is exceptionally hot and humid.
- Humidification system deploying water as cooling medium is environment friendly.

Spray air washers generally used in humidification plants consist of a chamber containing multiple banks of spray headers with spray nozzles, a tank for collecting spray water as it falls and an 'eliminator section' with PVC blades having 3 or 4 bends for removing droplets of water from the air after passing through the curtain of spray water, before discharging to the air ducts for distribution to production area. Air velocity, water spray density, spray pressure and other design criteria are optimised by each manufacturer, depending on the air-washer dimensions and spray header /nozzle sizes and configuration and eliminator design which are developed for different applications. Earlier designs of air washers were based on low air velocity, up to 3m/s (600 fpm). Some manufacturers have now developed high velocity air washers operating at 6m/s (1200 fpm) and even up to 9m/s (1800 fpm) air velocities which make the unit compact and facilitate shipment in pre-fabricated and assembled condition. However, high velocity air-washer designs require special eliminator design and construction and due to the additional pressure drop, the supply fans are to be selected at a higher static pressure

Resistance to air flow through the air washer varies with the type and number of inlet baffles, eliminator design, number of spray banks, air velocity, other components such as heating coil if any, dampers, etc. The total pressure drop can vary between 15 mm and 30 mm w. g depending on the overall size, air velocity, spray header and eliminator design.

Pre-fabricated air washer units with sheet metal casing and F.R.P-lined MS tank and other internal components in assembled condition are available in small and medium capacities up to 170,000 m/hr. (100,000 cfm). For large capacity air washers, casing may be constructed in masonry and water tank in R.C.C and components including spray system, eliminators, spray pumps, supply and return fans, dampers, rotary return air filters and water filters are assembled at site.

4.2 Supplying air to air washer plants

The supply air dampers or louvers regulate the amount of air fed to the air washer plant. They are made with light sheet metal which is rust proof. They can be opened or closed as shown in Fig. 4.4. It can be either operated manually or connected to a control unit, which monitors its opening.

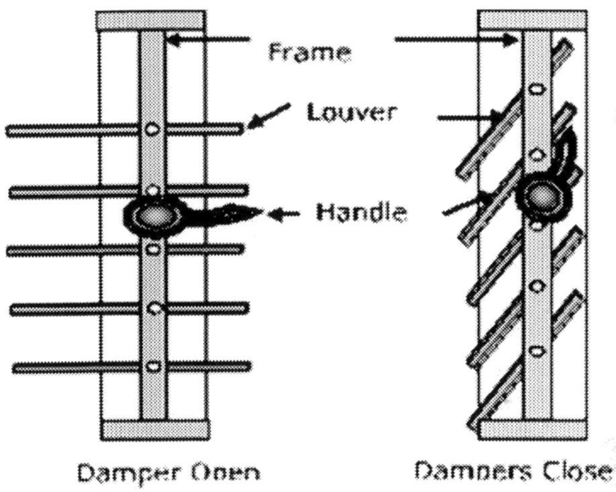

Figure 4.4 Supply air dampers

Separate set of dampers are installed so as to facilitate either taking only outside or recirculating the inside air. Sometimes a combination of outside air and inside air shall be made.

The louvers are made with light sheet metal or PVC. The size of louvers depends on the size of the windows or outlets. We can install either one set of louvers for one window or multiple sets, so that a part of air movement can be controlled.

VAV air-handling systems are equipped with routine configurations and control applications that are regularly taken for granted. For example, a single damper actuator often mechanically links and controls the outdoor and return air dampers of an air side economiser. Even if multiple actuators are used, they are often connected to the same output from an electronic controller and still work in tandem. Furthermore, when multiple actuators are controlled from different outputs, the control sequence is generally written so the outdoor air damper opens while the return damper closes at an equal rate. These control methods are traditional, proven and comfortable, but they do not necessarily represent the best way of controlling.

Sixty years ago, HVAC systems were constant-volume and variable-temperature systems. Fan sheaves needed to be adjusted and balancing dampers set so that the constant-speed fan could maintain a constant volume of air flowing at all times. Any variation in system resistance affected the amount of airflow in the system. Many research were conducted to determine the best way of sizing and selecting damper characteristics so the pressure drop in the mixing-box section of the AHU remained constant as the system went from minimum outdoor air to 100% outdoor air.
ASHRAE research determined that in a simple, draw-through, single-fan application, parallel-blade dampers for both the outdoor and return air, when modulated in equal increments but in opposite directions from return air to outdoor air, provided the most consistent system pressure drop at all positions of the two dampers.

Opposed-blade dampers exhibited almost four times the pressure drop with both dampers throttled 50% vs. one damper open 100% and the other damper closed. The parallel-blade outdoor and return air dampers, modulated together, provide the most linear response for constant-flow systems for HVAC systems with no return fan and almost equal pressure drop in the outdoor and return air paths. To open both dampers 100% when equal parts of outdoor air and return air are required is not a complex control algorithm. It merely requires sequencing the dampers.

Filters or strainers are installed in the supply airline to prevent unwanted materials such as leaf bits, fibres and dust particles coming along with air, which might spoil the equipment and chock the air path. Normally nylon or polypropylene mesh is used in case of recirculation of air, and wire mesh is used for outside air inlets. The size of mesh depends on the place of their installation. The mesh for fresh air is made with sturdy metal wires, and the openings are also bigger with areas of over 1 square inch. The purpose of this mesh is to restrict the direct entry of large-sized floating particles in outside air like leaves, papers bits and polythene pieces. It also protects the plant from birds.

The air from the production area is taken from specially designed underground trenches. Specially designed grills are provided to prevent heavy materials such as bobbins, cones, lumps of cotton and paper pieces from going in the trench.

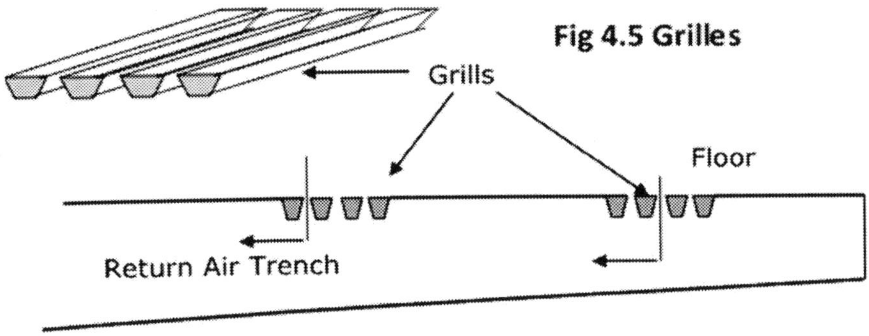

Figure 4.5 Grilles

In the return air ducts meshes of different sizes are used in tandem. The first set of mesh shall be coarser preventing heavy particles and lumps of fibrous dust entering the plant, whereas the subsequent meshes restrict the flow of small dust particles and micro dusts.

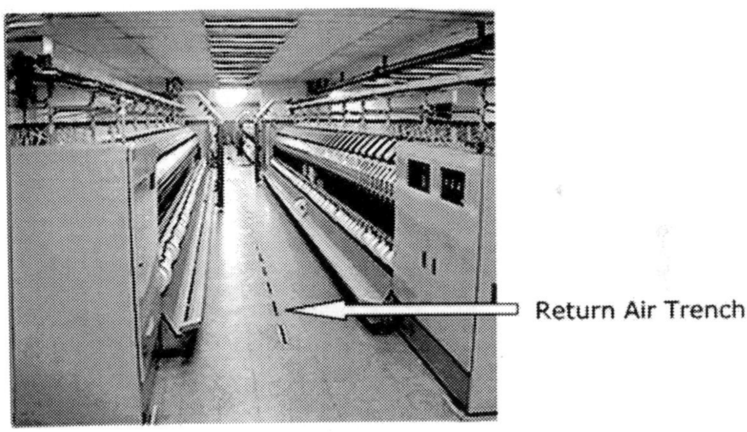

Figure 4.6 Return air trench

Floor grilles are the exhaust grilles fitted at different locations over the trench. The spent air along with floating fibre passes through this grille into the trench. The grilles should be painted with enamel paint. It is essential to specify aluminium with a baked enamel factory finish for grilles, registers and diffusers and not allow field painting or flat paint finishes.

Tim Carter suggests installing grill covers on the wall that has operating louvers to control the flow of return air depending on the situation.

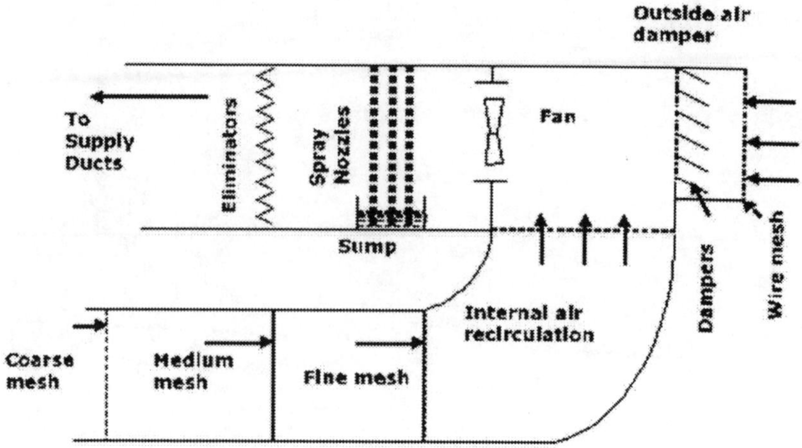

Figure 4.7 Typical air washer plant

In the return air system, contaminated air is returned to the AC, located outside of the room, through openings under the machines. Solids are removed by filtration and/or water sprayed into the air stream. An important part to be noted is majority of the solids in the room are carried upward by the heat generated by the machines, only about 30% to 40% of the total fall out is captured by this system. However, by placing the supply air diffusers in the top and return air suctions in the bottom and maintaining the quantity of air extraction balancing with the air inflow can reduce the problem of the solids going up.

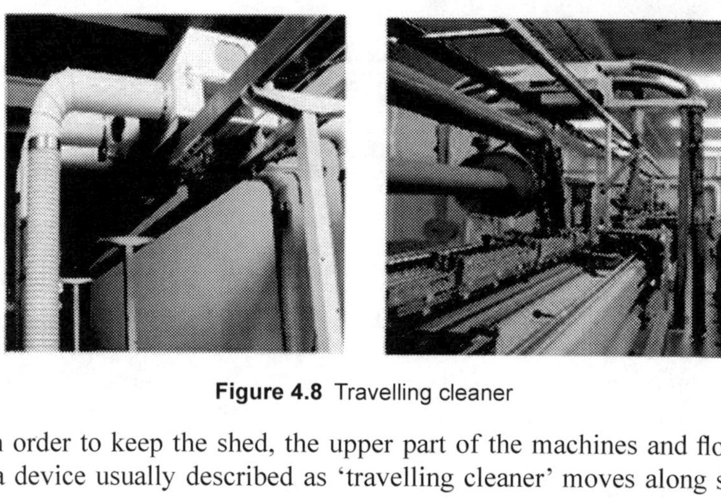

Figure 4.8 Travelling cleaner

In order to keep the shed, the upper part of the machines and floor lint free, a device usually described as 'travelling cleaner' moves along several machines, periodically blowing solids away from those parts and vacuuming off another 25% to 30% of the total. These travelling cleaners blow the air on the parts of the machine and make the accumulated fluff to become loose, and the sucks them. The sucked dust is collected in a bag and air is let out from the pores in the filter cloth by which the bag is made of. The remainder of the lint and dust is moved by air turbulence all over the room and clings to the machines, the ducts, the light fixtures, the ceiling and the walls. Its removal has to be performed by hand.

4.3 Air filter construction and cleaning system

The air fed to the air handling units should be filtered to have maximum efficiency. The function of these filters is to remove lint, dust, oil in some cases and other debris before they get into the water in the air washer units. There are a number of design factors that determine the CFM and air changes per hour (ACH). The most important factor is the selection and placement of filters within the system. Once the filter media is selected, the placement of the filters, both in terms of filter order and spacing within the system, must be considered as both contribute to the level of CFM that a unit can produce. There is no way to define the correct design of an air purification system, but evaluating the way in which airflow is created and restricted by the filter media is one way to better understand the ACH and the efficiency that can be achieved with the unit. The presence of pre-filters within air purification systems is critical to protecting the efficiency of the system. By removing larger particles from the air before it is cycled through the HEPA or HAPA filter media, higher airflow levels can be maintained for longer periods of time. Common dust concentrations exposed to filters in ventilation and air-condition systems are 0.45 to 1.0 mg/m^3 air in rural and suburban districts, 1.0 to 1.8 mg/m^3 air in metropolitan districts and 1.8 to 3.5 mg/m^3 air in industrial districts.

Ventilation filters may be classified and characterised after separation efficiency, i.e., ability to retain standard dust, blacking efficiency, i.e., capability to retain atmospheric dust, particle separation efficiency, i.e., capability to retain particles at fixed sizes, dust accumulation capacity, i.e., capability to keep standard dust before the pressure drop extend a certain value and airflow resistance and static pressure drop of the filters.

Filters can be grouped as air washers and dry filters. In air washers, the air is filtered by the water jets, whereas in dry filters, a porous medium such as felt, cloth, cellulose, glass, silk, etc., without adhesive liquid is used through which the air passes. The filtering may be by stationary filters or rotary filters with automatic cleaning. The stationary filters are either plain perforated sheets or zigzag sheets. Some of these units have moving paper media, while others have semi-permanent synthetic media. These are replaced depending on the usage. Automatic rotary air filters have a large rotating drum, made of perforated steel sheet and fitted with suitable filter media to arrest fluff and are fitted with drive mechanism and geared motor and necessary suction nozzles connected to a flexible suction hose for automatic cleaning and removal of the dust and fluff with centrifugal-type suction fan into a waste collection unit.

The performance of different filter media is normally as follows.

- Flat Panel type – disposable filters: Air velocity 0.1–1.0 m/s, Resistance 25–250 N/m2, Efficiency 20% to 35%
- Continuous roll – self-cleaning filters: Air velocity 2.5 m/s, Resistance 30–175 N/m2, Efficiency 25%
- Bag filters – Efficiency 40–90%
- HEPA (high-efficiency particulate air) Filters – Efficiency 99.97% for 0.3 μm particles and larger
- ULPA (ultra low penetrating air) Filters – Efficiency 99.9997 for 0.12 micron particles or larger
- Viscous filters-panel type –Cloth with viscous fluid coating – washable or disposable: Plates about 500 mm x 500 mm, Air velocity 1.5–2.5 m/s, Resistance 20–150 N/m2
- Viscous filters – Continuous roll – continuously moving, self-cleaning: Air velocity 2.5 m/s, Resistance 30– 175 N/m2
- Electrostatic precipitators: Cleaned automatically, Air velocity 1.5–2.5 m/s, Resistance negligible, Efficiency 30–40%
- Absolute: Dry panel with special coating – disposable or self-cleaning, Air velocity 2.5 m/s, Resistance 250–625 N/m2

Automatic cleaning is preferred as the generation of dust and fibre-fly varies and pressure drop in the filter is to be kept within the design limit to prevent reduction of outgoing air flow and also save energy consumed by the blower. Any reduction in outgoing return air flow is to be prevented to avoid buildup of fluff and fibre fly, as they will contaminate and spoil the yarn produced.

HEPA filtration technology which is recommended for cleaning air to make it safe for breathing was designed during World War II by the Atomic Energy Commission to remove and capture radioactive dust particles in order to protect the human respiratory system. HEPA filters are rated to remove up to 99.97% of all particles 0.3 μm in size or larger, which encompasses most non-viral airborne particles. HEPA is an acronym for 'High Efficiency

Particulate Air'. Basically, HEPA is a type of filter that can trap a large amount of very small particles that other vacuum cleaners would simply re-circulate back into the air. There are a couple of different categories of HEPA filters that can make understanding the abilities of filter confusing. HEPA filter media has important efficiency properties that are critical to successful air purification solutions. Not only is the filter media efficient at cleaning the air that passes through the filter, but it is designed in such a way that as particles are captured on the surface of the filter the cleaning efficiency of the filter actually increases before eventually decreasing as the filter becomes filled with particulates. There is also a new type of filter media called high airflow particulate arresting (HAPATM) that further improves on HEPA technology and can help systems achieve increased air circulation. Like HAPA, it is also rated to remove up to 99.97% of all particles 0.3 μm in size or larger. This filter media is constructed of a totally synthetic melt blown fibre material that is more durable than traditional HEPA filters. It has also shown greater depth loading capacity than HEPA-rated filter material, which means that more particles can collect on the filter surface before the efficiency of the filter begins to decrease. HAPA is less restrictive to airflow than HEPA, and systems utilizing HAPA filter media are capable of achieving higher CFM output than systems with HEPA filtration.

Filters are classified as follows by European Standards:

1. Basic Filters: Basic Filters in general are produced by using synthetic fibres. They are efficient for particles greater than 4–5 mm and with air speed more than 2.5 m/s. The pressure drop starts at approximately 50 Pa and final pressure drop approximately 150 Pa.

Table 4.1 European Standard Filter Class EU1 to EU4 – basic filters.

Class	Effect	Applications
EU1	Protects against insects and fibres Limited effect against larger pollen (<70%). Ineffective against smoke and blacking particles	Window units, heat exchangers, air heaters and fibre filters in textile industry
EU2	Effective against larger pollen (>85%) and larger atmospheric dust Limited effect against dust and blacking particles	Heating and cooling units in electrical transformers, garages, industrial halls and offices in industry
EU3	Effective against larger pollen (>85%) and larger atmospheric dust Limited effect against dust and blacking particles	Heating and cooling units in electrical transformers, garages, industrial halls, offices in industry
EU4	Limited effect against dust and blacking particles	In addition to EU3 kitchens and spray paint workshops

2. Fine filters: In general the fine filters are produced in glass fibres. They are efficient for particles bigger than 0.1 mm and air speed higher than 2 3 m/s. The pressure drop starts at approximately 50-100 Pa, and reaches final pressure drop at approximately 200-250 Pa.

Table 4.2 European Standard Filter Class EU 5 to EU 9 – finesfilters.

Class	Effect	Applications
EU5	Effective against pollen and finer atmospheric dust, Considerable effect against smoke. No effect against tobacco smoke	Churches, sport halls, department stores, schools, hotels, food stores
EU6	As EU5	As EU5
EU7	Effective against pollen and blacking dust	As EU6 and food industry, laboratories, theatres, hospital rooms, data rooms
EU8	Very effective against particles and blacking. Very effective against microbes. Effective against tobacco smoke.	Operating theatres, production rooms for fine optics and electronics. Hospital examination rooms.
EU9	As EU8	As EU8

3. Micro filters: Generally produced in glass fibres in combination with separators of paper, plastic or aluminium, efficient for particles greater than 0.01 mm, air speed higher than 0.5–1.0 m/s, start pressure drop approximately 250 Pa, final pressure drop according life span and economy. EU 10 to EU 14 is the classes available.

4.3.1 Stationary air filters

The initial filters used in the return air duct were just a flat wire mesh, which got replaced by nylon and polypropylene. As the volume of air required to be circulated was more and the air was dustier in cotton textile mills, there was a problem of frequent chocking up of the filters. The zigzag meshes were developed to provide more surface area for the air being filtered. Although, this reduced the frequency of cleaning, still the problems of improper cleaning, and progressive reduction of filtering were there. The development of rotary air filters, which are now very common in all modern plants ensures continuous cleaning of mesh, and always expose clean mesh for the air entering. This has improved the efficiency of air washer plants significantly.

The flat stationary filters, which are also termed as static filters, are used where the quantity of air moved is less. It can be seen in unit humidifiers installed above, the machines near the roof, trenches of the return air, opening provided to suck outside air in windows, etc. Apart from frequent chocking up of the mesh surface, the frequent cleaning makes the mesh weak. The problem of mesh tearing off was very high with flat stationary meshes. The zigzag surface made the mesh surface strong.

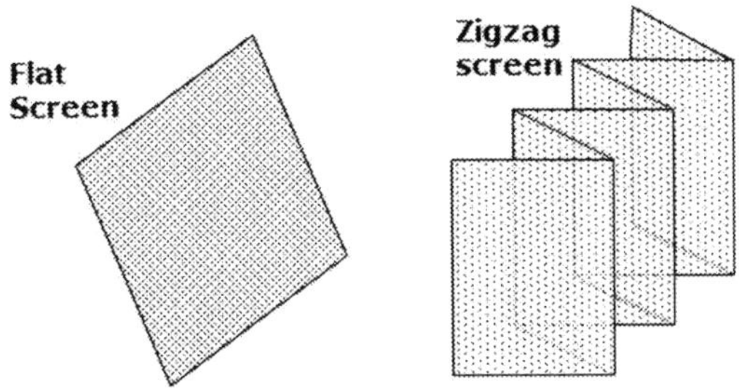

Fig 4.10 Stationary filter screens

For effective cleaning of the stationary screens, it is necessary to switch-off the plant. If the plant is switched off, the accumulated dust can be removed easily or else the force of suction makes the dust to cling on the surface. Stopping the plant and then entering the return air ducts to clean takes more time and is not affordable when the production is going.

4.3.2 Rotary air filters

In the rotary filters, the filter area gets cleaned on a continuous basis and hence has the best efficiency. The principle adopted is similar to the condensers used in Blow room for distribution of cotton to hopper feeders. The dusty air is sucked by a fan. The air on its way hits the rotary drum, which is a perforated cage, covered with a filter cloth.

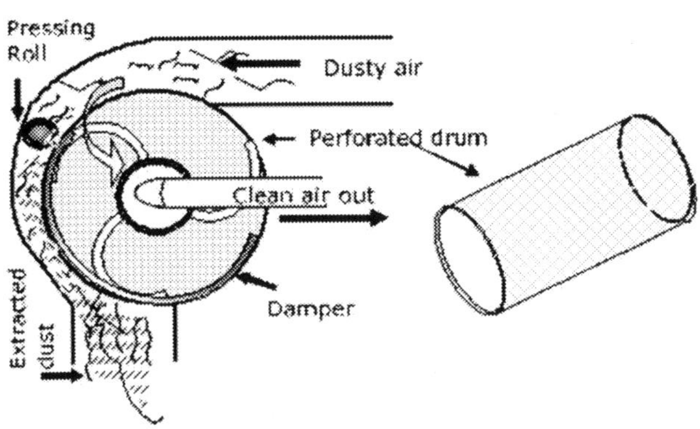

Figure 4.10 Rotary Air filter

The air is sucked from the inside of the cage and taken out by the side. The filter cloth allows only clean air to pass through the rotary drum. The fibrous dust collected on the filter cloth is taken out after pressing. A semi-circular damper prevents the fibrous dust from sticking to the rotary drum by obstructing the air movement. By this, the dust becomes loose. They are pressed by a pressing roller and collected separately. The clean surface of the drum is always getting exposed to the fresh dusty air being sucked. Therefore, the air which is let out of the filter is always clean and is uniform. As there is no requirement of manual cleaning like we have with stationary filters, the consistency is obtained in the air quality.

4.3.3 Conditioning the return air

Ronald J. Yeager and Gary D. Wolf (U.S. Patent 140760 dated 25th April 1995) developed a method and apparatus for conditioning the temperature and humidity of un-recycled ambient air without reheating. It provides a method and apparatus for conditioning the temperature and the humidity of un-recycled ambient air to a desired temperature and humidity by overcooling one part of the ambient air and bypassing another part of the ambient air and recombining the two parts to create air at the desired temperature and humidity. The apparatus comprises a cooling coil with variable cooling capacity and humidifier in parallel with a bypass duct. A portion of the incoming 100% outside air stream is routed through the cooling coil and humidifier and the remainder of the outside air is routed through the bypass duct. The cooled stream and the bypass stream then combine downstream of the exits of the cooling coil and bypass duct such that the combined air is within the conditioned window.

Some textile air washer systems are designed with temperature and humidity controls to automatically blend outside air with the in-plant air appropriate to the needs of the particular textile process being run in the plant. The mixture of air enters the actual washing section of the air washer unit after passing through solid cartridge-type filters.

4.3.4 Local dust collectors

These are small mobile units for collecting dust from the production units, where dust liberation is high because of the nature of the product being worked on particular machines. A small perforated drum is kept vertical, and an exhaust fan is fitted at the top. The dust liberated is collected on the surface of the drum, and the air is let off in the same work area. This type of units is used normally in reeling section, cutting section in garment industry where fleece cloth is in process, warping etc. The dust collected on the surface of the drum is cleaned manually at fixed intervals by stopping the fan.

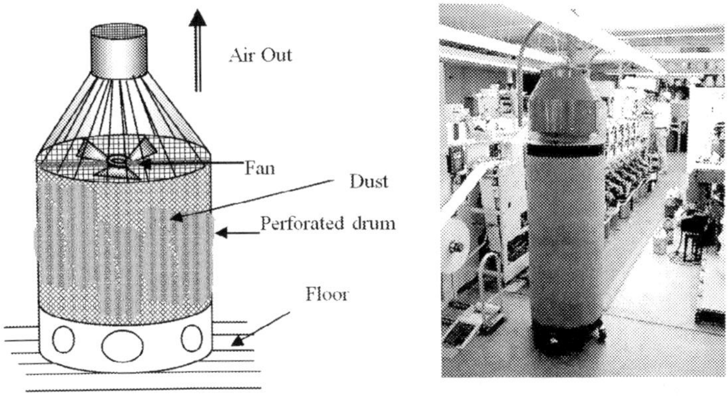

Figure 4.11 Local dust collectors

A number of commercial units are available, not only for use in textiles but also for different industries and organizations. The filtration capacity depends on the filter media.

4.3.5 Negative ionizers

High-density negative ionizer takes over where normal perforated filters cannot filter all micro particles. They operate by electronically generating a powerful stream of negative ions by around 70 trillion per second in intermittent pulses which 'wash' the air at a rate of nearly 100 feet per minute.

Air ions are molecules that have lost or gained an electron. They are present in normal air but are 'stripped' out when an air is subject to filtration and conditioning. Ions are produced by radioactive emission or by phenomenon called 'corona discharge' where a high voltage is applied to sharp point. All air ionization systems work by flooding the atmosphere with positive and negative ion, when ionized air comes in contact with a charge surface the charge surface attracts ions of the opposite polarity. As a result the static electricity that has built up on products, equipment and surface is neutralized. Typically, air is very insulative with a resistive exceeding 10E15 ohms/meter. By increasing the numbers of ions in the air it is possible to lower the resistivity of the air to 10E11 ohms/meter, thereby making the air more conductive. Conductive air can neutralise the static charge on every surface that it contents. The field from the charged surface attracts ions of the opposite polarity until the charged on the surface is neutralized. The static neutralizer is portable and uses a small fan to produce airflow. High voltages generated out of the supply line voltage are applied to stainless steel ion emitter points. These points produce intense alternating polarities, ionizing the air instantly in the fan airflow. The static eliminator-neutraliser is designed for use when handling sensitive electronic components where electrostatic discharge is a problem. The unit can also be used where static electricity causes problems such as attraction of drift to product, misalignment of small parts due to electrostatic 'jumping' and undesirable adhesion of plastic films due to electrostatic charge.

4.4 Type of fans used

A fan is typically a mechanical device that causes a movement of air, vapour and other gases in a given system. The fans are fixed on wall in between the air inlet and the spray nozzles. The air is blown on the water spray, which then hits the eliminators and passes into the distribution ducts. Diffusers are used to spread the humidified air uniformly in the work area.

For effecting the air flow, the fan develops a total pressure difference over the inlet and outlet air streams; the total pressure rise comprises two components, viz., i) static pressure that depends on the blade profile, number of blades, pitch, hub shape and so on, i.e., aerodynamic characteristics of the fan impeller and ii) dynamic pressure which develops due to velocity or kinetic energy imparted to the air stream. The fan static pressure is fully available to the user, but some of the fan velocity pressure is inevitably lost, although this loss appears as part of the system pressure drop.

There are two types, viz., axial flow fans and centrifugal fans used for blowing air into the ducts. In centrifugal fans, air is let through an inlet pipe to the centre or eye of the impeller from where it flows to a discharge pipe. In an axial-flow fan, with the runner and guide vanes in a cylindrical housing, air passes through the runner without changing its distance from the axis of rotation. Following are some of the terms used in Fans.

- *Aerofoil*: Airfoil also spelled as aerofoil, is a shaped surface, such as an airplane wing, tail, or propeller blade that produces lift and drag when moved through the air. An aerofoil produces a lifting force that acts at right angles to the air stream and a dragging force that acts in the same direction as the air Stream.
- *Lift:* If an aerofoil is to fulfill the desired function, it must experience a lift force, as well as, a drag force when in motion. The lift force arises because the speed at which the displaced air moves over the top

of the airfoil and over the top of the attached boundary layer is greater than the speed at which it moves over the bottom. The pressure acting on the airfoil from below is, therefore, greater than the pressure from above. The design of the airfoil, nevertheless, has a critical effect on the magnitude of the lift force.

- *Drag:* A fluid stream exerts a drag force (FD) on any obstacle placed in its path, and the same force arises if the object (aerofoil) moves and the fluid is stationary. How large it is and how it may be reduced are questions of obvious importance to designers who want to be certain that the structures will not collapse in the face of winds. It is conventional to describe drag forces in terms of a dimensionless quantity called the drag coefficient, defined irrespective of the shape of the body.
- *Selection of Fan:* Factors like cost optimisation, power rating and noise levels govern the selection of a fan suitable for a given application. Various combinations are possible to meet any given duty or operating requirements for best performance, lesser noise, power and cost characteristics. Once the volume of airflow and the static pressure of the system are known, it is possible to specify a fan. To select a fan one must consider the parameters like total airflow required, total operating pressure and fan installation space. Other considerations such as noise, reliability and operating environment should also be brought to bear on fan choice.

4.4.1 Axial flow fans

Axial flow fans, as the name indicates, draws the air and blows forward, which moves in the axis of the fan. There would be no centrifugal effect on the airflow generated. Guides or stator vanes serve to smoothen or straighten the airflow and improve efficiency. In general, an axial-flow fan is suitable for a larger flow rate with a relatively small pressure gain and a centrifugal fan for comparatively smaller flow rate and a large pressure rise. They are used to supply fresh air, to suck air from return air trenches, to suck air from rotary filters, to exhaust air out and so on. Depending on the purpose and the quantity of air handled, the size and materials of the fan are decided.

In a typical axial fan, the effective progress of the air is straight through the impeller at a constant distance from the axis. The primary component of blade force on the air is directed axially from inlet to outlet and thus provides the pressure rise by a process that may be called direct blade action. The blade force necessarily has an additional component in the tangential direction, providing the reaction to the driving torque: this sets the air spinning about the axis independently of its forward motion. The air delivering capacity of axial flow fans ranges from 100 to 500,000 cubic feet per minute (3 to 14,000 cubic meters per minute). Impellers usually have blades with cross sections matching those of an aerofoil. As compared to curved sheet blades, aerofoil can apply greater force to the air, thereby increasing maximum pressure and can maintain better efficiency over a wider range of volumetric flow. By increasing the thickness and curvature of the inner sections, the blades can be made stiffer; this limits flutter and allows the impellers to be run at higher speeds.

Figure 4.12 Axial flow fan

Normally high-efficiency axial flow fans, with aluminium impellers, adjustable pitch aluminium blades with direct drive totally enclosed motors are selected to deliver the design supply air quantity against the required static pressure, after considering pressure drop in fresh air damper, air washer internals, washer-dampers, supply air ducting, and supply air diffusers with volume control dampers. Design static pressure for supply air fan selection can vary from 40 mm to 55 mm w. g depending on the air washer length, spray headers and eliminators design, other internal components including heating coils, if any, design air velocity, etc. Return air fans are sized to recirculate up to 95% of design supply air quantity for each department, by considering the design static pressure required to overcome pressure drop in the return air floor grilles, masonry return air trench, rotary drum type return air filters, return air dampers, etc. Since the pressure drop in rotary drum filters, when dirty, can be as high as 30 mm w. g, the return air fan selection in mill humidification applications should account for up to 25 mm w. g. pressure drop in air filters and the return air fan selection and motor KW should cater to the required return air flow rate against a total static pressure of 50 to 55 mm w. g. after considering pressure drop in the floor grilles and return air trench.

Factors such as cost optimisation, power rating and noise levels govern the selection of a fan suitable for a given application. Various combinations are possible to meet any given duty or operating requirements for best performance, lesser noise, power and cost characteristics. Once the volume of airflow and the static pressure of the system are known, it is possible to specify a fan selection.

The axial flow fans are conventionally designed with impellers made of aluminium or mild steel. The grey area is the inconsistency in proper aerofoil selection and dimensional stability of the metallic impellers. This leads to high power consumption and high noise levels with lesser efficiency. The leading fan manufacturers in the world have been looking at FRP axial flow fans for higher energy efficiency. The improved design of FRP fan is aimed at higher lift to drag ratio and thereby increasing the overall efficiency. The new and improved aerodynamic fan designing, composite development, structural design combined with latest manufacturing process are also expected to result in consistent quality and higher productivity. Studies done by G. Srikanth, Sangeeta Nangia and Atul Mittal, for the project on *'Development of Energy Efficient FRP Axial-flow Fans'* launched by the Advanced Composites Mission of TIFAC showed a significant savings in power by using FRP Fans.

4.4.1.1 Advantages of axial flow FRP fans

The FRP fans offer certain critical advantages such as optimal aerodynamic design of fan impellers to provide higher efficiency for any specific application, reduction in overall weight of the fan, thereby extending the life of mechanical drive systems, requires lower drive motor rating and light duty bearing system, low power consumption resulting in appreciable energy savings. FRP fans fabricated by compression moulding/resin transfer moulding technique would have uniform dimensions and consistent quality and lower flow noise and mechanical noise levels compared to the conventional metallic fans.

Table 4.3 FRP Fan Test Results vs. Conventional Fans with Aluminium and Steel Impellers

Sl. No.	Type of FRP Fan	Flow Rate M^3/Sec	Total Pressure : mm of water gauge	Shaft Power KW
1.	Cooling Tower Fan	240.47	8.48	23.24
2.	Textile Mill Humidifier Fan	19.04	34.83	-
3.	Mine Ventilation Fan	48.60 to 81.00	92.83	89.63
4.	Radiator Cooling Fan for Railway Diesel Locomotives	49.76 to 60.21	88.56 to 102.98	74.95 to 78.60
5.	Air-heat Exchanger Fan	91.43 to 96.94	8.26 to 8.56	10.1 to 10.17

Contd...

4.4.2 Centrifugal humidifiers

In certain industrial and commercial environments, centrifugal humidifiers are used. In such devices, water is propelled outwardly by a rapidly rotating plate or other body to impinge on a surface where it is broken up into small droplets that are entrained in a stream of air and then discharged to the surroundings. Although the discharged air has a higher moisture content, the size of the droplets produced on the impingement surface is generally not sufficiently fine to be readily absorbed by the air. Instead, such humidifiers tend to produce a mist-like discharge. Not only does this result in a less than optimum overall humidity level but it also tends to create water build up on the surfaces of walls, equipment, furniture and so forth in the vicinity of the humidifier. More than merely an inconvenience, this promotes the growth of molds and other microorganisms that can pose health risks to workers in the area. Centrifugal humidifiers also have a tendency to become clogged with dust and other particles when they are used in industrial environments such as textile mills.

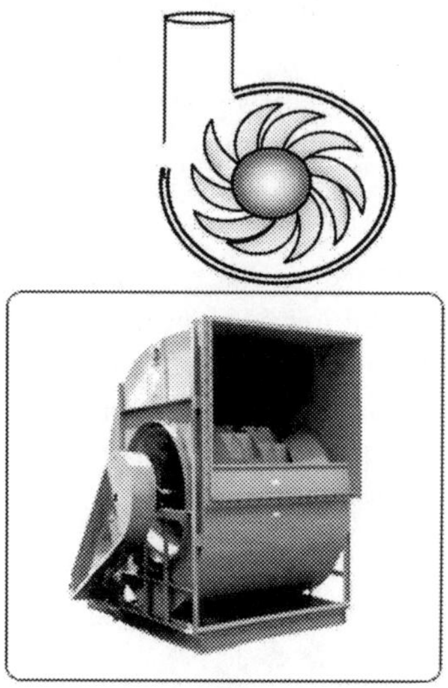

Figure 4.13 Centrifugal blower

Hashem, Feisal [United States Patent 6601778] in 2001 developed a centrifugal humidifier having a rotatable body for propelling water outwardly on to an impingement surface that break up the propelled water into droplets which are then entrained in a stream of air and discharged to the surroundings, wherein the impingement surface has a multiplicity of saw tooth ridges oriented with the more gradually inclined sides facing towards the direction from which the water is propelled by the rotatable body which promotes breaking up the water into finer droplets.

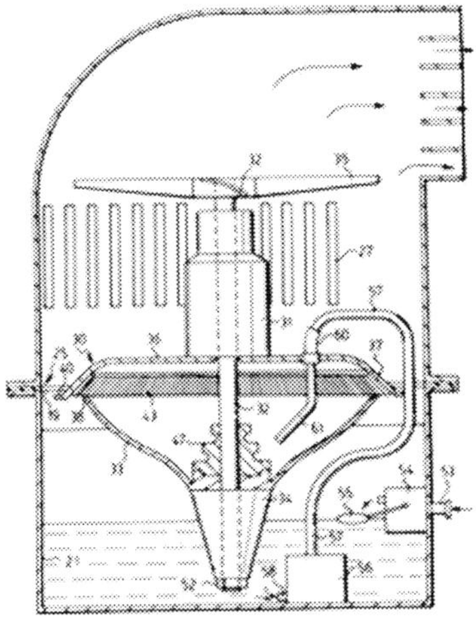

Figure 4.14 Hashem feisal centrifugal humidifier

Because of the saw tooth profile, the water tends to hit and bounce repeatedly on the first sides of successive ridges which promotes breaking up the water into finer droplets that are readily absorbed by the stream of air. The centrifugal humidifier therefor produce a fog-like discharge, rather than the mist-like discharge. This greatly reduces the problem of wetting nearby surfaces, and also greatly increases the throughput efficiency. This invention is claimed to eliminate or reduce the need for filters or similar moisture eliminators which are commonly used in other centrifugal humidifiers to reduce wetting. The saw tooth ridge impingement surface of also resists buildup of dust and other particles that can clog other centrifugal humidifiers.

4.5 Gravity louver damper

Gravity Louver damper also known as auto shutter is an arrangement provided at the delivery side of the axial fan and remain open when the fan is switched on and remain closed when the fan is switched off. This allows using less number of fans in a multiple fan system design without short circuit. Further when a motor trips and the fan is not able to run, the gravity Louver damper closes and the short circuit is prevented.

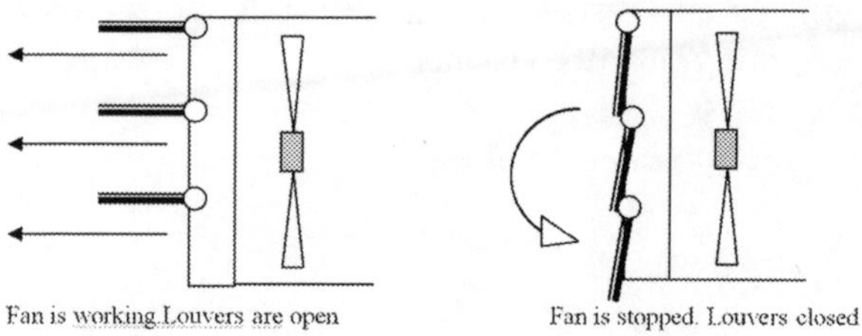

Figure 4.14 Gravity louvers

4.6 Eliminators

After passing through the spray section of the washer unit, air passes through mist eliminator blades to remove condensed moisture. These eliminator blades are constructed of stainless steel or synthetic materials in newer units, and galvanised metal in older ones.

Air supplied to the production area should have moisture in the form of atomised vapour, so that it could be absorbed easily by the fibres. The air should not come with water droplets, as they are heavy, and cannot be absorbed easily by the fibres. These droplets, if comes, deposit on the material and make it wet and stick them on the machine parts, walls, stresses, etc. The water droplets lead to rusting of machine parts. Eliminators arrest the droplets of water from moving along with the air carrying atomised water vapours. The eliminators normally get chocked with moist lint as the moist air hits the eliminator. It is very essential to clean the eliminators frequently. The eliminators should be designed in such a way that it is easy to open and clean. In olden days, removing of eliminator blades was a laborious job taking more than a day to open and re assemble. Now modular-type eliminator boxes are available. We can remove one module and fix another in its place. Hence, there is no need of stopping the plant for cleaning the eliminators.

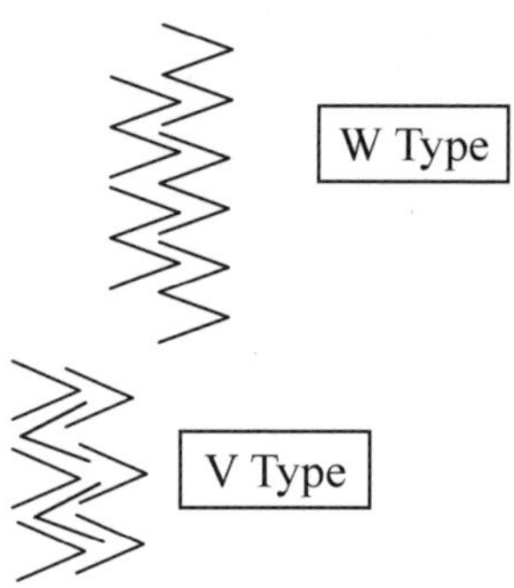

Figure 4.16 Eliminators

There are different types of eliminators available in the market with different shapes made of PVC, or sheet metals such as aluminium, galvanised iron, which are non-corrosive and non-rusting. Sheets in the form of W, V are arranged in such a way that the air has to hit, bend and move. The mesh pad eliminators are made by knitting stainless steel wires, which has a very high efficiency of 99% removal of liquid droplets 3 μm and larger. ACS, the introducers of mesh-type eliminators provide different types of eliminators.

ACS, the introducers of mesh-type eliminators provide different types of eliminators.

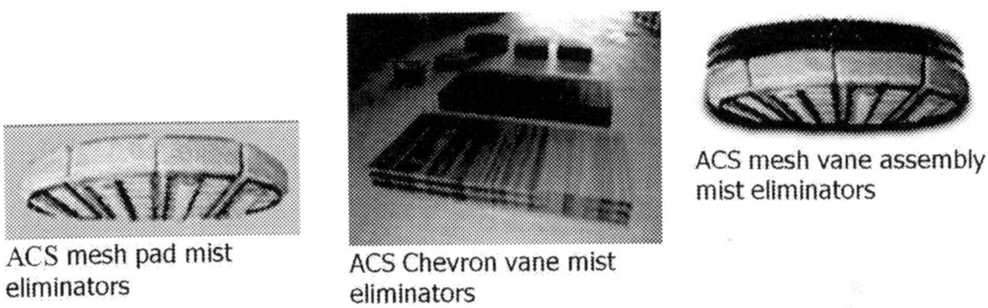

Figure 4.17 Knitted wire mesh mist eliminators.

After the mist eliminator section, some washer units are designed with steam reheat coils to temper the cooled air to suit a particular textile process.

Air volume dampers are provided for controlling the air input, and to divert air. By this, we can either take air from the inside of the department or from outside.

4.7 Water spray

4.7.1 Spray nozzle sets

The nozzles are the unit that atomise the water particles and help them mix with flowing air. The efficiency of humidification unit greatly depends on the fineness of the water particles in the spray. Normally two sets of nozzles installed one opposite to other are used in air washer plants for textile mills. The nozzles spray water against each other so that the incoming air must pass through a wall of water two or three feet thick. This does three things to the air:

1. Washes and cleans the air,
2. Saturates it with moisture, and
3. Controls the temperature of the air.

Spray sets is an assembly (bank) of two horizontal header pipes on which number of vertical branch pipes / riser pipes are fitted. The spray nozzles are fitted on to the riser pipes. The spray header is connected to the delivery of the pump. The pipes are evenly distributed in turn to distribute nozzles to create a uniform mist.

The spray nozzle headers are connected to a recirculating pump that continually picks up water from the air washer sump and pushes it through the nozzles. The capacity of this recirculating pump ranges from 250 to 1200 gallons per minute. The main chilled water pump located at the chilled water sump continually supplies chilled water to each individual air washer sump. Overflow and gravity take the chilled water back to the main refrigeration machine and chilled water sump. Capacities of air washer sump pans range from 600 to 2500 gallons. Spray nozzle systems typically use either high pressure water (1000 to 3000 psi) or a combination of compressed air and water (30–70 psi air) forced through a machined nozzle.

In old textile plants, the control of in plant air temperature was by regulating the temperature of the chilled water at the refrigeration machine. The humidity was controlled by the use of water and compressed air atomisers located in the overheads of the department requiring humidity. Some newer designs have automatic controls in each washer that automatically adjust the temperature of air washer sump water by letting more or less chilled water in. Automatic control valves are connected to the temperature and humidity sensing devices. This means that the amount of chilled water makeup can vary from one washer to another in a plant, but this variation is not large enough to disrupt adequate recirculation of treatment chemical being fed to the main chilled water sump. Air washer units have lint screens located across the front of the sump in addition to the chilled water sump screens and filters.

The spray nozzle selection is a very important aspect in the air washer plant. There are number of varieties of nozzles designed for different purposes. The nozzles might be hydraulic pressured or compress air. The nozzles might be flat spray nozzles, full cone spray nozzles, hollow cone spray nozzle, special cone spray nozzle, solid stream jet nozzle, special solid stream jet nozzle, etc. Droplet size in a given spray may vary from sub-microns to thousands of microns.

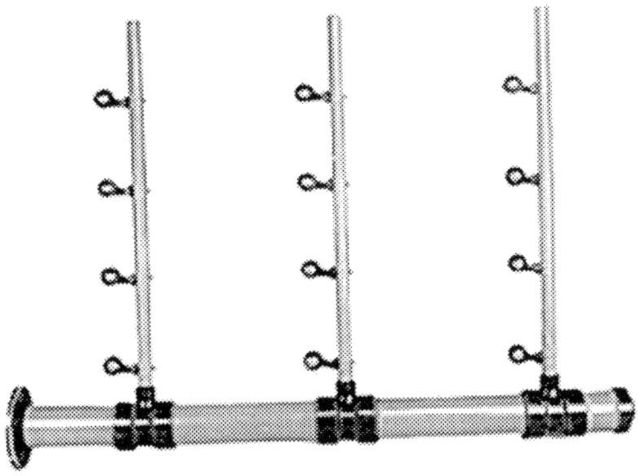

Figure 4.18 Spray nozzle bank

As the representative of droplet size, Sauter mean diameter are normally used in catalogue charts, which is a droplet size of spray generated at standard pressure for each nozzle and measured at the point that is most practical distance from nozzle orifice. There are many opinions on the classification of spray droplet sizes. 'The Mist Engineers', have classified the size droplets as shown in Fig. 4.18.a.

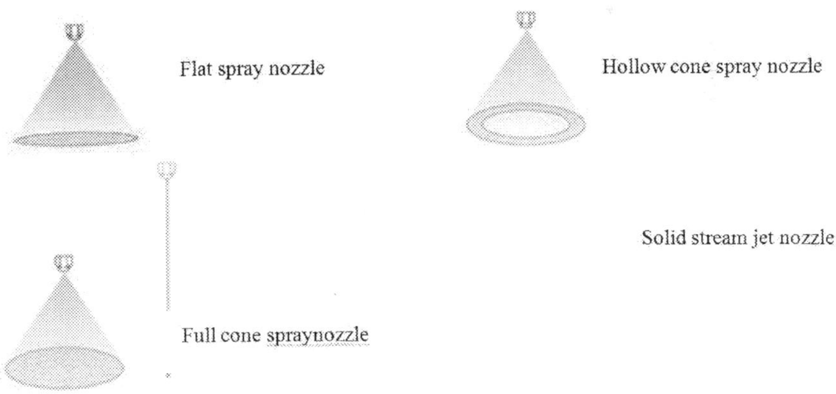

Figure 4.18.a Different types of nozzles

Single spray bank air washer is not normally selected for mill humidification due to its lower saturation efficiency. Double bank opposed spray type with G I header and stand pipe assembly and gun-metal / plastic spray nozzles are often used in air washers for mill humidification, since with this configuration it is possible to achieve 95% saturation efficiency with spray density of 3 gpm/ft^2 and air velocity not exceeding 3Mtr/Sec. The degree of saturation depends on the contact efficiency between air and water. A low velocity air flow is more conducive to higher saturation efficiency. However, manufacturers of high velocity air washers have now optimised their system for obtaining high saturation efficiency up to 95% with higher spray density.

4.7.1.1 Spray droplet size

Droplet size in a given spray may vary from sub-microns to thousands of microns. As the representative of droplet size, Sauter mean diameter is used in catalog charts, which is a droplet size of spray generated at standard pressure for each nozzle and measured at the point that is most practical distance from nozzle orifice. The droplet size plays an important role in evaporative distance. The sizes vary depending on the pressure and or maintenance of the system. When the spray nozzles are new and set up correctly, droplet sizes will range from 523 to 65,450 cubic microns. Both ultrasonic and steam humidifiers provide droplet sizes with volumes of approximately 0.52 cubic microns. High pressure spray nozzles (1000 to 3000 psi) have droplet volumes of approximately 1,767 cubic microns. Evaporative distances vary depending on field conditions – entering air temperature and relative humidity, air velocity, air volume, humidifier pulse rate, etc.

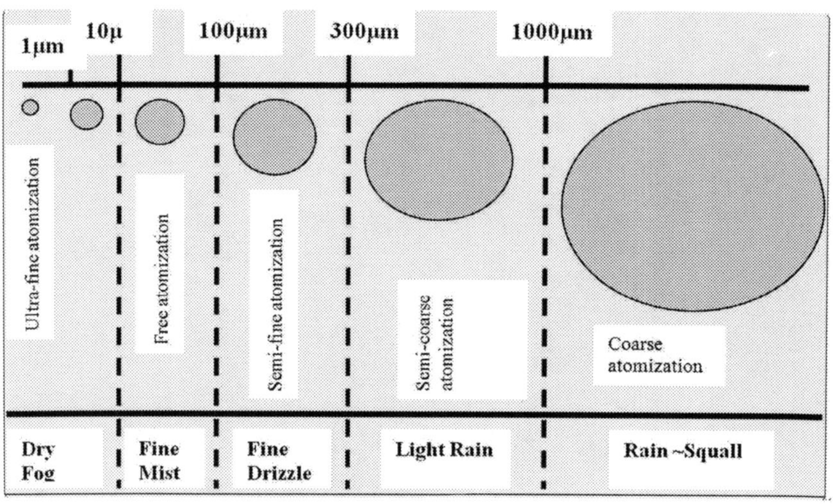

Figure. 4.19 Size of droplets

Droplet size plays a key role in achieving full evaporation. The smaller the droplet size the faster evaporation occurs. In ducted or AHU applications, large droplets may not evaporate and can collect on coils or inside of the ductwork. The spray nozzle's non-evaporated droplets will go to drain and therefore relative humidity set-points may not be achieved. Field experience reveals that 25% to 50% of spray nozzle water may go to drain.

4.7.2 Atomiser for humidification plants

Atomiser is a revolutionary product aimed at saving the precious energy hitherto wasted in the humidification plants. The concept of large number of small nozzles used in the air washer plants is sought to be changed drastically by introducing the revolutionary concept of atomiser. These atomisers change all the equations of humidification plants with respect to number of nozzles used; horsepower of the water pump and redefine them so as to conserve the precious energy. The problem of energy being lost as frictional losses in the small nozzles has been effectively addressed. Comparison of the atomizers with the existing (conventional) nozzles areas is as follows.

- Number of atomiser required per plant is limited to 6 to 20 numbers depending upon the size and condition of the plant.
- No replacement is required for atomisers which is otherwise the case with existing (conventional) small nozzles. Replacement is required due to choking as well as breakage during each cleaning time.
- No periodic maintenance required. It means savings in manpower and spare nozzles.
- Reduced frictional losses save energy (15% to 40% saving can be achieved depending upon the actual plant conditions).

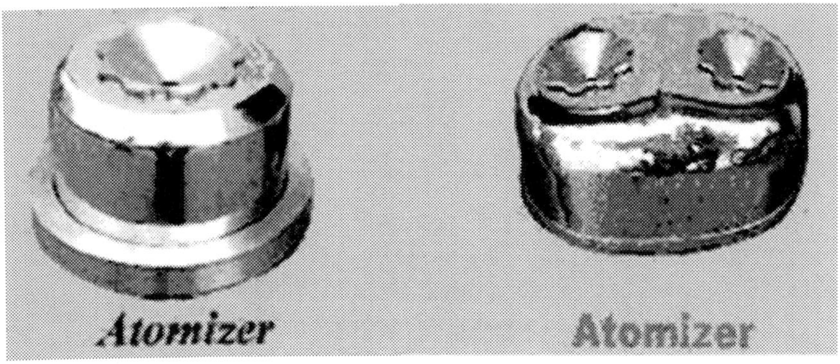

Figure 4. 20 Atomizer

4.7.3 Dry fog

The finest mist with its droplets' size distributed in a very tight range is called as dry fog. There is significant difference between each droplet size sprayed depending on the type of nozzle. Fig. 4.21 below shows the difference between size and distribution of droplet by the conventional pneumatic nozzle and by AKI Jet R under the same condition. It can also be seen that large sized droplets do bad things in humidification; they hit objects and make them wet.

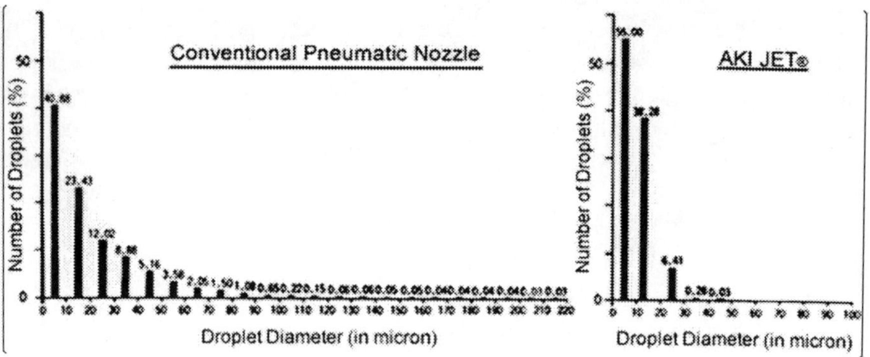

Figure 4.21 Droplet diameter

As Fig. 4.22 shows,, small droplets rebound from an object, but large droplets get burst and wet the object. This is just like how soap bubbles do. The dry fog does not wet the object.

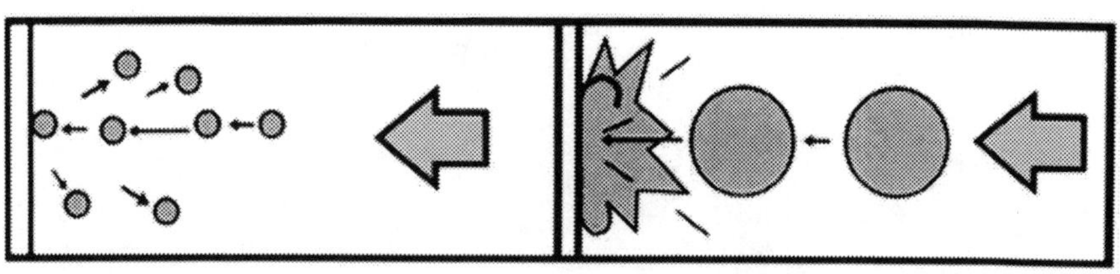

Figure 4.22 Concept of dry fog

Fig. 4.23 is of droplets collected by immersion sampling and droplet diameter distribution measured by the laser analyser for various humidifiers. AKI iet nozzle produces Dry fog with the maximum droplet diameter of 50 μm or less and Sauter mean droplet diameter of 5.9 microns.

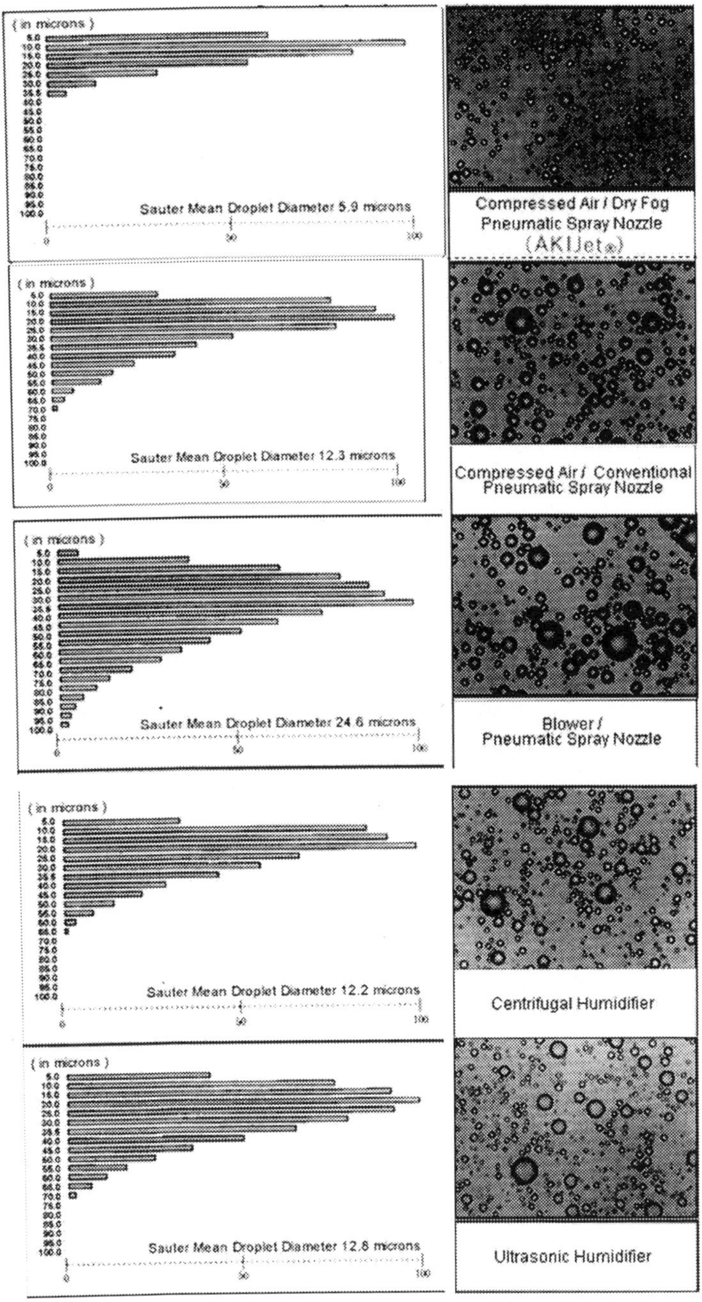

Figure 4.23 Spray droplet size by various humidifiers

Dry fog condition: Maximum droplet diameter 50 μ or less and mean droplet diameter 10 microns or less. Energy cost for dry fog humidification is only 20% of that of steam humidification as shown in Fig. 4.24.

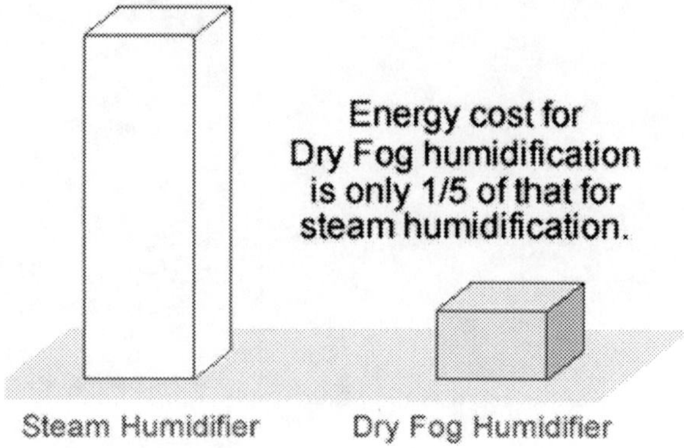

Figure 4.24 Energy cost comparison (Comparison of electric power consumption in humidifying 1 kg of water in 1 hour)

4.8. Water supply

4.8.1 Quality of water

Quality of water is very important for getting the effective results from humidification plants. Water should not be acidic as it can corrode the metal parts. The pH value should be within the range 7.5 to 8.5. Water, if hard, develops scales inside the pipes and chocks the nozzles. Total hardness should be preferably above 100 mg/l and within 250 mg/l. Demineralised water used in boilers should not be used in air washers as this may lead to corrosion. Moderate hardness helps to build a thin protective coating which reduces harmful effects of chlorides in coater. Because of evaporation of water drops, minerals tend to concentrate during re-circulation. Hence, it is necessary to bleed some water from the air washer tank and also do periodic cleaning to prevent growth of algae.

4.8.2 Filtering the water

The textile mill environment is with fibrous material, and loose fibres can be seen floating in the air, which settle on walls, open water in tanks, roof etc. The water fed to the air washer plant, if contaminated with fibres, choke up the nozzles and the force of spray reduces. The water shall not atomise, but big drops shall just fall down. The air will not get the required atoms of water. Therefore the water fed to the humidification plant should be free from fibres and floating materials. Hence, filtering water using either stationary strainers or rotary filters is practiced.

The purpose of the washer lint screens is to prevent fouling of the centrifugal spray pump. There are generally two banks of screens provided so that one can be left in place while the other is cleaned. The size of wire mesh in these duct work is sometimes designed into a unit to allow for simple heating of the air without washing. Large fans or blowers takes the air from the end of the washer unit and force it out through the plant duct work. The metal of construction is an important factor to be considered when recommending water treatment for textile air conditioner systems. On newer air washer units, areas that are regularly in contact with water are constructed of stainless steel, except for piping. *S*creens will vary from 1/4 to 1/8 inch mesh. Construction in new units is stainless steel, while old plants may be galvanised wire or copper. Some old washers have a small-scale drum filter inside the washer sump. The purpose of this filter is primarily to filter out lint during winter operation when some air washers operate autonomously without chilled water makeup.

In plants where refrigeration is not required during winter months, these washer units may operate autonomously, continually recirculating water from the sump up through the spray nozzles. When refrigeration machines are operating, air washer units are supplied with 40 to 50°F chilled water from the system's main chilled water sump. This sump is usually a large concrete structure containing from a few hundred gallons up to 50,000 gallons of chilled water. The total capacity of the chilled water system can vary between a few thousand gallons of water and over 100,000 gallons of water. Water in the chilled water system some time increases. When this occurs, the level in the sump necessarily rises until it hits the limit switch. Excess chilled water then flows to the cooling tower system as makeup. This procedure conserves both water and energy since it avoids wasting excess chilled water and increases the efficiency of the condenser unit. Chilled water sumps also contain some type of filter for suspended solids removal. These filters may be a simple wire mesh screen on a moving drum and a water spray device to remove suspended matter from the screen. Many potential fouling problems in the air washer systems can be avoided by removing suspended solids in the chilled water sump in this manner

There are some air washer units, particularly in synthetic plants, that do not use chilled water for the washing process. They continually re-circulate the same water through the spray nozzles, except for a small amount of makeup and blow down. Permanent cartridge-type metal filters both before and after the washing section are used to help keep the unit clean. These cartridge filters must be regularly steam cleaned. The eliminator sections of these washers are followed by steam reheat coils and chilled water coils to adequately control temperature and humidity.

Figure 4.25 Rotary water filter

4.9 Supply air distribution

Textile plants are generally served by uniform air distribution through sheet metal ducts running above the false ceiling in respective departments and taking care in the design to direct supply air through suitably positioned outlets to motor alleys and other points of concentrated heat loads. Plant duct work is either galvanised metal or aluminium. Some plants have the air conditioning duct work cast into the concrete structure of the building under the floor.

In the 'weave direct' (LoomSphere) system of humidification for a weaving shed, a separate air washer unit with independent supply fan and ducting system with branch ducts coming down to a level of approximately 0.8 to 1.2 m above the warp sheet of respective looms for directing the humidified air to the warp sheets are used to attain the desired high relative humidity condition of 78% to 80% at the warp sheet level of the looms. A separate air washer unit with separate supply fan and separate supply air ducting running above the false ceiling and ceiling diffusers at the false ceiling level are adopted to maintain around 65% R.H condition in the room itself.

4.9.1 Ducts and duct tape

The distribution of air uniformly or as required in the work area can be done by designing the duct dimensions suitably and joining them with suitable duct tapes. The ducts carry the humidified air from the plant and a common masonry plenum and run generally above the false ceiling. They are fabricated at site with 20 / 22/ 24 Gauge GI sheets and MS angles. They are generally supported from the roof purlins. The diffusers are fitted on the ducting which distributes the air uniformly, without directly blowing on the materials or creating turbulence. A recent development is using canvas as duct material. As canvas is porus, there is no need of any diffusers.

Figure 4.26 Canvas ducts for humidification

Duct tape, also called as 'Duck tape', is a cloth tape coated with a polyethylene resin on one side and very sticky rubber-based adhesive on the other. Unlike other tapes, the fabric backing gives duct tape strength yet allows it to be easily torn. Duct tape is also very malleable and can adhere to a wide variety of surfaces. While it was primarily designed for use in air ducts and similar applications, consumers have found a broad range of uses for this popular product. It can be used for a number of household repair jobs, as a fastener instead of screws or nails, and in car maintenance.

There are a variety of constructions using different backings and adhesives. One variation is black gaffer tape, which is designed to be non-reflective and cleanly removed, unlike standard duct tape. Another variation is heat-resistant foil (not cloth) duct tape useful for sealing heating and cooling ducts, produced because standard duct tape fails quickly when used on heating ducts. Duct tape is generally silvery gray, but also available in other colors and even printed designs.

Gaffer tape (also known as gaffer's tape or gaff tape as well as camera tape and spike tape for narrow, coloured gaffer tape) is a heavy cotton cloth pressure-sensitive tape with strong adhesive properties. It is widely used in theatre, photography, film and television production, and industrial staging work.

Berryhill, Robert A. in 1974 developed a piping arrangement for an air washer which enabled the air washer to maintain constant dew point of air passing from the leaving side of the air washer. Chilled water is sprayed from a first manifold and is used to cool air to a required wet bulb temperature. The air is then passed through subsequent manifolds having spray nozzles which further humidify the air until the air reaches a required dew point temperature in a substantially saturated state.

4.9.2 Diffusers

The distribution of air uniformly in the working area is done by diffusers. The diffusers are specially designed depending on the type of production unit, viz, spinning, weaving, knitting etc.

Figure 4.27 Diffuser

Air grilles are also used distribute the humidified air at various locations. They are of single deflection and double deflection type. They are available in GI / SS and extruded aluminium construction.

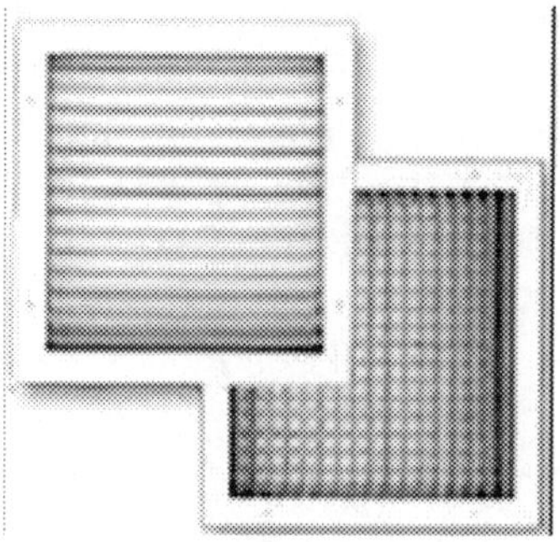

Figure 4.28 Air grilles

An apparatus for preventing the accumulation of moisture condensation on an air conditioning supply grill of a concentric diffuser in an air conditioning system having a return air duct was developed by William G. Suggs (US Patent No 595247 issued on 19th Aug 1997). This apparatus comprises one or more anti-sweat heater attached to air conditioning supply grill with one or more humidistat positioned within return air duct at least one relay and electrical wiring connecting each of humidistat to at least one of anti-sweat relays so that when one of humidistat senses an increased level of humidity in return air duct, one of relays may allow current to flow through anti-sweat heaters for warming of air conditioning supply grill and prevention of the accumulation of moisture condensation thereon.

There are different varieties of diffusers. While selecting a diffuser, one should ensure that there is no direct blowing of air on the materials working like sliver or roving ends as it can disturb them. The surface of the diffusers should be smooth and should not allow the fibrous materials to stick on them.

Figure 4.29 Cleaning of diffuser

Cleaning of the diffusers on a regular basis is very important to remove the sticky fibrous materials, or else it hinders the smooth flow of humidified air into the work area. Cleaning of diffusers which are in the ceiling level are generally neglected. A long stick fitted with a brush can be used to clean the diffusers when the machines are working. Compressed air can be used to clean the diffusers when the machines are stopped.

Fig. 4.29 shows different types of diffusers available in the market.

Figure 4.30 Different diffusers

Figure 4.31 Poor quality diffuser

4.10 Roto spray units

Roto-spray systems are similar in principle to the packaged air washer units, but are thought to be more efficient. These systems are housed in a cylindrical housing slightly larger than the supply duct work and are generally located on the roof of the textile plant. They have stationary spray nozzles and rotating eliminator blades that look like the compressor stages in a gas turbine engine. There is very little water in these units, and they are somewhat more difficult to treat from a water treatment standpoint because the same amount of dirt is concentrated in much less water. They are also not as accessible as a conventional air washer system because the access doors or manholes are bolted shut during operation. Roto-spray systems can be installed on the chilled water system as air washer units.

4.11 Chilling plants or cooling towers

Chilling plants cool the water to a very low temperature, which when circulated absorbs temperature from the air and cools it. In the same time it condenses the vapours in the air and dehumidifies it. Cooling towers are primarily used to cool water in air conditioning systems. They are also used in process industries to cool materials. There are two types of cooling towers: open and closed. In an open tower, the water to be chilled is open to the atmosphere and is cooled by evaporation as it cascades down through a structured packing material (called 'fill') designed to increase the surface area of falling water films, thereby increasing the evaporation rate. Induced-draft towers use a suction fan to pull air up through the fill ('counter-flow') or across the fill ('cross-flow').

Figure 4.32 Chilling plant

The warm water returning from the chillers is pumped to the top of the tower and distributed by spray nozzles over the 'fill'. The purpose of the fill is to expose as much of the water's surface as possible. As the water drops through the fill, fresh air from outside the cooling tower is forced through the tower. The water, cooled by partial evaporation, then falls into a basin at the bottom and is circulated back to the chiller to start the process all over again. Most cooling towers have the basic components of a pump to bring the water to the top of the tower, spray nozzles to distribute the water over the fill, a fan to move outside air through the tower the fill to expose the water surface to the air stream and a basin to capture and store the cooled water. Other components are also used in cooling towers such as float valves, filters and strainers, but these are the basics.

Figure 4.33 Coil-type evaporator

Figure 4.34 Fin-type evaporator

A single cooling tower can be used to supply chilled water to one or several chillers. For each chiller served by the cooling tower, there is a 'cell' in the tower. Each cell is independent and has its own fan, fill and basin. Each cell has an access door or panel for entry into the basin. In some older cooling towers, the access opening is under the water level of the basin which requires the basin to be drained before entry. This is not a problem with the vast majority of cooling towers in service today.

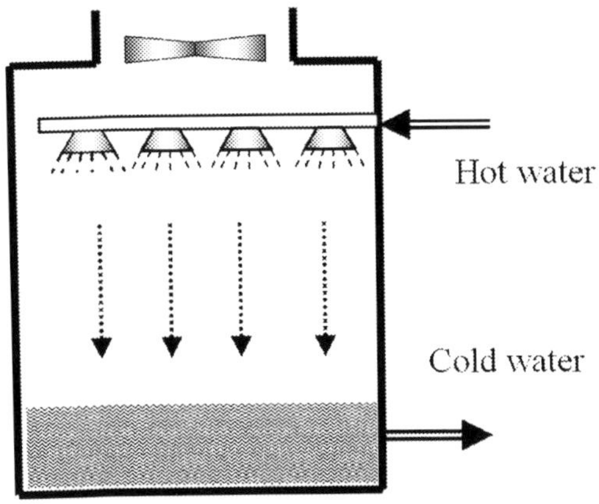

Figure 4.35 Induced draft counter flow tower

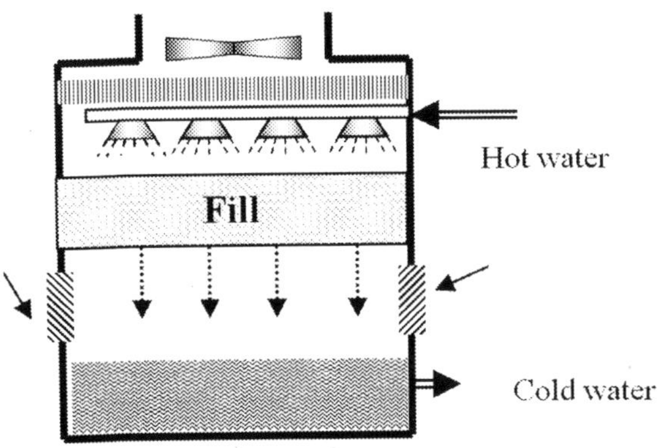

Figure 4.36. Induced draft counter flow tower with fill

Cooling towers come in four main types, viz 1) induced draft counter flow cooling towers, 2) forced draft counter flow cooling towers, 3) induced draft cross flow cooling towers and 4) forced draft cross flow cooling towers. In induced draft cooling towers, the fan is mounted at the top of the tower and is used to 'pull' air through the tower. In forced draft cooling towers, the fan is mounted on the side of the tower and is used to 'push' air through the tower. In counter flow cooling towers, the air enters at the bottom of the tower and travels up in the opposite direction of the water stream. In cross flow cooling towers, the air enters the side of the tower and travels perpendicularly to the water stream. The induced draft counter flow cooling tower is probably the most common. Cooling towers come in all sizes from as small as a minivan to as large as a three story building covering the area of a football field.

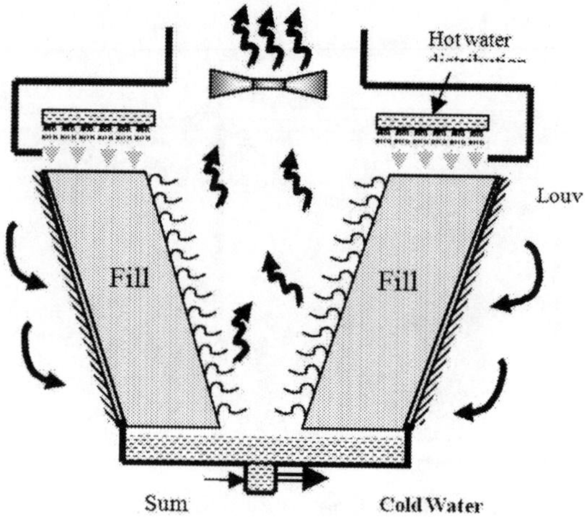

Figure 4.37 Induced draft, double flow, cross flow tower

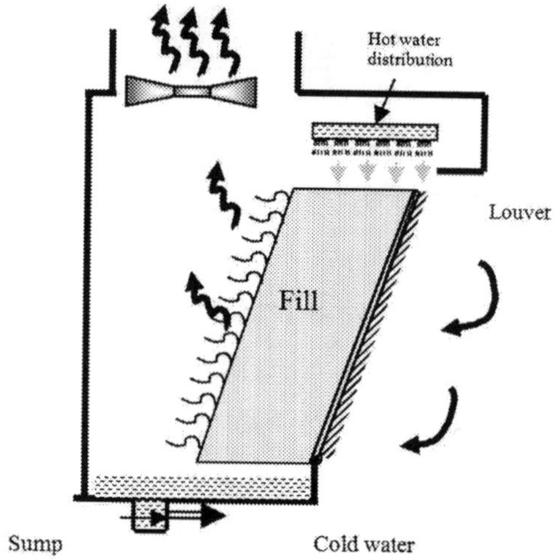

Figure 4.38 Induced draft, single flow cross flow tower

4.11.1 Open evaporative cooling towers with induced-draft air flow through fill

Several factors contribute to cooling tower efficiency, including proper sizing of the tower, the age of the tower, using more effective fill materials, using variable frequency drive fan motors and replacing old spray nozzles. The outside air that is drawn into the tower is contaminated with pollutants, dust, pollen, exhaust byproducts, insects and whatever else is suspended in it. As the water falls through the air stream, it captures these contaminants and carries them into the basin. This water, in turn, is delivered to the chiller. As the cooling water is heated in the chiller by absorbing heat from the refrigerant, the contaminants precipitate out of the water and adhere to the chiller tube walls, reducing the efficiency of the chiller. The result of all this is increased energy and maintenance costs. Heavier contaminants will tend to settle to the bottom of the cooling tower basin creating 'dead' areas. These dead areas form a shield between the basin surface and corrosion-preventing chemicals that are added to the water. This leads to under-deposit corrosion of the metal surfaces. Once corrosion begins, blisters form which release oxide particles which add more suspended solids to the condenser water stream. In addition, mud, dirt and debris in the basin provide an ideal environment for microbiological organisms, some of which secrete an acidic solution that can contribute to additional corrosion. Some of these microorganisms represent a health risk to maintenance personnel, and if released from the tower in the form of aerosols present a risk to the general public. Suspended solid accumulation can be minimised by the use of a side stream filter, often referred to as a sand filter. While these filters can remove some of the suspended solids, they do nothing about the settled debris in the basin. It is therefore very necessary to keep the cooling tower as clean as possible as well as to perform maintenance on the tower at regular intervals.

Counter-flow towers tend to be the most compact. This is because the upward flowing air stream is in contact with the entire cross section of water from top to bottom as it falls through the fill. Less space is needed because of their increased thermal efficiency and lack of the warm water distribution plenum needed by cross-flow towers. Their disadvantage, however, is the increased fan power resulting from air flowing directly against, rather than across, the falling water. Although bulkier, cross-flow towers consume less power and usually have a lower capital cost as well.

Forced-draft cooling towers use a bottom-mounted centrifugal blower to push air through the fill, as shown in Fig. 4.38. Centrifugal blowers are inherently quieter than the propeller-type fans of induced-flow towers, and by locating the blower beneath the unit, fan and motor noise is baffled, further reducing the noise of operation. Maintenance access is easier because all moving parts are located at the tower base. Corrosion is reduced as the blower handles dry, ambient air unlike the fans of induced-flow towers, which handle moisture-laden air. The major disadvantage of forced-draft towers is that they consume about twice the power of induced-flow towers for the same cooling capacity.

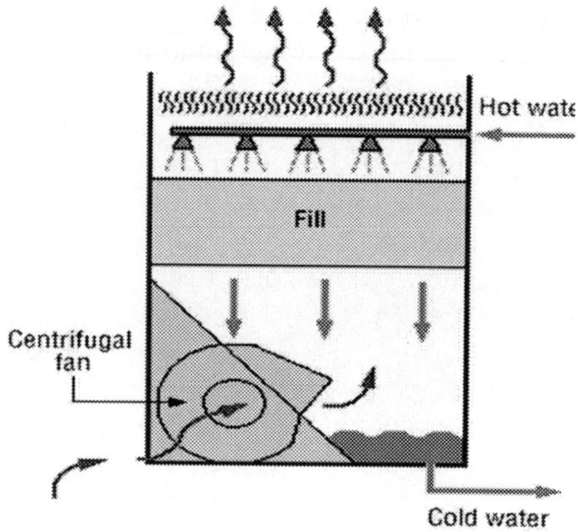

Figure 4.39 Open tower with forced-draft airflow

Cooling towers in textile plants are fairly standard, ranging from various types of old wooden field erected towers to modern metal package units. The cooling plant sizes vary from 300 to 2700 tons or higher per machine, depending on the size of the plant. Some cooling systems are designed with several package towers connected parallel to all of the refrigeration machines. In other cases, each refrigeration machine condenser unit is piped separately to its own cooling tower. The tonnage and number of air conditioning systems operating in a single plant depend on the combination of textile processes being accomplished and production capacity of the plant. Many textile processes require year-round refrigeration for proper temperature and humidity control, and these plants operate cooling towers 12 months of the year. Other plants vary cooling tower operations from six to eight months per year depending on the climate conditions of the plant. New textile plants are being designed with as few chilled water systems as is practical to maximise efficiency and reduce maintenance. In older plants, air conditioning systems were added on a piecemeal basis, resulting in several small systems in the same plant. It is important to know the number and location of the various cooling towers and chilled water systems and their interconnections so that appropriate chemical feed points can be selected for proper water treatment. For example, several refrigeration machines, each with its own individual cooling tower system, can be operated on a single large chilled water system, or each refrigeration machine can have its own individual chilled water circuit. These various combinations of equipment make feeding chemicals difficult and expensive in older plants since so many chemical feed points are necessary. Chilled water systems are piped around the average textile plant for use in air washers, process heat exchangers, and small office air conditioning units. By far, the largest user of chilled water is the air washer unit. A small textile plant may have only one refrigeration machine, one chilled water system and one air washer unit. Most textile chilled water systems have high level float switches on the chilled water sump. This switch controls a solenoid valve on interconnecting piping between the chilled water system and the cooling tower water circuit. During summer operation, when the chilled water is dehumidifying plant air, the volume of new textile plants are being designed with as few chilled water systems as is practical to maximise efficiency and reduce maintenance.

As plants get larger, more and more air washers are found. The average plant with 1200 to 2400 tons of air conditioning capacity may have six to eight air washers producing fairly low velocity air, or the same plant may have two to four high velocity air washer units. In plants where refrigeration is not required during winter months, these washer units may operate autonomously, continually recirculating water from the sump up through the spray nozzles. When refrigeration machines are operating, air washer units are supplied with 40 to 50°F chilled water from the system's main chilled water sump. This sump is usually a large concrete structure containing from a few hundred gallons up to 50,000 gallons of chilled water. The total capacity of the chilled water system can vary between a few thousand gallons of water and over 100,000 gallons of water. Sometimes, water increases in the chilled water system. When this occurs, the level in the sump necessarily rises until it hits the limit switch. Excess chilled water then flows to the cooling tower system as makeup. This procedure conserves both water and energy since it avoids wasting excess chilled water and increases the efficiency of the condenser unit.

Chilled water sumps also contain some type of filter for suspended solids removal. These filters have a simple wire mesh screen on a moving drum and a water spray device to remove suspended matter from the screen. Many potential fouling problems in the air washer systems can be avoided by removing suspended solids in the chilled water sump in this manner. Most textile plants use compressed air for controls, water atomisers, and other uses. Therefore, they need to deal with the problem of supplying cooling water to the air compressor heads, oil coolers and after coolers. The cooling water may come from the main cooling tower system in larger plants where there are several air conditioning units, or there may be a separate small tower provided for these units. Over the years, small textile plants have used once-through cooling water, but this practice is declining due to economics and discharge limitations. A few plants that are set up for once-through cooling water on air compressor systems conserve this water by sending it to one of the cooling towers during summer operation as makeup. If the cooling tower is not run during the winter months, this water again becomes once-through. When air compressor cooling water is utilised as makeup for a cooling tower system, a second sump float level control in the cooling tower sump controls the makeup from the air compressor units to avoid disturbing tower water concentrations and treatment chemical residuals. This second float is set at a higher level than the main makeup water valve so that it operates first when makeup is required. Water that is not needed when this float closes moves vertically a few feet from the float valve and then out a vented overflow line to waste.

4.11.2 Swamp coolers

In very dry climates, 'swamp coolers' are popular for improving comfort during hot weather. The evaporative cooler is a device that draws outside air through a wet pad. The sensible heat of the incoming air, as measured by a dry bulb thermometer, is reduced. The total heat (sensible heat plus latent heat) of the entering air is unchanged. Some of the sensible heat of the entering air is converted to latent heat by the evaporation of water in the wet cooler pads. If the entering air is dry enough, the results can be quite comfortable. These coolers cost less and are mechanically simple to understand and maintain.

4.11.3 Absorptive refrigeration

There is a process called absorptive refrigeration which uses heat to produce cooling. In one instance, a three-stage absorptive cooler first dehumidifies the air with a spray of salt-water or brine. The brine osmotically absorbs water vapour from the air. The second stage sprays water in the air, cooling the air by evaporation. Finally, to control the humidity, the air passes through another brine spray. The brine is re-concentrated by distillation. The system is used in some hospitals because, with filtering, a sufficiently hot regenerative distillation removes airborne organisms. Some use gas turbines to generate electricity. The exhausts of these are hot enough to drive an absorptive chiller that produces cold water. The cold water is then run through radiators in air ducts for hydronic cooling. The dual use of the energy, both to generate electricity and cooling, makes this technology attractive when regional utility and fuel prices are right. Producing heat, power, and cooling in one system is known as trigeneration.

Moriguchi, TetsuoYamaoka, Akira developed a (United States Patent 6718790 dated 13th April 2004) cooling device constructed by a water impermeable member defining a water passage, the water impermeable member being provided with a vapour permeable member which is permeable to water vapour and impermeable to water and has the form of a mesh and made of a material having water repellency. The cooling device has a long-life operation while preventing wear-out of the vapour permeable member. Further, optimally setting the size of an opening of the mesh member enables to securely dissipate water vapour outside of the cooling device through the vapour permeable member while efficiently suppressing water leak.

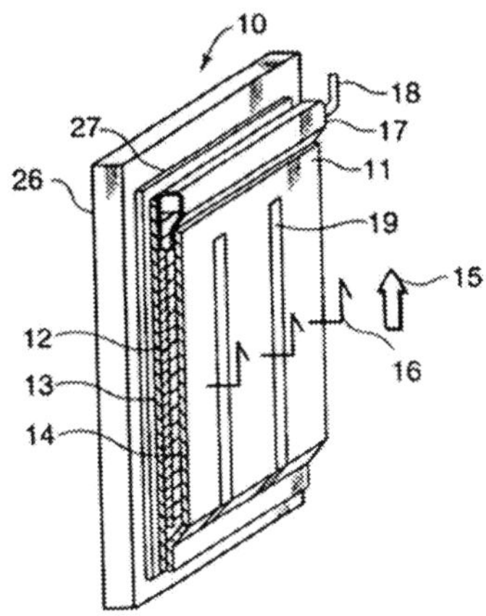

Figure 4.40 Cooling device

The cooling device includes a sheet-like vapour permeable member that is impermeable to water but permeable to water vapour, a supply channel for allowing water to pass by utilising a capillary phenomenon, and an impermeable member in the form of a plate or a sheet made of a material impermeable to any fluid and having heat conductivity. The impermeable member is arranged on the side of an object to be cooled as opposed to the vapour permeable member with the water passage. The cooling device is constructed in such a manner that water supplied to the water passage evaporates by heat exchange with water coming from the side where the object is located and that the water vapour is discharged outside of the cooling device through the vapour permeable member. The vapour permeable member is incorporated with a permeable membrane that has been developed and applied mostly in the fields of textile industries. Such a moisture permeable membrane is obtained by adhering a thin film formed with a multitude of micro-pores each having a size generally corresponding to one twenty-thousandth of water droplet or one thousand times as large as water vapour molecule onto a base cloth so as to provide the resultant permeable membrane with mechanical strength. This permeable membrane utilises a dimensional difference between gas molecule and liquid molecule acquired moisture permeability with use of such a physiochemical property that moisture is absorbed on a high-moisturised side with use of hygroscopicity inherent to a polyurethane film while water vapour is discharged on a low-moisturised side. However, in the cooling device incorporated with the aforementioned moisture permeable membrane, contact of the membrane with water for an extended period wears out the membrane, which hinders long-time use of such a membrane.

4.12 Temperature and humidity monitoring

The air after passing through the air washer and eliminator is sensed for the temperature. Depending on the requirement of the production area, the air can be either cooled or heated. Normally, for cooling, chilled water is added to the air washer, whereas for heating, the radiators or infrared heaters are provided after the eliminators. The temperature of outside and inside air is considered for making a decision for mixing the proportion of return air with the outside air. The return air is first filtered and then fed to air washer or thrown out to the atmosphere.

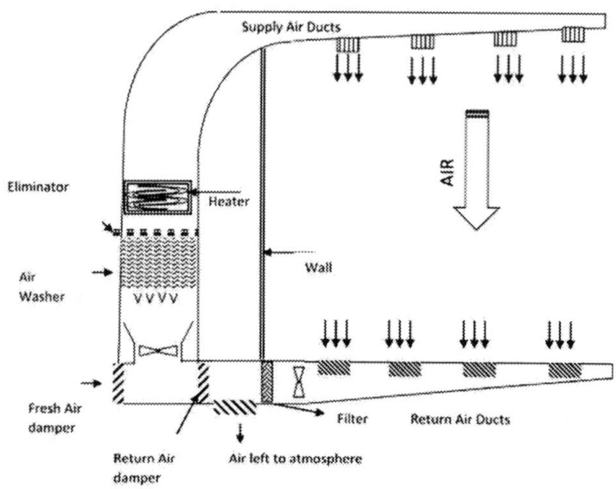

Figure 4.41 Centralized air conditioner

The humidity is monitored by increasing or decreasing the water pressure in the air washer and by judicially combining return air with outside air depending on the humidity conditions. Now various sensors are available, which monitor the opening and closing of the dampers and the water pressures to maintain the required humidity and temperature.

5

Humidification requirements in manmade fibre plants

5.1 Introduction

The manmade fibres include a variety of polymers, normally extruded through a spinneret, drawn and wound on spools. Texturizing is done if needed for thermoplastic filaments. Some of the manmade fibres are hygroscopic in nature, and their properties are influenced with absorption of water, whereas majority are not affected by moisture. The hydrophobic fibres are prone to generation of static charges in process; therefore, in manmade fibre manufacturing process from "spinning to weaving" also requires air conditioning with close tolerance in temperature and humidity.

Manufacturing in manmade fibre spinning can be divided into various sections such as quench box, take-up area, draw twisting area and textile area. In addition to the above manufacturing areas, the other rooms like invertor room, textile lab, instrumentation lab etc., also require air conditioning.

5.2 Quench box

The most critical air conditioning requirement in the manufacturing process of manmade fibre is the quench box. Molten polymer gets quenched and solidified as a fine filament yarn in the quench box. The quality of air-cleanliness, quantity, velocity across the quench screens, temperature, humidity and static pressure inside the quench box are all critical for good quality of the finished product and productivity.

Normally each extruder line has four quench stations. Each quench station requires 650 to 1000 cmh (400–600 cfm) in case of polyester and 900 to 1400 cmh (600–850 cfm) in case of nylon, to maintain an air velocity of 0.4 to 0.6 m/sec and 0.6 to 1.0 m/sec across the quench screen, respectively. The air quantity and velocity are indicative numbers and vary according to the process design and contour of the quench screens. The temperature of quench air varies from 16°C to 21°C and RH 65% ±5%. The quench air directly impinges on the hot molten polymer and hence should be clean for production of high-quality yarn; hence pre-filters, fine filters and HEPA filters are used in the supply air stream before the air enters the quench box.

Normally 100% fresh air is used in a quench air system. Recirculation in certain applications is considered if the ambient air temperature is high and no contamination of air is observed. Outside air is first passed through a coarse filter and then through a normal-water air washer. Air surrounding the factory is expected to have high levels of 'monomer' hence an air washer cleaning system is provided before conditioning the air as monomer gets dissolved in water. The air washer also helps to increase the humidity level particularly during the winter season. Air is then allowed to pass through a set of cooling coils, hot water coils and a humidifier section. In a quench system, the fan is kept before the cooling coil section, heating coil section, etc., and the system is a blow-through design against the normal draw-through design used in standard air conditioning. The fan power requirement for quench air is quite high compared to a normal air conditioning system due to the high static pressure requirement; it is therefore advisable to have the fan motor heat picked up by the air stream before entering the cooling coil. The filtration arrangement comprising microvee (fine) EU-9 and HEPA (superfine) EU-11/ EU-12 filters is provided after the air conditioning sections. The air is then taken through a special duct manufactured out of 6-mm MS sheet of welded construction so as to withstand high pressure. To maintain high cleanliness level of the supply air, the MS duct is sand blasted inside and painted with epoxy paint before installation. The MS duct is insulated outside with 75 mm-thick expanded polystyrene or equivalent material. The main quench supply air duct runs alongside the series of quench stations, normally below the quench boxes. Each quench box station is connected through a feeder duct with a damper for flow control. The quench supply air fan motor is provided with a variable frequency drive (VFD). With varying fan speed, a quench system is maintained at constant pressure in the quench box irrespective of pressure losses across the filters or due to closing down of any particular quench station. During the 1970s when VFDs were expensive, static pressure control was achieved through a variable inlet guide vane controlled by pneumatic actuators. These actuators were used to adjust the blower inlet guide vane based on the signal received from the static pressure sensor located in the common supply air duct. Even though many control methods are followed, the most common ones ar: as follows:

- Chille– wate– coil leaving dew point temperature controlling chilled water flow, with a 3-way mixing valve.
- Supply air temperature controlling hot water flow with a 3-way valve.
- Supply air relative humidity controlling steam humidifier valve.
- Static pressure sensor located at the end of quench duct controlling VFD of supply air blower motor.

The high-pressure quench air system needs a service corridor around the air washer for pressure equalisation to open the AHU doors for maintenance of the air handling system.

Machine air conditioning is required in filament spinning. The quench air outlet produces a laminar airflow with which maximum product quality is achieved in the spun-bond process. Luwa offers a comprehensive system including extraction, depending on requirements, which guarantees a speed variation of max. ± 0.05 m/s in the range of 0.3–0.7 m/s with minimum pressure loss.

Figure 5.1 Quench air outlet for filament spinning by auwa

5.3 Take-up area

A take-up room comes below the quench room floor. Take-up room air conditioning is a re-circulated air conditioning system. Temperature and humidity in a take-up area are also quite critical. The table below shows various conditions for polyester and nylon 6.

Table 5.1 Conditions for spinning nylon and polyester.

	Polyester	Nylon 6
Temperature	21–23°C	18–21°C
Relative Humidity	55–65%	45–55%
Room Pressure	Positive	Positive
Cleanliness of Air	Down to 5 μm	Down to 5 μm
Fresh Air	Normal	Normal

The take-up air distribution system is operated through special perforated registers at the ceiling level. Return air is collected at floor level through masonry floor trenches which are insulated from inside and treated outside by water proofing. The provision of special grilles for supply at the ceiling and floor return grilles enables close to unidirectional flow from top to bottom along the direction of filament movement to maintain the temperature and humidity conditions of air at the take-up point.

Aspirating type supply air diffusers are not used in the take-up area. Special heavy duty floor grilles are used for return air to take care of machinery movement over grilles. Masonry return air ducts are terminated in a corner of the take-up room from where sheet metal ducts are used to connect the return air back to the air handling system. The pressure differential between a quench air system and take-up room is maintained in such a way that quench air is at a marginally higher pressure compared to take-up so that air flow from quench air to take-up along the flow direction of filament is maintained.

Return air gets sucked through an axial flow return air fan and pumped to a return air plenum which has provision for 100% exhaust of return air or recirculation, based on ambient air conditions. Return air goes through an air washer, cooling coil, heating coil and humidifier before the supply air fan discharges it into a supply air plenum for filtration. Clean conditioned supply air is then allowed through GSS air ducting for distribution to the take-up room. The operating pressure of a take-up fan will be around 50 mm. The most common control scheme followed is as follows:

- Chilled water coil leaving dew-point temperature controlling chilled water flow with a 3-way mixing valve.
- Room air temperature controlling hot water flow with a 3-way valve.
- Room relative humidity controlling steam humidifier valve.
- Ambient air wet-bulb controlling energy enthalpy scheme, closing/ opening of exhaust dampers and fresh air dampers.

The control element mentioned under ambient air wet bulb controlling is for an energy enthalpy cycle. Whenever ambient air total enthalpy is lower than return air enthalpy, 100% outside air is used for air conditioning in order to save energy. For this purpose, ambient wet-bulb temperature is measured and whenever the ambient wet-bulb falls below the desired setting, an electric actuator opens the exhaust damper and allows all the return air to exhaust out. Simultaneously, the fresh air damper gets opened, and the damper connecting the return air to the mixing chamber is closed thereby allowing 100% outside air circulation.

5.4 Draw twist area

Draw twist air conditioning load is quite high due to the heaters fitted in the machine. Draw twist air conditioning does not fall under the critical category. A simple air conditioning system without any sophisticated controls is normally provided. Here also the energy saving enthalpy scheme taking 100% fresh air is used whenever ambient conditions are favourable. In draw twist, the yarn is heat set, elongated and twisted to give further strength to the yarn. The conditions required for draw twisting are 25°C to 27°C and RH up to 65% with cleanliness down to 5 microns.

5.5 Textile area

The textile area is where the yarn gets twisted through a ply twister, cable twister and sent as a finished product to tyre cord industries or woven through textile machinery. Air conditioning is not very critical for the textile area machinery or process. A few major manufacturers initially installed an evaporative cooling air washer for the textile area but subsequently converted this to air conditioning, basically to enhance human comfort. The sensible load in the textile area is quite high due to high equipment load. Manufacturers who converted to an air conditioning system from an air washer ventilation system observed substantial quality improvement in the final product and also good productivity improvement after air conditioning. The floor area for textile machinery is quite large and with very high equipment load, the air conditioning load is also quite high, which made these manufacturers to go in for ventilation in the initial startup stages. The temperature requirement for textile area ranges from 22°C to 25°C and RH of 55% to 70%. Air distribution is effected from the ceiling and return air gets collected through masonry floor ducts and floor grilles.

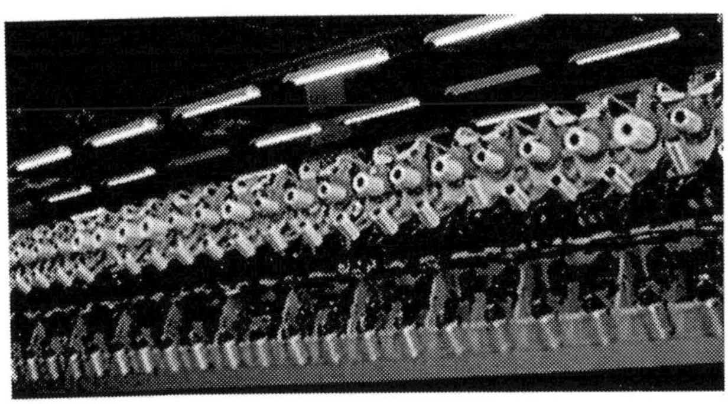

Figure 5.2 TexFog for synthetic twisting area developed by luwa

Since for synthetic textile manufacturing a high degree of humidity is crucial for hygienically perfect conditions, the air humidification must be optimal. Luwa developed a high-performance humidifying system TexFog, which uses only fresh water and does not require a water tank. Using TexFog is claimed to be perfectly safe from a hygiene point of view and fulfils the German hygiene regulation VDI 6022. As per Luwa it requires up to five times less energy than a conventional air washer and also allows a super saturation of the air.

5.6 Other areas

5.6.1 Invertor room air conditioning

A large number of process machines in the man-made fibre industry requires variable speed drive and a large number of invertor drives are used for motors driving rotary valves, extruder metering pumps, quench fans, finish applicators and tow collectors. These invertors are generally located in a common panel at two to three locations. Invertors dissipate a lot of heat during operation and need good ventilation or air conditioning for very reliable performance. Hence Invertor rooms are normally air conditioned, maintaining a temperature of around 23°C to 25°C and RH not exceeding 60%. A standard air conditioning system with air handling unit fitted with chilled water coil is used for invertor room air conditioning. Being a high sensible-load application, the system requires a higher dehumidified cfm compared to normal applications. No sophisticated controls are required for the invertor room air conditioning system.

5.6.2 Textile and instrumentation labs

Textile and instrumentation lab temperature requirement is 22°C to 25°C and RH of around 55%. While the temperature and humidity tolerance bandwidth are large for the instrumentation lab, the textile lab tolerance is very minimal. Even though both air conditioning systems use standard air handling units, the control scheme for a textile lab is quite elaborate with a 3-way mixing valve for chilled water coil, 2-way valve for hot water coil, dehumidifier etc. An instrumentation lab normally has a cooling coil controlled by a 3- way mixing valve. Both the systems are re-circulated type and the AHUs are located in a separate room adjacent to the lab area.

5.6.3 Refrigeration plant

The refrigeration requirements for all the processes are combined and a 'centralised refrigeration plant' to take care of all the individual requirements. Normally, a separate utility building adjacent to the manufacturing area houses the refrigeration plant along with other utilities such as air compressors, boilers, generators, UPS etc.

The refrigeration requirement for a Nylon plant is comparatively high compared to a PSY plant. The refrigeration requirements for PSF is125 to 170 ton / metric ton, whereas for Nylon it is 600 to 650 ton / metric ton per day. The above refrigeration load estimates will vary depending on the location of the plant (city weather conditions), design of quench chamber, fresh air or re-circulated air for quench or other areas, take-up room size machinery and volume and whether textile facility is included in the overall facility or not.

In the 1960s and early 1970s, when centrifugal chillers were not common, a battery of water-cooled reciprocating chillers used to serve the refrigeration requirement. Since the process and inside room temperature requirement between quench take up and other areas are different, refrigeration plants were grouped to operate at two levels of chilled water leaving condition, viz., 5°C and 7.5°C to make the system energy efficient. After the development of centrifugal chillers, gradually most of the plants replaced both reciprocating and low energy centrifugal chillers with energy efficient centrifugal chillers of 0.65 KW/TR. The scene changed from a battery of reciprocating chillers of 100/120 TR capacity to a few energy efficient centrifugal chillers of 600/800 TR. This change gave a substantial energy benefit to the man-made fibre industry. During the same period, a few organisations went in for captive power plants with diesel generating sets or a gas turbine. These organisations installed absorption chillers and produced refrigeration through the cogeneration route. The chilled water produced in the refrigeration plant is pumped to various air conditioning systems through constant speed centrifugal pumps. Since the plant room is located away from the manufacturing area, where air washers are located, extensive chilled water piping is common for this industry. All utility piping used to get clubbed and laid on pipe racks from the utility area to the manufacturing area.

The air conditioning scheme normally used is a system with a normal-water air washer and a closed circuit chilled water coil. In the initial period of evolution of this industry, a chilled water air washer was extensively used, for cooling and dehumidifying instead of a chilled water coil plus air washer to save cost. The piping system used to be slightly different for chilled water air washer application. Chilled water pumps located in the plant room would supply chilled water to the make-up of the air washer through a 2-way control valve. The amount of chilled water allowed into the air washer is based on the process load requirement of the specific air washer. The overflow water from the air washer would then be collected into a common piping header by a gravity flow arrangement up to the plant room. The return piping from the air washer being gravity return, proper care (over-sizing) had to be done. The chilled water being exposed to the air stream normally picked up impurities such as dust particles, monomers and these used to settle down in the chiller. For maintenance, chillers for such applications usually had a drain provision between baffle plates. The open gravity return system had a lot of operating problems such as overflow of hot well during shutdown, collection of impurities in the circulation water, clogging of piping etc. Hence, new plants which came up during the late 1970s avoided chilled water air washers and had a closed-circuit cooling coil with a separate normal water air washer.

5.6.4 texturising

The extremely high-energy requirements in false twisting on the one hand, and the homogeneous temperature and humidity needs in the bobbin creel area on the other, create a need for targeted machine air conditioning. The air is guided directly to the bobbins via low-induction slide vane outlets. A simultaneously controlled extraction of the heat produced by the process guarantees optimal running of the machine

Figure 5.3 Luwa machine air conditioning

6 Humidification for nonwoven plants

6.1 Processes in nonwoven plant

The nonwovens are produced by mechanically entangling a fine web of fibres either by water jet, air jet or punching by needles. The nonwovens production is not as simple as we talk. As the fibres are not bonded fully while in manufacturing are affected greatly by static charges. Normally, a number of webs produced from different carding machines are laid one above the other and then passed through the jets or needle punches. A slight variation in humidity can cause the webs to jump up or break, which greatly affects the web density.

Figure 6.1 Lap preparation

Uneven GSM is one of the major problems being faced by the nonwoven manufacturers. Uneven humidity not only affects the web density, i.e., GSM, but also the web thickness. This poses the problem of uneven height of pads when equal number of pads are counted and filled in bags. The uneven heights bring a suspicion among the customers about the number of pads and the quality of pads.

The carding process is difficult, especially when bleached cottons are processed for producing absorbent cotton pads, balls or buds. As the cotton loses its natural wax in the bleaching process, lubrication properties are lost and fibres become sticky. Higher moisture makes the cotton lumpy, whereas a lower humidity results in lapping on cylinder and doffers. It is very difficult to get a good working on cards, even with a low doffer speed of 7 to 10 RPM. The lapping on cylinders, doffers and flats are so high that there are frequent breakdowns. Because of this reason, the number of absorbent cotton manufacturers prefers having old and obsolete carding machines rather than latest high-speed machines. However, if they can provide correct humidity and temperature, it is possible to get a good working with least breakdowns and the advantage of latest carding machines can be taken.

In water entangling system of nonwoven manufacturing, after water jets, the web is dried using hot air driers. The nonwoven web needs to be monitored with uniform moisture. This demands a controlled temperature and humidity in the web former, pad cutting, ball making, bud making and packing areas.

Cotton bud making requires highly sophisticated cards and auto-leveler draw frame. As the hank of sliver is very fine (approximately Ne10), high sophistication is required in temperature and humidity. These cards and draw frames normally do not work continuously because of their high productivity compared to bud making machines. If a bud-making machine works 24 hours a day and 7 days a week, for the blow room and cards few hours is sufficient. This type of stopping and restarting the machines also create problem in maintaining humidity and temperature at pre-determined levels.

In order to prevent the hot air drier making the complete department hot, it is needed to exhaust the hot air by providing hoods above the drying area coupled with exhaust fans. Similarly, the steam escaped out from bleaching kiers needs to be guided out of the room as it increases humidity in adjacent opening and baling machines. There is a need to have suitable exhaust system to prevent steam entering the processes of lap preparation, etc.

Apart from the need of good working, maintaining a germfree atmosphere is essential as normally the absorbent cotton nonwovens are used for cosmetic and medical purposes. The water sumps needs periodic cleaning and anti-microbial treatment. We cannot just use normal soft water as we use in spinning and weaving plants. Periodic checking of water inlet and humidified air entry points for the presence of micro-organisms is practiced. All doors shall be of insect proof and shall be held closed with spring pressure.

6.2 Air distribution needs in nonwoven industry

Air distribution plays a major role in a nonwoven production unit. In order to guarantee the quality of the product to be manufactured, very specific prerequisites must be fulfilled. In particular in the preparation phase and when the nonwoven is formed, the following points are important:

- Consistently high humidity of the material for optimal processing
- Conditioned air with correct dosage, at the production critical positions on the machine
- Air humidity level that is still hygienically safe
- Efficient cleaning of the air, taking into account the maximum workplace concentration. Stabilizing and bonding need constant conditioning

Luwa has developed modular construction of the TAC® system for air engineering processes in a nonwoven manufacturing plant designed considering the characteristics of the various production machines. Since in a nonwoven manufacture a high degree of humidity is crucial for hygienically perfect conditions, the air humidification must be optimal. TexFog designed for this purposes uses only fresh water and does not require a water tank. Therefore, this is safe from a hygiene point of view, which fulfills the German hygiene regulation VDI 6022. It is a low energy consuming and claimed to operate with five times less energy compared to a conventional air washer and also allows a super saturation of the air.

Figure 6.2 TexFog – hygienic and energy-saving humidification

6.2.1 Flow master

The FlowMaster displacement air outlet is suitable for local conditioning (spot air-conditioning) in manufacturing halls with an irregular local heat load distribution. It is a displacement air outlet distributing the supply air at a low speed near the floor, with little induction and turbulence. The air that has been heated up by machines and humans is removed through the extraction openings from the room or directly as machine extract air.

Figure 6.3 Flow master

6.3 Material air conditioning and spot air conditioning

For nonwoven plants, two types of air conditioning, viz., spot air conditioning and material air-conditioning are required. With spot air conditioning, maintaining the required air parameters is guaranteed in work zones. With material air conditioning, it is ensured that the material to be processed is surrounded by air in an optimal condition at every process step. In other process steps, as in bonding and converting, comfort air conditioning is the most important aspect.

In case of opening bleached cottons after drying, wet cotton openers are used. The cotton opened is dried in hot air driers and then goes for further opening. In that position, maintaining humidity is very critical. Normally moisture sensors are used to predict the moisture in the dried cotton, and dry fog humidification is used. The spot humidifiers are fixed inside the hoppers. The uniformity in moisture content is very essential in the laps to get uniform bulkiness of the pads and correct height in the bags. Therefore, material humidification with precision is very essential.

6.4 Cleaning of the air

A multi-level filter technology with at least one first automatic filter step is required, depending on the waste or contaminants present. Depending on the process, separate air filters are required. For example, where cotton nonwovens are produced, the raw cotton is first processed in regular blow room with 4 to 6 cleaning points, like a normal coarse count spinning mill. The material is then fed to a series of cards through chutes. The dusty air is to be removed and filtered in blow room as well as in carding. This requires heavy duty filters. To avoid contamination of dust, the pre-opening units are normally housed in a separate building. The opened cotton is fed to a cake making machine, where cotton is combined with water and pressed with press foot in a perforated stainless steel drum. After bleaching and drying, the cotton is made into bales, which are used for further processing of preparation of lap. In the bleached blow room and carding, the concentration is on opening and not on cleaning. Here we need not extract the dust as done in pre-opening. Air is extracted from the blow room and carding, filtered and reused.

Automatic panel filter for exhaust air containing dust and fibres with cleaning nozzles is available, which needs only minimal space. Luwa provides a cleaning robot having a variable number of nozzles and the variable speed depending on the dust content.

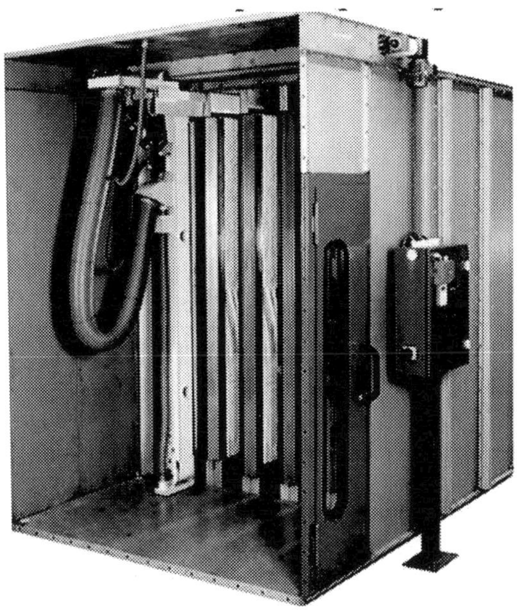

Figure 6.4 Cleaning robot

Figure 6.5 Multidrum-VAC by luwa

Multi VAC is a simpler automatic filter unit suitable for applications with a smaller air volume. This has number of filters in drum shape with different porosity, so that wastes are segregated and sent for recirculation

6.5 Recycling and disposal of wastes

Fibres recycling with waste separators is used for retrieving good fibres from bleach carding and bleach blow room area.

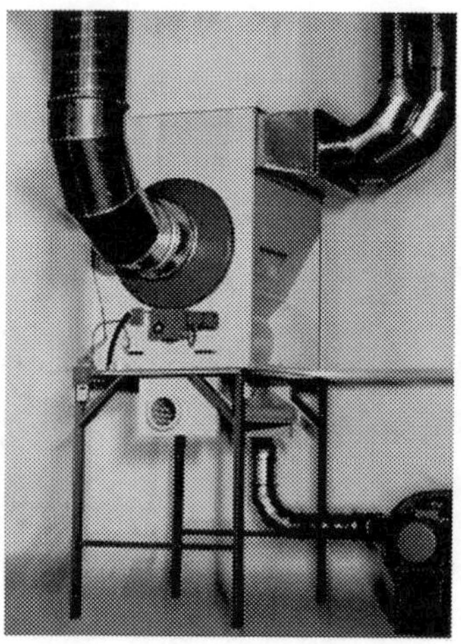

Figure 6.6 Compacting unit

Figure 6.7 Rotary pre-filter by luwa

Rotary pre-filters are used in pre blow room and carding to extract dirty non-reusable dust. The wastes collected on drums are sucked by a compacting unit, where they are pressed into small bales for despatch.

6.6 Compact filter unit CFU

Compact, stand-alone type of filters are used to collect non-usable clean short fibres. They are collected and disposed for papermaking industry.

Figure 6.8 Compact filter unit CFU

7
Concept of total air control

7.1 Introduction

TAC, the total air control is the concept given by Luwa for processing traditional textiles such as in spinning, weaving and knitting mills. It comprises solutions for all air engineering processes, including regulating, controlling and services. TAC covers air treatment in entire production facilities, specific areas and immediate machine environments, machine and plant cleaning, appropriate disposal of collected particles and fibres as well as a control system and various after sales services. The impact of climate control on the interplay of people, machines and materials is unquestioned. Without a satisfactory indoor climate, industrial production is unlikely to be entirely successful. Regardless of the extreme conditions of the outside climate, the TAC aims at providing required humidity, temperature and sufficient clean air circulation. Customised air is the concept of TAC which is tailor made for the process. Optimum air quality and custom solutions for each production process decisively affect productivity and product quality.

The plants are designed by studying the process in detail. Intimate knowledge of the production processes is very essential for the humidification plant designers. The engineers should work closely with production experts who contribute their knowledge of their specifications and the intricacies of production processes. The production experts and their process know-how interface with humidification specialists directly builds in the process a solid bridge between air treatment technology and process know-how. The Competence Centers of the humidification plant manufacturers provide instant access, anywhere in the world, to the extensive industry and process know-how. This factor alone is of immeasurable benefit to customers in all project phases, from initial consulting and conceptualising, planning and execution, to coordination, installation and certification, etc. It also includes customer training and after sales services. The plant manufacturers take care of planning and execution through qualification, validation and servicing. Design-build concept saves resources and valuable time. Further electronic controls to monitor the temperature, humidity and air flow are very essential, and they are integrated part of the solutions. Training the technicians to monitor the plant is also a part of the package of TAC. Modular total solution for all air control processes, TAC, comprises of air treatment, air conduction, cleaning, waste handling, controls and services.

Now a number of other companies are also working towards giving complete solution.

TAC offers increased productivity through:

- Tailor-made solutions adapted to any production machine.
- Stable climate in the work zones.
- Efficient dust removal.
- Thorough machine cleaning.
- Robust extraction and disposal systems.
- Unique regulating and controlling with branch-specific software.
- Precise and dust-insensitive transmitters.
- Worldwide service organization.

Reduction of operating costs is achieved through:

- Air conducting concepts with reduced air and water requirements.
- Customer-specific plant engineering.
- Application of energy-efficient components.
- Appropriate regulating and controlling.

Air in the right place is the very specific pre-requisite that must be fulfilled in order that the quality of the product to be processed is guaranteed with consistent climate, precisely regulated supply of treated air, optimally placed exhaust air openings in the floor, efficient cleaning of the machines and reliable controlling and regulating of temperature and humidity. In order to achieve even temperature and humidity and at the same to avoid fibre fly and dust development, the air flow, in particular the supply and removal of the air plus the cleaning, plays a crucial role. According to the manufacturing process, a room, machine or combination air conditioning system is to be foreseen.

7.2 Different individual solutions

Different individual solutions for airflow and air conditioning for different manufacturing processes in a textile mill by Luwa are, for instance:

- Baffle plate air outlet for ring spinning
- RotorSphere air outlet for rotor spinning
- LoomSphere air outlet for weaving machines
- AirBell air outlet for knitting mills

RotorSphere air outlet systems are specially designed for Rotor Spinning. The logic considered is that compared to the ring spinning process, rotor spinning is basically a closed process. This requires less air and can be run with a smaller air conditioning plant. So investment is kept at a low level. In addition, there are savings with the fan and pump performance as well as water consumption. In addition, with RotorSphere a much improved homogeneity of the micro-climate is achieved along the rotors and the feed cans situated between them. RotorSphere also ensures that there is a good working atmosphere for the factory staff.

Figure 7.1 Rotor sphere air outlet for rotor spinning

Since ring spinning, friction spinning and air jet spinning are an open process, a room air conditioning system is to be foreseen. To keep the number of thread breakages low and increase productivity the air above the machine must be distributed optimally. In the work zone, i.e., the drafting system, there must be a predominance of vertical air movements so that the humidity can reach the yarn optimally and the dust produced is pushed downwards. There, the dust is taken up by the exhaust air system and removed from the spinning hall.

Figure 7.2 Baffle plate air outlet for ring spinning

The patented air outlet LoomSphere developed by Luwa allows direct air conditioning of the weaving machines at the warp sheet, without demanding for the maintenance of 85% humidity all over the shed. This new concept is the main feature of the air conditioning concept for weaving mills. It guarantees the desired conditions regarding humidity and temperature on the warp.

Figure 7.3 LoomSphere air outlet for weaving machines

Since in a nonwoven manufacture a high degree of humidity is crucial for hygienically perfect conditions, the air humidification must be optimal. Luwa offers the high-performance humidifying system TexFog, which only uses fresh water and does not require a water tank. Using TexFog is safe from a hygiene point of view and fulfills the German hygiene regulation VDI 6022. It requires up to five times less energy than a conventional air washer and also allows a super saturation of the air. For nonwoven plants, two types of air conditioning are provided, they are spot air conditioning and material air-conditioning. With spot air conditioning, maintaining the required air parameters is guaranteed in work zones. With material air conditioning, it is ensured that the material to be processed is surrounded by air in an optimal condition at every process step. In other process steps, as in bonding and converting, comfort air conditioning is the most important aspect.

Figure 7.4 Tex Fog – hygienic and energy-saving humidification for nonwoven plants

7.3 Cleaning and disposal of air

The Flow Master displacement air outlet is suitable for local conditioning (spot air conditioning) in manufacturing halls with an irregular local heat load distribution. Flow Master is a displacement air outlet which distributes the supply air at a low speed near the floor, with little induction and turbulence. The air that has been heated up by machines and humans rises and is removed via the extraction openings from the room or directly as machine extract air.

Figure 7.5 Flow master displacement air outlet

Regulation and reduction of fibre fly and dust are important for the quality of the products to be processed. High-performance travelling cleaners and room air changes counteract these problems. In production fringe areas – near creels, transport systems, casings or interim storage – additional cleaning measures are necessary. Vacuum cleaning plants are very common in textile mills. Ceiling cleaner Circulaire, developed by Luwa considerably reduces cleaning work and increase the production time. Waste collection from the point of origin on a continuous basis and compacting them for transporting outside is a very important aspect of total air control.

Figure 7.6 Travelling cleaner STC on ring spinning machines

Figure 7.7 Travelling cleaner LTC with loom sphere air outlet in a weaving mill

The travelling cleaner is an important part of an efficient air conditioning concept for regulating and controlling the air which is reliable and economical

7.3.1 Automatic panel filter APF – cleaning robot with variable cleaning for individual applications

A multi-level filter technology with at least one first automatic filter step is required, depending on the waste or contaminants present. Luwa offers an automatic panel filter for exhaust air containing dust and fibres, which needs only minimal space. Thanks to the high filtering degree, the requirements of the international norms for residue dust amounts (MAC values) are fulfilled. The filter medium is automatically cleaned by means of a nozzle belt. Having a variable number of nozzles and the variable speed of the robot, depending on the dust content, is a new feature.

Figure 7.8 Cleaning robot

7.3.2 Multidrum-VAC – compact and powerful

The simpler automatic filter unit Multi drum-VAC is suitable for applications with a smaller air volume. As pioneers for efficient and environmentally friendly recycling or disposal concepts, Luwa offers a wide range of products for extracting and compacting dust, fibres and dirt particles. Fibre recycling is carried out with our waste separator, and the usable fibres are recycled directly to the production unit after extraction. Non-usable waste is separated out with the automatic rotary pre-filter unit. The wide product and system range comprises individual products right through to fully automatic plants.

Figure 7.9 Multi-drum VAC

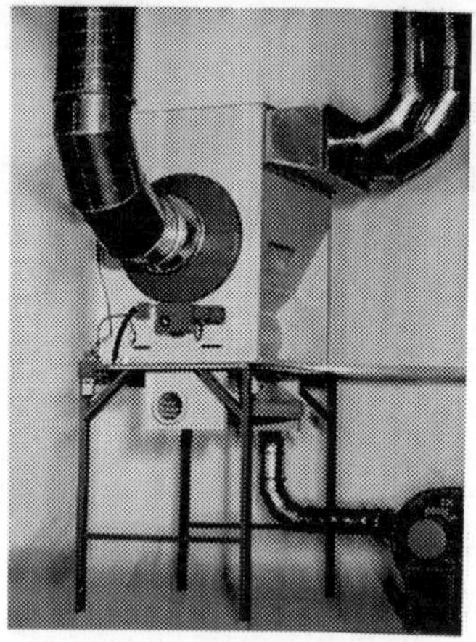

Figure 7.10 Good fibre recycling with waste separator WSB

Figure 7.11 Rotary pre-filter unit RPF for elimination of not reusable waste

Compact stand-alone filter units are operating and maintenance friendly, are used for collecting clean short lint from the blow room and carding after bleaching.

Figure 7.12 Compact filter unit

7.4 Controls

7.4.1 Transmitters

Luwa offers a range of transmitters especially developed for the textile industry. They are fast-reacting, precise and have particularly high-performance temperature and humidity sensors.

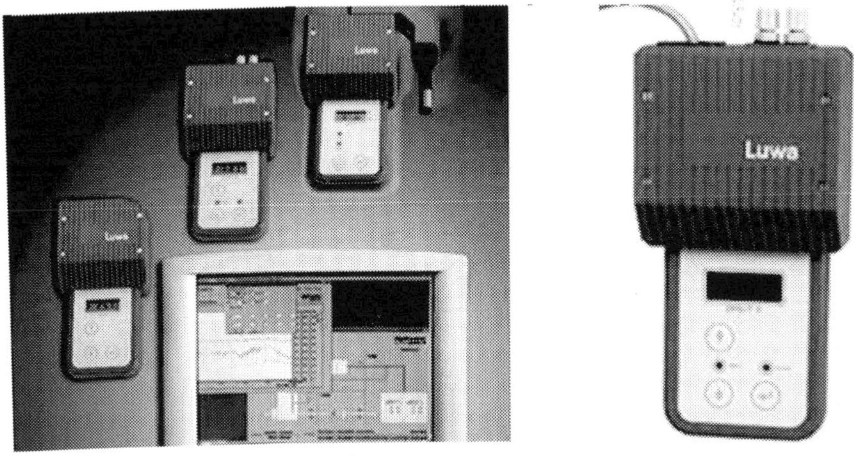

Figure 7.13 Control units

The speed and precision of the sensors help to ensure that the air conditions are always in an optimal range. The permanent and precise registering of temperature and humidity values as well as of pressures therefore allow energy-optimised permanent operation.

DigiControl has an important task within our total concept. It is the automatic system for regulating and controlling, monitoring and optimising of all air engineering processes. It consists of harmonised hardware and software. It records, registers and stores all relevant data.

7.5 After sales service – trouble-free operation

After sales service is very important to ensure that the plants perform to optical efficiency. The quality and reliability of equipment have no effect unless it is handled properly by trained personnel it cannot perform. The manufacturer can improve his design by getting proper feedback from the users, for which he needs to have continuous interaction. The concept of after sales service addresses both the requirements, i.e., training the people to handle the equipments properly and to understand the problems and requirements. Hence in the concept of total air control servicing by the manufacturer takes an important role. Regular maintenance contributes considerably to a reduction in operating costs. Expert checking of the plant and adjustment of the set values guarantees irreproachable, trouble-free operation and lengthens a plant's life. The service contract includes regular checks of the plants regarding condition, function, operating costs and safety. Maintenance and function-relevant cleaning ensures the machine in always ready for operation condition. As per the contract, the manufacturer carry out maintenance tasks, breakdown remedying, conversion, tele-service and repair work, training and instruction of the operating staff according to requirements. Also the manufacturers make their know-how available for ssoptimising and advising about existing plants from other manufacturers. The main focus is on operating safety and reduction of operating costs.

8

Localized humidification control

8.1 Need for localized humidification control

The conditioned air supplied by a humidification or air conditioning plant is normally by diffusers fitted to the ducts in the roof level, which is intended to spread the air uniformly in the working area. The return air trenches suck the air and hence the air shall be moving from the diffusers to return air trenches. But our purpose of humidifying the department is to help the material in process to absorb the humidity from the air supplied. It is more important for the working materials to get exposed to the humidity rather than the complete department. The human beings working in the production area require a humidity range of 45% to 60% for comfortable working, whereas the working material depending on the material and process requires different humidity and temperatures. Sometimes, we need to dehumidify to get correct results. Hence, it shall be useful if one can develop systems to provide humidity or dehumidify at the spot where the material is working. Spot humidifiers are one such development.

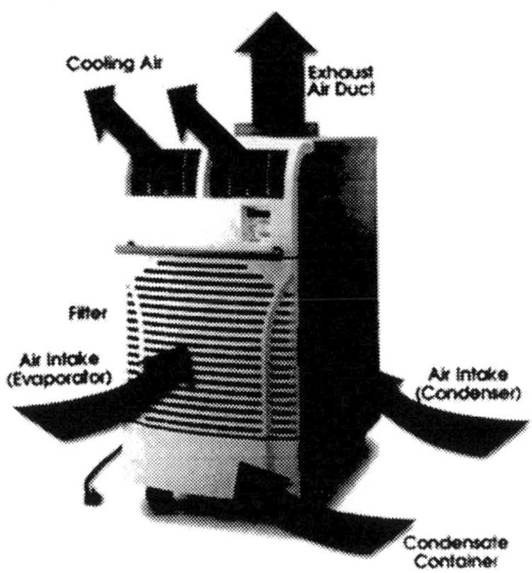

Figure 8.1 Spot coolers

Energy costs continue to escalate while, at the same time, utilities are frequently unable to provide all of the electricity that industry needs. When these circumstances are combined with the need to control capital expenses, the spot cooling becomes the choice. In a wide variety of applications, spot coolers are the answer to controlling both energy and capital expenses.

Spot cooling is a method of cooling overheated areas within a larger area, as opposed to providing general cooling by means of a remote centrally controlled air conditioning system. In operation, spot cooling is an extremely efficient means of cooling people, processes and equipment because it directs a localized stream of cool air exactly where it is needed.

The concept of spot cooling is not a new idea, but the use of spot coolers to accomplish localised cooling within work environments and equipment rooms is a recent development. Room air or ambient air is drawn in through the evaporator intake in front. The air is cooled as it passes over the evaporator coils and leaves through the cool air discharge. The condenser intake on the side of the unit draws in air to cool the condenser coils. This warm air is then discharged out of the condenser exhaust on top.

8.2 Use of heating lamps

In a number of cases, we need to keep a small area in the production department either dry or hot depending on the nature of the material in processes. This cannot be done by central plants. The use of the heating lamps along with suitable partitioning shall be useful. The classical examples are the use of heating lamps above the creel of draw frames, combers and speed frames during rainy season, use of heating lamp on the warp sheet of a loom while weaving cotton PVA-doubled yarns for loops in super soft terry towel making, conditioning chambers, etc.

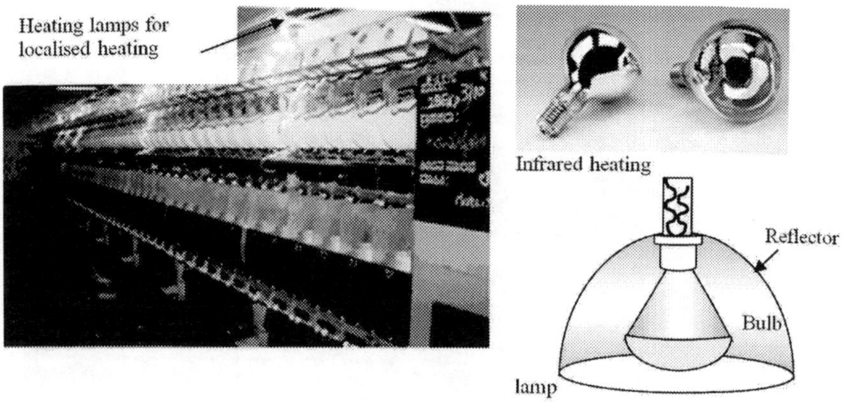

Figure 8.2 Heating lamps

While installing heating lamps, it is necessary to understand the purpose and adjust the distance between the lamp and the machine to get the required results. If the lamps are very near, the metal parts of the machine also become hot, which is not desired. The heat from the lamp should not get wasted by radiation; hence, use of reflectors in the back is suggested. As the bulbs shall be very hot, it can break if comes in contact with water droplets. The reflector also protects the bulb from the possible water droplets.

8.3 Subsystem humidification

Subsystem humidification of air conditioning instead of a block system helps improvement of loom efficiency, the comfort and the hygienic conditions within the weaving room and reduction of the energy needed for the conditioning of the air in the weaving room. These goals are obtained by precisely controlling the moisture content of the yarn being fed into each loom by removing solids and heat generated by the weaving process at their source and by improving the handling of the air in the weaving room. One may to understand that warp yarns after sizing usually are dried by the hot air cylinders in slashers to 5–6% moisture. But during stops, part of the yarn is over dried to 1–2% moisture. Most warp yarn breaks are concentrated in these parts. Filling yarn moisture falls within the 3–7% range, depending mainly on the spinning technology employed. Sometimes, yarn conditioning prior to weaving is required. Yarn exposure to the room air during the weaving process is as short as 15 minutes or less. The weaving process would run best with a very high room humidity level, in the 85–95% range, but operator comfort imposes a limit, usually 55–65%. Room temperature is normally kept within 24–26°C. range. Therefore, the concept of localised humidification is beneficial. Recently, two companies (LUWA-Bahnson and LTG Air Engineering) introduced systems projecting cool nearly saturated air from about 800 to 1500 mm over the shed on the yarn. Better loom performance has been reported. Similarly, trials done by SITRA (Patent No 174964 dated 2 Jul 1990) at Coimbatore indicated that inducing humidified air at the creel of ring frame is economical and gives equally good results.

Loom Sphere is a localised humidity unit developed by Luwa, placed as close as possible above the warp, producing a low-turbulence air flow. Normally Loom Sphere diffuser is installed approximately 800 mm above the weaving loom, depending on the harness and beam change. This arrangement results in the high number of air changes of 80 to 150 per hour required in the weaving zone to extract the dust created there. The broad air diffuser with integrated filter creates a laminar air flow that penetrates into the weaving zone without mixing with the dry and dusty ambient air. This concept with laminar air flow enables high air humidity of greater than 80% at the critical areas of the loom.

This system has two essential advantages. There is no room air because of the laminar air flow, and the air that flows slowly guarantees that there is sufficient humidity on the warp. The warm outlet air is collected directly under the weaving machine, together with dust and fiber fly, and is removed. This direct air conditioning in combination with a reduced air volume for the room is the ideal TAC combi air-conditioning. The TAC combi air conditioning is claimed to offer a high degree of efficiency and comfort with low investment and subsequent low running costs.

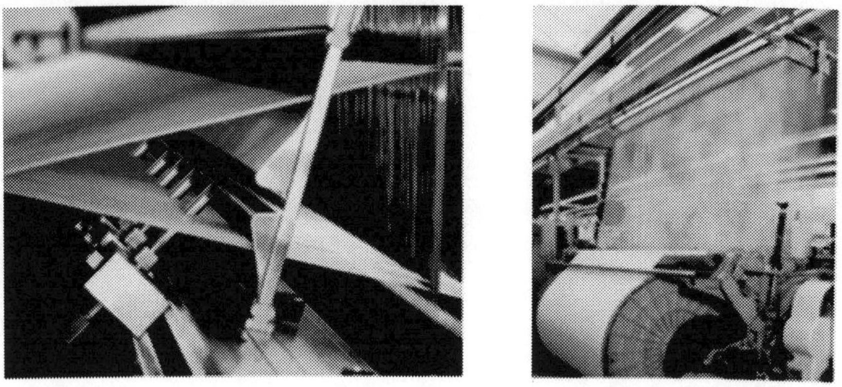

Figure 8.3 Loom sphere humidification

Table 8.1 Comparison of conventional and loom sphere humidification systems.

	Conventional System	Loom Sphere System
Relative humidity – room	80%RH	65%RH
Relative humidity at Warp	<70%RH	>80%RH
Air volume of humidification plant	100%	60%
Air change – room	45–60 per hour	30–35 per hour
Air change at warp level	45–60 per hour	80–150 per hour
Power consumption	100%	60%

System developed by Shofner U.S. Pat. Nos. 5,676,177 & 5,910,598, Chern U.S. Pat. No. 4,966,017 and Vignoni U.S. Pat. No. 5,275,350 has following subsystems performing a definite task separately. The necessary moisture required by the yarn to perform best in the loom is metered exactly and directly on the yarn. Thereby, yarn breakage is reduced and weaving room air humidity can be lowered, improving human comfort. Lint and dust generated by the weaving process are removed at their source. Air contamination is lowered, thereby improving hygienic conditions and reducing air filtration requirements and making travelling cleaners unnecessary.

Heat generated by the weaving process is partly removed by water cooling the lubricating oil of the weaving machine, thereby reducing the weaving room heat load. Large peripherally located air conditioning units are replaced by smaller units, distributed over or under the roof of the room, each serving the area of 4–24 looms. Due to subsystems, filtration and cooling requirements are greatly reduced. Air humidification is not needed. The fan and coil conditioning system is best suited for this application. Subsystems are operated on each loom separately, thereby increasing production flexibility. Loom efficiency is increased as yarn breaks are lowered to less than half. Conditioned air will be projected on the aisles of work place. Room air will be collected for recycling at a higher level. Thereby turbulences, usually produced by ascending warm air, are prevented. Under floor air ducts are no longer necessary. Due to subsystems, less air will have to be turned over. Due to decentralisation of AC units, said air will be transported over shorter distances. Thereby, substantial amounts of energy are saved. As lesser air ducts are needed, the investment costs are reduced. By elimination of underground ducts and openings in the floor, replacement or rearrangement of weaving machines is made easier.

8.4 Air bell air outlet

Air bell air outlet for knitting mills developed by Luwa is another example of localised humidification. This is used for air conditioning of circular knitting machines. Air bell has the effect that the air flow above the machine forms a sort of bell. Fly and dust are eliminated directly with the outlet air and cannot move to other circular knitting machines. With this, the work area is freed from fly and dust and fault frequency is reduced by up to 50%.

Figure 8.4 Air bell

8.5 Exhausting air

Providing ducts and exhaust fans to collect the dry or humidified air from selected parts of the machine also helps in providing the clean work area to the material in process and preventing disturbance to adjacent processes. The air suction units like Pneumafil provided in ring frames, speed frames, draw frames, carding, etc., not only concentrates on preventing the dust flying out but also guides the air movement. The conditioned air fed to the production department through the diffusers at the top spreads all over, whereas we need it to interact with the material being processed. As the return air system is sucking air from the department, the conditioned air is also sucked out wasting all efforts for conditioning. The system of extracting air from the machines directs the wind flow towards the machines and the material in process.

Figure 8.5 Flow master by luwa

Figure 8.6 Collection of dusty air from machines

The Flow Master displacement air outlet developed by Luwa is an example for local conditioning in manufacturing halls with an irregular local heat load distribution. Flow Master is a displacement air outlet which distributes the supply air at a low speed near the floor, with little induction and turbulence. The air that has been heated up by machines and humans rises and is removed via the extraction openings from the room or directly as machine extraction air.

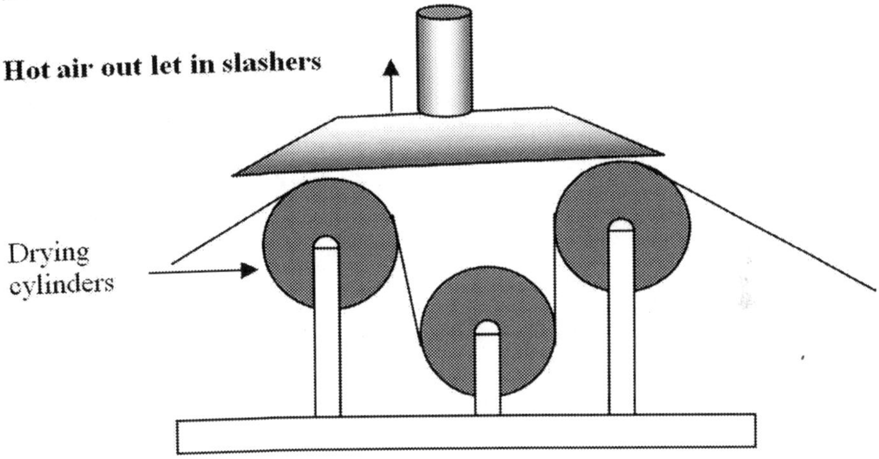

Figure 8.7 Hot air hoods

In sizing, for drying the sized yarns either steam heated cylinders or infrared lights are used. While drying, the moisture in the yarn evaporates and forms a column surrounding the yarn. This reduces the drying efficiency. Therefore, it is needed to continuously remove the humid air from the point of drying. Hoods are provided on the top of the drying surface, and the air is exhausted by using exhaust fans. If this is not done, it not only reduces the drying efficiency of the drying units but also allows the humid air to escape freely in the work area. This humid air moves up and after reaching the roof the water condenses and falls as big drops all over. These drops shall be very big and if fall on the dried size beam create havoc by developing stickiness. Therefore, even though one says control of humidity is not needed for sizing, monitoring and exhausting of the hot humid air are very essential.

Some mills have exhaust fans on the walls. These are provided to throw out hot air from the working area. Where humidification is required and provided by various means, the exhaust fans tend to throw the humidified air out. Therefore, it does more harm than good. It is always advisable to provide suitable ducting and collect the hot air from the point of generation and guide them out by installing exhaust fans at suitable location.

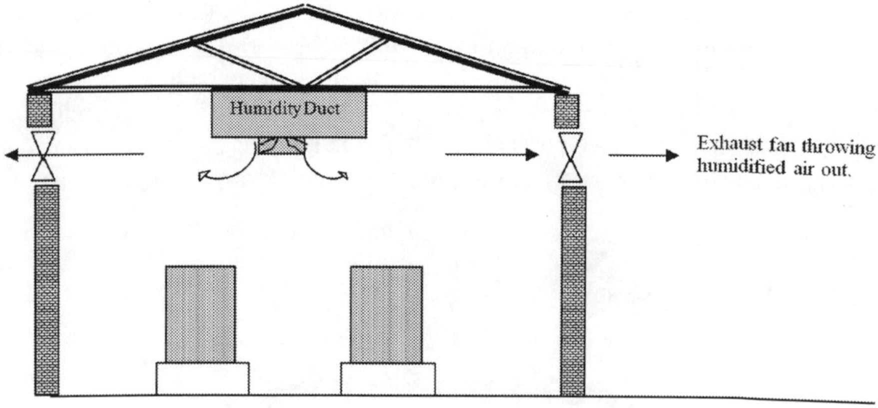

Figure 8. 8 Wrong positioning of exhaust fans

8.6 Machine air conditioning

In machines where the speeds are very high like texturizing, ring frames, etc., the generation of heat is very high. Further, the ballooning throws out the moisture from the fibres in process. Therefore a specialised treatment of air and humidity is required at those points, whereas in the creel, as the materials are almost stationary, they tend to absorb more moisture. The treatment needed for creel is therefore to be different compared to actual spinning area.

Figure 8.9 Machine air conditioning

The extremely high-energy requirements in false twisting on one hand and the homogeneous temperature and humidity needs in the bobbin creel area on the other create a need for targeted machine air conditioning. The air is guided directly to the bobbins via low-induction slide vane outlets. A simultaneously controlled extraction of the heat produced by the process guarantees optimal running of the machine.

8.7 Super soft terry towels weaving

Towels have the basic objective of absorbing moisture from the body. More the absorbency, the towel is called as of superior quality. Normal methods used to make the towels soft are using lower twists in the pile yarn, using softeners while dyeing, having a higher loop height etc. One of the recent methods is to use the pile yarn made by doubling cotton with a water-soluble filament. Once the towel is washed, the water-soluble filament dissolves, and the pile yarn shall be almost of zero twist, giving a super soft feel and maximum absorbency. PVA filament, Solvoran is an example. The production of such towels poses a peculiar problem. The PVA needs a highly dry weather to avoid partial dissolving of PVA films while weaving, whereas the ground beam made up of 100% cotton requires a very high humidity of 85% and above. We need to keep the pile beam area hot and dry by using heating lamps, whereas ground beam needs to be humidified using mist humidifiers. This is a classic example of localised humidification control.

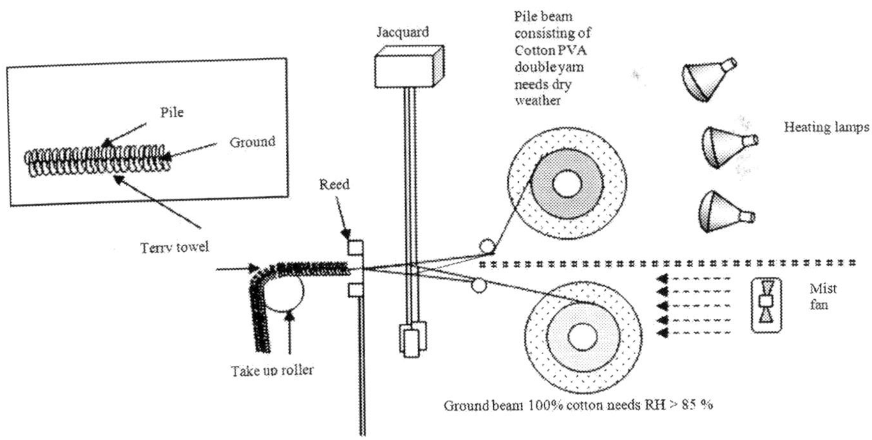

Figure 8.10 Terry loom manufacturing Super-soft Towel with PVA cotton-doubled yarns

Localized humidity controls are as important as centralised controls. The total air control concept given by Luwa has number of units those are basically localized units designed specifically for the purpose.

8.8 Yarn conditioning plants

Humidification needs are not only for getting a good working during productio, but also to ensure that the material is made suitable for further working. During spinning and winding, the yarn package runs at a very high spee, because of which the moisture in the yarn is thrown out. The normal moisture in ring spun cops and cone winding shall be around 4.5% to 5.0%. To have a good working in warping or knitting, the yarn should have moisture of above 7%. The commercial moisture regain considered for overseas market is 8.5%, whereas in the mills, it is very difficult to get that moisture content. To get the advantage of moisture content, people started keeping high humidity in winding section by installing mist humidifiers. This resulted in the rusting of machine parts, but the moisture in the cone did not increase. The cones were remaining dry as the cones were put in polythene bags immediately after doffing to avoid handling stains. Then the system of keeping cones in a humidified room for 24 hours was started, which only added the space problem, but the cones did not absorb moisture as they were tightly built to have higher yarn content on cones and to have higher container capacity to reduce transportation costs.

Figure 8.11 Cones kept in humid room

The yarns made for hosiery are normally waxed, which also is a hindrance for the yarn to absorb moisture. By keeping cones in a humidified room for a long time gave higher moisture content in the outer surface of cones, but not uniformly inside the cone. Some even tried subjecting cones to direct steam at different pressures. The imbedded air in between the fibres and yarns had no place to come out, which was also a hindrance to moisture absorption. There used to be condensation of water droplets at the surface of cones and in the polythene bags, but the moisture was not entering the core of the yarn. It increased the problems rather than reducing.

The yarn conditioning plants and auto claves were designed to give moisture to the yarn by force.

The cones are kept in an enclosed chamber, which shall first be made vacuum by a vacuum pump. By this, the trapped air between fibres and yarns are taken out. Afterwards, live steam would be allowed to enter the chamber. As the air has been removed, the water molecules can enter the cavity in the fibres, and the absorption of moisture becomes more effective.

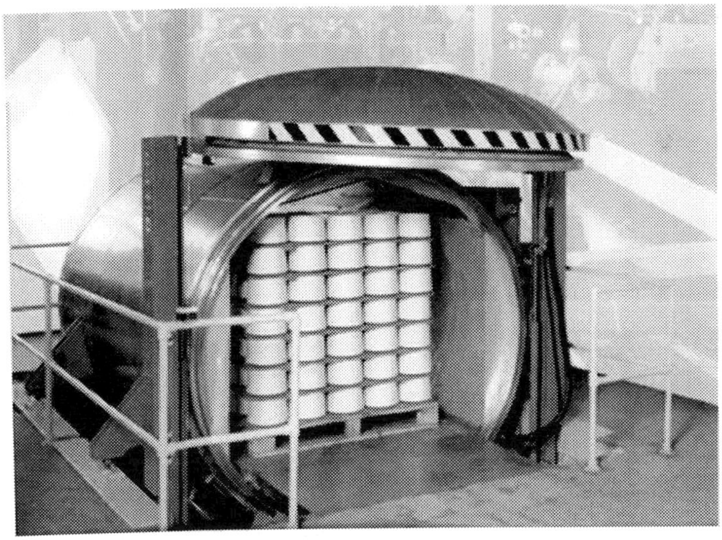

Figure 8. 12 Yarn conditioning machine

8.9 Static elimination

Static charges develop because of the friction between either fibres or between fibres and the machine parts. The humidification makes the air conductive and reduces the static charges. However, the generation of static charges shall not be uniform throughout the production area. Therefore, it shall be advantageous to attack the places of static generation and eliminate the charges rather than increasing humidity throughout the working area. Various devices are available for arresting static charges; out of them air ionizers have a direct link to air handling.

Air ions are molecules that have lost or gained an electron. Ions are present in normal air but are "stripped" out when an air is subject to filtration and conditioning. They are produced by radioactive emission or by phenomenon called 'corona discharge', where a high voltage is applied to sharp point. All air ionisation systems work by flooding the atmosphere with positive and negative ion, when ionised air comes in contact with a charge surface; the charge surface attracts ions of the opposite polarity.

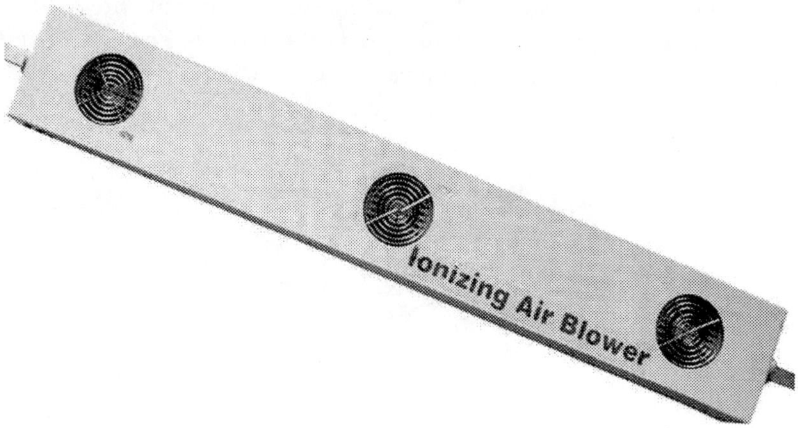

Figure 8.13 Static eliminators

As a result, the static electricity that has built up on products, equipment and surface is neutralized. Typically, air is highly insulative with a resistive exceeding 10E15 ohms/meter. By increasing the numbers of ions in the air, it is possible to lower the resistance of the air to 10E11 ohms/meter, thereby making the air more conductive. Conductive air can neutralise the static charge on every surface that it contents. The field from the charged surface attracts ions of the opposite polarity until the charged on the surface is neutralised. Different types of air ionisers are available with built in blowers which can be installed very near to the source of static charges.

9

Maintenance of humidity

9.1 Understanding the calculations

The effect of music is not only by the use of correct instrument but also by its proper use. The role of instrumentalist is equally important as the instrument itself. Similar is the case with air-handling systems. Proper use of the plant and maintaining it are very important to get the required results. It must be noted that one cannot go on changing or updating a humidification plant in any industry due to its huge cost and complete shutting down of the factory might be essential.

Meeting ventilation and comfort requirements in multiple-space air-handling systems are challenging. The factory and building heating and cooling loads are not static; therefore, systems need to be flexible to respond to the continually changing outdoor conditions and occupant activities. To optimise comfort and minimise energy usage, the HVAC control system must also be dynamic. An effective way to assure that the comfort, working and ventilation needs for all areas of the factory are met is to continuously monitor system operation. The system must adjust control parameters as necessary to meet the needs for each specific operating condition. This type of system level control can accomplish objectives while keeping operating energy costs in check. The key to optimising a system is communicating controls on the HVAC equipment.

The heating of the spray water or the air prior to the air washer creates a condition that allows the air passing to the plant to hold a higher level of moisture. The amount of moisture that is required for the plant use will dictate the degree that the air or spray water will need to be heated. In dealing with the problems of maintaining constant inside air conditions one has to take note of the variations in outside air conditions and also the heat generated in the working area. This needs many time consuming and tedious calculations. One might have to use psychrometric chart that graphically represents the interrelation of air temperature and moisture content and is a basic design tool for building engineers and designers. Psychrometric charts are an aid to understand the properties of air. The science of air–water interactions is called

Psychrometrics, which is the measurement of thermodynamic properties in moist air. As a problem-solving tool, Psychrometrics excel in clearly showing how changes in heating, cooling, humidification and dehumidification can affect the properties of moist air. Psychrometric data are needed to solve various problems and processes relating to air distribution. Most complex problems relating to heating, cooling and humidification are combinations of relatively simple problems. The psychrometric chart illustrates these processes in graphic form, clearly showing how changes affect the properties of moist air. The psychrometric chart is a graphical representation of the thermodynamic properties which impact moist air. It consists of eight major components:

1) Humidity ratio values are plotted vertically along the right-hand margin, beginning with 0 at the bottom and extending to 0.03 at the top. Absolute humidity can be indicated by several parameters, including pressure and dew point temperature.
2) Enthalpy, or total heat, is plotted with oblique lines, at intervals of 5 Btu/lb., of dry air, extending from upper left to lower right.
3) Dry-bulb temperature lines are plotted vertically at 1°F intervals. The readings are given in *x* axis. In some charts, kinetic energy of air molecules is plotted in *x* axis.
4) Wet-bulb temperature lines are indicated obliquely and fall almost parallel to enthalpy lines. They are shown at 1°F intervals.
5) Relative humidity lines curve cross the chart from left to right at intervals of 10%. They begin at the bottom at 10% and end at the top with the saturation curve (100%).
6) Volume lines indicating cubic feet per pound of dry air are plotted at intervals of 0.5 cubic foot.
7) Two-phase region includes a narrow, cross-hatched area to the left of the saturation region indicating a mixture of condensed water in equilibrium.
8) The protractor at the upper left of the chart contains two scales. One is for the ratio of enthalpy difference. The other is for a ratio of sensible heat to the total heat. The protractor establishes the angle of a line on the chart along which a process will follow.

A study of these charts help in measuring and determining the properties of both outside and inside air. It also helps in establishing the conditions of air that will be most suitable and comfortable in a given situation. The charts are created for different atmospheric pressures. With the development of online sensors that study the level of temperature and humidity and send signals for monitoring the plant to get the correct temperature and humidity, the need for the textile technologists to understand the principles and operations of humidification and air condition is not felt. However, when there is a breakdown, the technicians become helpless and try various means without understanding the basics and land into more troubles. It is therefore necessary to have some basic concepts on the design and operation of air-handling units.

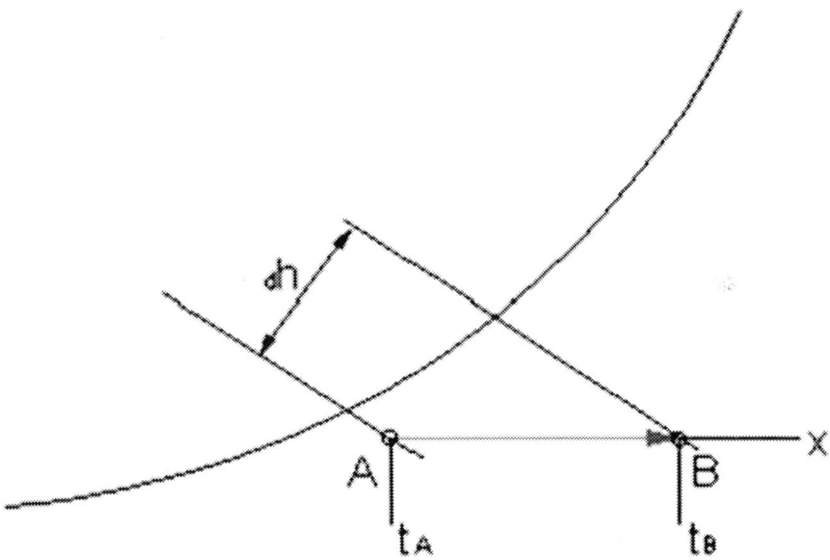

Figure 9.1 Psychrometric chart

The atmosphere is a mixture of air; mostly oxygen and nitrogen and water vapour. Psychrometric is the study of moist air and the change of air conditions. The Mollier diagram graphically represents the interrelation of air temperature and moisture content and is a basic design tool for building engineers and designers. Humidity ratio is the ratio of mass of water to mass of dry air in the sample. Since temperature is energy and absolute humidity is a mass measurement, it is evident that temperature and absolute humidity are completely independent quantities.

The most common changes of air conditions are

- Heating of air
- Mixing of air
- Cooling and dehumidifying of air
- Humidifying by adding steam or water

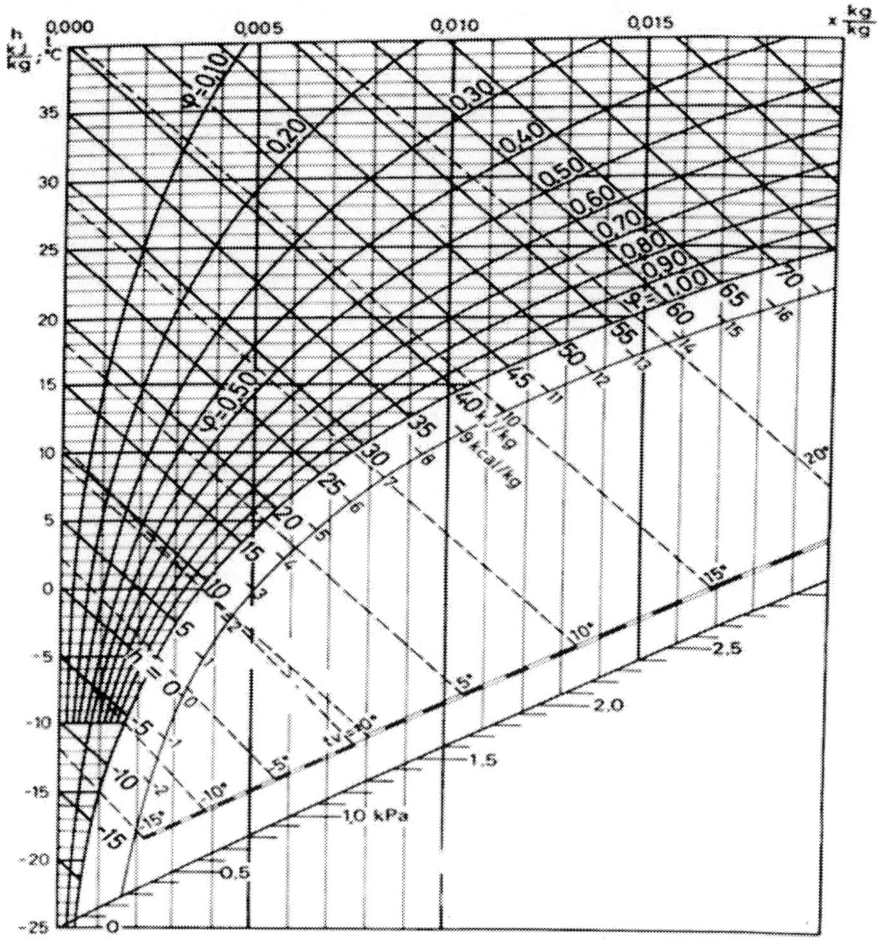

Figure 9.2 Moiler diagram

9.1.1 Heating of Air

The process of heating of air can be expressed in the Mollier diagram as explained below.

Heating of air moves the air condition from A to B along a constant specific humidity — x - line. The supplied heat – dH – can be read in the diagram as shown.

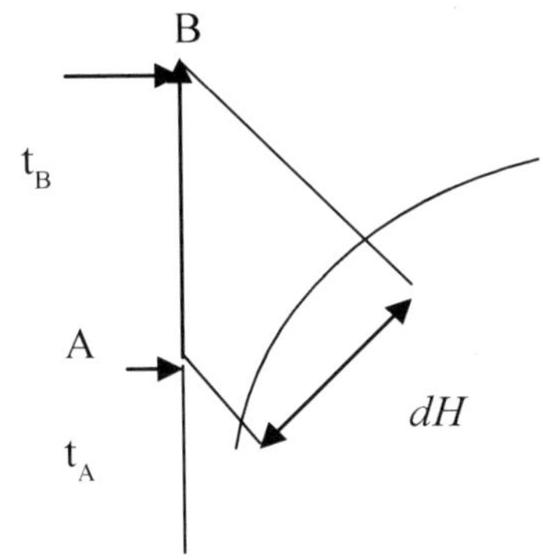

The heating process of air expressed can also be expressed in the psychrometric chart as:

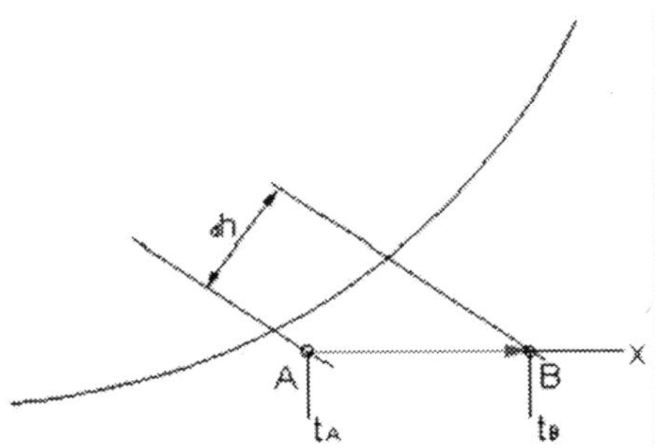

9.1.2 Mixing of air of different conditions

When mixing air of a condition A and air of a condition C, the mixing point will be on the straight line in point B. The position of point B depends on air volume A and air volume C.

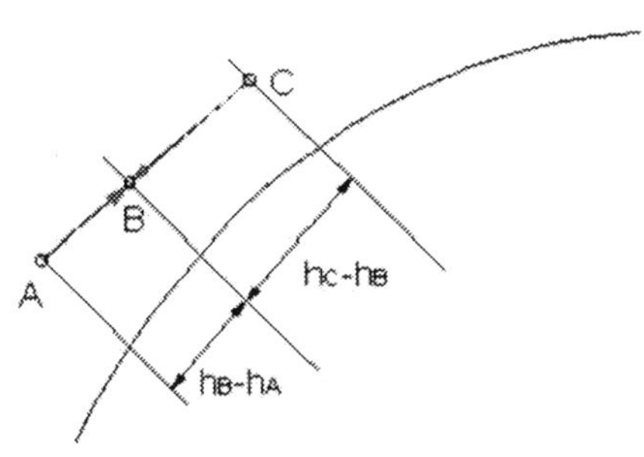

The heat balance for the mix can be expressed as: $QA\ hA + QC\ hC = (QA + QC)\ hB$ (1)

whereQ = volume of the airand h = heat of the air. The moisture balance for the mix can be expressed as: $QA\ xA + QC\ xC = (QA + QC)\ xB$ (2)

where x = water content in the air

When hot air is mixed with cold air, the result may be fog. When the mixing point is below the saturation line, parts of the moisture in the air condensates as small droplets floating in the air. The fog process can be expressed in the Mollier diagram as:

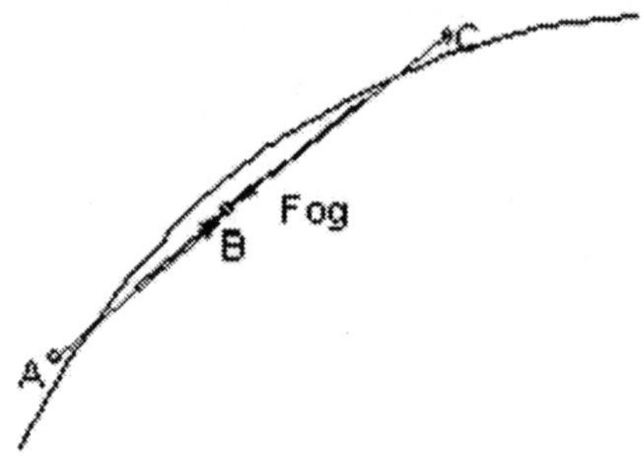

9.1.3 Cooling and dehumidifying air

With a cold surface in the air, the air may be cooled and even dehumidified. If the temperature on the cooling surface is higher than the dew point temperature – t_{DP}, the air cools along the constant specific humidity - x - line. The cooling process of air can be expressed in the Mollier diagram as:

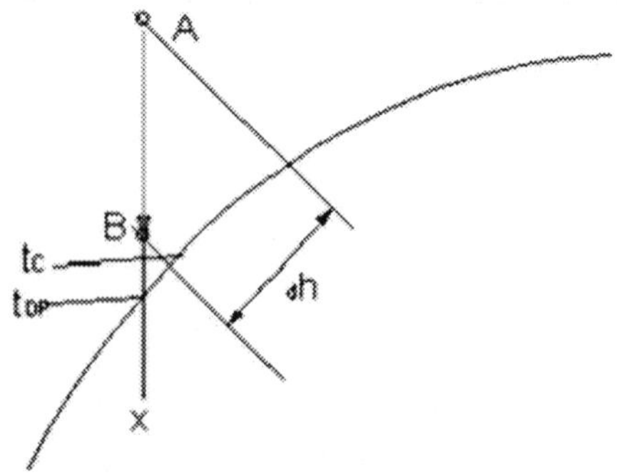

It the temperature on the cold surface is lower than the dew point temperature –t_{DP}, the air cools in the direction of a point C as shown here.

Vapour in the air condensates on the surface, and the amount of water condensate will be x_A - x_B.

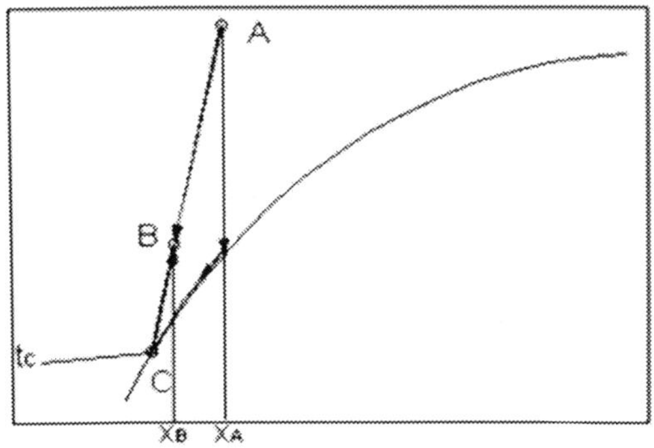

9.1.4 Humidifying, adding steam or water

If water is added to air without any heat supply, the air condition will change adiabatic along a constant enthalpy line – h. The dry temperature of the air will decrease as shown in the Mollier diagram here. If steam is added to the air, the air condition will change along a constant dh/dx line as shown.

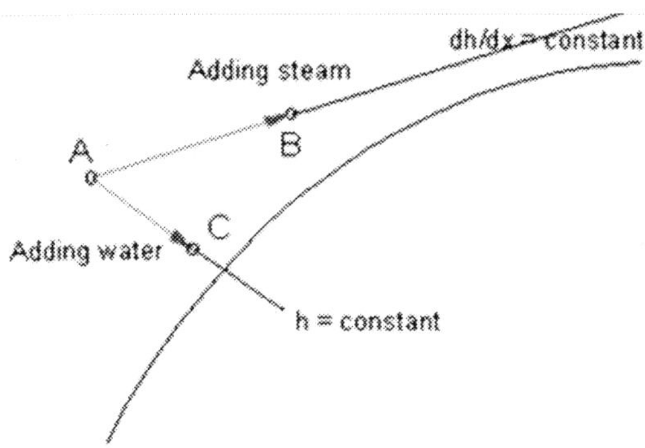

9.1.5 Humidifying air by adding water

If water is added to air without any heat supply, the air condition will change adiabatic along a constant enthalpy line – *h*. The dry temperature of the air will decrease as shown in the Mollier diagram above.

The added water can be expressed as: $mw = v\,\rho\,(xC - xA)$ *(1)*

where mw = mass of added water (kg/s), v = volume flow of air (m3/s), ρ = density of air – vary with temperature, 1.293 kg/m3 at 20oC (kg/m3) and x = specific humidity of air (kg/kg). If the water added is at the same temperature as the air, the change in enthalpy is zero.

9.1.6 Humidifying air by adding steam

If steam is added to the air, the air condition will change along a constant *dh/dx* line for steam as shown above. When adding saturated steam at atmospheric pressure, the constant line *dh/dx* = 2,502 kJ/kg (the evaporation he

at of water at atmospheric pressure). When adding saturated steam at atmospheric pressure, the temperature rise is very small; in general less than 1oC. For practical purposes, the process of adding saturated steam at atmospheric pressure approximates the horizontal temperature line. The water vapour added can be calculated using (1).The enthalpy added is found using the Mollier diagram.

9.1.7 Calculation of humidification heat load and supply air quantity

Cooling capacity for a typical textile air conditioning system is provided by a standard Carrier or refrigeration machine; sizes run from 300 to 1200 tons or higher per machine, depending on the size of the plant. Total plant tonnage will run from 300 tons in a very small plant to 15,000 to 18,000 tons or higher for a very large textile plant. In large plants, the total tonnage may be designed with a single cooling tower circuit and a single chilled water system; but in most cases, there will be several smaller systems operating independently. The tonnage and number of air conditioning systems operating in a single plant depend on the combination of textile processes being accomplished and production capacity of the plant. The room sensible heat load calculations are worked out as the sum total of heat dissipation from the machines' operating kW load in various departments, lighting load, solar heat gain through insulated roof, occupancy load and heat gain from supply air fan motors. Temperature of humidified air leaving the air washer based on outside dry-bulb and wet-bulb temperatures and design saturation efficiency is calculated as follows and plotted on the psychrometric chart.

Leaving Air Temperature = (Entry Air DBT – Entry Air WBT) x % saturated efficiency

The line of supply air temperature rise in the department due to heat gains is plotted horizontally, starting from the point of 'leaving air temperature condition' after humidification in air washer, till it reaches the design relative humidity line for the corresponding department.

The required supply air quantity is calculated as follows:

Supply Air Quantity in CFM = (Room sensible heat BTU/Hr)/ (1.08 x Temp raise of humidified air in °F)

During winter, when the outside temperature is lower than inside temperature, there shall be a loss of heat, which can be calculated as

$Ht = (Ar.Ur + Aw.Uw + Awi.Uwi)(Ti\text{-}To)$.

where Ht is the transmission loss in kcal/hr., A is the area of surface in m2, U is the overall heat transmission coefficient for winter condition in kcal/hr.m2.°C. Ti is the dry-bulb temperature in °C, and To is the outside dry-bulb temperature in °C. The suffixes r, w and wi represent roof, walls and windows.

Considerable amount of heat is generated while the production is going on because of the running machinery, electrical installations and the people working. The net heat gain is calculated by using the formula: $Hn = Hi - Ht$, where Hn is the net heat gain, Hi is the inside heat and Ht is the heat loss.

Production machines are driven by electric motors. Electrical power consumed by motors is converted into mechanical energy which is transmitted to the production machine through a shaft connecting them together. The electricity consumed finally results in generation of heat. Hm the heat load generated by machines is given as the sum of heat from the motors ($H1$) and the heat from the production machines (H2). The Hm is calculated by the formula Hm = (PR-fL860)/n where PR is the rated capacity of motor in KW, fL is the load factor, i.e. ratio of actual load of motor to the rated capacity and n is the efficiency of motor at actual load. Normally, $H1$ is about 15 to 20% of the heat load from machines.

Most of the textile machines are equipped with pneumatic systems to keep the material clean or to collect the materials back for reprocessing. The power consumed by the pneumatic system is also converted into heat and adds to the heat load. Heat load due to this pneumatic system is calculated by equation $Ha = PR$. fL860, where Ha is the heat load of pneumatic system in kcal/hr., PR is the KW of motor driving the fan for pneumatic system and fL is the load factor of the motor. This heat load can be avoided from the total heat by discharging it out without allowing it to get mixed with the conditioned air. It may be done by either overhead ducts or underground ducts with forced suction.

There shall be infiltration of heat because of various openings in the building. The heat carried away by infiltration is calculated by equation $Hinf = M(Ho\text{-}Hi)$, where $Hinf$ is the rate of heat loss by infiltration, M is the mass of infiltering air in kg/hour., and Ho and Hi represent the enthalpy of the outside air and inside air.

The threshold level (R) of air change equivalent to balance the heat loss and heat gain to maintain uniform temperature is worked out by the formula $R = Hn / M\ (Ho\text{-}Hi)$, where M is the mass of air in the department

9.1.8 Air changes per hour

The air movement in the form of infiltration and exhilaration from the building influences the relationship between temperature and relative humidity. Number of air changes per hour is decided by dividing the product of 60 minutes and total CFM of the fans and the total volume of air in cubic feet to be conditioned.

$$\text{Number of Air Changes per Hour (ACH)} = \frac{\text{Airflow of Air Cleaner in Cubic Feet per Min (CFM)} \times 60\text{ Min}}{\text{Cubic Footage of Area } (L \times W \times H)}$$

The equation $Q = 3.39H/(TD1-TW2-1.3)$ gives the capacity of the humidification plants, where Q is the air circulation rate in m3/h, H is the heat load in kcal/h, $TD1$ is the psychrometrically available dry-bulb temperature in the department in °C and $TW2$ is the wet-bulb temperature of outside air in °C. The relation between $TD1$ and $TW2$ is fixed by the RH to be maintained. Therefore, the Q is dependent only on the heat load. If heat load is reduced by any means, the required capacity of the plant reduces without affecting the desired RH. Smaller the plant size lower would be the capital cost as well as running cost. If some means are provided to avoid incidence of direct radiation from sun to roof, heat due to sun effect shall be nil. This can be done by false roof over the main roof, i.e. double sheeting with a small gap of 3 to 6 inches. Another method is to spray water on the roof to eliminate the rise in surface temperature. In this case, the heat instead coming to the department from roof, shall be going out from the department.

9.2 Operation and problems associated with air washer systems

There are fundamental differences between summer and winter operation of textile air washer systems. A basic understanding of these differences is necessary in order to fully comprehend the treatment problem presented. Dehumidification occurs when the water that is spraying from the nozzles is cooler than the wet-bulb temperatures of the incoming air. In this case, the incoming air is moist and contains a higher level of moisture than the plant requires. By spraying this air with colder water, the incoming air is cooled down. This cooling process also removes moisture from the incoming air. Dehumidification usually occurs during the hotter humid months of the late spring, summer and very early autumn. During the period of time that the air washer acts as a dehumidifier, the air washer will take on the water that has been stripped from the air. This effect causes the dissolved solids level in the air washer to drop. It also creates a surplus of water in the air washer system that is removed from the air washer through the system overflow. As the solids level in an air washer drops, the Langelier Index will also drop. The index is used to indicate the tendency of the system water to be either scale forming or corrosive. During the period of time that the air washer acts as a humidifier, the water in the system will tend to cycle up and establish a positive Langelier Index. This is scale forming and is treated as such. During the humidification process, the water can become lower in solids and dissolved contaminants than the system make up water, this causes the water to have a lower Langelier Index and become corrosive. At this point, the chemical treatment program will need to be changed to protect the system from the potential corrosion that could occur at this time. The chemical treatment of an air washer must be looked at as two systems operating in one piece of equipment during different times of the year.

9.2.1 Summer operation

Considerable amount of heat is generated from electricity used for driving the electric motors and providing lighting, the mechanical friction of the machine parts, heat generated by the people working etc., which all added to the already higher temperature in the air entering in the department during summer. Because of the direct radiation received from the sun, the surface on the top of the roof goes much higher than the atmospheric dry bulb temperature. Hence, the temperature exceeds the maximum permissible limits. During summer, most textile air washer systems dehumidify plant air. The condensation from this dehumidification process drops into the air washer sump and enters the chilled water system. It may add condensation in quantities equal to or greater than the volume of the entire chilled water system. There will be no raw water makeup to the chilled water system; instead, overflow from the chilled water sump goes either to the sewer or to the cooling tower through a solenoid valve. Condensation of this magnitude directly affects the parameters to be considered for proper water treatment. For example, a textile plant with 30 PPM total hardness in raw water may find chilled water hardness on a hot day to be 5 to 6 PPM or lower. The pH will generally drop to 6.5 to 7.0, and alkalinity will similarly drop. Dissolved solids tend to stay the same or to increase slightly depending upon the degree of process contamination.

9.2.2 Winter operation

During winter, the dry-bulb temperature inside the department shall be higher than that of the outside. Therefore, winter operation of air washer systems is just the reverse of that of summer operations. Humidification of plant air generally takes place, and the air washer evaporates water just like a cooling tower in summer. Air washer sumps are each equipped with a float-controlled makeup line so that they can be operated independently during this humidification time of the year. During this time, each washer allowed to dry, after which high pressure water or steam easily removes them. Physical inspection of the air washer unit during operation is the only way to determine the effectiveness of the chemical treatment program, and the point at which a shutdown for cleaning will be required.

During winter, when heat losses from the mill buildings are more than the internal heat loads, heat from external source is needed to maintain the required temperature. The heat losses under equilibrium conditions can be divided into the transmission losses through walls, windows and roof and the losses due to infiltration of outside cold air which leaks in through various openings of the building. Both sensible heat and latent heat are required to raise the temperature to required level. When starting from cold, substantial additional heat and moisture are required to raise the temperature, the rate of which depends on the heat capacity of the building structure and material content in the department and the time in which the heating is to be achieved.

In a building, there shall be several openings such as cracks or crevices in the roof and dampers, broken glass panes, open doors, exhaust fan openings, etc. The main doors are frequently kept open for the movement of materials and men. Through one or more such openings cold air enters the department and warm air leaves out carrying heat from department. It is suggested to install air curtains in the doors where material movements are there; however, care should be taken to ensure that the sliver cans and bobbin trolleys are covered to avoid disturbing of materials while paring below the air curtains.

9.2.3 Problems encountered in air washer operations

9.2.3.1 Microbiological growth:

Air washers provide an ideal environment for the growth of microorganisms because of the process contaminants and soluble oils that feed them. Most deposits from air washer units contain dirt and debris from the process involved, corrosion products and some crystalline particulate matter. The most important part of the deposit is the microbial growth or slime masses. These are the most difficult to measure. They result in the very sticky slime that combines with the dirt and debris, corrosion products and crystalline matter to form hard encrusted deposits above the water level and thick slimy masses below the water on metal surfaces inside the washer. When microbial growths combine with process contamination, the resulting deposit in some cases is like a separate organic chemical that is almost impossible to remove. Controlling the growth of microorganisms in a chilled water or air washer system is the key to an effective treatment program. Microorganism growth can cause odours, carryover by blocking air passages, encrustation and corrosion under deposits. In addition, they can cause air washer sump screens to plug, which results in overflow of solids into the sump recirculating pump and the subsequent plugging of spray nozzles and into the fan. Deposits occur inside the fan shroud, disturbing air flow. The resulting solids get carried through the fan deposit in the duct work. In extreme cases, deposits develop on the fan blade itself, unbalancing the fan so that it must be shut down and cleaned.

Hygiene is of paramount importance when releasing water into an atmosphere as any viruses or bacteria in the water could potentially be inhaled by people in the vicinity. Modern humidification systems incorporate a variety of hygiene features, but the most effective type should combine both flush cycles and a form of silver ion dosing. The flush cycles will ensure that water cannot stagnate in the pipes and allow bacteria to form. Any cold water humidification system should typically auto-flush at least every 24 hours.

Silver ion dosing is a relatively new development in hygiene control in humidifiers. As silver is effective against over 650 types of bacteria and virus, it provides added reassurance by eliminating any organisms in the water before they enter the system. Silver also has a residual effect throughout the pipework. In the past, humidifiers typically used to incorporate UV sterilization, but this can potentially allow viruses to enter the system 'shadowed'" by particles in the water, or allowed in by UV bulbs that have dulled with age.

Regular servicing is also an important aspect of hygienic humidification. No matter what hygiene features a system has, inspections should be carried out by a competent individual from time to time to ensure optimum and hygienic performance. Spray systems, such as the JetSpray from Condair, incorporate self-cleaning nozzles that can reduce maintenance to just an annual check.

9.2.3.2 Oil:

Lubricating oils (oily smoke) that are atomised into the air steam due to heat generated from high-speed machinery operations can foul the system. The buildup of lint in the system will also cause system fouling. Oil and other materials picked up from the plant air can be tremendous nutrients to feed microorganism growth. Even though pre removal of oil may be accomplished with the filters, some oil will be present in the washer and will cause sticky surfaces and increased microbiological growth, and will generally increase the fouling tendency of the unit. Additionally, if certain amines are used in the microbicide program, oil can be coagulated in the washers and chilled water sump to form an extremely sticky slime that will quickly collect fibers and other suspended solids. These problems can be controlled by using an oil emulsifier to pull the oil into the water and then allow it to be removed with the bleed or overflow, and by low level foaming of the sump water. This foaming will float the lint and allow it to be removed with the bleed or the overflow. If left untreated, the oil in a system can film out and buildup to a point that the system is forced to shut down. It can form jelly-like buildups on the eliminator sections of the air washer which will stop air and moisture flow.

9.2.3.3 *Foam:*

This problem is generally caused by the chemicals being added for treatment, but may be the result of impurities being cleaned from plant air or of high solids in the washer water. Severe foam can overflow the sump pan of the washer onto the floor, spreading slime and dirt, and producing hazardous walking conditions. It can also be sucked out of the eliminator section of the washer.

9.2.3.4 *Carryover of solids:*

Carryover of solids is caused by microorganism growth on eliminator blades which disrupts air flow and allows solids to pass into plant duct work, and by foaming which also sends solids from the washer sump into the plant. Extremely high levels of dissolved solids in the washer sump water can also result in carryover. These solids can cause a variety of problems in the plant, from spotting of product to disrupting the temperature and humidity controls. Their worst damage is done in the plant duct work. Most textile plant duct work contains a mat of lint and fibre on its inside surfaces. This mat builds up over a period of time and is removed on a regular basis. When this mat becomes wet with solids from the washer, it becomes encrusted with the dirt and salts that are present in the washer water. Severe corrosion results, particularly if the ducts are aluminium and one of the treatment chemicals in the water is phosphate.

9.2.3.5 *Corrosion:*

Corrosion potential is most severe in textile air washer and chilled water systems during summer operation as the water in the systems de-concentrates. Chiller tube sheets and heads are the most vulnerable areas for corrosion. Idle chilled water lines in winter are also vulnerable if they are left full of water. Lint, dirt, fibre, oil and microbiological deposits from the air washers cause most of the corrosion problems on the chiller tube sheets. They combine in some cases to totally slime areas of the tube sheet. Severe pitting occurs underneath deposits of this nature. Corrosion in any metallic system is controlled by the passivation of the metallic surfaces that come into contact with the water. Carbon dioxide, sulfur dioxide and other corrosive gasses scrubbed from the air stream can cause serious problems when dissolved in the water. They form dilute mild acids which will readily attack the unprotected metallic surfaces of the system. Nozzle loss is very characteristic of this condition. To correct this condition, corrosion inhibitors are used which lay down a thin monomolecular coating on the metal surface, while increasing the pH of the system to a level that is close to 9.0. This neutralises the dilute acids, puts a protective coating on the metal surface and retards the rapid growth of bacteria in the system. Corrosion is characteristic of conditions during the period when the air washer acts as a dehumidifier. At this point, the water that is washed from the air will dilute the natural buffers that are found in the water, and allow the pH of the system to slowly drop. The result is corrosion and the perfect conditions for bacteria growth. It is normally recommended to clean both the tube sheet and the heads thoroughly and to coat them with several coats of suitable epoxy paint. There is little corrosion potential inside the air washer units themselves, except for the mild steel or galvanised piping and any other mild steel structures present. Except for the insides of piping, the best preventive measure for corrosion remains to be a good coating of paint.

9.2. 3.6 Encrustation

This is the name given to a deposit when it occurs above the water level in a dry area of an air passage in an air washer unit. This is a very difficult deposit to remove. Neither strong acids nor caustic seem to be effective as cleaning agents. In some cases, a chisel is the only way to remove these deposits. These deposits are composed of lint, dirt and debris, fiber and process contamination. They are held together by the organic slime from the microbiological growth. When a deposit of this nature dries out, it adheres to air washer surfaces like epoxy glue.

9.2.3.7 Odours

Most odours in textile plant air washer systems result from the chemicals used in treating the system. For example, chlorine compounds, polychlorophenates, and some organic sulphur compounds, used to kill bacteria, result in severe odour problems in plant air. Odour problems in air washer systems generally do not result from gases inside the textile plant. Most of these gases are dealt with using some form of smoke abatement.

9.2.3.8 Biological control

An air washer puts air out for people to breath, which must be clean and free of odours. This eliminates many of the biocides as the sulphur based carbamates for use in this type of system. To control buildup of slime use of rapid oxidizers with low odour levels as bromine and hydrogen peroxide is used in conjunction with a cationic or nonionic biocide. Usually this includes the use of one of the polyquat biocides or glutaraldehyde. The biocides should be fed twice each week in a shock type basis, with the oxidizers being fed continuously or as a shock weekly, based on the conditions of the system and the oxidizer being used.

9.3 Efficient use of humidification plants

It is not only the installation of plant, but the operation of the plant to get the best performance is very important. The performance includes providing the required temperature and humidity within the time limit at the lowest possible cost. The humidifiers consume huge power in a textile mill. Therefore proper usage of humidification plants is very essential. By monitoring suitably the air and water, it is possible to get the required conditions, whereas to get it at the lowest cost is important. The power consumption is one of the major factors. The survey done by SITRA on energy effectiveness reveals the following.

- By running humidification plant scientifically production can be improved. Concentrating at the spray dwell time of humidified air in the spray chamber can improve the plant performance which gives a boost in output yarn produced.
- Usage of high-efficiency FRP fans instead of M.S. or aluminum blades consume less power for the given air output.
- Spot capacitor at the motor terminals inside humidification plant premises is more important and this aspect is often neglected in many mills.
- The use of false ceiling suppresses the heat entering the premises from the top. Ventilating the roof or the attic by force like the lateral high volume low-pressure exhaust fans or by self-propelled roof extractors, can bring down the room temperature by 5°C.
- Generally, the roof heating by solar radiation causes 50% of the heat load in any premises. Insulation of false ceiling can avoid the same. If the roof of humidification plant is concreted, SITRA suggests of covering the open terrace in patches with the standard asbestos sheet at a height of say 2 feet above the terrace. This can be done

above humidification plant chamber, MCC, compressor house and wherever open terrace on RCC roof. This is meant to avoid the solar heat load. The secondary insulation, i.e. under deck insulation, the attic fan helps to remove the stale hot air under ceiling and insulate false ceiling from solar heat load. This method of attic ventilation and cross ventilation inside premises is low cost energy saving type compared to high cost cooling of equipments and premises.

Following are some of the common humidification design features for energy saving.

a) Loom Sphere (Weave-Direct): This system of loom humidification provides high RH required at warp sheet level only and results in overall reduction of total humidified supply air quantity by almost 25%, thereby achieving significant saving in energy cost.

b) Return air system: Return air trenches designed with minimum corners and bends and given a smooth finish reduce air resistance. Rotary return air drum filters with effective suction fan for continuous fluff removal will help to keep the return air filters clean and reduce air resistance, thereby reducing power consumption of return air fan motor.

c) Under-deck insulation: Generally, industrial roofing is of asbestos cement sheets or metal sheets (GI/aluminum), and under-deck insulation with minimum 50 mm thick resin bonded glass wool of 32 kg/m3 density is necessary to reduce the heat gain through the roof and minimise the attic temperature rise within the false ceiling.

d) Drive motors for textile machinery: Since the connected textile machinery loads are very high, only high efficiency drive motors should be selected to reduce the heat gain and the heat load, thereby reducing the supply air quantities to the departments being humidified thus minimising supply fan motors' power consumption.

e) Use of PVC air inlet louvers and eliminators: With PVC blades and moderate air velocity (not exceeding 600 fpm for low velocity air washer system and not exceeding 1200 fpm for high velocity air washer system) are desirable to reduce the air resistance in the air washer with consequent reduction in design static pressure for supply air fan selection so as to reduce fan motor power consumption.

f) Use of variable speed drive (Invertor control) on air washer water circulation-cum-spray pumps, helps to achieve power saving on part load operations. However, variable speed drive for fan motors is generally not adopted since any reduction in air flow can slow down the removal of fluff and other waste materials and contaminate the department causing quality deterioration of the yarns / fabrics inside.

g) Higher saturation efficiency of air washer: Saturation efficiency is defined as the ratio : (Entering Air DB – Leaving Air DB) / (Entering Air DB – Entering Air WB)

   Improving the saturation efficiency reduces the temperature of air leaving the air washer thereby increasing the delta 'T' (temperature rise of saturated air due to heat pick up in the department) for a specific heat load and relative humidity condition in the department with consequent reduction in required supply air quantity and fan motor power consumption.

h) *Good design and maintenance of air filters*: to minimize fluff carry over into the air washer unit, proper spray density in air washers and proper sizing and cleanliness of spray nozzles to improve water distribution also helps improve saturation efficiency of an air washer for reducing supply air flow-rate and bring down energy cost.

i) *Use of digital controls* in conjunction with electronic temperature and humidity sensors in humidification plants for getting feedback of temperature and relative humidity conditions in each department of the mill and processing of such information by digital controller for automatic control of the position of outside air exhaust air, return air and by-pass air dampers' positions will maintain stable inside condition in the departments with added benefit of power saving at part load conditions.

9.3.1 Efficiency rating (SEER)

Some countries set minimum requirements for energy efficiency. The efficiency of air conditioners is often rated by the seasonal energy efficiencyoratio (SEER). The SEER rating is calculated by dividing the total number of BTUs of heat removed from the air by the total amount of energy required by the air conditioner in watt-hours. The higher the ratio, the more energy efficient the air conditioner is. On a power basis, the SEER ratio relates the cooling power of the air conditioner (in BTU per hour) to the electrical power consumption (in watts). SEER is very similar to the "coefficient of performance" commonly used in thermodynamics, except COP is a unit-less parameter. (To convert SEER to COP, multiply by 0.293 or (1055 / 3600). In theory, a SEER of 13 is equivalent to a COP of 3.8. This means that 3.8 units of heat energy are pumped per unit of work energy.)

9.3.2 Works practice

In a number of cases, especially in winter and rainy seasons, it is seen that steam is used for heating the department although there is enough amount of heat to maintain the desired level of temperature and humidity inside the departments, and no special efforts are necessary to fresh air from outside. The following factors are normally seen as responsible to the failure of reducing the steam for heating.

a) Recommended measures are not fully implemented: One of the important actions taken is closing the dampers on fresh sir port and exhaust port. These dampers, though closed, allow lot of fresh air in or exhaust hot air out through the crevices between the blades of the dampers. It is suggested to cover these dampers with tarpaulin curtain or PVC curtain to stop leakages. However, due to many practical problems, this is not implemented.

b) Disagreement to work the departments at low temperatures is another factor for high steam demand. Higher the dry-bulb temperature higher will be the heat losses and higher heat and moisture requirement. It is therefore desirable to work the department at as low temperature as possible with the required level of RH. It should be noted that RH is more important for good working rather than the temperature.

c) Use of steam for the department has become a tradition in some cases and is difficult to change that mid set. It is difficult to break a belief that department does not work smoothly without steam during winter.

9.3.3 Typical operating procedures followed in a textile mill

Purpose:

1. To maintain the humidification plants so as to get the required temperature and humidity to the production plant with least consumption of power and water.
2. To protect hygiene of the people working in the department by controlling growth of micro-organisms in the water tank and in the humidification ducts.

The performance of a spinning or weaving mill depends directly on the performance of humidification plant as the strength and elongation properties, generation of static electricity etc., depend on the relative humidity of the working area. As the humidification plants work with water, there are possibilities of them working as breeding area for micro-organisms and gives a foul smell. Timely maintenance is the only remedy for improving performance of the plant and preventing growth of micro- organisms. The plants are maintained on weekly, monthly, six monthly and yearly bases with well- defined checklists.

The maintenance of humidification plants is done in teams. Separate teams work for cleaning the water tanks and refilling water, ducts cleaning, return air trench cleaning, opening fan bearings and greasing, cleaning the nozzles in nozzle bank, cleaning eliminators, checking the doors and so on. All works shall be done simultaneously to avoid the plant from stopping for a long time, as it affects production and quality of the yarns and fabrics.

Procedure

a) The maintenance activities of humidification plant are planned by discussing with the production personnel, so that critical fibres/ designs are not worked at that time.

b) A board indicating that the plant is under maintenance and should not be started shall be put whenever the plant is taken on maintenance or stopped for repairs.

c) The entry point of fresh air is checked, and the big objectionable particles such as leafs, feathers of birds, plastic sheets an, papers are taken out.

d) The doors of the humidification plant are checked and ensured that they are intact, and safety systems are in place as needed.

e) The air washer is cleaned thoroughly including the removal of all solids and bacterial growth. Then the system is drained and flushed to remove solids from isolated areas. Next the system is charged with the proper treatment products based on the time of year and atmospheric conditions.

f) The old stagnant water is removed from the water tank, and the tank surfaces are cleaned. Any fungus or other micro-organism growth if found is removed.

g) The water filters are checked for its condition and replaced if needed.

h) The air filters in the return air system are checked for its condition and replaced if needed.

i) Ensuring proper atomisation of water is very necessary. If fibre or dust is clogging to the water nozzle, it will noteatomise. The nozzles are cleaned with nylon bristle.

j) The conditions of eliminators are checked. If eliminators are blunt, or chocked, it condenses water and solids will enter the humidity duct.

k) The return air damper and rotary air filter are cleaned.
l) The choked up floor grills of the return air system are cleaned using a metal brush.
m) Inside the return air trench are cleaned with hard fibre brush.
n) The inside of the humidification ducts are cleaned by using compressed air and scrubbing.
o) The diffusers are cleaned from both inside and outside by using proper brushes.
p) The water tank is filled with fresh water, and the chemicals are applied as needed.
q) It is necessary to ensure that all members of the cleaning gang have come out of the ducts, return air trenches and other parts of the humidification plant before starting the plant.
r) A senior member checks and ensures that all the cleaning and maintenance tools taken inside have been brought out.
s) Before starting the positions of dampers and bye pass, whether they have been set as per the requirement of the production departmenteis checked.
t) The fans are started and checked to ensure that there is no vibration or the blades touching any parts.
u) The water pump is started and ensured that all nozzles are working and water is getting fullydatomised.

9.4 Keeping AHUs clean

Keeping the air-Handling Units clean is very important for getting the best performance. Some of the average run textile mills are equipped with non-automatic dust filtration and humidification plants. The dust-laden air is discharged into dust chambers and filter rooms. The dust and fly are not removed completely from the air which is humidified and recirculated. Workers in these mills are exposed to the risk of breathing air polluted with dust and fly and contracting respiratory ailments, byssinosis (lung disease) chronic bronchitis, etc. Unfortunately, the managements of these textile mills are more concerned with the maintenance of the required atmospheric conditions of temperature and humidity rather than the reduction of dust and fly in the departmental air and maintenance of clean and hygienic environments for the workers.

Air is cleaned by an air washer when particulate matter impinges on the wetted surfaces of the eliminator blades. Air is also cleaned, when the sprayed water generates enough ionisation to neutralise the charges on the particles in the service area so that they can be dropped out into the sump as they are drawn through the air washer. The cleaning of the air will remove lint, oil, and other airborne particles from the air. This builds up the level of the solids in the water which must be removed either from over flow or bleed. Chemical treatment can improve particle removal and prevent nozzle clogging from excessive solids buildup.

As the water droplets are broken up at the nozzles, ionisation occurs. This places a charge on the water droplet which allows it to attract charged particles that are in the air. This will allow the removal of any of the airborne solids from the air. It can also cause the buildup of charges in the air that exits the air washer. These charges can cause static electricity to build in the air, which can cause textile processing problems if not neutralised by an antistatic treatment program.

Air washer systems should be inspected regularly to insure that wetted components are free of dirt, scale, oil, debri, and microbiological fouling. Preventative maintenance and the use of chemical treatment programme can virtually eliminate the problems that are caused by these contaminants, which reduce excessive man-hours for cleaning up the unit and down time. The suggested method is cleaning the air washer thoroughly including the removal of all solids and bacterial growth. Then draining and flushing the system to remove solids from isolated areas. Next is to charge the system with the proper treatment products based on the time of year and atmospheric conditions. Finally, the system must be tested regularly to maintain the proper levels of products in the system during the different operating conditions.

9.4.1 Maintenance of filter

Filter performance is very important for effective operation of air-Handling units. This is evaluated by different criteria. It is desirable that filters, or filter media, be characterised by low penetration across the filter of contaminants to be filtered. At the same time, there should be a relatively low-pressure drop, or resistance, across the filter.

A depth media filter is a deep, relatively uniform density media, i.e. a system in which the solidity of the depth media remains constant throughout its thickness. If the percent solidity of the depth media is sufficiently high, relatively large particles will tend to collect in only the upstream portions of the media, tending to load on the front end of the media, and not penetrate very deeply. Those that block the air passage ways are termed retained solids, i.e. retention of solids within the filter media. Penetration, often expressed as a percentage, is defined as: Pen = $C / C\delta$, where C is the particle concentration on exit through the filter, and $C\delta$ is the particle concentration before entry through the filter. Filter efficiency is defined as: 100% minusnpenetration. Because it is desirable for effective filters to maintain values as low as possible for both penetration and pressure drop across the filter, they are rated according to value termed alpha (α), which is the slope of penetration versus pressure drop across the filter. Steeper slopes or higher alpha values are indicative of better filter performance. Alpha is expressed according to the following formula: $\boldsymbol{\alpha} = 100 \log (C/C\boldsymbol{\delta})\ \Delta P$, wher, $\boldsymbol{\Delta P}$ is the pressure drop across the filter.

A filter will have reached its lifetime, when a limiting pressure drop across the filter media is reached. For any specific application, the limiting pressure drop will be the point at which the filter needs replacement, as set forth in specifications applicable to the system, or through regulatory requirements.

9.4.2 Cleaning coil

Cooling coils inside AHUs function in a dark cool environment and with copious condensation in over humid atmosphere provide a moist garden for the propagation of mold and other micro-organisms. These organisms multiply to huge concentrations creating a hidden bio-film of mold deep inside the HVAC system. These microbial contaminants travel from the coil through the air stream. Drain pans below the cooling coils are equally affected. Dust and debris which pass through inefficient filters or bypass poorly installed filters combine with the microbial contaminants forming a black sludge that helps choke the coil between fins cutting down air quantity supplied by the blower due to increase in pressure drop across the coil.

Chemical cleaning, although can prevent the problem to some extent, cannot be carried out every day or every month because of the extensive and messy procedure involved in the cleaning process. However, formation of mold and collection of dust and dirt goes on as long as the AHU keeps operating. Most maintenance people will resort to chemical cleaning once or twice a year and usually the work is entrusted to sub-contractors who employ poor labour that does not mind being exposed to chemicals required for the task. The AHU must be shut down for several hours for the cleaning work to be effectively completed and after start up, traces of the chemical are carried by the air stream into human-occupied spaces.

For many years, the design and manufacture of air-handling units has been neglected, as it was considered a very simple product best left to sheet metal duct work fabricators. Very often galvanised sheet construction was downgraded to mild steel and rust was a common problem with the "insides" of the unit and the debris that would collect inside the coil. With a framework of extruded aluminium, double skin insulated panels and stainless steel drain pans, AHU design and construction havs progressed to a point when only specialist manufacturers are now involved. But the physical location of these units, with adequate service space all around, is still limited by the constraints of space. Under such circumstances, maintenance staff can hardly be blamed if coils have not been cleaned and drain pans never touched after installation.

Recently, there have been significant strides in the development of UVC light production sources, those penetrate the outer structure of the cell and alter the DNA molecule. This prevents replication, causing cell death. Germicidal effectiveness of UVC is directly related to the dose applied, and the dosage is the integral product of time and intensity. A high intensity for a short period of time and low intensity for a long period of time are nearly reciprocal and are equal in killing power, therefore, the energy required to destroy microorganisms is given as microwatt-seconds (or micro-joules) per square centimeter. Independent testing performed by Rapid Precision Testing Laboratories, Cordova, Tenn. has shown that when compared to the older generation of UVC lamps, high-output UVC emitters specifically designed for HVAC use only 1.75 times the electrical power yet produce 2.5 to 6 times the output at temperatures ranging from 32 to 90 F. Produce over five times the output at conditions most often found in HVAC systems. Produce a wider band width for more killing power at a broader spectrum. Provide a significant increase in output per inch (arc length) of tube (foot-print) as well as an increase in tube life, meaning that fewer tubes can be used for a longer period of time. Significant life-cycle cost reductions over conventional designs are the end result.

A typical dirty cooling coil

A clean coil after installation of UVC lamps

Figure 9.3 Cleaning coil using UVCslamps

Scientific research has concluded that maximum germicidal effectiveness occurs at about 265 nm. There are no low-pressure, mercury vapour UVC devices currently available that produce this optimum spectral line. However, some high-output emitters can deliver a broader band of total output (250 to 260 nm), so in addition to the benefit of increased output, they are closer to the optimum wave length, thus further enhancing available effectiveness.

Finned cooling coils and air cooled condensers comprise two major heat exchangers in all room air conditioners, most packaged and split air conditioners as well as most packaged chillers used in central plant AC systems. Keeping such finned heat exchangers in prime condition without any substantial loss of heat exchange capacity is the responsibility of the company that maintains the AC. Performance of finned cooling coil surfaces is largely dependent on two factors: maintaining the design airflow across the coil surface and keeping the finned surface clean and corrosion free

Vaniklin-WB, a chemical cleaner invented by Mr. Yashwant Jhaveri for cleaning of air conditioner coils, is first diluted with water in the required proportion specified by the manufacturer, and then liberally sprayed over the coil surface using a hand sprayer, of the type commonly used at home for green plants, or a larger foot-operated spray pump, in case of large air-handling units. If the coil surface can be conveniently accessed from both, the air entering and leaving sides then the chemical spraying should be carried out from both sides so as to reach deep inside the coil fins. After about ten minutes (will vary depending on how dirty the coil is) of spraying, rinse the coil with fresh water till the foam goes away or till the chemical is washed off. Small room air conditioner coils can be thoroughly cleaned by immersing the entire AC (after disconnecting and removing the fan motor, control panel and the electrics) in a tank containing the diluted chemical, provided the AC can be brought to a workshop for general overhaul and painting. A noticeable increase in air quantity after cleaning was observed.

Air washers are periodically shut down and washed out manually to help control the severe fouling and deposition problems. The frequency of shutdown and washout depends on the type of textile process and the severity of the problem. It may be done every week in some plants, while in other plants five to six week intervals between washouts may be standard. In some cases where fouling potential is minimal, washers operate three or more months between washouts. Most textile plants do this maintenance during scheduled shutdown periods. The actual cleaning is difficult while the unit is in operation, except for daily screen cleaning. Some deposits will not come off wet.

9.4.3 Duct cleaning

Cleaning both return air ducts and supply air ducts is very important. The air ducts can accumulate deposits of floating micro dust and fibres, construction dirt, dust, smoke, insects and other air borne pollutants and can become an ideal breeding ground for mold spores, mildew, pollen, bacterial colonies and other health-threatening microorganisms. Higher humidity allows moisture to mix with dust in the system that breeds microbial contamination. Condensate drain pans and other system components often become contaminated with microbial slime, which in extreme cases has led to very serious levels of contamination such as legionella (Legionnaire's disease).

before cleaning

before cleaning

A portable vacuum unit with HEPA filters for 'source remove' of dust and debris from ducts

A portable air compressor to operate the forward and reverse motion brushes for dislodging debris from walls of ducts

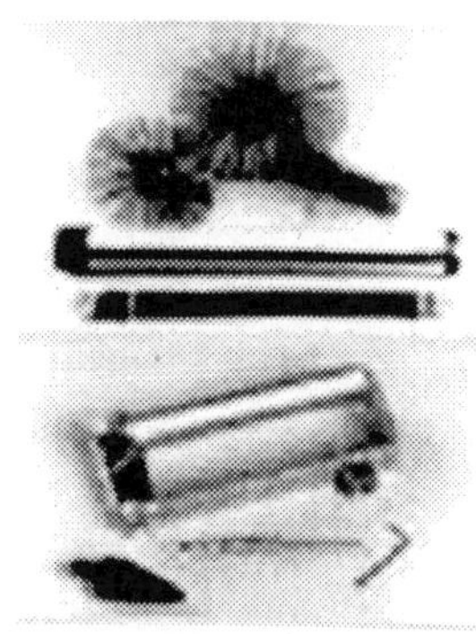

Brushes for inserting inside the ducts with extension rods (Top). Hole outer for duct holes used to connect vacuum unit (Bottom)

Figure 9.4 Air distribution duct cleaning

Mold, fungi, dust, slime, all need to be reduced or eliminated from air conveyance systems to promote acceptable indoor air quality. A clean, well-maintained system works better and lasts longer. Good filtration of the outside air and the return air are essential to keep clean air entering the air-handling unit. Housekeeping is also important to prevent the mold from spawning in areas other than the air-handling system.

Most ductwork is concealed above false ceilings and is out of sight, and not given attention. The method adopted to clean them is normally 'source removal' which involves the physical removal of contaminants from inside the ducts. To achieve proper cleaning, the duct work must be put under a vacuum, after which the contaminants are dislodged from the inner walls of the duct work using high pressure compressed air. The portable power vacuum unit is the single most important piece of equipment required. It should be portable, lightweight and have a small 'footprint' so that it can be easily loaded and unloaded from transport vehicles, moved up or down narrow stairways and fit into small places. The blower inside the unit should be capable of delivering 1800 to 2500 cfm for light work and 6000 cfm or more for large projects. The static head, which is a measure of the vacuum unit's ability to overcome the internal static resistance from filters and external static resistance of the ductwork, should be a minimum of 100 to 150 mm. Vacuum units should have three progressive filtration stages to ensure that the air is almost perfectly clean before discharge into the air conditioned space. The first stage coarse pre-filter should be cleanable and reusable. The second-stage intermediate filter should be 65 to 85% efficiency to capture most of the remaining visible contaminants. The final-stage HEPA filter will have a minimum efficiency of 99.97 percent at 0.3 μm.

The National Air Duct Cleaners Association of USA suggests the duct cleaning procedure as follows:

- PHASE 1: The air-handling unit is shut down and will remain off through the entire cleaning process. A negative pressure machine, equipped with a HEPA filter, is attached to the ductwork. The interior of the air-handling unit blower section and ductwork is vacuumed with a backpack vacuum cleaner, equipped with a HEPA filter. It is necessary to install access doors in the ductwork and in air-handling units to facilitate the cleaning and eventual fogging of the biocide. Without HEPA filtration, the mold spores would pass through the vacuum and go back into the duct, or the space served by the duct. The use of the negative pressure machine will allow the capture of airborne particles created by the vacuuming process. This initial vacuum phase is to remove debris, dust and dirt from the system and to expose the maximum surface area of the mold growths, so that the biocide can be most effective.

- PHASE II: The coils and drain pans are cleaned with a non-acidic, biodegradable cleaner. Portable pressure washers and a wet-vac are used to clean into the depth of the coil and to remove any excess water. Un-clog pan tabs are installed in the drain pan to prevent sludge from returning. Filters are also changed at this time.
- PHASE III: All grills in the ductwork and the air-handling units are sealed with plastic and tape. This sealing allows the biocide fog to fill the unit and the duct and to remain there while it kills the mold. On larger duct systems, the fogging machine will be placed into the duct, through the access doors, to assure that the entire length of duct is treated. Throughout the fogging process, all workers will wear breathing masks, goggles, and gloves. During the fogging, and for 24 hours thereafter, the area served by the system must be kept off limits to everyone. These 24 hours is for the biocide to kill the mold and dry out. Once dry, the residue of the biocide consists of nonvolatile compounds, and is harmless.
- PHASE. IV: With the grills still sealed, the negative pressure machine is attached to the ductwork. By now dead mold particles are dislodged from the duct. This is done with either rotary brushes or compressed air-driven skipper balls. The dislodged particles are vacuumed up with backpack vacuums, equipped with HEPA filters. The particles that become airborne due to the vacuuming process are drawn into the negative air machine, which is also equipped with a HEPA filter. The removal of the dead particles is crucial, as they will tend to give off an ammonia vapour as a by-product of the chemical reaction that occurs when the mold is killed.
- PHASE IV-A (Optional): Using the access doors to gain entry, the interior of the ductwork will be painted with a fungicidal, protective coating, which is EPA registered paint, custom designed for air conveyance systems. Once applied, it dries to a white, elastic finish that will allow for movement without splitting. This coating will prevent microbial growth and will not support bacteria. The fungicidal ingredient, which is EPA registered will assure cleaner air for occupants.
- PHASE V: The access doors are closed, the plastic and tape are removed from the grills and the air handler. Each grill is cleaned or vacuumed as necessary, and the air-handling unit is vacuumed. The system can at this time be returned to service. Used plastic, tape and any debris are cleaned up from the work area.

Site preparations to be taken before starting the work are as follows:Duct cleaning should be scheduled during periods when the area is unoccupied to prevent exposure to loosened particles of dust and dirt.

- The duct layout drawing should be studied so that the routing, size and number of bends, dampers, etc., are clearly understood. In case a drawing is not available, a sketch showing the routing and size should be prepared.
- If the duct is very long, an inspection door should be provided in the main duct, if not already provided
- In case of long ducts having a number of branches, cleaning should be done by isolating a section at a time.
- Place dust protection covers on furnishings, work stations, etc., to prevent damage.

9.4.4 Cleaning the supply water sump

Cleaning of the sump water at regular intervals is very important, as it is normally contaminated with microorganisms. Amplification of microorganisms can be controlled either by the use of traditional chemical biocides such as chlorine or to use dissolved ozone, a powerful biocide capable of destroying bacteria and viruses as well as oxidising many organic and inorganic compounds. It is also a de-scaling agent. Owing to its high oxidation potential; ozone offers to purify the air stream indirectly by spraying ozone-sanitized fresh water into the air stream rather than introducing gaseous ozone directly into the air. Concentrations of dissolved ozone in the liquid phase can be controlled to yield efficacious microorganism control while keeping the ozone concentration in the supply air stream below threshold limits of 0.1 PPM.

A complete equipment survey, line diagram and a thorough understanding of equipment operation are necessary to recommend effective water treatment programs for air washer and chilled water systems. Treatment problems must be categorised relative to their severity. In most cases, treatment problems will be present specifically to operating conditions in the plant. The selection of an effective microbicide and dispersant is generally the best place to start in developing a water treatment program for air washers and chilled water systems. Choose one that will kill the organisms that predominate in particular system, one that will minimize foaming, one that is stable at the temperatures of the chilled water in the system and one that will not coagulate or precipitate with other solids or oils in the system water. Feed rates for the selected microbicide must be below the threshold at which odour problems begin. In some cases, it is desirable to feed microbicide continuously at low levels rather than slug dosages at high levels. The dispersant selected should be able to handle primarily microbiological deposits, but in systems with heavy oil concentrations, a strong oil dispersant may be appropriate. The oil dispersant should not foam excessively or present odour problems. Generally, anti-foam for use in this type of environment should be an emulsifiable product stable to any temperature to be encountered in the chilled water system, free of odour-causing materials; effective for the pH range found in the system and has been evaluated and found to be effective in combating the type of foam encountered.

Periodic slug feeding of the antifoam in small dosages normally control foam in chilled water and air washer systems. Because of the differences between textile mill plants, selection of a corrosion inhibitor for air washer and chilled water systems is not easy. The most important part of controlling corrosion is keeping the system clean. Molybdate, zinc and phosphates are among the materials used for corrosion control in textile systems. Application technology for each of these materials should be considered in relation to the operating parameters of the washer or chilled water system to be treated. Scale control is not a factor, except in the cooling tower system. This is because of the tremendous dilution of chilled water during summer operation and the fact that there are no heat transfer surfaces for the concentrated air washer water during winter operation.

9.4.5 Quality of water

Soft water provides for easier maintenance of a humidifier. When hard water is evaporated, the mineral residue consists of a hard scale which normally requires some drastic treatment (such as chipping or acid) for its removal. When soft water is used, the residue is commonly called soft and can usually be removed by flushing the unit with water or going over the surface with a brush. It should be remembered that softening water does not reduce the total amount of minerals present; ion exchange softening merely converts the calcium and magnesium minerals to sodium minerals. Many humidifiers today automatically accomplish periodic flushing with fresh soft water to keep the mineral concentration down, and the unit operating satisfactorily.

Use of Dispersants: The proper use of dispersants and biological control agents allows the system water to remain cleaner. This improves the atomisation at the nozzles which creates a finer droplet and a more uniform spray pattern. This is in part due to reduction in the surface tension of the water itself, coupled with the control and the removal of the potential system foulants. Because reduced surface tension allows the spray nozzles to break up water into a finer spray, the increased water surface thus exposed to the air stream promotes higher evaporation of the air stream. This finer spray pattern also assists the air washer in scrubbing the air more efficiently, which eliminates more airborne particles. The lower surface tension or 'wetter' water also keeps the circulating system clean and prolongs operational time between clean-ups.

Control of Scale: Scale is controlled in the air washer system by using a blend of dispersants and scale modifiers. These products distort the crystalline structure of the scale and suspend the scale in the water which can be removed through the bleed. This improves the operation of the heat exchange potential of the system, maintains a cleaner spray nozzle and stops down time from scale buildup. The scaling potential for an air washer occurs when the system is humidifying the air. Humidification allows the air washer to act similar to a cooling tower, where the water cycles up the solids in the water and eventually creates a scaling condition. At this time, the system can be treated very much like a cooling tower in both the areas of biological control and scale prevention.

A modification of the pan-type humidifier uses wicks to increase the surface of water exposed to the air and thus increase the evaporation rate. The wicks in such humidifiers are particularly susceptible to clogging due to scale. When this occurs, the wicks must be replaced. When soft water is used, however, the mineral deposit can be re-dissolved by soaking the wicks in fresh soft water. There is also another type of humidifier that physically sprays a fine mist of water into the air. If minerals are present in the water, they settle out of the air as a fine powder. Depending on the mineral concentration, the amount of water evaporated and the use of the humidified air, a wide variety of problems may be encountered.

9.5 Monitoring humidity

A number of monitoring units are developed to constantly measure and understand the temperature and humidity. The disk charts and the temperature and humidity charts show the history of temperature and/or humidity as reported by a host (sensor).

The charts support both Fahrenheit and Celsius when viewing temperature trends. The charts also give you the option to use several different display modes, time frames and chart types.

The motion tracking page shows when motion was detected by the motion sensor and includes a chart showing motion activity.

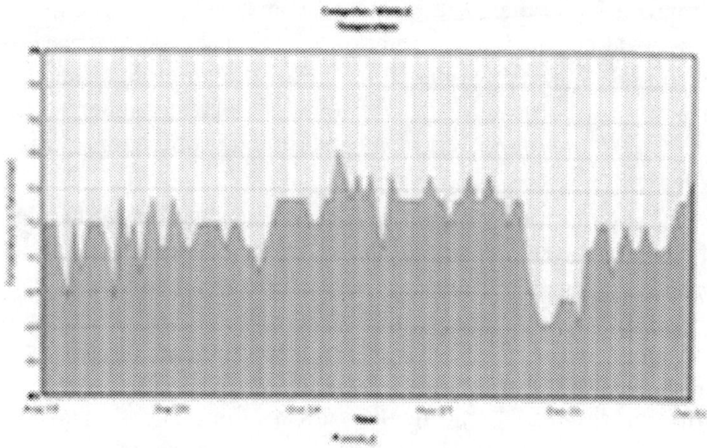

Figure 9.5 Motion tracking (v2.71+)

Humidity sensors are gaining more significance in diverse areas of measurement and control technology. Manufacturers are not only improving the accuracy and long-term drift of their sensors they are also improving their durability for use in different environments and simultaneously reducing the component size and the price.

9.5.1 Calibrating calibrating

relative humidity instruments accurately and efficiently is a daunting challenge, as the need for humidity monitoring and control proliferates. As a critical environmental factor, the accurate measurement of relative humidity is rapidly becoming a quality and regulatory concern in various applications. In response, many companies are scrutinising their humidity calibration methods to ensure adherence to required standards. Along the way, many are finding that their processes are inefficient, incomplete or incapable of providing reliable results. Others remain unaware of the effectiveness of their processes and may be surprised that the high accuracy they expected may not exist.

Many secondary relative humidity sensors, which are based on electronics and conductivity, claim of accuracies ranging from ±1 to 4%RH. This performance is impressive compared to the state of the industry 3 decades ago when horse hair hygrometers at ±10% RH were the norm. However, their calibration systems are not clear. All secondary RH sensors are dependent on accuracy in calibration in order to deliver specified performance. While the best RH sensor may claim an intrinsic accuracy of ±1%, calibration errors can render them useless. All secondary RH sensors require calibration to a higher standard, not only during production but also as part of ongoing maintenance to correct for drift. The element of time to the measurement of humidity is an extremely important factor in accurate calibrations. Good calibration cannot save a bad sensor, but improper calibration will cause an otherwise excellent system to produce erroneous and potentially damaging results. Ms. Veriteq Work developed calibration units by producing a time-based record of temperature and relative humidity that can be graphically displayed on a computer. The advantage of such a record is that it details the level of thermal and RH stability during calibration and, if multiple units are tested, may also indicate the homogeneity of the test chamber. Thus, data loggers make it easy to determine when, or if, a steady state has been achieved. Conversely, they also make it obvious when a calibration has been performed hastily or in unacceptable conditions.

RH is more difficult to measure than most water-related or atmospheric properties such as temperature, pressure, flow, volume, mass or level. The complexity begins with the broad range of moisture conditions RH sensors must operate in. For example, a sensor rated to measure 1 to 90% RH from −40 to 70°C, must perform in humidity conditions ranging from 1,000 parts per billion to 200,000,000 parts per billion. The dynamic range this represents is 200,000:1, a figure that would challenge the linearity of most sensors. RH is also a temperature-dependent variable. Its value can change significantly with even slight variations in temperature and without any change in water content. For example, a 1°C variance in temperature at 20°C and 50% RH can introduce an error of ±3% RH, an enormous variance in a calibration process. At 90% RH, even a 0.2°C variance will result in a ±1% RH error. These temperature effects highlight the importance of thermal stability, a condition that is often difficult to achieve in a calibration environment. The most significant calibration challenge is that RH testing must be carried out in air, which is not reliable and uniform. RH sensors measure only the water in the immediate layer of gas contacting the surface of the sensor, a fact that emphasises the need for stable and homogenous environmental conditions. Air is a poor thermal conductor, and the temperature at any given point can be affected by thermal currents and temperature gradients that make such conditions not only difficult to achieve but also time consuming.

All secondary RH sensors are non-linear devices with temperature dependencies. Most hygrometric sensors work by changing their electrical properties with variations in humidity and temperature. Because of this non-linearity and temperature dependency, it is necessary to validate sensor operation at multiple RH and temperature points, ideally covering the range of the intended application. This requirement is not only difficult to achieve without expensive equipment but is also time consuming. Unless questioned or challenged, many flawed humidity calibration processes may remain unchanged indefinitely. To avoid this, companies need to conduct periodic reviews of existing equipment and procedures and to look further at the effects and results of their calibration processes. The following questions will prove helpful in carrying out these reviews:

Reference standards: Whether the standards used are traceable to national standards, for examaple, NIST or NPL? Have they been validated? Do the standards have an uncertainty sufficiently low to justify the final calibration accuracy claimed? Have all the elements of uncertainty been considered in determining the reference standard's accuracy statement?

Procedures: Have the procedures been validated? Do they clearly specify adequate times to allow for thermal and humidity stability? Are there provisions to ensure uniformity of the chamber environment? Are multiple relative humidity and temperature calibration points tested? Are the procedures written, followed and periodically reviewed?

Instruments and sensors: Are the calibrated instruments trusted by those who use and rely on them? Do they need calibration more often than expected? Are the as-found calibration conditions frequently out-of-spec? Do the instruments require significant adjustments for each calibration? Are the instruments or sensing devices reasonably interchangeable or are there 'unexplained' differences?

Products and processes: Are there recurring problems with product quality or efficiency that relate to temperature or humidity sensitivity?

9.5.2 CO_2 sensors

CO_2 sensors can be used in a variety of ways to bring more ventilation air into a building or room when the CO_2 level rises either above a fixed set point or above a maximum differential over the outdoor CO_2 level. The typical approach is to set a minimum airflow set point for average or minimal occupancy and override the minimum set point as the CO_2 level increases. This increase in airflow over requirements of the thermostat usually requires reheat to compensate for more supply air than is necessary to satisfy the thermostat. There might be enough outdoor air in the building, but is to be properly distributed throughout the envelope. This is done by activating the fan in a parallel-fan box that serves the room when a CO_2 sensor in the space indicates an under-ventilated condition. The parallel fan acts as a transfer fan, moving additional unmitigated air into the overpopulated space. If adding the transfer air does not reduce the CO_2 level in the space, the minimum primary airflow to the space can be increased. However, before introducing more primary air to a given space, the first step should be to introduce transfer air from less occupied areas into the space, because this air has already been treated and can be used for dilution of contaminants with a minimal energy penalty.

9.5.3 Temperature and humidity control in a confined area

Adrian O. Bartley, III, William E developed a method (Ref United States Patent 4183224) and apparatus for precise temperature and humidity control within a confined or determined chamber. The method involves determining a desired condition in a controlled space. The essence of the concept is to provide humidity control by the addition of moisture at all operating conditions and, in effect, never having to control dehumidification. The system is claimed to be much more controllable than previous approaches, which as a result achieves energy conservation. In addition, the method for heating or cooling does not affect the humidity, and humidity is added after all the desired temperature relationships are achieved, which enables the system to eliminate control into the dehumidification mode. The process includes a higher air flow rate than normally utilised and is claimed to improve the precision of the humidification and temperature control process. The apparatus is arranged in a neat configuration to achieve better mixing of steam humidification with dry heated air, the positioning of the apparatus to achieve the desired air flow characteristics through the room and more uniformity with substantial elimination of temperature variations by lights, or people entering or leaving. The apparatus includes structure to cool the conditioned air without stripping moisture. The system is more energy conservative because only a small percentage of makeup air is added to a large quantity of circulated air with the makeup in effect controlling the temperature and humidity of the circulated air.

Arima M, Hara E.H, Katzberg J.D in Communications, Power, and Computing Conference IEEE in May 1995 described a HVAC system controller which employs a control algorithm using either fuzzy logic reasoning or rough set theory. The controller deduces the appropriate control outputs from sensor readings. The system controls temperature and humidity at the reference point by adjusting the flow of hot water into a heating coil for heating operation and the flow of chilled air through an air duct for cooling operation. To control humidity, the controller turns on and off a humidifier.

Programmable controller PLC has taken the microprocessor as a foundation and has taken industry control device, i.e., the computer technology, the automatic control technology and the communication technological development. It has a simple structure, the convenient programming, reliable higher merit which is now widely used in the commercial run and in automatic controls. Statistics have indicated that programmable controller in the industrial automation installations applies mostly one kind of equipment. PLC is in the black-white control logic foundation, with the 3C technology (Computer, Control, and Communication) unifies and unceasingly develops perfectly. At present, from the small-scale single plane sequential control, it has developed including situation and so on process control, position control all controls domain. PLC has become an integral part of almost all machines where process monitoring is essential.

9.6 Problems of not getting required conditions

In spite of having good air-handling systems in place, we might at time get an unsatisfactory result. The reasons are many.

- Someone holds the doors open for extended periods of time, or during system startup after a power failure or system shutdown.
- The basic design configuration of the HVAC apparatus did not feature inherent RH control or even positive control during abnormal or upset conditions.
- The rooms at the ends of the duct runs have no air flow. Speeding up the fan to overcome the additional static friction from the first air outlet to the last air outlet would be less expensive than replacing the ductwork with larger ductwork, but would cause additional fan noise in the rooms closest to the fan rooms.
- Added additional machines in the space available increased the load on AHU.

Careful design can prevent under or oversized equipment, which can shorten equipment life or create uncomfortable conditions. Oversized cooling systems can result in poor dehumidification; although the system lowers the temperature quickly, it does not run long enough to dehumidify the area. There are several important design principles that are essential. The heating and cooling loads for each room are to be calculated using an approved method, determining the amount of both sensible and latent cooling needed on the hottest summer day and heating required on the coolest winter day. It should consider the amount of solar gain through windows, heat transmitted through walls heat exchanged through air leaking into the building, etc. The sensible heat is more important than the absolute heat. The ratio of sensible to total load is called the sensible heat ratio. Dehumidification is accomplished by the latent capacity of the equipment. In well-insulated buildings, latent load tends to make up a larger portion of total capacity than it does in poor insulated buildings. Since equipment is designed for a predetermined sensible heat ratio, designers are often faced with difficulty finding equipment that meets their design calculations. The designers will select equipment that matches or slightly exceeds the design latent load and allow slight under sizing of the sensible capacity. In this manner, good dehumidification can take place and provide sufficient cooling comfort. After load calculations and equipment selection, the designer should focus on duct design. The best duct system will be sealed with approved sealing materials (typically mastic or aluminum tape) to limit air leakage, duct length and the number of bends will be minimised, and ducts will be installed within the conditioned envelope of the home. Return grille sizes and locations are important along with correct type and number of supply diffusers to ensure proper air circulation. Some programmable thermostats will start the compressor prior to the air handler fan operation, thus allowing the coils to start dehumidifying early in the cooling cycle. Two speed compressors allow the air conditioner to operate at higher efficiency and lower energy costs during most of the cooling days. Other technologies are available that will reduce humidity levels in very energy-efficient homes during periods when air conditioning is not needed, such as energy recovery ventilators (ERV) and dehumidifiers. ERVs are similar to heat recovery ventilators (HRVs) which preheat incoming ventilation air with outgoing conditioned air. ERVs take the efficiency a step higher by recovering latent and sensible energy from the air stream. Commercial cooling systems often have desiccant systems to dehumidify the conditioning air. Desiccants are materials that attract moisture and can be dried or regenerated by adding heat supplied by natural gas, waste heat or the sun.

9.7 Basic humidity control preparation

9.7.1 Fresh Air Unit

Fresh air units control room pressure and amount of fresh air more accurately according to operation and even with variation on air balance by room pressure sensor. They could be designed to treat outdoor air to become neutral air (air, which has the same temperature and humidity as the room). Therefore, variations from outdoor conditions do not affect much.

Figure 9. 6 Fresh air unit

9.7.2 Air distribution

Good air distribution contributes to good humidity control and avoids dead spot. The designer should consider location of supply and return air where the supply air could have low air temperature and high relative humidity at the air diffuser. Heat loss and heat gain estimates are part of a design procedure that flows from system selection decisions, the actual load calculations, to equipment selection procedures, to placement and selection of air distribution hardware, to duct routing and airway sizing. Load calculations affect every aspect of the system design procedure. The calculations must be as accurate as possible.

- Equipment capacity that matches the size of the applied heating and cooling loads will deliver comfort, efficiency and reliability over the entire range of operating conditions.
- Heating and cooling loads determine the total air delivery requirement (blower CFM) and the air flow requirement for each room (room CFM). This airflow information is then used to select supply air outlets and to size the duct runs.
- Load information is also used to estimate purchased energy requirements and to estimate annual operating cost. In this regard, the energy and operating cost estimates will only be as accurate as the load estimate.

In humidity control, the designer should be aware of the followings:

i. most cooling coil does not perform accurately as specified,
ii. there are always leaks in the air-handling unit and air duct,
iii. apparatus dew point does not stay as specified,
iv. air flow does not stay as specified,
v. calculation could be wrong
vi. assumptions could be wrong

Therefore, a good humidity control design should be more foolproof than conventional comfort air conditioning. Previously, designer who has such awareness will apply overcooling and reheat system. Though the system could safeguard the designer and provide the required result, but it is an energy eater.

9.8 Guidelines to select supply outlets and return grilles

Supply outlets (grilles and registers) should be of the appropriate style and size for the application and be in an appropriate location for the application.

- Supply outlets should not produce objectionable noise. Design guides and manufacturers' information establish limits for face velocity.
- Supply outlets should provide the appropriate throw for the installed location. Floor outlets should throw the supply air to the ceiling; ceiling outlets should throw to the wall, etc. Size depends on product performance, the supply CFM value and the face velocity limitation.
- Supply air should not be blown directly into the occupied zone.
- Floor outlets that blow air straight up the exposed wall are best for cold climate heating; and if properly selected, adequate for cooling.
- Ceiling outlets are best for cooling, but will not warm slab or exposed floors during the winter.
- The relation between supply CFM, throw, face velocity and drop is established by manufacturer performance data. Performance is very sensitive to size and devices that appear to be generally similar can have substantially different performance characteristics.
- A low resistance return path should be provided for every room that receives supply air – a wall opening with no door, a transfer grille or a ducted return. Door undercuts are not acceptable.
- Return grilles shall be the correct size for the grille flow rate. Filter grilles have a lower face velocity than plain grilles.

- The location of the return grille does not affect room air patterns which are controlled by the supply outlets and will not have a significant effect on pockets of stagnate air. Low returns do 'pull' warm air down to the floor and high returns do not 'pull' cool air up into the occupied zone.

9.9 Guidelines to size the duct runs

The resistance (inches water gauge of static pressure) of the longest circulation path (longest supply run plus longest return run) shall be compatible with the performance of the blower that is supplied with the heating-cooling equipment. Airway sizes that are compatible with the blower performance shall be increased if airflow velocity creates a potential noise problem. All systems shall have adequate provision for balancing airflow. The length of the longest circulation path and the available static pressure determine the friction rate used for airway sizing. The length of the circulation path includes the straight runs and the equivalent length of the fittings along the path. One fitting can add from 5 feet to more than 60 feet to the length of the path. External static pressure is determined from the equipment manufacturer's blower performance data, preferably for medium-speed operation. The available static pressure equals the external static pressure minus the pressure drop through all the air-side devices in the circulation path. Refer to blower table footnotes and manufacturer pressure drop data for devices that were not in place when blower performance was laboratory-tested by the equipment manufacturer. Accessory or after-market filters (or any device) that produce a substantial increase in system resistance shall not be installed if the blower cannot accommodate the increased resistance by speed change. An arbitrary increase in system resistance may cause low airflow to rooms, a high temperature rise across a furnace heat exchanger, or low suction pressure at the cooling coil. The room heat loss and heat gain estimate and the heating and cooling factors determine the design value for room airflow. Airway size is determined by sectional flow rate and the design friction rate value. The friction chart or duct slide rule used for airway sizing shall be technically correct for the type of duct material. Airway velocities shall not exceed specified design limits. Branch (run out) ducts shall be equipped with a hand damper (for balancing).

9.10 Power consumption

The humidification and air control in textile industry contribute for nearly 15% of the power consumed. The number of trials is being made to reduce the power consumption as the power cost is increasing and is next only to the raw material cost. Selection of a suitable plant and having systems to operate a part without affecting the working are considered as very important. There are many factors that are to be considered when calculating total power consumption. When comparing evaporative humidification to electric steam humidification, major operating cost differences are easily noted. When comparing ultrasonic humidification to spray nozzle humidification in ducted or AHU applications, the operating costs can be similar. Spray nozzle systems may use less electricity but in most cases are less efficient in evaporation. The spray nozzle humidifiers may operate longer in order to get the same amount of evaporation as ultrasonic humidifiers. Also, the unevaporated water from spray nozzles will go to drain. (This is treated water – RO/DI – and has cost associated with its production.)

There are often claims that some humidification systems require significantly less energy to operate than others. While some energy savings between different types of humidifiers exist, large differences in overall facility energy consumption many times do not materialise. Often times, when two different humidification systems are compared, only the energy required to operate the humidifiers is considered. This is not the entire picture. Different humidification systems affect the entire HVAC system energy usage differently. Mr. Bryce Dvorak discusses the power consumption issues as follows.

9.10.1 The two schools of humidification

Two types of humidification familiar are isothermal and adiabatic. With isothermal humidification, steam is injected into an air-handling unit or directly into the space to be humidified. Water can be boiled with electric resistance, natural gas, or some other heat source to make steam. In adiabatic method, water is atomised into very small droplets.

Figure 9.7 Pressurised water jets

This can be done by forcing pressurised water through very tiny nozzles (high-pressure humidification) or by vibrating a pan of water at very high frequencies (ultrasonic humidification). The droplets are then injected into the air-handling unit or directly into the space that is to be humidified. The atomised water droplets are then evaporated into the air to increase the relative humidity.

9.10.2 Evaporation

Many companies are advertising that the adiabatic humidification technologies are extraordinary energy savers. The fact is, when strictly comparing only the humidifier, an isothermal (boiling) unit will always use more energy than an adiabatic (evaporating) unit of equal capacity. But the energy implications are significantly more complex than this. The small droplets of water introduced by an adiabatic humidification system are not evaporated into the air for free. The water droplets absorb heat from the air to evaporate. We have all experienced heat absorption from evaporation. When we sweat, the moisture on our skin evaporates absorbing heat from our body to keep us cool, just as the water droplets absorb heat from the air during adiabatic humidification. Pound for pound, evaporating water droplets require nearly as much energy input as boiling.

9.10.3 Energy Impacts

Adiabatic humidification can offer an advantage if humidification is required at the same time that cooling is needed. Some type of cooling is needed most of the year for many air-handling systems. This type of humidification will provide a free cooling effect along with the humidifying. However, during the heating season adiabatic humidification can result in increased space-heating energy. More heat may be required to maintain the desired space temperature. Isothermal humidification, on the other hand, does not significantly affect the temperature of the surrounding air.

There may be several reasons to choose one type of humidification over the other, including energy usage. Every system should be considered case-by-case to determine what is right for your application. When analysing the energy consumption of a humidification system, it is important to look at the whole picture. Focusing just on the humidifiers can lead to some misleading results.

9.10.4 Normal practices to reduce power consumption

The developments done to reduce power consumption include the fan blades made by light materials, reducing the resistance to the movement of air by selection of proper dampers and providing continually clean surface of filters to the air, ensuring that the water is atomis0ed to the thinnest possible, the path of air in the ducts is smooth, the ducts are always kept clean, the return air grills are not allowed to get chocked, preventing leakages of humidified air, ensuring that heat generated by the machines are properly utilised and disposed etc.

9.11 Ionisation of air using chemicals

Fluff generation in spinning section is very common, which not only leads to health problems for the men working but also creates working problems and mars efficiency. Fluff accumulation increases yarn imperfection by 30 to 40% and objectionable faults by 30 to 80%. It also increases the invisible loss and sweeping wastes eventually reducing the profitability. The fluff increases breakages in ring frames and fly frames reducing the production, increasing the soft wastes and further liberation of fly. Spun-in fluff increases the breaks in winding, warping, weaving and knitting operations resulting in reduced efficiency, increased hard wastes, defective fabric and increased costs. About 50% of the breakages in winding are attributed to spun in fluff and loose fluff in the department. High fly liberation reduces the concentration and stamina of the people working apart from leading to lung-related diseases.

One of the reasons for the liberation of dust is the drying up of cotton while being processed in the machines because of the speed, forced air movement, heat generation by moving parts and so on. While cotton in the mixing before feeding to blow room has 7–7.5% moisture, the card sliver has only 3–4% while coming out of the machine. Similar is the case with all the machines. Cotton with low moisture develops positive charges while moving or while getting rubbed with other fibres. These positive charges lead to further fly liberation.

The automobile pollution and the generator exhausts contain very fine black dust. They remain at very low level during dry winter season due to fog and get sucked by the humidification plants and enter the production area. The air washer fails to restrict them as the particles are very small in size. Once they enter the production area, they settle on the material and leave their mark.

One of the systems followed from a long time is to spray lubricating and hygroscopic materials like cotton spray oil or antistatic chemicals on the cotton mixing and allowing the mixing to get conditioned for 8 to 24 hours, depending on the type of spray used and the humidity conditions maintained in the department. However, in recent years, after the development of Bale Pluckers, where the bales are directly fed to blow room, the concept of mixing cotton in bins and spraying lubricants has become redundant. We need some other system to provide moisture to the cotton and reduce the generation of positive charges.

Teechnol, a specialty chemical developed by Texcell, Mumbai, is added to the water in humidification plant rather than spraying on cotton mixing. The air washer plant atomises Teechnol along with water and liberates wet ions into the existing air stream, which are carried as positively charged and negatively charged water clusters. Brownian movement holds the charged particles in suspension. They travel through the enclosed environment and move by magnetic attraction to oppositely charged surface to immediately neutralise static charge as they develop. It also agglomerates the fine dust into bigger particle, which can be removed by filtration. For a tank of 10,000 Litre capacities, 7 Kgs of Teechnol is used on the first da, and reduced on the subsequent days to 2-3 Kgs. The exact quantity, however, depends on the fibres, the humidity conditions, the count spun and the speed of the machines working. The controlled studies by using Teechnol have shown the following results for Ne 31s Carded.

	Before	**After**
Ends down %	1.65	0.51
Clearer waste/5 hours	900 Gms	550 Gms
Winding efficiency%	77.00	84.30
Winding production Kgs	384	454
Clearer cuts/100 KM	64.6	48.4
Yarn breaks per 100 KM	68.6	38.5
Hard waste%	1.45	1.14
Yarn U%	16.3	15.57
Thin /100 KM	700	554
Thick / 100 KM	204	90
Neps / 100 KM	1112	980
Total imperfections	2016	1624

The most important thing to remember when considering humidity control is to get good advice from an expert with experience in the textile industry. There are many different issues involved with humidification of textile plants and as the issues described above demonstrate, getting it wrong can be an expensive mistake.

10 Auxiliary units to make humidification units effective

10.1 Building designing to have effective control of humidity

Building design place a very important role in helping maintaining the temperature and humidity not only at required level but also at the optimum costs, mainly the power consumption. Heating, ventilating and air conditioning (HVAC systems) contribute for significant power cost in textile industry. The use of high-performance HVAC equipment can result in considerable energy, emissions and cost savings (1–40%), but if the building is not supporting, there shall be losses due to leakages, interference of outside environments, etc. Whole building design coupled with an 'extended comfort zone' can produce much greater savings (40–70%). In addition, high-performance HVAC can provide increased user thermal comfort and contribute to improved indoor environmental quality (IEQ). Positive room pressure is crucial since infiltration will bring in humidity. Therefore, the designer should analyse air balance on fresh air, exhaust air and leakage so that room pressure will be positive during anytime of operation.

The power required for the supply, pumping and exhaust system of the humidification plant of a spinning department depends on many factors. The most important factors are power for driving machinery, lighting and heating load, number of people inside, temperature gradient and relative humidity. The usual procedure is to train the back-propagation neural network (BPN) with the available data and, once it is trained, BPN will be used for inferring. Some other investigators have used different neural network architectures. S. Rajasekaran, R. Prakasam, S. Dhandapani and Vasanthakalyani David developed a method of prediction of humidification power in textile-spinning mills using functional and neural networks. In this general methodology to build and work with functional network (FN), an alternative to neural network paradigm is given. This functional network architecture is claimed to be successfully applied to predict power required for the humidification plants.

Studies by SITRA indicate that humidification system accounts for about 15% of the total energy consumption in a mill. The humidification plants are to be judiciously operated based on the outside conditions in different seasons. In multi-fan system, the non-working supply fan should be isolated to avoid recirculation of air inside the supply plant room. The pneumafil air must be collected in a separate duct and a positive exhaust must be provided for the same to avoid backpressure in the pneumafil system. The capacity of the exhaust and pumping system should be 80% and 45% of the supply system respectively to avoid excess energy consumption. Interconnection of plants is to be avoided for better performance. Provision of rotary filters for exhaust systems, V-type filters for supply system, PVC clamp-on water nozzles, Attic area (space between roof and false ceiling) ventilation, proper design of ducts and grilles, separate energy meter for plants are some of the other measures for conservation of energy. The problems observed are over-designed humidification plants, non-compliance of maintenance plan and lack of training for the operatives.

10.1.1 False ceiling

False ceiling is an additional roofing made of fireproof non-conducting material to protect the work area from direct exposure to the outside temperature and humidity. This roof in non-load bearing, and hence no equipment can be installed on this. Light sheets of asbestos or similar materials is supported by frame work of inverted '**T**' angles, hanged by support of thin rods fitted to purlins.

Figure 10.1 False ceiling

The false ceiling designing is a very important aspect of humidity control. It greatly reduces the volume to be controlled resulting in huge savings in humidification costs, which is up to 30%.

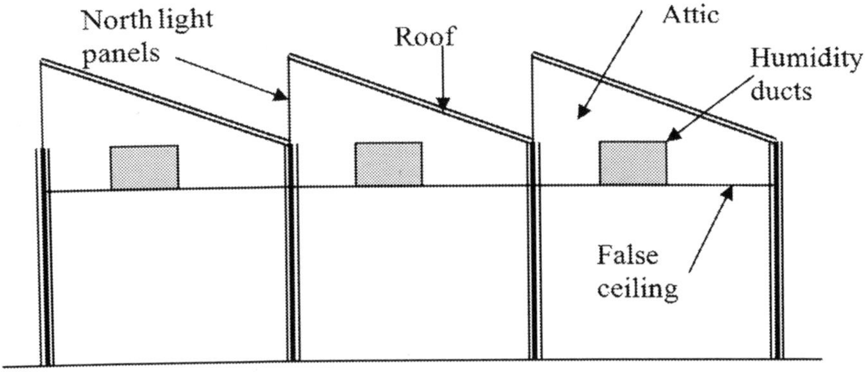

Figure 10.2 Placement of false ceilings

The false ceiling needs proper closing of gaps to avoid fluff and micro-dust entering and accumulating in the attic area. The fluff and dust accumulation in the attic area is a major threat for textile mills as it is vulnerable for fire accidents. Even though proper care is taken to ensure no electrical installations are above the false ceiling, there are incidences of fires because of the very high temperature in attic area, as there is no circulation of air. It is therefore suggested to have wider "**T**" angles to hold the false ceiling sheets to give full support to the sheets and no air gap is left out.

Figure 10.3 Fixing false ceiling

10.1.2 Insulation of ducts

Insulation of ducts is important to avoid loss or gain of heat during air transmission. The attic space generally has high temperature in the afternoons between 50° and 60°C (122 to 140°F), whereas the duct carrying the humidified air shall be around 25–28°C (77 to 82°F). The ducts are therefore insulated with either mineral wool or glass wool. Insulation is also required to be provided over the false ceiling and under the roof to arrest the roof heat entering into the department which otherwise reduces the humidification plant capacity. Following are the guidelines as per section 15880 of San Diego State University, which can be used by all.

a) Ductwork carrying unheated or tempered air (less than 80°F, i.e. 26.6°C) shall be uninsulated.
b) Concealed warm air duct should have 1 inch (25.4 mm) thick fibreglass blanket type insulation.
c) Exposed warm air ductwork should have 1 inch (25.4 mm) thick fibreglass semi-rigid board type insulation with suitable external surface for finish application. Leave ductwork un-insulated only when serving the room in which it is located.
d) Insulate supply ducts as noted above for warm air ductwork, with the addition of vapour barrier jackets with properly specified materials and installation.
e) Insulation, jackets, facings, adhesives, coatings and accessories shall be fire hazard rated in accordance with the requirements of UL 723 and shall have a maximum flame spread of 25 maximum fuel contributed and smoke developed rating of 50. Flameproofing treatments subject to deterioration due to moisture or humidity are not acceptable. All materials shall be UL listed.

Some mills have adopted the concrete load-bearing beams as humidification ducts. In this case, no additional insulation is required.

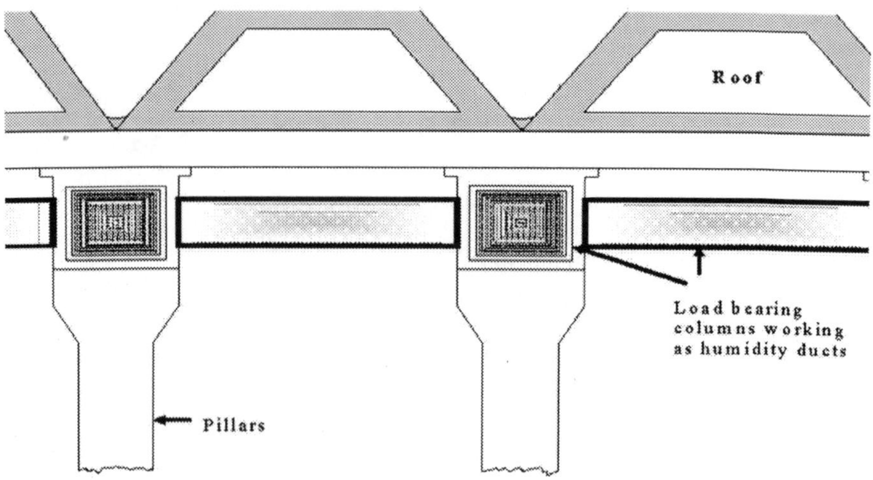

Figure 10.4 Load-bearing columns/ducts

10.1.3 Roof maintenance and humidity control

The radiation from the rooftop is very high. A number of methods are followed depending on the situation to reduce the effect of radiation from roof. Some provide two layers of roof sheet with a gap of around 3 to 6 inches between them. This prevents direct radiation of outside temperature to the working area. It helps in all the seasons like summer, rains and cold winter. The principle used is similar to a vacuum flask, but here is no vacuum between the two layers of roof sheets. Some people put an insulation between two layers of roofing.

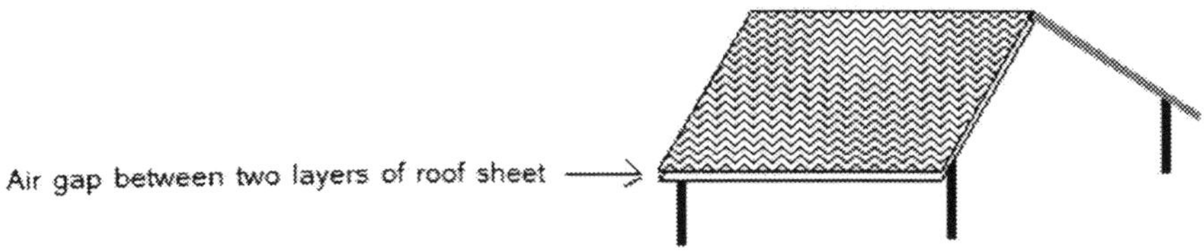

Figure 10.5 Air gap between two layers of roof sheet

During hot summer, one of the methods to eliminate the rise in surface temperature due to sun effect is spraying water on the roof. The water temperature goes down due to evaporative cooling and the water reduces the surface temperature of roof by contact. Theoretically, the surface temperature goes down as low as the wet bulb temperature of the surrounding atmosphere. This drop in surface temperature down to wet bulb temperature means that the surface temperature is reduced below the dry bulb temperature inside the department. Therefore, instead of heat flowing into the department through the roof it would be going out from the department.

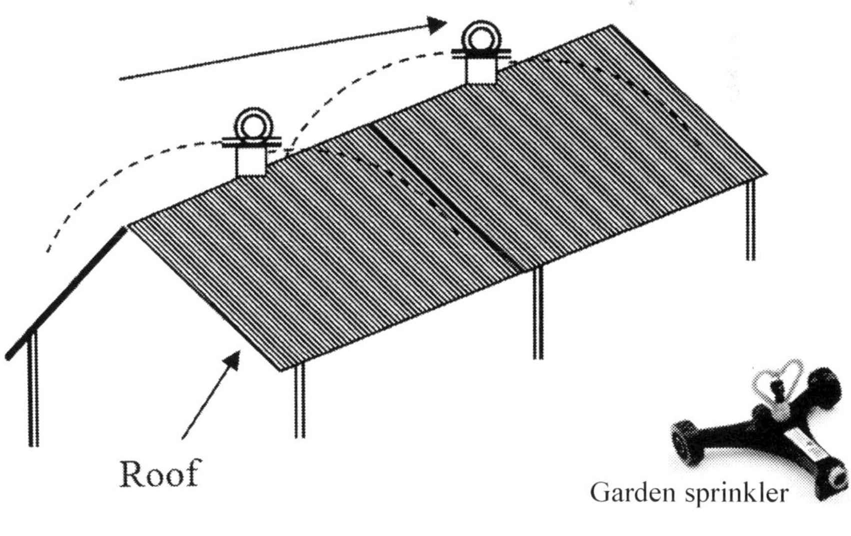

Figure 10. 6 Roof coolers

The heat transfer coefficient of the roof is generally higher when heat is flowing upwards as compared to the heat going down through the roof from outside to inside the shed. With roof cooling as the heat flows outside, it results in higher heat transfer coefficient for the roof, which is advantageous and provides a comfortable working condition.

For spraying water on the roof mills install roof coolers on the top of the roof. They are simple garden sprinklers, which rotate due to the pressure of the water being sprayed. It wets the roof and insulates the working area from high outside temperature.

The roof cooling using sprinklers is a crude method, and there is no control on the water sprayed. It cannot spray uniformly on all the places. Further the spray often gets disturbed because of choking in the nozzles as a result of floating fibers getting contaminated.

10.1.3.1 Painting rooftop

Use of proper paint to the roof can reduce the radiation loss. M.R. Vidhyaprakash, Chief Executive of Excel CoolCoat says that up to 60% cost of humidification could be saved by painting the roof with Excel CoolCoat as it reflects 90% of surface heat back to atmosphere, whereas the surface of roofing remains comfortable to touch even in peak summer afternoons. For buildings with existing humidification plant and false ceiling, an attractive 15% savings can be achieved in maintaining RH in the area of running humidification pump. This system is being used in textile mills also. More the sunshine more will be the difference in temperature. The temperature was recorded 8 months from the date of coating, proving the life. Considerable improvement in working condition and labour efficiency is noted. Table 10.1 gives the actual readings. The temperatures were recorded at around 2.30 pm. (Ambient temperature: −39°C (102 °C)). The roof was of asbestos. The temperature indicated is the exact temperature of asbestos roof sheets which was recorded with the aid of laser-guided infrared noncontact thermometer.

Table 10.1 Effect of painting roofs.

	Uncoated		Coated	
	°F	°C	°F	°C
Recording No. 1	132	56	107	42
Recording No. 2	129	54	105	41
Recording No. 3	132	56	105	41

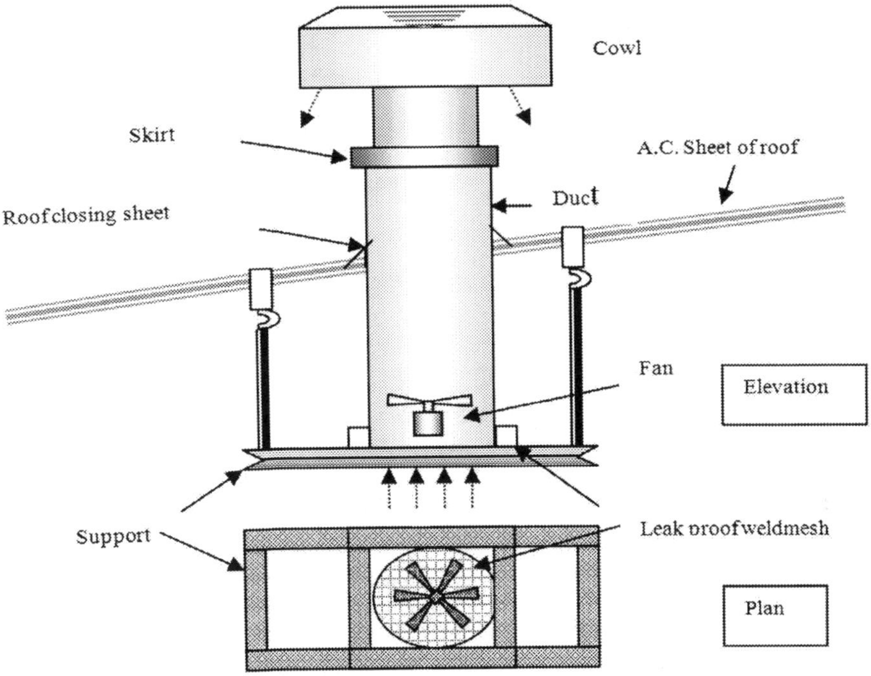

Figure 10.7 Roof extractor

Motorised roof extract fans by Blowtech Engineers

Powerless roof exhaust fans

10.1.3.2 Roof extractors

An axial fan assembly located at the truss level hanging from the roof purlins is known as a roof extractor. They could exhaust large amount of the heat/ smoke accumulating at roof level by displacing large quantity of hot air and thereby inducing large quantity of fresh air through windows / ventilators in a factory. They are ideally suited in foundries, press shops, forge shops, heat treatment shops and large-sized factories. Normally, roof extractors were without positively driven fans, but now they have come with positively driven fans which are more effective in removing hot air.

10.1.3.3 High-volume low speed fans

A high-volume low-speed (HVLS) fan is a type of mechanical fan greater than 7 feet (2.1 m) in diameter. They are generally ceiling fans although some are pole mounted. HVLS fans move slowly and distribute large amounts of air at low rotational speed. HVLS fans work on the principle that cool moving air, breaks up the moisture-saturated boundary layer surrounding the body and accelerates evaporation to produce a cooling effect. Ceiling fans produce a column of air as they turn. This column of air moves down and out along the floor.

Called a horizontal floor jet, this deep wall of horizontal moving air is relative to the diameter of a fan, and to a lesser degree, the speed of a fan. Once the floor jet reaches its potential, it migrates outwards until it meets a side wall or other vertical surface. Under ideal conditions, an 8 feet (2.4 m) diameter fan produces a floor jet of air approximately 36 inches (910 mm) deep. A 24 feet (7.3 m) diameter fan produces a floor jet 108 inches (2,700 mm) deep, tall enough to engulf a human standing on the floor.

Figure 10.8 A six-blade HVLS fan

10.1.4 Enclosing humidification plants

Enclosing of the humidification plants installed outside the main building enhances humidifier application flexibility. Fully insulated enclosures operate in temperatures from −40°F to +122°F (-40°C to +50°C) and offer supplemental heating and cooling systems that maintain required operation conditions. For example, DriSteem Humidifier Company offer outdoor enclosures for use with gas-to-steam (GTS), liquid-to-steam (LTS), steam to steam (STS) and vaporstream (electric) humidifiers.

Enclosures can be installed on the ground or on the roof. Outdoor enclosures are ideal for facilities that have limited interior space. Factory constructed and assembled enclosures are available. The outdoor enclosure is normally shipped complete with humidifier preinstalled and tested, ready to easily connect to gas, water, steam or electricity. It is proved that the outdoor enclosure provides reliable operation under extreme conditions. The enclosures should have easy access for service. Steal enclosure doors are used to provide full access to all internal components. The doors feature stainless steel hinges and latches operable from both the exterior and interior of the unit.

The enclosing protects in both cold and hot climates. To ensure complete safety and operation in all climates, the outdoor enclosure has supplemental heating and cooling systems that automatically maintain required operation conditions. DriSteem humidifiers housed in outdoor enclosures operate properly at temperatures ranging from −40°F to 122°F (−40°C to 50°C). The enclosure is constructed of heavy-duty galvanized steel and is fully insulated. Gaskets on doors ensure a tight seal.

10.1.5 Vapour barrier

Humid vapour could penetrate through the wall, ceiling and floor. This is influenced more by the difference in vapour pressures between outside and inside. Therefore, effective vapour barrier should be part of wall, ceiling and floor. A vapour barrier is any material used for damp proofing, typically a plastic or foil sheet, that resists diffusion of moisture through wall, ceiling and floor assemblies of buildings to prevent interstitial condensation and of packaging. Technically, many of these materials are only vapour retarders as they have varying degrees of permeability. Simple material like aluminium foil is an excellent vapour barrier but one should be cared to seal the joint between the sheets. Water proof paint is also a good vapour barrier. Cold room panel is both thermal and vapour insulation and convenient for retrofit work.

A vapour barrier stops more vapour transmission than a vapour retarder. A vapour barrier is usually defined as a layer with a permeance rating of 0.1 perm or less, while a vapour retarder is usually defined as a layer with permeance greater than 0.1 perm but less than or equal to 1 perm.

10.1.6 Vapour retarders

Vapour retarders help slowing the diffusion of water vapour through a building assembly. During the winter, a vapour retarder on the interior of a wall will slow down the transfer of water vapour from the humid interior into the cool stud bays. During the summer, a vapour retarder on the exterior of a wall will slow down the transfer of water vapour from damp siding towards the cool stud bays.

The fundamental purpose of a roof assembly is to keep water from entering a building through the roof. Low-slope roof assemblies, when properly designed and constructed, perform this function well. However, moist air within a building can enter a roof assembly and condense into water. The climate in which a building is located significantly will affect the type, direction of flow and degree of moisture migration and vapour drive that will occur into and out of a building.

Vapour drive from a building's interior to exterior is likely to be strongest when the exterior temperature and relative humidity are low and the interior temperature and relative humidity are high. These conditions will occur most often during winter months in cold climate regions. Warmer interior air exerts a higher vapour pressure than cooler outside air. Roof assemblies create a barrier between these areas of differing vapour pressures. If a roof assembly is not sufficiently insulated, warm, moist air will rise into the roof assembly and may cool to its dew-point temperature, causing condensation.

There are three primary methods used to prevent moisture from accumulating in low-slope roof assemblies. They are use of vapour retarders, ventilation of interior space and self-drying roof assembly design. For vapour retarders to perform, the temperature at vapour retarders' level must be warmer than the dew-point temperature. To ensure that sufficient insulation must be provided and installed above the vapour retarder. Joan P. Crowe of NRCA technical services explains a graphical method of designing the vapour retarders and the insulation considering the R values. Dew-point temperature can be determined by using a simplified version of the ASHRAE psychrometric chart.

Table 10.2 Psychrometric chart.

R.H%	Design Dry-Bulb (Interior) Temperature in Degree F														
	32	35	40	45	50	55	60	65	70	75	80	85	90	95	100
100	32	35	40	45	50	55	60	65	70	75	80	85	90	95	100
90	30	33	37	42	47	52	57	62	67	72	77	82	87	92	97
80	27	30	34	39	44	49	54	58	64	68	73	78	83	88	93
70	24	27	31	36	40	45	50	55	60	64	69	74	79	84	88
60	20	24	28	32	36	41	46	51	55	60	65	69	74	79	83
50	16	20	24	28	33	36	41	46	50	55	60	64	69	73	78
40	12	15	18	23	27	31	35	40	45	49	53	58	62	67	71
30	8	10	14	16	21	25	29	33	37	42	46	50	54	59	62
20	6	7	8	9	13	16	20	24	28	31	35	40	43	48	52
10	4	4	5	5	6	8	9	10	13	17	20	24	27	30	34

Along the top of the table, locate the design dry-bulb (interior) temperature column, and the relative humidity row along the left side of the table. The dew-point temperature is at the intersection of the design dry-bulb temperature column and relative humidity row. Finally, locate the dew-point temperature value on the temperature gradient line. If the dew-point temperature falls within the insulation that is placed above the vapour retarder, condensation should not occur on the bottom side of the vapour retarder. If it falls below the insulation, additional insulation is required.

10.1.7 Curtain

The required temperature and humidity ars different from section to section in a textile mill, and the outside temperature and humidity shall be seldom same as the required levels in the production departments. It is therefore practiced to insulate the plant and buildings. However, doors are needed for movement of materials and men. The doors when open can allow the humidified air to go out or allow outside air to infiltrate creating a disturbance in the working. The use of door closers is not very convenient considering the frequency of material and men movement, and the visibility requirement. Use of curtains can solve the problem to some extent. Two types of curtains are popular, viz. Split polyurethane curtains and Air curtains.

10.1.7.1 Split polyurethane curtains

Split polyurethane curtains although cannot prevent infiltration can restrict the speed of infiltration and reduce the impact to some extent. It is cheap and easy to install and hence is popular and used widely.

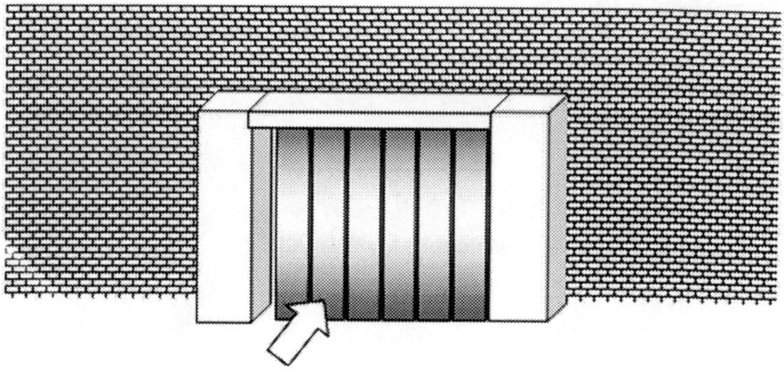

Figure 10. 9 Split polyurethane curtain

10.1.7.2 Air curtain

An air curtain is a fan-powered device used for separating two spaces from each other. This is intended to help keep outside air out and avoid cold draught by mixing in warm air from the air curtain.

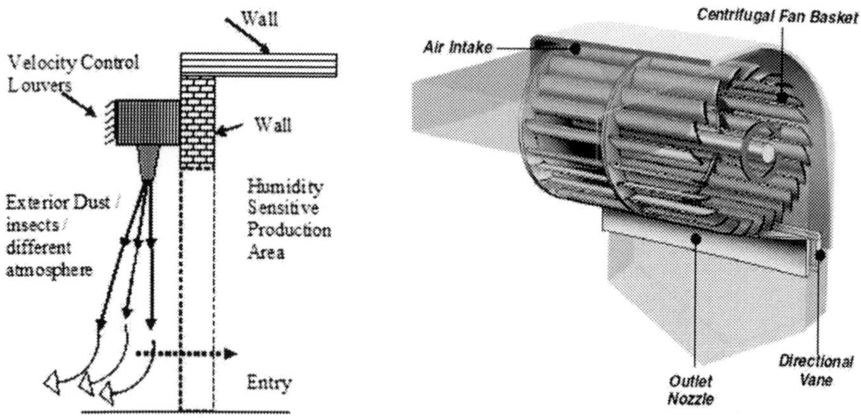

Figure 10.10 Air curtain

The most common configuration for air curtains is a downward facing fan mounted over an opening. Normally this opening is an entrance to a building or a section in a humidity sensitive production plant. Airflow through a door depends on wind forces, temperature differences and pressure differences. The fan must be powerful enough to generate a jet that can reach the floor. Air curtains can be used to save energy by reducing the energy transfer between two spaces, although a closed and well-sealed physical door is much more effective. Alternatively, a combination can be utilised; when the door is opened air curtain turns on, minimizing air flow from inside to outside and vice versa.

Air curtain creates an invisible barrier of high velocity air to separate different environments. Air enters the unit through the intake. It is then compressed by scrolled fan housings and forced through a nozzle which is directed at the open doorway. The system utilises centrifugal fans mounted on direct driven, dual-shafted motors. The result is a uniform air screen across the opening with enough force to stop winds up to 25 mph. It is important to note that when there is an in-draft caused by an exhaust system in the building (negative pressure), the performance of any air door will be heavily affected.

10.1.8 Windowless constructions

The old mills normally had the window area as much as 25% of the wall area, and also large north-light roofing in the top. This was mainly due to take as much advantage as possible of natural light. The present trend is to reduce the window area to as much low as possible or to have windowless constructions. The following are the main reasons.

a) The overall heat transfer coefficient for 3-mm-thick glass window is 5.37 kcal/hr.m^2.°C and for 6-mm-thick glass 5.03 kcal/hr.m^2.°C, whereas for a brick wall of 45cm thick, it is as low as 1.17 kcal/hr.m^2.°C. Heat transmission is thus reduced by about 75% by avoiding window area and having a brick wall. The cost of providing artificial light is comparatively very less compared to the cost of cooling/heating the air.
b) Too much window area gives too vivid a view of the outside to the operating personnel and causes distraction of them.
c) Workers usually keep the windows open that allows lot of outside air to come in. In a department without windows, it is easy to control the ambient conditions.
d) Window panes get broken while opening and closing and need maintenance. If broken windows are not repaired, conditions around the broken window are affected due to outside window.

10.1.9 Underground return air duct for motors and pneumafil

The heat load of motors is normally 15 to 20% of the total machine load. If a way could be found to discharge the heat generated by motors outside the conditioned space, would result in significant reduction in heat load and also the operating costs of humidification. Underground return ducts are recommended for this.

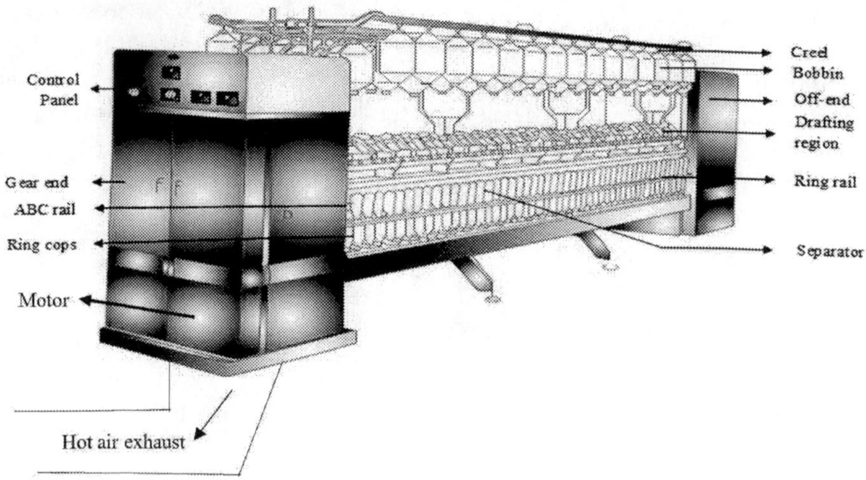

Figure 10.11 Underground return air duct for motors

The driving motor of each ring frame is enclosed, and openings are provided in the bottom so that the air can be drawn out through return air ducts. This prevents the heat from getting discharged into the department. This concentrated release of hot air is of great advantage, as the maximum heat load in a spinning mill is coming from ring frame motors. The exhaustion of hot air also helps keeping the motors cool and reduces power consumption. The air circulated to cool electrical winding inside the motor is discharged at great velocity. This high velocity air creates lot of turbulence and keeps the fluff, etc., floating into atmosphere. Isolation of motors helps keeping the atmosphere clean. Similarly, providing underground ducts to exhaust pneumafil air shall help in reducing the heat load of the department, directs humidified air to the point of spinning at each spindle and reduces dust level in the department.

11
Air conditioning units

11.1 History of air conditioning

The term air conditioning commonly refers to the cooling and dehumidification of indoor air for thermal comfort. In a broader sense, the term can refer to any form of cooling, heating, ventilation or disinfection that modifies the condition of air. An air conditioner is an appliance, system or mechanism designed to extract heat from an area, typically using a refrigeration cycle but sometimes using evaporation. Glenn Elert refers an air conditioner as a mechanical system for controlling temperature (by providing cool air), humidity (by providing dry air) and ventilation (by providing fresh air). It is a combined process that performs many functions simultaneously. It conditions the air, transports and introduces to the space being conditioned. It provides heating and cooling from its central plant or rooftop units. It also controls and maintains the temperature, humidity, air movement, air cleanliness, sound leve, and pressure differential in a space within predetermined limits for the comfort and health of the occupants of the conditioned space or for the purpose of product processing. An air conditioning, or HVAC&R, system is composed of components and equipment arranged in sequence to condition the air, to transport it to the conditioned spac, and to control the indoor environmental parameters of a specific space within required limits. Most air conditioning systems perform the following functions:

- Provide the cooling and heating energy required.
- Condition the supply air, that is, heat or cool, humidify or dehumidify, clean and purif, and attenuate any objectionable noise produced by the HVAC&R equipment.
- Distribute the conditioned air, containing sufficient outdoor air, to the conditioned space.
- Control and maintain the indoor environmental parameters such as temperature, humidity, cleanliness, air movement, sound leve, and pressure differential between the conditioned space and surroundings within predetermined limits.

The first air conditioning probably started with early man when he moved inside a cool, dark cave to escape the summer heat and to protect self from the winter cold. Humans have continually searched for ways to keep cool in the summer. The ancient Egyptians circulated aqueduct water throughout the walls of houses which cooled them. As this was a luxury due to water usage being expensive only the wealthiest could afford it. An earlier form of air conditioning was invented in Persia (Iran) thousands of years ago in the form of wind shafts on the roof, which caught the wind and passed it through water and blew the cooled air into the building. The buildings had their own cisterns and wind towers cooling buildings during the hot season. Cisterns were generally large outdoor pools which collected the rain water; wind towers used windows to catch the wind and internal vanes directed the air flow down in to the building over the cistern tanks and out through a downwind cooling tower. The cistern of water evaporated which in turn cooled the air within the building. As early as the times of the ancient Egyptians, Romans and Greeks, wet fabric mats were hung in the windows and doorways of dwellings to cool the indoor environment. Wind blowing through the mats evaporated the water thereby cooling the air entering the building. The same technique was later used in the royal palaces of India. Today, we find the benefits of air conditioning to be much more than personal comfort; it is necessary for improving productivity in the workplace, making possible good medical/ health service and preserving fresh and frozen foods.

In 1820, British scientist Michael Faraday discovered that compressing and liquefying ammonia could chill air when the liquefied ammonia was allowed to evaporate. In 1842, Florida physician Dr. John Gorrie used compressor technology to create ice, which he used to cool air for his patients in his hospital. He hoped eventually to use his ice-making machine to regulate the temperature of buildings. He even envisioned centralized air conditioning that could cool entire cities. Though his prototype leaked and performed irregularly, Gorrie was granted a patent in 1851 for his ice-making machine. His hopes for its success vanished soon afterwards when his chief financial banker died; Gorrie did not get the money he needed to develop the machine. Dr. Gorrie died impoverished in 1855 and the idea of air conditioning faded away for 50 years.

One of the first uses of air conditioning for personal comfort was in 1902 when the New York Stock Exchange's new building was equipped with a central cooling as well as heating system. Alfred Wolff, an engineer from Hoboken, New Jersey, who is considered the forerunner in the quest to cool a working environment, helped design the new system, transferring this budding technology from textile mills to commercial buildings. Later in 1902, the first modern, electrical air conditioning was invented by Willis Havilland Carrier (1876–1950). His invention differed from Wolff's in that it controlled not only temperatur, but also humidity for improved manufacturing process control for a printing plant in Brooklyn, New York. This specifically helped to provide low heat and humidity for consistent paper dimensions and ink alignment. In 1906, Stuart W. Cramer of Charlotte, North Carolina, was exploring ways to add moisture to the air in his southern textile mill. Cramer coined the term "air conditioning" and used it in a patent claim he filed that year as an alternative to "water conditioning", then a well-known process for making textiles easier to work. He combined moisture with ventilation to actually "condition" and to change the air in the factories, controlling the humidity so necessary in textile plants. Willis Carrier adopted the term and incorporated it into the name of the company he founded in 1907, The Carrier Air Conditioning Company of America. Later, Carrier's technology was applied to increase productivity in the workplace, and the Carrier Engineering Company, now called Carrier (a division of United Technologies Corporation), was formed in 1915 to meet the new demand. Later still, air conditioning use was expanded to improve comfort in homes and automobiles. Residential sales didn't take off until the 195's. The Royal Victoria Hospital, Belfast, Northern Ireland, is a landmark building in building engineering services (built in 1906) and lays claim to being the first "Air conditioned building in the world".

Air conditioning, or manufactured air, as it was first called, was originally considered to be simply controlling humidity. Textile mills had a higher production rate if the inside humidity could be managed. Then temperature control was added and circulated air with controlled humidity and a constant temperature.

The first air conditioners and refrigerators employed toxic or flammable gases such as ammonia, methyl chlorid, and propane which could result in fatal accidents when they leaked. Thomas Midgley, Jr. created the first chlorofluorocarbon gas, Freon, in 1928. The refrigerant was much safer for humans but was later found to be harmful to the atmosphere's ozone layer. "Freon" is a trade name of DuPont for any chlorofluorocarbon (CFC), hydrogenated CFC (HCFC, or hydrofluorocarbon (HFC) refrigerant, the name of each including a number indicating molecular composition (R-11, R-12, R-22, R-134). The blend most used in direct-expansion comfort cooling is an HCFC known as R-22. It is to be phased out for use in new equipment by 2010 and completely discontinued by 2020. R-11 and R-12 are no longer manufactured in the US, the only source for purchase being the cleaned and purified gas recovered from other air conditioner systems. Several non-ozone depleting refrigerants have been developed as alternatives, including R-410A, known by the brand name "Puron". Innovation in air conditioning technologies continues, with much recent emphasis placed on energy efficiency and for improving indoor air quality. As an alternative to high global warming refrigerants, such as R-134a in cars' and R-22, R-410a in residential air conditioning, natural alternatives like CO_2 (R-744) have been proposed.

11.2 Application of air conditioning

Air conditioning systems vary considerably in size and derive their energy from many different sources. Popularity of residential air conditioners has increased dramatically with the advent of central air, a strategy that utilises the ducting for both heating and cooling. Commercial air conditioners have changed a lot in the past few years as energy costs rise and power sources change and improve. The use of natural gas-powered industrial chillers has grown considerably, and they are used for commercial air conditioning in many applications. Air conditioning applications are broadly divided into two applications; Comfort and Process.

11.2.1 Comforts applications

Comfort applications aim to provide an indoor environment that remains relatively constant in a range preferred by humans despite changes in external weather conditions or in internal heat loads. Some claim that comfort air conditioning increases worker productivity but is disputed; one counter argument being that apparent increases in productivity can be explained as resulting from workers perceiving that their employer shows an interest in their welfare. Comfort air conditioning makes deep plan buildings feasible. Without air conditioning, buildings must be built narrower or with light wells so that inner spaces receive sufficient fresh air. Air conditioning also allows building to be taller since wind speed increases significantly with altitude making natural ventilation impractical for very tall buildings. The various sectors of the economy using comfort air conditioning systems are as follows:

(a) *The commercial sector* includes office buildings, supermarkets, department stores, shopping centers, restaurants and others. Many high-rise office buildings use complicated air conditioning systems to satisfy multiple-tenant requirements. In light commercial buildings, the air conditioning system serves the conditioned space of only a single-zone or comparatively smaller area. For shopping malls and restaurants, air conditioning is necessary to attract customers.

(b) *The institutional sector* includes such applications as schools, colleges, universities, libraries, museums, indoor stadiums, cinemas, theatres, concert halls and recreations centres. For example, one of the large indoor stadiums, the Super dome in New Orleans, Louisiana, can seat 78,000 people.

(c) *The residential and lodging sector* consists of hotels, motels, apartment houses and private homes. Many systems serving the lodging industry and apartment houses are operated continuously, on a 24 hour, 7-day-a-week schedule.

(d) *The health care sector*_encompasses hospitals, nursing homes and convalescent care facilities. Special air filters are generally used in hospitals to remove bacteria and particulates of sub-micrometer size from areas such as operating rooms, nurserie, and intensive care units. The relative humidity in a general clinical area is often maintained at a minimum of 30% in winter.

(e) *The transportation sector* includes aircraft, automobiles, railroad cars, buses and cruising ships. Passengers increasingly demand ease and environmental comfort, especially for long distance travel. Modern airplanes flying at high altitudes may require a pressure differential of about 5 psi between the cabin and the outside atmosphere.

11.2.2 Process applications

Process applications aim to provide an indoor environment to suit a process being carried out that remains relatively constant despite changes in external weather conditions or in internal heat loads. Although often in the comfort range, it is the process that determines conditions not human preference. Process applications include the following:

(a) *In textile mills* proper control of humidity increases the strength of the yarn and fabric during processing. For many textile manufacturing processes, too high a value for the space RH% can cause problems in the spinning process. On the other hand, a lower RH% may induce static electricity that is harmful for the production processes. The fluff generation is high with low RH%.

(b) *Cleans rooms:* Many electronic products require clean rooms for manufacturing such things as integrated circuits, since their quality is adversely affected by airborne particles. RH% control is also needed to prevent corrosion and condensation and to eliminate static electricity. Temperature control maintains materials and instruments at stable condition and is also required for workers who wear dust-free garments. For example, a class 100 clean room in an electronic factory requires a temperature of 72 ± 2°F (22.2 ± 1.1°C), a RH% at 45 ±5, and a count of dust not to exceed 100 particles/ft^3 (3531 particles/m^3).

(c) *Precision manufacturers* always need precise temperature control during production of precision instruments, tool, and equipment. Bausch and Lomb successfully constructed a constant-temperature control room of 68 ± 0.1°F (20 ± 0.56°C) to produce light grating products in the 1950s.

(d) *Pharmaceutical products* require temperature, humidit, and air cleanliness control. If the temperature or humidity crosses certain limits, there are chances of deterioration of the products. High-efficiency air filters must be installed for most of the areas in pharmaceutical factories to prevent contamination.

(e) *Warehouses:* Modern refrigerated warehouses not only store commodities in coolers at temperatures of 27 to 32°F (-2.8 to 0°C) and frozen foods at −10 to 20°F (-23 to -29°C, but also provide RH control for perishable foods between 90 and 100%. Refrigerated storage is used to prevent deterioration. Temperature control can be performed by refrigeration systems, but the simultaneous control of both temperature and RH in the space can only be performed by process air conditioning systems.

(f) *Hospital operating theatres* in which air is filtered to high levels to reduce infection risk and the humidity controlled to limit patient dehydration. Although temperatures are often in the comfort range, some specialist procedures such as open heart surgery require low temperatures (about 18°C, 64°F) and others such as neonatal relatively high temperatures (about 28°C, 82°F).

(g) *Facilities for breeding laboratory animals*: Since many animals normally only reproduce in spring, holding them in rooms at which conditions mirror spring all year can cause them to reproduce year round.

(h) *Aircraft air conditioning*: Although nominally aimed at providing comfort for passengers and cooling of equipment, aircraft air conditioning presents a special process because of the low air pressure outside the aircraft.

In both comfort and process applications, the objective is not only to control temperature but also other factors including humidity, air movement and air quality

11.3 Different types of air conditioners

11.3.1 Dehumidifier

Some air conditioning units dry the air without cooling and are classed as dehumidifiers. They work like normal air conditioning units except that they contain a heat exchanger which is placed between the intake and exhaust. To achieve a comfort level in tropical humid areas, they contain convection fans, but only consume about 1/3 of the electricity. Dehumidifiers are also preferred by people who do not like the draft created by air coolers. Dehumidifiers are discussed in detail in a separate chapter.

11.3.2 Individual room air conditioning system

Individual room or simply individual air conditioning systems employ a single, self-contained air conditioner, a packaged terminal, a separated indoor-outdoor split uni, or a heat pump. The heat pump extracts heat from a heat source and rejects heat to air or water at a higher temperature for heating. Unlike other systems, these systems normally use a totally independent unit or units in each room. Individual air conditioning systems can be classified into two categories: Room Air Conditioner (window-mounted) and Packaged Terminal Air Conditioner (PTAC), installed in a sleeve through the outside wall. An evaporator fan pressurises and supplies the conditioned air to the space. In tube-and-fin coil, the refrigerant evaporates, expands directly inside the tube, and absorbs the heat energy from the ambient air during the cooling season; it is called a direct expansion (DX) coil. When the hot refrigerant releases heat energy to the conditioned space during the heating season, it acts as a heat pump. An air filter removes airborne particulates. A compressor compresses the refrigerant from a lower evaporating pressure to a higher condensing pressure. A condenser liquefies refrigerant from hot gas to liquid and rejects heat through a coil and a condenser fan. A temperature control system senses the space air temperature (sensor) and starts or stops the compressor to control its cooling and heating capacity through a thermostat. The difference between a room air conditioner and a room heat pump, and a packaged terminal air conditioner and a packaged terminal heat pump, is that a four-way reversing valve is added to all room heat pumps. Sometimes room air conditioners are separated into two split units: an outdoor condensing unit with compressor and condense, and an indoor air handler in order to have the air handler in a more advantageous location and to reduce the compressor noise indoors. Individual air conditioning systems are characterised by the use of a DX coil for a single room. This is simple and direct way of cooling the air.

Most of the individual systems do not employ connecting ductwork. Outdoor air is introduced through an opening or through a small air damper. Individual systems are usually used only for the perimeter zone of the building.

11.3.3 Evaporative-cooling air conditioning system

Evaporative-cooling air conditioning systems use the cooling effect of the evaporation of water to cool an air stream directly or indirectly. It could be a factory-assembled packaged unit or a field-built system. When an evaporative cooler provides only a portion of the cooling effect, then it becomes a component of a central hydronic or a packaged unit system. An evaporative-cooling system consists of an intake chamber, filter(s), supply fan, direct-contact or indirect-contact heat exchanger, exhaust fan, water sprays, recirculating water pum, and water sump. Evaporative-cooling systems are characterised by low energy use compared with refrigeration cooling. They produce cool and humid air and are widely used.

11.3.4 Refrigeration air conditionin

Refrigeration air conditioning equipment is used to reduce the humidity of the air processed by the system. The relatively cold evaporator coil condenses heater vapour from the processed air sending the water to drain and removing water vapour from the cooled source which in return lowers the relative humidity. In the vapour compression refrigeration cycle, heat is transferred from a lower temperature source to a higher temperature heat sink. Heat naturally flows in the opposite direction this is due to the second law of thermodynamics; work is required to move heat from cold to hot. A freezer works in much the same way; it moves heat out the interior into the room in which it stands. A compressor driven by a motor creates the most common refrigeration cycle. Evaporation occurs when the heat is absorbed and condensation occurs when heat isdrealised, air conditioners are designed to use a compressor to cause pressure changes between two compartments, and also they actively pump a refrigerant around. The refrigerant is pumped into the low-pressure compartment, where, despite the low temperature, the low pressure causes the refrigerant to evaporate into vapour, which takes the heat with it. In the second compartment the refrigerant vapour is compressed and forced through another heat exchange coil, condensing into a liquid, rejecting the heat previously absorbed from the cooled area. The heat exchanger in the condenser section is cooled most often by a fan blowing outside air through it, but in some cases, it is possible to cool it by other means like water fountain.

11.3. 5 Desiccant-based air conditioning system

In desiccant-based air conditioning system, latent cooling is performed by desiccant dehumidification and sensible cooling by evaporative cooling or refrigeration. Thus, a considerable part of expensive vapour compression refrigeration is replaced by inexpensive evaporative cooling. A desiccant-based air conditioning system is usually a hybrid system of dehumidification, evaporative cooling, refrigeration, and regeneration of desiccant. In this system, there are two airstreams: a process air stream and a regenerative air stream. Process air can be all outdoor air or a mixture of outdoor and re-circulating air. Process air is also conditioned air supplied directly to the conditioned space or enclosed manufacturing process, or to the air-handling unit, packaged unit, or terminal for further treatment. Regenerative air stream is a high-temperature air stream used to reactivate the desiccant. A desiccant-based air conditioned system consists of rotary desiccant dehumidifiers, heat pipe heat exchangers, direct or indirect evaporative coolers, DX coils and vapour compression unit or water cooling coils and chillers, fans, pumps, filters, controls, duct, and piping.

11.3.6 Thermal storage air conditioning system

In a thermal storage air conditioning system or simply thermal storage system, the electricity-driven refrigeration compressors are operated during off-peak hours. Stored chilled water or stored ice in tanks is used to provide cooling in buildings during peak hours when high electric demand charges and electric energy rates are in effect. A thermal storage system reduces high electric demand for HVAC&R and partially or fully shifts the high electric energy rates from peak hours to off-peak hours. A thermal storage air conditioning system is always a central air conditioning system using chilled water as the cooling medium. In addition to the air, water and refrigeration control systems, there are chilled-water tanks or ice storage tanks, storage circulating pump, and controls.

11.3.7 Clean-room air conditioning system

Clean-room or clean-space air conditioning systems serve spaces where there is a need for critical control of particulates, temperature, relative humidity, ventilation, noise, vibration and space pressurisation. In this the quality of indoor environmental control directly affects the quality of the products produced in the clean space. It consists of a recirculating air unit and a makeup air unit both include dampers, pre-filters, coils, fans, high-efficiency particulate air (HEPA) filters, ductwork, piping work, pumps, refrigeration system, and related controls except for a humidifier in the makeup unit.

11.3.8 Space conditioning air conditioning system

Space conditioning air conditioning systems are also called space air conditioning systems. They have cooling, dehumidification, heatin, and filtration performed predominately by fan coils, water-source heat pump, or other devices within or above the conditioned space, or very near it. A fan coil consists of a small fan and a coil. A water-source heat pump usually consists of a fan, a finned coil to condition the ai, and a water coil to reject heat to a water loop during cooling, or to extract heat from the same water loop during heating. Single or multiple fan coils are always used to serve a single conditioned room. Usually, a small console water-source heat pump is used for each control zone in the perimeter zone of a building, and a large water-source heat pump may serve several rooms with ducts in the core of the building. Space air conditioning systems normally have only short supply ducts within the conditioned space, and there are no return ducts except the large core water-source heat pumps. The pressure drop required for the recirculation of conditioned space air is often equal to or less than 0.6 in. water column (WC) (150 Pa). Most of the energy needed to transport return and recirculating air is saved in a space air conditioning system, compared to a unitary packaged or a central hydronic air conditioning system. Space air conditioning systems are usually employed with a dedicated outdoor ventilation air system to provide outdoor air for the occupants in the conditioned space. Space air conditioning systems often have comparatively higher noise level and need more periodic maintenance inside the conditioned space.

11.3.9 Unitary packaged air conditioning system

Unitary packaged air conditioning systems can be called, in brief, packaged air conditioning systems or packaged systems. These systems employ either a single, self-contained packaged unit or two-split units. A single packaged unit contains fans, filters, DX coils, compressors, condenser, and other accessories. In the split system, the indoor air handler comprises controls and the air system, containing mainly fans, filter, and DX coils; and the outdoor condensing unit is the refrigeration system, composed of compressors and condensers. Rooftop packaged systems are most widely used. Packaged air conditioning systems can be used to serve either a single room or multiple rooms. A supply duct is often installed for the distribution of conditioned air, and a DX coil is used to cool it. Other components can be added to these systems for operation of a heat pump system; i.e., a centralised system is used to reject heat during the cooling season and to condense heat for heating during the heating season. Sometimes perimeter baseboard heaters or unit heaters are added as a part of a unitary packaged system to provide heating required in the perimeter zone. Packaged air conditioning systems that employ large unitary packaged units are central systems by nature because of the centralised air distributing ductwork or centralised heat rejection systems. They are characterised by the use of integrated, factory-assemble, and ready-to-use packaged units as the primary equipment as well as DX coils for cooling, compared to chilled water in central hydronic air conditioning systems. Modern large rooftop packaged units have many complicated components and controls which can perform similar functions to the central hydronic systems in many applications.

11.3.10 Central hydronic air conditioning system

Central hydronic air conditioning systems are also called central air conditioning systems. In a central hydronic air conditioning system, air is cooled or heated by coils filled with chilled or hot water distributed from a central cooling or heating plant. It is mostly applied to large-area buildings with many zones of conditioned space or to separate buildings. Water has a far greater heat capacity than air. The following is a comparison of these two media for carrying heat energy at 68°F (20°C):

	Air	Water
Specific heat, Btu/lb • °F	0.243	1.0
Density, at 68°F, lb/ft^3	0.075	62.4
Heat capacity of fluid at 68°F, Btu/ft^3 • °F	0.018	62.4

The heat capacity per unit volume of water is 3466 times greater than that of air. Transporting heating and cooling energy from a central plant to remote air-handling units in fan rooms is far more efficient using water than conditioned air in a large air conditioning project. However, an additional water system lowers the evaporating temperature of the refrigerating system and makes a small or medium size project more complicated and expensive.

11.3.11 Air system

An air system is sometimes called the air-handling system. The function of an air system is to condition, transport, distribute the conditioned air, recirculating the air and to control the indoor environment according to requirements. The major components of an air system are the air-handling units, supply/

return ductwork, fan-powered boxes, space diffusion device, and exhaust systems. An air-handling unit (AHU) usually consists of supply fan(s), filter(s), a cooling coil, a heating coil, a mixing bo, and other accessories. It is the primary equipment of the air system. AHU conditions the outdoor/recirculating air, supplies the conditioned air to the conditioned space and extracts the returned air from the space through ductwork and space diffusion devices. A fan-powered variable-air-volume (VAV) box, often abbreviated as fan-powered box, employs a small fan with or without a heating coil. It draws the return air from the ceiling plenum, mixes it with the conditioned air from the AH, and supplies to the conditioned space. Space diffusion devices include slot diffusers mounted in the suspended ceiling; their purpose is to distribute the conditioned air evenly over the entire space according to requirements. The return air enters the ceiling plenum through many scattered return slots. Exhaust systems have exhaust fan(s) and ductwork to exhaust air from the lavatories, mechanical rooms, and electrical rooms.

11.3.12 Water System

The water system includes chilled and hot water systems, chilled and hot water pumps, condenser water syste, and condenser water pumps. The purpose of the water system is (1) to transport chilled water and hot water from the central plant to the air-handling units, fan-coil unit, and fan-powered boxes and (2) to transport the condenser water from the cooling tower, well water, or other sources to the condenser inside the central plant.

11.4 Design for air conditioning system

Designing an air conditioning plant needs more concentration as the system design determines the basic characteristics. After an air conditioning system is constructed according to the design, it is difficult and expensive to change the design concept.

Engineering Responsibilities: The normal procedure in a design includes initiation of a construction project by owner or developer, selection of design team, setting of the design criteria and indoor environmental parameters, selection of conceptual alternatives for systems and subsystems; preparation of schematic layouts of HVAC&R, preparation of contract documents, working drawings, specifications, materials and construction methods, commissioning guidelines, review of shop drawings and commissioning schedule, operating and maintenance manuals, monitoring, supervision and inspection of construction, supervision of commissioning: testing and balancing; functional performance tests, modification of drawings to the as-built condition and the finalisation of the operation and maintenance manual. It is necessary for the designer to select among the available alternatives for optimum comfort, economics, energy conservation, noise, safety, flexibility, reliability, convenience and maintainability. Experience, education and judgment are important in the selection process. Factors requiring input from both the architect and the mechanical engineer include shape and the orientation of the building, thermal characteristics of the building envelope, especially the type and size of the windows and the construction of external walls and roofs, location of the ductwork and piping to avoid interference with each other, or with other trades, layout of the diffusers and supply and return grilles, minimum clearance provided between the structural members and the suspended ceiling for the installation of ductwork and piping and location and size of the rooms for central plant, fan rooms, duct and pipe shafts.

If the architect makes a decision that is thermally unsound, the HVAC&R engineer must offset the additional loads by increasing the HVAC&R system capacity to reduce the energy consumption. HVAC&R and other building services must coordinate utilization of daylight and the type of artificial light to be installed, the layout of diffusers, grilles, return inlet, and light fitting in the suspended ceiling, integration with fire alarm and smoke control systems, electric power and plumbing requirements for the HVAC&R equipment and lighting for equipment rooms and coordination of the layout of the ductwork, piping, electric cables, etc.

The use of safety factors allows for the unpredictable in design, installation and operation. An HVAC&R system should not be over designed by using a greater safety factor than is actually required. The initial cost of an over designed HVAC&R system is always higher, and it is energy-inefficient. When an HVAC&R system design is under pressure to reduce the initial cost, some avenues to be considered are selecting an optimum safety factor, minimizing redundancy, conducting a detailed economic analysis for the selection of a better alternative, calculating the space load, the capacity of the syste, and the equipment requirements more precisely and adopting optimum diversity factors based on actual experience data observed from similar buildings.

11.4.1 Raw materials

Air conditioners are made of different types of metal. Frequently, plastic and other nontraditional materials are used to reduce weight and cost. Copper or aluminium tubing, which are critical ingredients in many air conditioner plants, provide superior thermal properties and a positive influence on system efficiency. Various components in an air conditioner will differ with the application, but usually they are comprised of stainless steel and other corrosion-resistant metals. Self-contained units that house the refrigeration system will usually be encased in sheet metal that is protected from environmental conditions by a paint or powder coating. The working fluid, the fluid that circulates through the air-conditioning system, is typically a liquid with strong thermodynamic characteristics such ae Freon, hydrocarbons, ammonia or water.

Refrigeration cycle: In the refrigeration cycle, a heat pump transfers heat from a lower temperature heat source into a higher temperature heat sink. Heat would naturally flow in the opposite direction. This is the most common type of air conditioning. A refrigerator works in much the same way, as it pumps the heat out of the interior into the room in which it stands. The most common refrigeration cycle uses an electric motor to drive a compressor. A refrigerant is pumped into the cooled compartment (the evaporator coil), where the low pressure and low temperature cause the refrigerant to evaporate into a vapour, taking heat with it. In the other compartment (the condenser), the refrigerant vapour is compressed and forced through another heat exchange coil, condensing into a liquid, rejecting the heat previously absorbed from the cooled space. This cycle takes advantage of the universal gas law $PV = nRT$, where P is pressure, V is volume, R is the universal gas constant, T is temperature and n is the number of moles of gas (1 mole = 6.022×10^{23} molecules).

Refrigeration air conditioning equipment usually reduces the humidity of the air. The relatively cold (below the dew point) evaporator coil condenses water vapour from the processed air, sending the water to a drain and removing water vapour from the cooled space which lowers the RH. In food retailing establishments' large open chiller cabinets act as highly effective air dehumidifying units. Some air conditioning units dry the air without cooling it. They work like a normal air conditioner, except that a heat exchanger is placed between the intake and exhaust. In combination with convection fans they achieve a similar level of comfort as an air cooler in humid tropical climates, but only consume about 1/3 of the electricity.

11.4.2 Design

All air conditioners have four basic components; a pump, an evaporator, a condense, and an expansion valve. All have a working fluid and an opposing fluid medium as well. Two air conditioners may look entirely dissimilar in size, shape and configuration, yet both function in basically the same way.

Most air conditioners derive their power from an electrically-driven motor and pump combination to circulate the refrigerant fluid. Some natural gas-driven chillers couple the pump with a gas engine in order to give off significantly more torque.

As the working fluid or refrigerant circulates through the air conditioning system at high pressure via the pump enter an evaporator and changes into a gas state, taking heat from the opposing fluid medium and operating just like a heat exchanger. The working fluid then moves to the condenser and gives off heat to the atmosphere by condensing back into a liquid. After passing through an expansion valve, the working fluid returns to a low pressure state. When the cooling medium passes near the evaporator, heat is drawn to the evaporator. This process effectively cools the opposing medium, providing localized cooling where needed. Early air conditioners used Freon as the working fluid, but because of the hazardous effects Freon has on the environment, it has been phased out. Recent designs have met strict challenges to improve the efficiency of a unit, while using an inferior substitute for Freon.

11.4.3 The manufacturing process

Creating encasement parts from galvanized sheet metal and structural steel: Most air conditioners have raw material in the form of structural steel shapes and sheet steel. As the sheet metal is processed into fabrication cells or work cells, it is cut, formed, punched, drilled, sheared and/or bent into a useful shape or form. The encasements or wrappers, the metal that envelopes outdoor units, is made of galvanized sheet metal to provide protection against corrosion. Galvanized sheet metal is also used to form the bottom pan, face plate, and various support brackets throughout an air conditioner. This sheet metal is sheared on a shear press in a fabrication cell soon after arriving in production hall. Structural steel shapes are cut and mitered on a band saw to form useful brackets and supports.

Punch pressing the sheet metal forms: From the shear press, the sheet metal is loaded on a CNC (Computer Numerical Control) punch press. The punch press has the option of receiving its computer programme from a drafting CAD/CAM Computer-Aided Drafting Computer-Aided Manufacturing) program or from an independently written CNC program. The CAD/CAM program will transform a drafted or modelled part on the computer into a file that can be read by the punch press, telling it where to punch holes in the sheet metal. Dies and other punching instruments are stored in the machine and mechanically brought to the punching arm, where it can be used to drive through the sheet. The NC (Numerically Controlled) press brakes bend the sheet into its final form, using a computer file to program itself. Different bending dies are used for different shapes and configurations and may be changed for each component.

Some brackets, fin, and sheet components are normally outsourced to other facilities or companies to produce large quantities. They are brought to the assembly plant only when needed for assembly. Many of the brackets are produced on a hydraulic or mechanical press, where brackets of different shapes and configurations can be produced from a coiled sheet and unrolled continuously into the machine. High volumes of parts can be produced because the press can often produce a complex shape with one hit.

Cleaning the parts: It is essential that all parts are clean and free of dirt, oil, greas, and lubricants before they are powder coated. Various cleaning methods are used to accomplish this necessary task. Large solution tanks filled with a cleaning solvent agitate and knock off the oil when parts are submersed. Spray wash systems use pressurized cleaning solutions to knock off dirt and grease. Vapour degreasing, suspending the parts above a harsh cleansing vapour, uses an acid solution and will leave the parts free of petroleum products. For additional corrosion protection, many parts will be primed in a phosphate primer bath before entering a drying oven to prepare them for the application of the powder coating.

Powder coating: Before brackets, pan, and wrappers are assembled together, they are fed through a powder coating operation. The powder coating system sprays a paint-like dry powder onto the parts as they are fed through a booth on an overhead conveyor. This can be done by robotic sprayers that are programmed where to spray as each part feeds through the booth on the conveyor. The parts are statically charged to attract the powder to adhere to deep crevices and bends within each part. The powder-coated parts are then fed through an oven, usually with the same conveyor system, where the powder is permanently baked onto the metal. The process takes less than 10 minutes.

Bending the tubing for the condenser and evaporator: The condenser and evaporator both act as a heat exchanger in air conditioning systems and are made of copper or aluminium tubing bent around in coil form to maximise the distance through which the working fluid travels. The opposing fluid, or cooling fluid, passes around the tubes as the working fluid draws away its heat in the evaporator. This is accomplished by taking many small diameter copper tubes bent in the same shape and anchoring them with guide rods and aluminium plates. The working fluid or refrigerant flows through the copper tubes and the opposing fluid flows around them in between the aluminium plates. The tubes will often end up with hairpin bends performed by NC benders, using the same principle as the NC press brake. Each bend is identical to the next. The benders use previously straightened tubing to bend around a fixed die with a mandrel fed through the inner diameter to keep it from collapsing during the bend. The mandrel is raked back through the inside of the tube when the bend has been accomplished.

Tubing supplied to the manufacturer in a coil form goes through a uncoiler and straightener before being fed through the bender. Some tubing will be cut into desired lengths on an abrasive saw that will cut several small tubes in one stroke. The aluminium plates are punched out on a punch press and formed on a mechanical press to place divots or waves in the plate. These waves maximize the thermodynamic heat transfer between the working fluid and the opposing medium. When the copper tubes are finished in the bending cell, they are transported by automatic-guided vehicle (AGV) to the assembly cell, where they are stacked on the guide rods and fed through the plates or fins.

Joining the copper tubing with the aluminium plates: A major part of the assembly is the joining of the copper tubing with the aluminium plates. This assembly becomes the evaporator and is accomplished by taking the stacked copper tubing in their hairpin configuration and mechanically fusing them to the aluminium plates. The fusing occurs by taking a bullet or mandrel and feeding it through the copper tubing to expand it and push it against the inner part of the hole of the plate. This provides a thrifty, yet useful bond between the tubing and plate, allowing for heat transfer.

The condenser is manufactured in a similar manner, except that the opposing medium is usually air, which cools off the copper or aluminium condenser coils without the plates. They are held by brackets which support the coiled tubin, and are connected to the evaporator with fittings or couplings. The condenser is usually just one tube that may be bent around in a number of hairpin bends. The expansion valve, a complete component normally outsourced, is installed in the piping after the condenser. It allows the pressure of the working fluid to decrease and re-enter the pump.

Installing the pump: The pump is normally a purchased unit. Designed to increase system pressure and circulate the working fluid, the pump is connected with fittings to the system and anchored in place by support brackets and a base. It is bolted together with the other structural members of the air conditioner and covered by the wrapper or sheet metal encasement. The encasement is either riveted or bolted together to provide adequate protection for the inner components.

11.4.4 Quality control

Quality of the individual components is always checked at various stages of the manufacturing process. Outsourced parts must pass an incoming dimensional inspection from a quality assurance representative before being approved for use in the final product. Usually, each fabrication cell will have a quality control plan to verify dimensional integrity of each part. The unit will undergo a performance test when assembly is complete to assure the customer that each unit operates efficiently.

11.4.5 Design of the control system

Controlling and maintaining the indoor environmental parameters within predetermined limits depends mainly on adequate equipment capacity and the quality of the control system. Energy can be saved when the systems are operated at part load with the equipment's capacity following the system load accurately by means of capacity control. Modern air conditioning control systems for the air and water systems and for the central plant consist of electronic sensors, microprocessor-operated and controlled modules that can analyze and perform calculations from both digital and analog input signals. Control systems using digital signals compatible with the microprocessor are called direct digital control (DDC) systems. Outputs from the control modules often actuate dampers, valve, and relays by means of pneumatic actuators in large buildings and by means of electric actuators for small projects. Thermostats control the operation of HVAC systems, turning on the heating or cooling systems to bring the building to the set temperature. Typically the heating and cooling systems have separate control systems (even though they may share a thermostat) so that the temperature is only controlled "one-way".

There are rapid changes in HVAC&R controls from conventional systems to energy management systems, to DDC with microprocessor intelligenc, and then to open-protocol BACnet, etc. The designers need to keep a pace with the developments. It was normal that many designers preferred to prepare a conceptual design and a sequence of operations and then to ask the representative of the control manufacturer to design the control system. Only about one-third of the designers designed the control system themselves and asked the representative of the control manufacturer to comment on it.

Designer should be able to prepare the sequence of operations and select the best-fit control sequences for the controllers from a variety of the manufacturers that offer equipment in the HVAC&R field. The designer may not be a specialist in the details of construction or of wiring diagrams of controllers or DDC modules, but should be quite clear about the function and sequence of the desired operation, as well as the criteria for the sensors, controllers, DDC module, and controlled devices.

If the HVAC&R system designer does not perform these duties personally, preparation of a systems operation and maintenance manual with clear instructions would be difficult. It would also be difficult for the operator to understand the designer's intention and to operate the HVAC&R system satisfactorily.

Drawings: The layout of an HVAC&R system and the locations and dimensions of its equipment, instruments, ducts, pipes, etc., are best shown and illustrated by drawings. HVAC&R drawings consist of mainly the following:

- *Floor plans.* System layout including plant room, fan rooms, mechanical room, ductwor, and pipelines are illustrated on floor plans. Each floor has at least one floor plan. HVAC&R floor plans are always drawn over the same floor plan of the architectural drawing.
- *Detail drawings.* These drawings show the details of a certain section of an HVAC&R system, or the detail of the installation of certain equipment or the connection between equipment and ductwork or pipeline. Standard details are often used to save time.
- *Sections and elevations.* Sectional drawings are helpful to show the inner part of a section of a system, a piece of equipment or a device. They are especially useful for places such as the plant room, fan roo, and mechanical room where lots of equipment, ductwor, and pipelines are found. Elevations often show clearly the relationship between the HVAC&R components and the building structure.
- *Piping diagram.* This diagram shows the piping layout of the water system(s) and the flow of water from the central plant to the HVAC&R equipment on each floor.
- *Air duct diagram.* This diagram illustrates the air duct layout as well as the airflow from the air-handling unit or packaged unit to the conditioned spaces on each floor through space diffusion devices.
- *Control diagrams.*
  - *Specifications*: Detailed descriptions of equipment, instruments, ductwork and pipelines, as well as performances, operating characteristic, and control sequences are better defined in specifications. Specifications usually consist of the legal contract between the owner and the contractor, installer

or vendor and the technical specifications that specify in detail the equipment and material to be used and how they are installed.

– *Codes and Standards*: Codes are generally mandatory state or city laws or regulations that force the designer to create the design without violating human safety and welfare. State and city codes concerning structural integrity, electrical safety, fire protectio, and prevention of explosion of pressure vessels must be followed. Standards describe consistent methods of testing, specify confirmed design guideline, and recommend standard practices. Conformance to standards is usually voluntary.

Air conditioner manufacturers face the challenge of improving efficiency and lowering costs. Because of the environmental concerns, working fluids now consist typically of ammonia or water. New research is under way to design new working fluids and better system components to keep up with rapidly expanding markets and applications. The competitiveness of the industry should remain strong, driving more innovations in manufacturing and design.

11.5 Central air conditioning

Central air conditioning, commonly referred to as *central air* (US) or *air-con* (UK), is an air conditioning system which uses ducts to distribute cooled and/or dehumidified air to more than one room, or uses pipes to distribute chilled water to heat exchangers in more than one room, and which is not plugged into a standard electrical outlet. With a *split system*, the condenser and compressor are located in an outdoor unit; the evaporator is mounted in the air handling unit. With a *package system*, all components are located in a single outdoor unit that may be located on the ground or roof.

Central air conditioning performs like a regular air conditioner but has several added benefits. When the air handling unit turns on, room air is drawn in from various parts of the house through return-air ducts. This air is pulled through a filter where airborne particles such as dust and lint are removed. Sophisticated filters may remove microscopic pollutants as well. The filtered air is routed to air supply ductwork that carries it back to rooms. Whenever the air conditioner is running, this cycle repeats continually. Because the central air conditioning unit is located outside the main building, it offers a lower level of noise indoors than a free-standing air conditioning unit.

11.6 Health Implications

Poorly maintained air conditioning systems (especially large, centralized systems) can occasionally promote the growth and spread of microorganisms, such as Legionella pneumophila, the infectious agent responsible for Legionnaire's disease, or thermophilic actinomycetes. Conversely, air conditioning (including filtration, humidification, cooling, disinfection, etc.) can be used to provide a clean, safe, hypoallergenic atmosphere in hospital operating rooms and other environments where an appropriate atmosphere is critical to patient safety and well-being. Air conditioning can have a positive effect on sufferers of allergies and asthma. In serious heat waves, air conditioning can save the lives of the elderly. Although many people superstitiously believe that air conditioning is unconditionally dangerous for one's health, especially in areas where air conditioning is not common, this belief is unsupported by fact; properly maintained air conditioning systems do not cause or promote illness. As with heating systems, the advantages of air conditioning generally far outweigh the disadvantages.

Conditions favourable to the growth of microorganisms are sometimesfound in mechanical ventilation and air conditioning systems. B. P. Ager and J. A. Tickner suggested preventive measures such as inclusion of the cleaning of water reservoirs and inspection of filters. Biocidalwater treatment is a possible additional control measure to minimize growth of microorganisms in cooling towers.

12
Dehumidification

12.1 Introduction

As providing humidity to the required level is important for getting good working, ensuring that humidity does not cross the specified limits is also important. When the humidity is more than requirement, we get problems like lapping of cotton on the drafting rollers in draw frames, speed frames, ring frames, etc., sagging of webs in cards and combers, dust and trash particles not getting removed properly in blow room, and so on. A very high humidity can also cause health problem among the people working. In summer as the temperature also shall be high, higher humidity causes sweating of the people making it inconvenient to work. The problem is more in coastal areas where the humidity is normally on a higher side. There is a need to know the method of reducing humidity when it is more in order to achieve optimum conditions. Dehumidifiers come to help in such cases. They are also preferred by people who do not like the draft created by air coolers.

12.2 Reducing the humidity

Normally in the land locked areas of hot countries, the requirement is of increasing the humidity as the outside is normally dry compared to the humidity required for working. The humidity is increased by spraying water on the flowing air. Sometimes the humidity inside becomes more than required because of excess spray of water and improper ventilation. The internal RH% will be more compared to outside humidity. In those cases, outside air is taken in and the humidity is reduced.

In cases where the outside humidity is more, the air washer is stopped and the departmental air is re-circulated. It can help to some extent, but by this the temperature in the department increases, and people feel uneasy due to lack of fresh air. If the air temperature is high, chilling can bring down both humidity and temperature. Chilled water spray on hot humid air attracts the moisture in the air and condenses them, which is collected by the eliminators and sent down to sump. When the air temperature is low along with higher humidity, heating the department helps in reducing the RH%. Different types of dehumidifiers are available, which are used to reduce humidity and bring it to required level.

12.3 Types of dehumidification

Different methods are followed for dehumidification. If the outside air is dry and cool, then circulating it without spraying water and letting out the return air can reduce humidity quickly. However, if the outside is more humid, then we need to remove the moisture from the air, which is let inside the department. Drying the air with cooling is required when outside air is hot and humid. The air is passed over a set of cooling coils, where the moisture from the air gets condensed on the cooling coils. The cooled and dehumidified air can then be let into the working area.

Some air conditioning units actually dry the air without cooling. They work like normal air conditioning units except that they contain a heat exchanger which is placed between the intake and exhaust. In some cases, desiccators are used filled with hygroscopic agents, on which the humid air is passed. The moisture from air is absorbed by the desiccator. To achieve a comfort level in tropical humid areas they contain convection fans, but only consume about 1/3 of the electricity.

12.3.1 Desiccators type

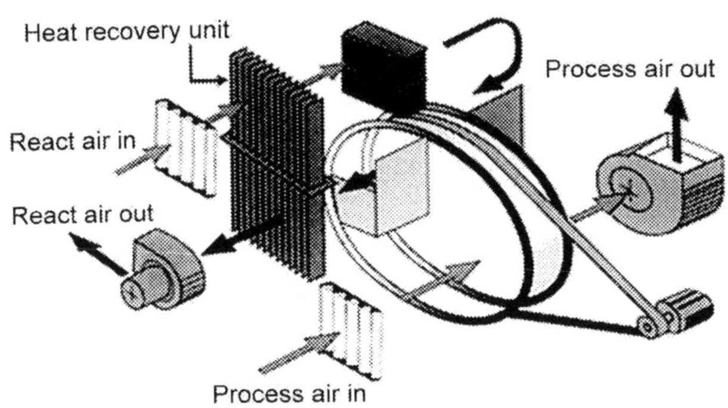

Figure 12.1 Desiccators-type dehumidifier

In this system,moisture-laden air enters through the process inlet and passes through the desiccant rotor, where the moisture is absorbed by the desiccant. The dry, dehumidified air is then delivered to conditioned area through the process outlet. The rotor then rotates to the reactivation sector and the moisture is removed from the desiccant rotor, by blowing hot air through the reactivation outlet to the atmosphere. The hot desiccant rotor is cooled by a small portion of the process air before rotating back to the process sector. Thus, the absorption activity is continuous.

12.3.2 Energy Recovery Ventilators

An energy recovery ventilator (ERV) is a type of mechanical equipment that features a heat exchanger combined with a ventilation system for providing controlled ventilation into a building. This type of equipment was introduced as 'air-to-air' heat exchangers in the colder regions of the U.S., Canada, Europe and Scandinavia. In these areas, tightlybuilt modern houses were developing problems with indoor air quality and excessive humidity during the winter. The air-to-air heat exchanger brought in fresh outside air to combat these problems, and preheated it at the same time. These products are now called heat recovery ventilators or HRVs. ERVs are available in wall mount or duct-connected models. There are four main types of heat exchanger cores, that is, cross flow plate cores, counter flow flat-plate cores, heat pipe core and rotary wheel core.

The efficiency of ERVs refers to the amount of temperature adjustment of the incoming fresh air that the outgoing stale air can accomplish in the heat exchanger. In the winter, the outgoing heated stale air will pre-heat the incoming fresh air to some extent, according to the unit's efficiency. The opposite occurs in the summer with the outgoing air pre-cooling the incoming air. A unit with humidity regulation will remove humidity from humid summer fresh air and add humidity to the incoming dry winter air.

ERVs take the efficiency a step higher by recovering latent and sensible energy from the air stream. Dehumidification equipment is available for incorporation into the overall HVAC system and can be controlled by a programmable thermostat/humidistat. A dehumidifier that incorporates fresh air exchange is less expensive than an ERV system to install, but does not offer energy recovery.

12.3.3 Open coil duct heaters

The use of bulky finned tubular heaters earlier greatly restricted airflow, which decreased performance. The bulkiness of the heater created a lot of weight and added to the overall product size. The life of the heater was just marginally acceptable. The heater system needed to be made more efficient in order to support the overall goals of the project. An open coil duct heater is more compact, lighter and more efficient. It extends life, improves performance and lowers cost in the following ways: The heater reduces obstruction by 50%, which significantly decreases back pressure, increases drying performance and lowers the strain on the blower. While the process air temperature was intentionally increased, their efficiency helped lowering the actual circuit temperature by 50%. This multiplies the heater life and gives their customer a new advantage.

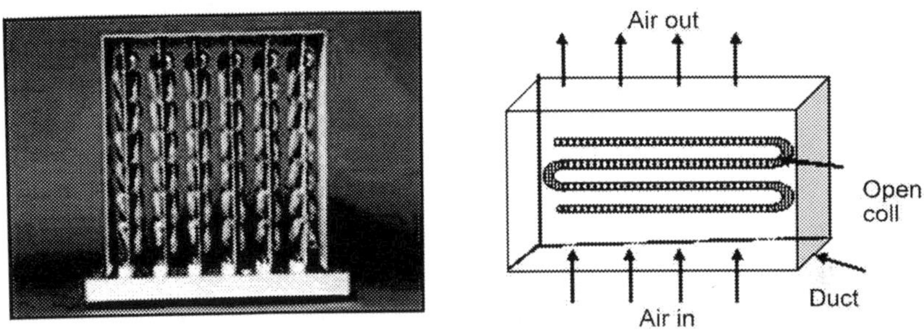

Figure 12.2 Open coil duct heater

12.3.4 Space heaters

When the humidity is more and the temperature is low, normally which happens if there is continuous raining outside for a long time, the method used is to heat the space either by closed steam pipes or by heating lamps. This shall increase the temperature. At higher temperature, the moisture holding capacity of air increases, and hence the relative humidity shall drop, although absolute humidity shall remain more or less same, excepting for the reduction due to expansion of air. As normally outlet shall not be there for the air while space heaters are used, the air cannot expand because of the increase in temperature, instead the pressure of air shall increase.

Figure 12.3 Propane space heater

Figure 12. 4 Radiant vent-free space heater

Figure 12.5 Portable electric space heater

Figure 12. 6 Diesel space heater

The space heating by steam pipes demands installation of a separate steam boiler, which requires high cost for installation and maintenance. Hence, this system is followed in big textile mills, where steam is produced on a regular basis for the processes such assizing, wet processing and so on. The spinning units thatdo not require steam normally adopt electrical heaters of coil type or infrared lights.

Figure 12. 7 Infrared heater

Infrared lights have advantage that they can be fitted in localizedareas. The other advantages are as follows:

- Infrared heater is nonpolluting and does not burn, whereas steam boilers normally burn some fuels.
- Heater emits 91.7% of the energy it uses.
- The infrared heater generates heat instantly, whereas incase of boilers we need to wait for hours together for the pressure to build up. There is no heat accumulation after it's turned off whereas with steam pipe, it takes hours to cool down.

- The halogen tubes will reduce the energy bill to half with halogen FIR radiant energy. More economical and effective compared to conventional electric heaters.
- There is minimum heat loss with the infrared heater because warm air doesn't rise. Heat is not conveyed by convection, it is transmitted like sunshine.
- Even if vinyl is put very close to the heater it will not burn or melt.
- All functions can be made remote controlled.

12.3.5 Chilling

When both the temperature and humidityare high, the method followed is to cool the air, wherein the moisture gets condensed and falls down. As the air cools, its water holding capacity shall also reduce. Cooling towers are used to cool water below the dew point temperature of atmospheric air and the chilled water is either circulated in cooling coils or sprayed on the stream of water in an air washer.

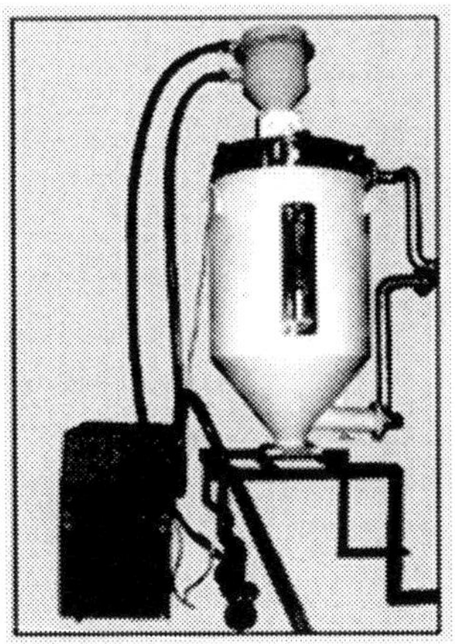

Figure 12.8 Chilling unit

13
Designing HVAC

13.1 What is HVAC?

The term HVAC refers to the three disciplines of *Heating*, *Ventilating* and *Air Conditioning* in air management and a fourth discipline, *Controls* which pervade the entire HVAC field. Controls determine how HVAC systems operate to meet the design goals of comfort, safety and cost-effective operation.

- *Heating* can be accomplished by heating the air within a space, or by heating the work area directly by radiation.
- *Ventilating* maintains an adequate mixture of gases in the air, controls odours and removes contaminants from occupied spaces. It can be accomplished passively through natural ventilation or actively through mechanical distribution systems.
- *Air-conditioning* refers to the sensible and latent cooling of air. Sensible cooling involves the control of air temperature while latent cooling involves the control of air humidity. Air is cooled by transferring heat between spaces, such as with a water loop heat pump system, or by rejecting it to the outside air via air-cooled or water-cooled equipment. Heat can also be rejected to the ground using geothermal exchange. Cool air is not comfortable if it is too humid. Air is dehumidified by condensing its moisture on a cold surface or by removing the moisture through absorption. In dry climates, humidification may be required for comfort instead of dehumidification. Evaporative humidification also cools the air. In such climates, it is possible to use radiant cooling systems, similar to the radiant heating systems.
- *Controls* ensure comfort, provide safe operation of the equipment and enable judicious use of energy resources. HVAC systems are sized to meet heating and cooling loads that historically occur only 1% to 2.5% of the time. It is the function of the controls to ensure that the systems perform properly, reliably and efficiently during those conditions that occur 97.5% to 99% of the time.

Each HVAC discipline has specific design requirements and also opportunities for energy savings. However, energy savings in one area may augment or diminish savings in another. This applies to interactions between components of an HVAC system, as well as between the system and the lighting and envelope systems. Therefore, understanding how one system or subsystem affects another is essential to get the best of the opportunities for energy savings. This design approach is known as *whole building design.*

13.2 Fundamental of energy and resource-efficient HVAC design

Consider all aspects of the building simultaneously: Energy-efficient, climate responsive construction requires a whole building perspective that integrates architectural and engineering concerns early in the design process. For example, the evaluation of a building envelope design must consider its effect on cooling loads and day lighting. An energy-efficient building envelope, coupled with a suitable lighting system and efficient, properly sized HVAC equipment will cost less to purchase and operate than a building whose systems are selected in isolation from each other.

Decide on design goals as early as possible: A building that only meets energy code requirements will often have a different HVAC system than one that uses 40% less energy than the code. The difference is likely to be not only component size but also basic system type.

'Right Size' HVAC systems to ensure efficient operation: Safety factors for HVAC systems allow for uncertainties in the final design, construction and use of the building, but should be used reasonably. Greatly oversized equipment operates less efficiently and costs more than properly sized equipment. It is unreasonable and expensive to assume a simultaneous worst-case scenario for all load components (occupancy, lighting, shading devices, weather) and then to apply the highest safety factors for sizing.

Consider part-load performance when selecting equipment: Part-load performance of equipment is a critical consideration for HVAC sizing. Most heating and cooling equipment operate at their rated efficiency only when fully loaded. However, HVAC systems are sized to meet design heating and cooling conditions that historically occur only 1% to 2.5% of the time. Thus, HVAC systems are intentionally oversized at least 97.5% to 99% of the time. In addition, most equipment is further oversized to handle pick-up loads and to provide a factor of safety. Therefore, systems almost never operate at full load. In fact, most systems operate at 50% or less of their capacity.

Shift or shave electric loads during peak demand periods: Many electric utilities offer lower rates during off-peak periods that typically occur at night.

Whenever possible, design the systems to take advantage of this situation. For example, energy management systems can shed non-critical loads at peak periods to prevent short duration electrical demands from affecting energy bills for the entire year, or, off-peak thermal ice storage systems can be designed to run chillers at night to make ice that can be used for cooling the building during the next afternoon when rates are higher.

Plan for expansion, but don't size for it: A change in building use or the incorporation of new technologies can lead to an increased demand for cooling. But, it is wasteful to provide excess capacity now as it may never be used or the equipment could be obsolete by the time it is needed. It is better to plan equipment and space so that future expansion is possible like adequate size of mechanical rooms and use of modular equipment.

Commission the HVAC systems: Commercial HVAC systems do not always work as expected. Problems can be caused by the design of the HVAC system or because equipment and controls are improperly connected or installed. A part of commissioning involves testing the HVAC systems under all aspects of operation, revealing and correcting problems, and ensuring that everything works as intended. A comprehensive commissioning program will also ensure that operation and maintenance personnel are properly trained in the functioning of all systems.

Establish an Operations and Maintenance (O&M) Program: Proper performance and energy-efficient operation of HVAC systems can only be ensured through a successful O&M program. The building design team should provide systems that will perform effectively at the level of maintenance that the owner is able to provide. In turn, the owner must understand that different components of the HVAC system will require different degrees of maintenance to perform properly.

13.2.1 Design recommendations

Consider all aspects of the building simultaneously. The building should incorporate as many features as possible that reduce heating and cooling loads, for example:

(a) In skin-load dominated structures, employ passive heating or cooling strategies (e.g., sun control and shading devices, thermal mass).

(b) In internal-load dominated structures, include glazing that has a high cooling index.

(c) Specify exterior wall constructions that avoid thermal bridging.

(d) Detail the exterior wall constructions with air retarder systems.

(e) Incorporate the highest R-value wall and roof construction that is cost-effective.
(f) Design efficient lighting systems.
(g) Use daylight dimming controls whenever possible.
(h) Specify efficient office equipment.
(i) Accept life-cycle horizons of 20 to 25 years for equipment and 50 to 75 years for walls and glazing.

Decide on design goals as early as possible. It is important that the design team knows where it is headed long before the construction documents phase. Emphasise communication between all members of the design team throughout the design process.

(a) Develop a written "Basis of Design" that conveys to all members of the project goals for energy efficiency. For example, such a design might highlight the intent to incorporate day lighting and the attendant use of high-performance glazing, suitable lighting controls and interior layout depending on the process.
(b) Establish a quantitative goal for annual energy consumption and costs.
(c) Clarify goals to meet or exceed the minimum requirements of codes or regulations during schematic design.

"Right Size" HVAC systems to ensure efficient operation:

(a) Accept the HVAC safety factors and pick-up load allowance stated in ASHRAE/IES 90.1-1989 as an upper limit.
(b) Apply safety factors to a reasonable baseline. It is unreasonable to assume only the extremes while designing. Safety factors should be applied to a baseline that was created using reasonable assumptions.
(c) Take advantage of the new generation of dependable computerized analysis tools, to reduce uncertainty and eliminate excess over sizing. Hour-by-hour computer simulations can anticipate how building design and operation affect peak loads. Issues such as diversity, pick-up requirement, and self-shading due to building geometry can be quantified. As uncertainties are reduced, over sizing factors can be reduced or applied to a more realistic baseline.

Consider part-load performance when selecting equipment. Select systems that can operate efficiently at part-load. For example: Variable volume fan systems and variable speed drive controls; variable capacity boiler plants (e.g., step-fired (hi/lo) boilers, modular boiler plants, modulating flame boilers); condensing boilers operate more efficiently (9-–96%) as the part-load decreases to the minimum turn-down ratio; variable capacity cooling plants (e.g., modular chiller plants, multiple compressor equipmen, and variable speed chillers); variable capacity cooling towers (e.g., multiple cell towers with variable speed or two speed fans, reset controls); variable capacity pump systems (e.g., primary/secondary pump loops, variable speed pump motors; an, temperature reset controls for hot water, chilled wate, and supply air.

Shift or shave the load. Investigate the utility company's rate structure; negotiate for a favourable rate structure and take advantage of the on-peak and off-peak rate differences. Use energy management controls systems to avoid unnecessary peak demand charges (peak shaving and demand limiting). Explore thermal storage systems (e.g., thermal ice storage). Examine alternate fuel sources for systems (e.g., district steam vs. natural gas vs. fuel oil; steam or natural gas chillers; dual fuel boilers).

Plan for rightgsizing. "Right sizing" means, avoiding systems, which have more capacity than currently required. This concept extends to accommodating for planned expansion and not providing excess capacity today. It includes providing the physical space required for additional equipment: boilers, chillers, pumps, cooling towers and designing distribution systems that can easily accept additional equipment, and can be expanded in future. The result is savings in first cost and operating cost.

Commission the HVAC systems. ASHRAE Guideline 1 recommends a comprehensive commissioning protocol for HVAC equipment. However, there is a demand for more general Total Building Commissioning (TBC) concepts.

Establish an operations and maintenance program.

(a) Specify systems that can be properly maintained by the owner, based on the owner's stated resources. See WBDG Functional/Operational.
(b) Provide as part of construction, contract system interfaces to allow personnel to easily monitor and adjust system parameters.
(c) Make systems control, operatio, and maintenance training part of the construction contract.
(d) Include complete documentation regarding operation and maintenance of all equipment and controls systems as part of the construction contract.
(e) Establish a written, comprehensive operation and maintenance program, based on the requirements of the facility, equipmen, and systems installed. See WBDG Sustainable O&M Practices.

13.2.2 Types of HVAC systems

13.2.2.1 Heating systems

1. *Boilers* are used to generate steam or hot water fired by natural gas, fuel oi, or coal. The following boilers have combustion efficiencies between 78% and 86%.
   - *Fire tube steel boilers* in which hot gases from the combustion chamber pass through tubes that are surrounded by water, typically do not exceed 25 million Btu/hr. (MMBtu/hr.), but capacities up to 70 MMBtu/hr. are available.
   - *Water tube steel boiler*, thah pass hot combustion gases over water-filled tubes, which range from small, low pressure units e.g., around 10 MMBtu/hr., to very large, high-pressure units with steam outputs of about 300 MMBtu/hr.
   - *Cast iron boilers* are used in small installations (0.35 to10 MMBtu/hr.) where long service life is important. Since these boilers are composed of pre cast sections, they can be more readily field-assembled than water tube or fire tube boilers. At similar capacities, cast-iron boilers are more expensive than fire tube or water tube boilers.
   - *Condensing boilers* achieve higher system efficiencies by extracting so much heat from the flue gases that the moisture in the gas condenses. The gases that remain can often be vented directly to the outside, simplifying and reducing the cost of breeching. They are typically fired with natural gas and operate between 95% and 96% combustion efficiencies. They also operate more efficiently than non-condensing boilers at part-load. Condensing boilers are available in capacities between 0.3 and 2 MMBtu/hr, and can be connected in modular installations.
2. *Furnaces* can be used for residential and small commercial heating systems. Furnaces use natural gas, fuel oi, and electricity for the heat source. Natural gas furnaces are available in condensing and non-condensing models. The cooling can be packaged within the system, or a cooling coil can be added. When direct expansion systems with coils are used, the condenser can be part of the package or remote.
3. *Heat pumps* are devices that add heat to or extract heat from a conditioned space. Both refrigerators and air conditioners are types of heat pumps that extract heat from a cooler, conditioned space and reject it to a warmer space (i.e., the outdoors). Heating can be

obtained if this cycle is reversed: heat is moved from the outdoors to the conditioned space indoors. Heat pumps are available in two major types: conventional packaged (air-source) and water-source (conventional or geothermal).

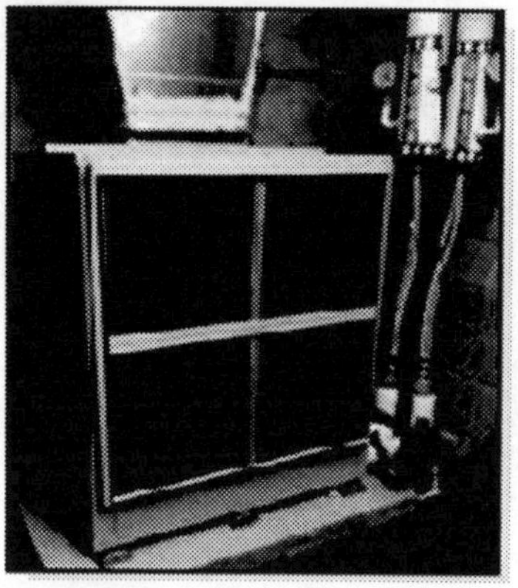

Figure 13.1 Geothermal heat pump unit

13.2.2.2 Heating controls

Four controls are explained here out of which the first three controls increase energy efficiency by reducing on/off cycling of boilers, whereas the fourth improves the efficiency during operation.

1. ***Modulating flame:*** The heat input to the boiler can be adjusted continually (modulated) up or down to match the heating load required. Modulating flame boilers have a minimum turn-down ratio, below which the boiler cycles off. This ratio is 25% for most boilers, but some can be turned down to as low as 10%.
2. ***Step-fired***: The heat input to the boiler changes in steps, usually high/low/off. Compared to steady-state units, the capacity of the boiler can come closer to the required heating load.
3. ***Modular boilers***: This is assembling of groups of smaller boilers into modular plants. As the heating load increases, a new boiler enters on-line, augmenting the capacity of the heating system in a gradual manner. As the heating load decreases, the boilers are taken off-line one by one.
4. ***Oxygen trim*** systems continuously adjust the amount of combustion air to achieve high combustion efficiency. They are usually cost-effective for large boilers that have modulating flame controls.

13.2.2.3 Ventilation systems

Ventilation systems deliver conditioned air to occupied spaces. Depending on the building type, ventilation air may be comprised of 100% outside air, such as in a laboratory building, or some mixture of re-circulated interior air and outside air. In commercial and institution buildings, there are a number of different types of systems for delivering this air.

1. ***Constant air volume (CAV)*** systems deliver a constant rate of air while varying the temperature of the supply air. If more than one zone is served by a CAV system, the supply air is cooled at a central location to meet the need of the zone with highest demand. The other zones get overcooled or, if comfort is to be maintained, the air is reheated at the terminal units. CAV systems with reheat are inefficient because they expend energy to cool air that will be heated again. CAV systems with reheat, however, provide superior comfort in any zone. Constant airflow reduces pockets of 'dead' air, and reheat provides close control of the space temperature.
2. ***Variable air volume (VAV)*** systems vary the amount of air supplied to a zone while holding the supply air temperature constant. This strategy saves fan energy and uses less reheat than in a CAV system. VAV systems, however, can have problems assuring uniform space temperature at low airflow rates. At times, the minimum airflow required for ventilation or for proper temperature control may be higher than is required to meet the space load. When this occurs reheat may be required.
3. ***Low-flow air diffusers*** in VAV systems help maintain uniform air distribution in a space at low airflows. These devices can be passive or active. Passive low flow diffusers are designed to mix the supply air with the room air efficiently at low flow. Active diffusers actually move the outlet vanes of the diffuser to maintain good mixing at low flow. Active diffusers can also be used as VAV terminal units.
4. ***Fan-powered VAV*** terminal units provide a method to improve air distribution at low load conditions. These units combine the benefits of a VAV system, by reducing central fan energy and reheat energy, with the benefits of a CAV system, by maintaining good airflow. There are two major types; series and parallel. Series fan-powered units maintain constant airflow to the zone at all times, whereas, parallel fan-powered units allow the airflow to the zone to vary somewhat, but do not allow the airflow in the zone to drop below a desired level. Both allow the central fan to throttle down to the minimum airflow required for ventilation.

5. ***Raised floor air distribution*** delivers air low in the space, at low velocity and relatively high temperature compared to traditional plenum mounted distribution systems. Delivering air through a series of adjustable floor-mounted registers permits room air to be stratified with lower temperatures in the bottom portion of the room where people are located and high temperatures towards the ceiling. This system type is attracting increasing interest because it has the potential to save energy and to provide a high degree of individual comfort control. These systems have historically used constant-volume air delivery. Now VAV systems that are more easily designed, installed, and operated with raised floor plenum systems are available.

13.2.2.4 Ventilation system controls

Ventilation control systems have become more complex now a days but when installed and maintained properly are more dependable. They have the advancements as explained below.

1. ***Direct digital control (DDC)*** systems using digital-logic controllers and electrically operated actuators are replacing traditional pneumatic controls. Pneumatic systems use analog-logic controllers and air-pressure actuators. DDC systems are repeatable and reliable, provide accurate system responses, and can be monitored from a central computer station. DDC systems also require less maintenance than pneumatic systems. However, pneumatic controllers can be less expensive than electric actuators. Hybrid systems use a combination of digital logic controllers and pneumatic actuators.
2. ***CAV systems*** with controls to reset the supply air temperature at the cooling coil provide the warmest air possible to the space with the highest cooling load. This reduces reheat throughout the system. However, the temperature should be no higher than is necessary to properly dehumidify the air. Another option to reduce reheat is to use a bypass system. Bypass systems work like variable volume systems at the zones, but have constant airflow across the central fan.
3. ***VAV systems*** can now be designed to serve areas with as little as six tons of cooling loads. Inlet vanes or, better yet, variable speed fans are used to control air volume. In systems that have supply and return fans, airflow monitoring stations are used to maintain the balance between supplies and return airflow.
4. ***CO_2-based control systems*** control the amount of outside air required for ventilation. These systems monitor the CO_2 in the return air

and modulate the outside air damper to provide only the amount of outside air required to maintain desired levels. There must be a minimum amount of outside air even when the spaces are unoccupied. Alternately, detectors of volatile organic compounds (VOC) can supplement the CO_2 monitoring system.

13.2.2.5 Chillers and condensers

1. Chillers. In large commercial and institutional buildings, devices used to produce cool water are called chillers. The water is pumped to air handling units to cool the air. They use either mechanical refrigeration processes or absorption processes.

Mechanical refrigeration chillers may have one or more compressors. These compressors can be powered by electric motors, fossil fuel engines or turbines. Refrigeration systems achieve variable capacity by bringing compressors on or off line, by unloading stages within the compressors, or by varying the speed of the compressor. There are different types of compressors. Reciprocating compressors are usually found in air-cooled direct expansion (DX) systems for residential and small commercial systems with capacities of 10 to 200 tons. Multiple compressors can be employed in a single system to match part load conditions. Scroll compressors are available in 1 to 15 ton range. Multiple compressors can be found in water chillers with capacities of 20 to 500 tons. Scroll compressors require less maintenance than reciprocating compressors. Rotary screw compressors are used in chillers with 70 to 500 tons capacity. Centrifugal compressors are used in chillers with capacities of 100 to 7,000 tons. Centrifugal chillers are the most efficient of the large-capacity chillers.

Absorption chillers are heat-operated devices that produce chilled water via an absorption cycle. Absorption chillers can be direct-fired, using natural gas or fuel oil, or indirect-fired. Indirect-fired units may use different sources for heat: hot water or steam from a boiler, steam from district heating, or waste heat in the form of water, air, or other gas. Absorption chillers can be single-effect or double-effect, where one or two vapour generators are used. Double-effect chillers use two generators sequentially to increase efficiency. Several manufacturers offer absorption chiller/heater units, which use the heat produced by firing to provide space heating and service hot water.

Evaporative coolers, also called swamp coolers, are packaged units that cool the air by humidifying it and then evaporating the moisture. The equipment is most effective in dry climates. It can significantly reduce the peak electric demand when compared to electric chillers.

Figure 13.2 Cooling tower

Typical full-load operating efficiencies for chillers are noted below:

- Small air-cooled electric chillers have 1.6-1.1 kW/ton (COP of 2.2 to 3.2).
- Large and medium-sized air-cooled electric chillers have 0.95-0.85 kW/ton (COP of 3.7 to 4.1).
- Similar water-cooled electric chillers have 0.8-0.7 kW/ton (COP of 4.4 to 5.0). Lower values such as 0.6-0.5 kW/ton chillers (COP of 5.9 to 7.0) may indicate energy efficient equipment, but part-load performance should also be examined.
- The COP of absorption units is in the range of 0.4-0.6 for single-effect chillers, and 0.8-1.05 for double-effect chillers.
- Engine-driven chillers attain COP of 1.2 to 2.0.

2. Condensers are heat exchangers that are required for chillers to reject heat that has been removed from the conditioned spaces. They can be either air-cooled or water-cooled. Water-cooled condensers often rely on rooftop cooling towers for rejecting heat into the environment; however, it is possible to reject the to the ground or river water.

- *Air-cooled condensers* are offered on smaller, packaged systems (typically from less than one ton to 120 tons). They are initially less costly than water-cooled condensers, but do not allow the chiller to operate as efficiently.

- *Water-cooled condensers* use water that is cooled directly from the evaporative condenser or indirectly via a cooling tower. The lower temperature achieved by evaporating water allows chillers to operate more efficiently.
- A *waterside economizer* consists of controls and a heat exchanger installed between the cooling tower water loop and the chilled water loop. When the outdoor air temperature is low and/or the air is very dry, the temperature of the cooling tower water may be low enough to directly cool the chilled water loop without use of the chiller, resulting in significant energy savings.

13.2.2.6 Air-conditioning equipment controls

1. Controls that significantly affect the energy efficiency of chillers include variable speed drives, multiple compressors and water temperature reset controls.
    (a) *Variable speed drives* achieve good part-load performance by matching the motor output to the chiller load, and by cycling off at a lower fraction of capacity than constant-speed chillers.
    (b) *Multiple compressors* achieve a closer match of the load than single-compressor chillers by sequencing the compressors as needed.
    (c) *Water temperature reset controls* raise the water temperature as the demand decreases, allowing for more efficient chiller operation.
2. Strategies that significantly affect the energy efficiency of cooling towers include the use of variable-speed or multiple-speed fans, wet-bulb reset strategies, where the temperature of the cooling water is adjusted according to the temperature and humidity of outside air (instead of maintaining it constant) and fans and pumps that use variable frequency drive (VFD) controls to reduce energy use at part-load
3. Integrated chiller plant controls use monitoring and computational strategies to yield the minimum combined energy cost for the chillers, cooling towers, fans and pumps. This approach can be significantly more effective (though more difficult to implement) than optimising the operation of each piece of equipment independently.

13.2.2.7 Heat recovery

Air is blown across copper coils to reject heat from this air-cooled condenser. This is an important component of many HVAC systems.

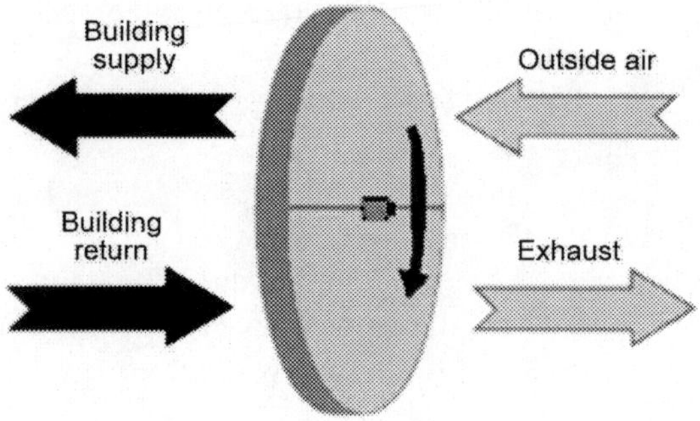

Figure 13.3 Enthalpy recovery wheel

Types of heat recovery include mainly air to air heat exchanger and desiccant wheels.

(a) *Air-to-air heat exchangers* transfer heat or coolness from one air stream to another. They are usually classified as one of the following:
- Plate heat exchangers, with 60%-75% efficiencies
- Glycol loop heat exchangers, with 50%-70% efficiencies (including pump energy use)
- Heat pipe heat exchangers, with efficiencies as high as 80%

(b) *Desiccant wheels* retrieve both sensible and latent heat, with efficiencies as high as 85%. Desiccant dehumidification of the air is achieved by inserting a rotating wheel in the air stream that needs to be dried. The desiccant extracts moisture from the air stream. The wheel then rotates, exposing the moist part to another air stream that dries (or regenerates) the desiccant material. Two typical methods of regeneration are Energy recovery wheels and Gas fired desiccant dehumidification.
- Energy (Enthalpy) recovery wheels are located in the outside intake and the exhaust air streams. The exhaust air regenerates the desiccant.
- Gas-fired desiccant dehumidification packages are located in the outside intake air stream or in the entire supply air stream. Outside air is heated by the gas furnace and is blown over the wheel to regenerate the desiccant.

(c) Other forms of heat exchange include:
- Indirect evaporative cooling (IDEC), which uses water-to-air heat exchange to pre cool air.

- Electric heat recovery chillers receive up to 50% of rejected heat, usually though split or multiple condensers.
- Absorption chiller/heaters can use a fraction (typically 50%) of the heat input for cooling and the rest for heating.
- Gas-fired, engine driven chillers retrieve much of the heat rejected (usually 20–50%).

13.2.2.8 Cogeneration

Cogeneration is a process in which electric power is generated at the facility where the waste heat is recovered to produce service hot water, process heat or absorption cooling. Currently, packaged cogeneration systems between about 60 to 600 kW are widely available. Extensive research and marketing efforts are underway for smaller systems (as low as 4 kW).

13.2.2.9 Fuel cells

Fuel cells use chemical processes to generate electricity. The heat generated by fuel cells can also be recovered, as in cogeneration. Currently, the minimum size for a fuel cell in building applications is 200 kW. Note that fuel cells need continuous, full-load operation.

13.2.4.10 Design and analysis tools

Building energy simulations allow the system designer to compare different HVAC systems and control strategies. These tools vary in their scope and level of complexity. Some tools analyse individual components of HVAC systems (e.g., motors) under simplified assumptions regarding the component use (e.g., annual hours of operation). Other tools simulate entire buildings, including energy gains/losses through the building envelope, energy gains from internal loads and energy used by the HVAC systems to maintain user-prescribed space conditions (e.g., temperature, humidity, ventilation rates). The latter tools require expertise and experience to obtain accurate results due to the detailed input required. Some building simulation packages have reduced input requirements. The trade-off is that these tools are typically not as accurate, since the programs use defaults or assumptions to replace the user inputs. However, simplified tools can be used early in the design process to investigate the influence of HVAC system selection on energy efficiency strategies such as day lighting

13.2.2.11 Under floor air distribution

It covers the systems which make use of the floor void, created by the access floor, directly as a plenum for the distribution of the conditioned air. They are classified in the category of Under Floor Air Distribution – UFAD. Liebert Hiross of Italy developed 2 modules Hiross C and Hiross V, which generate conditioned air to be delivered into the under floor supply plenum and draw air back through grilles or space into return under floor plenum thus reducing the need for a false ceiling. This is stated to be offering up to 10% reduction of floor-to-ceiling height and 5-7% of building construction costs. This is said to be suitable for both open plans and cellular offices space and in height-limited space.

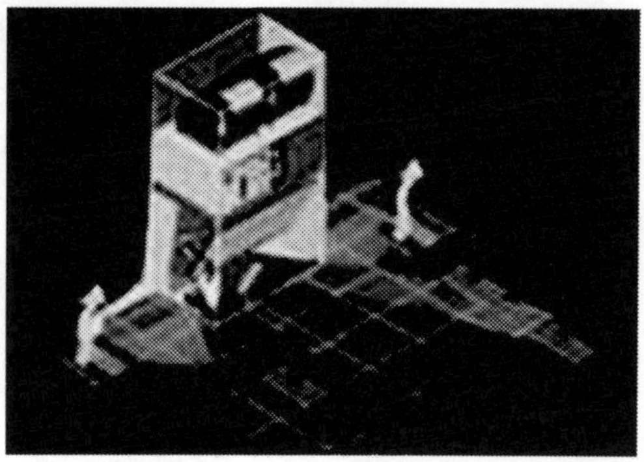

Figure 13.4 Hiross C – conditioned air module

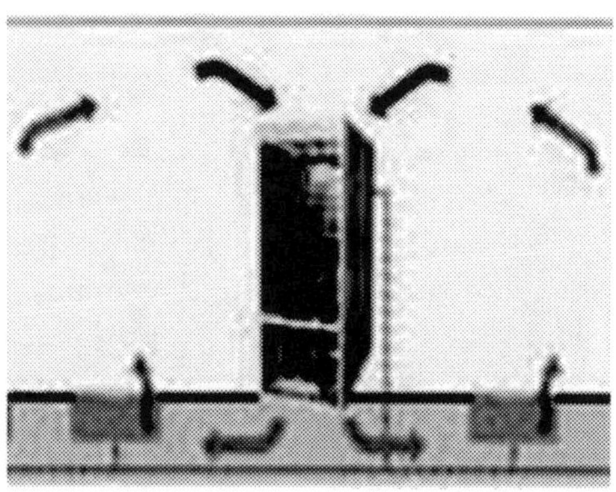

Figure 13. 5 Hiross V – variable air flow unit

Now softwares are developed for designing the ducts. DSD developed by Dynacomp, Inc. is menu-driven and interactively sizes air duct systems using the static-regain, constant-velocity, or equal-friction methods of ASHRAE, either independently or together.

DSD includes these features:

- Calculates pressure required at blower and the drop in each section and path.
- Handles round, rectangular and oval ducts with variable roughness; galvanised, fiberglass or flex duct.
- Allows mixed use of the three sizing methods: static regain for trunks and main branches and equal friction for run outs (for example).
- Altitude and humidity corrected.
- Calculates path and section drops as inputs are made; sizes of sections and settings of dampers may be reviewed and changed so as to balance the path drops during the design process.
- Prints out losses (in detail) for straight sections and for each fitting, and prints out a list of duct and fitting requirements.
- Handles exhaust as well as blowing applications.

14

Air pollution control in textile industry

14.1 Introduction

The textile industry is known for pollution problems that need to be resolved. The textile industry is not a single entity but encompasses a range of industrial units which use a variety of natural and synthetic fibres to produce various fabrics. The industry is very complex in nature as far as varieties of products, process and raw materials are concerned. During production, the cloth has to pass through various processes and chemical operations such as sizing, desizing, scouring, mercerizing, bleaching, dyeing, printing and finishing. A number of dyes and auxiliary chemicals are used to impart desired quality in the fabrics. The wastewater of the industry is highly alkaline in nature and have high BOD and chemical oxygen demand (COD) and contains high concentration of TDS and alkalinity. It can cause environmental problems unless it is properly treated before disposal. The terms bio chemical oxygen demand (BOD), COD and total dissolved solids (TDS) are defined as below.

- BOD = Bio Chemical Oxygen Demand. It is a measure of the amount of oxygen that bacteria will consume while decomposing organic matter under aerobic conditions. BOD is determined by incubating a sealed sample of water for five days and measuring the loss of oxygen from the beginning to the end of the test. Samples often must be diluted prior to incubation or the bacteria will deplete all of the oxygen in the bottle before the test is complete.
- COD = Chemical Oxygen Demand. It does not differentiate between biologically available and inert organic matter. It is a measure of the total quantity of oxygen required to oxidise all organic material into carbon dioxide and water. COD values are always greater than BOD values, but COD measurements can be made in a few hours while BOD measurements take five days.
- TDS = Total Dissolved Solids. It is determined by evaporating a known volume of water and weighing the residue. Waters with TDS are unpalatable and potentially unhealthy.

The industry also generates air pollution. Processing of fibres prior to and during spinning and weaving generates dust, lint etc., which degrades working environment in the industry. Dust may cause respiratory diseases in workers. A chronic lung disease, byssinosis is commonly observed among workers exposed to cotton, flax and hemp dust. Besides this, there are a number of process operations including spinning weaving that produce noise to the tune of 90 dB (A).

The industry specific standards for textile industries have been evolved and notified in India under the Environment (Protection) Act, 1986 which are to be compiled by the unit. The industries are being persuaded to install necessary pollution control measures. In order to assist the industries and regulatory agencies, a comprehensive industry documents is published by CPCB, i.e. Central Pollution Control Board.

CETP: For small industries operating in clusters, the scheme 'Common effluent treatment plant' (CETP) is being enforced. Such scheme is already operational in number of industrial areas. Under the scheme, grant up to 50% of the cost of the CETP is provided by State and Central Government.

Eco Mark Scheme: The Central Pollution Control Board (CPCB) has developed criteria for eco-labelling of textile materials and the same is being executed by BIS.

14.2 Source of air pollution

Smoke, odour and dust are the main pollutants responsible for air pollution. The major air pollution problem occurs during the finishing stages, where various processes are employed for coating the fabrics. Coating materials include lubricating oils, plasticisers, paints and water-repellent chemicals; essentially, organic (usually hydrocarbon) compounds such as oils, waxes or solvents. After the coatings are applied, the coated fabrics are cured by heating in ovens, dryers, tenter frames, etc. This results in the vaporisation of the organic compounds into high molecular weight volatile organic (usually hydrocarbon) compounds (VOCs), which take the form of visible smoke and invisible but objectionable odour. In terms of actual emissions, the industry must also deal with larger particles, principally lint.

Smoke is basically made up of tiny solid or liquid particles of VOCs less than one micron in size that are suspended in the gaseous discharge. When the smoke/gas mix goes up the stack, the Environmental Protection Agency (EPA) measures the density or opacity of the emission using an official transparent template, and labels the opacity as ranging from 0 to 100%. The more opaque the smoke, the more visible it is. Opacity of smoke is actually related to the quantity, rather than the weight, of particles present in the gas. For a given particle size and stack diameter, percentage of opacity is a function of the grain loading or weight of particles, measured as grains per cubic foot (1 lb. = 7,000 grains).

The problem with odour is that it isn't practically measurable. It is a sensation originated by interaction between molecules of hydrocarbons and nerve fibres in the human olfactory membranes. The human nose can differentiate between 4,000 different odours. The odours associated with textile plant emissions are usually caused by hydrocarbons with molecular weight less than 200 and fewer than 15 carbon molecules. These odorous molecules attach themselves to the particles of smoke and can be carried great distances from their point of origin. When air pollution equipment has abated the smoke, the odour molecules have no vehicle to carry them. That is why the installation of smoke abatement equipment sometimes solves the odour problem, too. Now relatively inexpensive equipments are also available to address the problem directly.

14.3 Air pollution control

Selection of an air pollution control (APC) system should involve several orderly steps including analysis of chemicals involved, evaluation of existing exhaust system, consideration of exhaust stream pre cooling, consideration, comparison and selection of optimum APC equipment. Analysis of chemicals helps in deciding the VOCs to be abated.

14.3.1 Analysis of existing exhaust system

It is important to evaluate the mechanical parameters of the existing exhaust system to be sure it will properly complement the new APC equipment. These are some major considerations:

1. In order to keep smoke out of the working area, pressure inside the heating equipment should be negative. Investigate all openings in the body of the equipment and eliminate those that are not necessary for the process.
2. Exhaust gas volume should be adequate to absorb moisture evaporated from the textiles and also to bring the total VOC content below 25% of lower explosive limit.
3. Exhaust system ductwork from heating equipment to cooling section of the APC requires special attention. The ductwork should be of solid steel construction, welded and tested for gas and liquid tightness. The welded seams should be located at the top of the duct.

4. It is important to prevent condensation of VOCs in the exhaust system and subsequent dripping of condensate back into the oven. Velocity in the duct should be sufficient to keep the droplets moving. The gas temperature should remain high until the gas enters the cooling section of the APC system and the ductwork should be kept as short as possible. It is suggested to insulate the entire run of exhaust system ductwork.
5. If the heating equipment has several gas-collection zones with separate exhaust blowers, ductwork design should incorporate a manifold and provisions for pressure adjustment to ensure adequate exhaust from each zone. A separate 'zero pressure' point, located between discharge of the zone blower and the manifold, will prevent the central blower from influencing exhaust rates from individual zones.

14.3.2 Consideration of exhaust stream pre-cooling

For the APC system to capture VOCs, the vapours are to be condensed; therefore, cooling the air stream is the first step. The efficiency of the cooling/condensation process depends on the boiling temperatures of the specific VOCs, their concentration in the exhaust gas and operating temperature of the APC device. The operating temperature of the APC cooling section should be selected to ensure effective condensation while minimising capital and operating costs of the cooling system. In textile applications, most VOCs condense at the saturation temperature of the exhaust stream, which usually ranges between 110°F and 160°F. Sometimes, as in a textile printing operation with paint based on a low-molecular-weight VOC, it is necessary to reduce gas temperature to 100°F or below to achieve reasonable efficiency of condensation.

14.3.3 Cooling equipment

Cooling can be accomplished indirectly with fin and tube heat exchangers, or directly by injecting cooling liquid in the pre scrubbing section of an APC system. The appropriate cooling design should be selected on an individual basis, taking into consideration the composition of the exhaust, amount of solids and the plant's operating experience with similar systems. Tubular heat exchangers are rarely used because of the difficulty of cleaning if tubes become clogged. An evaporative cooling system provides effective cooling with minimum maintenance, simply by spraying cooling water into the exhaust gas. This method, however, produces water droplets contaminated by lint and oil, consequently adding to total inlet loading of the system.

14.3.4 Fibre filter

A fibre filter typically consists of glass or polyester fibres from 5 to 10 microns in diameter, packed between flat or cylindrical screens. Particles of condensed VOCs are collected on the surface of the fibres by inertial impaction, direct interception and Brownian movement or diffusion. Removal efficiency of fibre filter rises with density and thickness of the fibre bed, but the pressure drop through the bed and probability of plugging also rise. The fibre filter, or fibre bed mist eliminator, is seldom recommended for textile industry applications, because the high flammability of VOCs and dry lint in the gas stream usually precludes the use of dry filtration devices. There are other disadvantages as well. Fibre filter usage is limited to rare applications where the product of VOC condensation is free-flowing oil, and no waxes or lint are present.

14.3.5 Wet scrubber

Like fibre filters, wet scrubber utilise impaction, interception and diffusion as collection techniques. The important difference is that the filtering role is performed by a scrubbing liquid rather than fibres, removing the potential for fire or explosion. The scrubbing liquid is usually water, which is sprayed as droplets which become targets for precipitated particles and absorbed gases. The droplets are larger than the particles, and become heavier and heavier as they collect impinged particles, so they are relatively easy to remove. In textile applications, simple wet scrubbers are successfully used to remove lint and other solids above about 5 microns in size, but not for VOC smoke abatement.

14.3.6 High-energy venturi scrubber

Condensed VOC droplets are typically smaller than one micron. To collect them from the exhaust gas stream, the optimum target droplets must be in the range of 10 to 15 microns in size; much smaller than the drops generated in the simple wet scrubber. To generate these tiny droplets in the quantities required for efficient scrubbing, a substantial amount of electric energy is required. The most common wet scrubber for sub-micronic particle removal is the high-energy venturi scrubber (HEVS). The HVES is essentially a gas-atomising device relying on sheering and impaction forces to break water into an evenly distributed cloud of very fine droplets at high density. The density raises the probability of collisions between sub micronic droplets of condensed VOCs and the larger (but still tiny) droplets of recycling liquid. The droplets are produced in the venturi throat of the scrubber, where velocity of the moving gas can be as high as 400 feet per second. This is about eight times higher than the velocity of the exhaust gas in the duct, resulting in high pressure losses and high-energy requirements for operating the system.

A venturi scrubber is designed to effectively use the energy from the inlet gas stream to atomise the liquid being used to scrub the gas stream. A venturi scrubber consists of three sections: a converging section, a throat section and a diverging section. The inlet gas stream enters the converging section and, as the area decreases, gas velocity increases. Liquid is introduced either at the throat or at the entrance to the converging section. The inlet gas, forced to move at extremely high velocities in the small throat section, shears the liquid from its walls, producing an enormous number of very tiny droplets. Particle and gas removal occur in the diverging section as the inlet gas stream mixes with the fog of tiny liquid droplets. The inlet stream then exits through the diverging section, where it is forced to slow down.

Venturis can be used to collect both particulate and gaseous pollutants, but they are more effective in removing particles than gaseous pollutants.

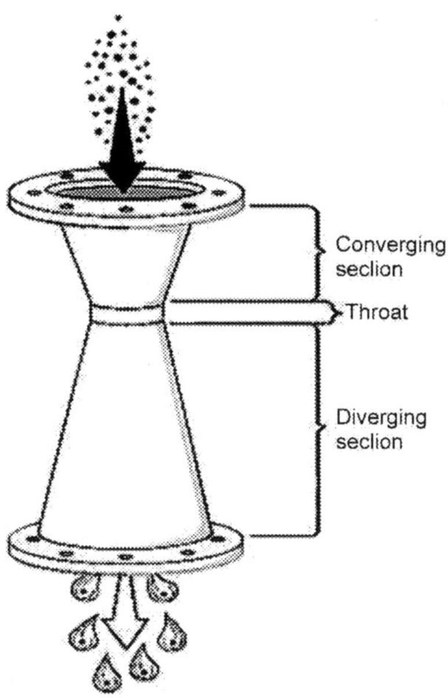

Figure 14.1 Venturi scrubber

14.3.7 Packed bed scrubber

Wet scrubbers are more successful in applications where the basic problems are odour and relatively large particles of lint and other solids, rather than smoke. To provide odour removal in textile operations, basic, acidic or oxidising agents are added to the recycling liquid. Successful removal agents include potassium permanganate, sodium hypochlorite, chlorine dioxide, ozone and dichromate. In order to provide intimate contact between scrubbing liquid and odour producing gas, packed bed scrubbers are employed. The packed bed utilises specially shaped plastic or ceramic elements which are highly permeable to gas flow. However, the conventional packed bed presents its own problem. Solid particles that are not soluble in water like lint will clog conventional packing. For textile odour remediation, two specialised scrubbers can be used:

- In a multi-rod deck scrubber, gas and scrubbing liquid compete for space between solid tubular structures. The result is highly turbulent intermixing, with little chance of clogging.
- In a multi-channel scrubber, channels are continually created by interaction of gas, scrubbing liquid and the packing of solid glass or ceramic spheres (marbles). The scrubbing action removes not only absorbed gas, but also solid particles trapped in the scrubbing liquid; the rotating marbles are highly resistant to clogging.

In these scrubbers, efficiency of removal of particles above 2 microns is better than 90%, and substantial odour reduction is accomplished through the mass transfer capability of the rods or marbles flooded with scrubbing liquid.

14 3.8 Optimised scrubber/quencher

Where the problem is basically odour-producing vapours and larger solid particles like lint, a preferred treatment is a low-pressure-drop scrubbing system combined with a quencher for pre-cooling the gas stream, and arrangements for separating out pollutants from the scrubbing liquid. One such system utilises a venturi-rod design with open type spray nozzle. In a typical situation, gas at 300°F to 400°F enters the scrubber/quencher where it's cooled to full saturation temperature while particles of lint are trapped in the scrubbing liquid. VOCs condense in the quencher to become sub micronic droplets of oil. These droplets, together with larger droplets of scrubbing liquid mixed with solid particles, must be separated from the gas stream.

14.3.9 Selecting a separator

Separation is required for all particle collection systems, whether designed for sub-micronic or larger particles, or both. In selecting the separation method for a textile plant APC system, gravity-based settling chambers are immediately ruled out, as they are seldom effective with particles under 50 microns, and almost all the polluting particles emitted by textile finishing processes fall into this range. Cyclonic separators are suitable for larger particles and scrubbing liquid. However, the inlet velocity required for separating particles under 5 microns often results in excessive energy consumption and wear on the equipment. Since tiny particles usually respond well to electrical forces, electrostatic precipitation is particularly effective in removing sub-micronic particles. For instance, electrical force exerted on an electrically charged particle of 0.1 micron can be more than a million times the force of gravity on that particle. So where the hydrocarbon aerosol droplets of VOCs are concerned, electrostatic separation often follows the scrubber stage in an APC system for textile plant exhaust gas. Electrostatic precipitators (ESPs) offer high capture efficiency of sub-micronic particulate by using electrical forces to remove suspended particles from the gas stream. Because an electrostatic precipitator (ESP) uses electricity only to charge particles in a passive air stream, it requires much less energy than a Venturi scrubber. Three steps are involved in electrostatic precipitation: charging, collection and removal. In all cases, a corona discharge electrically charges the suspended particles, which are collected on electrodes. Removal strategies vary.

14.3.9.1 Dry electrostatic precipitators

In a dry precipitator, the electrodes are periodically washed off to remove accumulated particles for disposal. Despite that washing, dry ESPs are prone to particle buildup on the walls of the collection chamber. The buildup acts as insulation, reducing the overall efficiency of the equipment. Dry ESPs can be used alone for removal of free flowing particulates, but must be combined with scrubbers or similar equipment to capture VOCs.

14.3.9.2 Wet electrostatic precipitators

In a wet electrostatic precipitator (wet ESP) a liquid, usually water, continuously washes particle buildup off the collection surface. The wet ESP is widely used for applications where the gas to be treated is hot, has a high moisture content and contains sticky particles or sub-micronic particles that can't be treated by other methods, which is typical textile industry condition. In most cases the gas is pretreated in a scrubber. The wet ESP follows to complete particulate removal with a high level of efficiency. In a conventional 'up-flow' wet ESP, the gas stream flows co-current to the water. This requires frequent cleaning because bottom water sprays are not completely effective in reaching the top of the chamber where sub-micronic particles accumulate. The 'down flow' wet ESP overcomes these problems, but does require mist elimination following treatment.

Proctor & Gamble has created a self-cleaning electrostatic air filtration system that rivals a HEPA filter in efficiency. It has substantially lower power requirements, produces only minimal back pressure, and can be used for applications where HEPA filters cannot. This air filtration technology has gone through two proof-of-concept stages. In the first, a single nozzle was shown to remove as much as 99% of the particulate from the air stream. In the second, an array of nozzles was tested, which allowed P&G researchers to devise a method whereby a multi-nozzle system uses the same voltage as does a single nozzle system. This electro-statically charged spray technology rivals HEPA filters in its ability to remove particulates from the air, but uses substantially less power to move the air through the filtering droplets, and produces only minimal back pressure on the air stream. One advantage of using a charged non-aqueous liquid as the filtering medium is that the liquid is continuously cleaned to maintain efficiency. While designed for home use, the technology can scale up to large Heating, Ventilation, Air Conditioning and Refrigeration (HVAC) systems or down to personal respirators, and encompasses both the hardware and the specialised fluid that it uses.

14.9.3 Water cooled jacketed wet ESP for sub-micronic sizes

This is a modification on the wet ESP design that is well suited for removal of sub-micronic particles. The jacketed, water chilled vertical-tube wet ESP increases removal efficiency and reduces operating costs and maintenance, by harnessing temperature differential to create particle movement perpendicular to the direction of the gas flow. The design combines a quencher to cool the gas stream with an up-flow, vertical tube wet ESP. It integrates variables of velocity, particle size and temperature to maximise removal of sub-micronic particles. The hot, contaminated inlet stream is cooled from 300°F to 140°F in a quencher. Then a low-energy scrubber removes particles larger than 2 microns. The pretreated gas stream passes into a multitude of tubes containing ionizing electrodes. Each electrode has a number of sharp points which give off the corona discharge that mobilises the particles. When the particles are charged to the point of saturation, they migrate toward the walls of the collecting tubes. Within the double-walled, water cooled collection chamber several processes take place which greatly increase the effectiveness of the device.

1. *Thermophoresis* (heat transmission): Inside the collection chamber, cold water is circulated in a closed-loop system. When gas temperatures on opposite sides of a particle are different, the gas molecules on the warmer side have a higher kinetic energy than the molecules on the cooler side. The hot, active gas molecules will strike the surface of the particle with more force propelling the particle toward the cooler temperature zone. This process is Thermophoresis. Particle velocity depends on the temperature gradient, the relative thermal conductivities of the gas and the particle, and the density and viscosity of the gas. Thermal forces for particles of 0.1 micron can be 100 times greater than those for particles of one micron.
2. *Diffusiophoresis* (diffusion transmission): Diffusiophoresis occurs when the partial vapour pressure of water in a carrier gas is greater than the water vapour pressure at the surface of the water droplets in a scrubbing liquid. In the Condensing WESP, water vapour will condense onto the water droplets, creating a bulk motion of gas and entrained particles toward the droplets. Because the droplets are larger than the particles, they are more easily removed by either electrical or mechanical means. Diffusiophoresis can improve the efficiency of collecting fine particles, when compared to a conventional wet ESP.
3. *Condensing action:* In the Croll-Reynolds brand of water cooled jacketed wet ESP Condensing WESP, charged particles, both liquid droplets and solids, are subject to electrical forces in addition to thermophoresis and diffusiophoresis. They migrate towards the positive (grounded) wall of the collection tubes. The interior of the tubes is cooled by circulating water, and the cool tube surface condenses the water vapour. This process deposits a wet film on the collection surface which tends to flush particles down the wall and out of the system. The wet film also discourages sticking of oils and waxes, and reduces the need for system shutdown for cleaning. By condensing water vapour from the contaminated gas, the Condensing WESP design can cut external water needs by thirty percent compared with conventional wet ESPs. The continuous presence of condensate also reduces corrosion.

Corona Power in Condensing WESP: Corona power of an electrostatic precipitator is the product of the operating voltage and the operating corona current. The efficiency of a wet ESP is directly proportional to its corona power. In operation, corona power is limited by internal sparking between ionizing and collecting electrodes. Since efficiency is directly proportional to the electrical power conveyed to the moving gas, each time a precipitator sparks, the voltage and consequently the particulate collection efficiency is reduced. For a given equipment design, gas flow condition and particle concentration, sparking rate is a function of gas-to-liquid ratio of entrained liquids, quality of the wet ESP components and the appropriate selection of high-voltage power supply components to match wet ESP geometry and the physical and chemical nature of the suspended particles. Consequently, the power level and efficiency of a wet ESP can be increased by proper design of its automatic voltage control system. The industry in general seems satisfied with 40% conduction, i.e. the ability to deliver 40% of available power to the gas. However, with proper selection of the transformer rectifier, automatic voltage control and automatic current-limiting power systems can be designed to deliver over 86% conduction, the maximum theoretically possible.

14.4 Use of UV disinfestations

Ultraviolet light is part of the spectrum of electromagnetic energy generated by the sun. The full spectrum includes, in order of increasing energy, radio waves, infrared, visible light, ultraviolet, x-rays, gamma rays and cosmic rays. Most sources of light generate some UV. For air disinfection, UV is generated by electric lamps that resemble ordinary fluorescent lamps. Germicidal UV is of a specific type (253.7nm wavelength) known to kill airborne germs that transmit infections from person to person within work area. This is aimed at the upper room air so that only airborne microbes are directly exposed. Room occupants are exposed only to low levels of reflected UV − levels below that known to cause eye irritation. UV does not prevent transmission of infections (e.g. colds) by direct person to person contact.

UV exposure can be harmful, or harmless, depending on the type of UV, the type of exposure, the duration of exposure and individual differences in response to UV. There are three types of UV:

(a) *UV-C* – Also known as 'shortwave' UV, includes germicidal (253.7 nm wavelength) UV used for air disinfection. Unintentional overexposure causes transient redness and eye irritation, but does not cause skin cancer or cataracts.

(b) *UV-B* – A small, but dangerous part of sunlight. Most solar UV-B is absorbed by the diminishing atmospheric ozone layer. Prolonged exposure is responsible for some type of skin cancer, skin aging and cataracts (clouding of the lens of the eye).

(c) *UV-A* – Long-wave UV, also known as 'black-light', the major type of UV in sunlight, responsible for skin tanning, generally not harmful.

UV air cleaners and ultraviolet water purifiers utilise germicidal UV (UVC) light to enhance the quality of life and well-being through creating a better and healthier indoor environment. The UV disinfection systems literally sterilise microorganisms. UVC reduces or eliminates germs such as mold, viruses, bacteria, fungi and mold spores from the indoor air of homes, offices and commercial buildings, ensuring a higher indoor air quality.

Studies done by North Carolina Department of Environment, Health and Natural Resources on the air washer portion of a textile air handing system indicated that UV light, which destroys all microorganisms (algae, bacteria – including Legionnaires Disease Bacteria, fungi, molds, virus particles), adds nothing to the water, does not alter the pH or cause scaling or other deposits, and requires no special handling or processing. The UV unit can be utilised on both cooling tower and air washers without extensive installation. Further as there is reduction in operation and maintenance cost, no discharge of toxic chemicals, and no extensive maintenance to keep the UV unit operating efficiently, a short payback period and substantial savings are possible.

14.5 An integrated approach for the textile industry

The best way to reduce pollution is to prevent it in the first place. Some companies have creatively implemented pollution prevention techniques that improve efficiency and increase profits while at the same time minimising environmental impacts. This can be done in many ways such as reducing material inputs, re-engineering processes to reuse by-products, improving management practices and employing substitution of toxic chemicals.

Croll Reynolds Air Pollution Control Systems Westfield developed Condenser WESP for controlling the pollutions, specifically for textile industry. The powerful combination of liquid scrubbing, thermophoresis and diffusiophoresis effects and high-voltage electrostatic precipitation in a condensing water chilled jacketed wet ESP is an effective method of submicron particle removal. In the textile industry, a Condensing WESP can help obtaining zero opacity, satisfying state and federal regulations for textile exhaust. The wet ESP is also an excellent candidate for operation as part of an energy-efficient, low-maintenance integrated system. By combining a Condensing WESP with a relatively inexpensive, low-energy scrubber which acts as the air distribution device and odour abatement section, large and small particles are effectively removed from textile-plant exhaust.

Table 14. 1 Selection of control technology for textile industry applications.

Process conditions requirements	Wet ESP	Scrubber	Dry ESP	Fiber Filter
Captures particles under one micron	yes	yes	yes	yes
Handles moist gas	yes	yes	no	no
High efficiency for submicron particles with low-energy consumption	yes	no	yes	no
Particles collected in liquid	yes	yes	no	no
Recommended for flammable chemicals	yes	yes	no	no

What to look for in an advanced wet ESP system for textile plant APC:

- Wide-spaced electrodes or large-diameter tubes allowing operation at 40,000 volts or higher.
- Rigid ionising electrodes that ensure system maintainability and increase longevity.
- Continuous cleaning action by liquid film on the walls of collecting electrodes to prevent accumulation of stick oils and waxes, ensuring high removal efficiency with minimum maintenance.
- High-voltage system with solid-state controls capable of delivering maximum available electric power to the exhaust gas.
- Pre-scrubbing section located in the bottom section of the wet ESP to minimise physical size and capital expense for the APC system.

Appendix – 1

Some of the commercial humidification plants

The principles of various systems are explained in previous chapters; however, understanding shall be difficult without seeing the plants in operation and working with them. A number of manufacturers have given various equipment and one needs to understand and select what is suitable to him. Here an attempt is made to introduce some of the different models working and the claims made by the manufacturers about their equipment. It is practically impossible to cover all the humidification units available on earth.

1 Samarth air washers – Ahmedabad, India

Tube axial flow fan

Eliminators

Spray bank

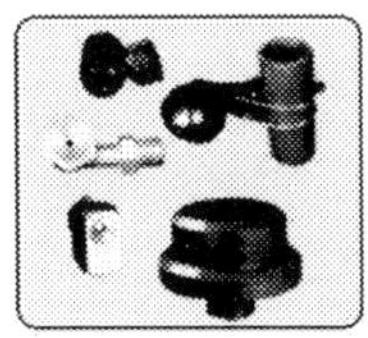

Nozzles

Axial flow fanshave air flow capacity varying from 3,000 CFM to 90,000 CFM and static pressure up to 60 mm WG. They are high-efficiency fans with aerofoil shape, and die cast aluminium adjustable angle blades, suitable for direct drive or 'V' belt drive with standard electric motor. Air washer plants are supplied with different types of eliminators like 'W' shape, 'C' shape, Luwa type, 5 bend 6 pass PVC eliminators with corrosion resistance frame and fixing items in different size and shape to suit any plant with excellent water retention and minimum pressure drop. Air washer can have single, double or triple bank spray system to suit required saturation efficiency. Nozzles are of Non-Clog type. High-efficiency clamp with PVC/polycarbonate body and SS dome and clamps. Stand pipes are of PVC/GI with GI main header pipes and butterfly valves with pot strainer. All air washers piping are in PVC materials, which are non-corrosive and have long life. Samarth manufactures Luwa type PVC clamp on nozzles with SS clamp, all type brass nozzles (Atomiser) suitable in air washer plant. PVC / AL / GL / Polycarbonate material with louvers type or 1″ dia. holes are offered for varied requirement and performance. Corrosion resistant frame and service door is provided as per requirement.

2 Cool Air Mechanical – Atlanta, USA

2.1 Coolair humidifiers

Coolair supply and fit an extensive range of off-the-shelf and made to measure humidifiers and dehumidifiers suitable for places and to suit different outside conditions. Refrigeration is designed by understanding requirements of coldroom and industry's close-control environment.

Coolair humidifier

Air cooling unit

Custom built package air washer units in various models of capacities from 3,000 cfm to 40,000 cfm are available. The design incorporates important components like air filter section and air washer system including of water pump and water elimination arrangements in both model, i.e., spray bank type and cell type. The unit requires no civil works and provides facility of being installed within very short time.

Refrigeration

3 Chang Shu Fan Co Ltd [Rtex Marketing Consultancy Service]– China

Chang Shu Fan Co. Ltd, located in Jiang Su province, is specialised in textile fan. The company has been focusing on exploring and developing air treating machinery and has produced and supplied energy saving humidification and dust filter, including turnkey project of systems. The products cover up to 300 kinds of axial fan, centrifugal fan, exhausting fan, mist spring fan, textile supplementary fan and energy saving and dust removing unit.

Company has made innovations and applied in the textile industry widely and made achievements like ASFU Energy Saving Humidification System jointly developed with Dong Hua University (former China Textile University), Design Institute of Textile Ministryand Beijing No. 2 Cotton Mill. This system is praised as a vital innovative in the fields of textile humidification and considered to be a replacement of older ones due to 30% energy saving annually as well as reaching international standards.

4 AMCO – American Moistening Company – Pineville, NC,USA

4.1 Compressed air humidification

The compressed air humidification by AMCO is claimed to control plant humidity to within ±2%RH. Self-cleaning atomisers eliminate clogging, Gravity-fed system prevents leaking or dripping. The unit uses tap water and compressed air and produces 'fog like' mist which is easily absorbed in atmosphere, easy to install, economical to operate with low maintenance costs.

Operations: The humidity control cabinet, in conjunction with the remote mounted RH sensor, continually monitors plant RH. The system is automatically activated when the RH falls below targeted levels. The controller activates the air control unit, allowing 28 psi of air into the system. This mixes with gravity-fed water in the atomiser chamber and produces a very fine mist which is absorbed in the plant's atmosphere. When desired RH levels are reached, the system is automatically shut off.

Engineering and Installation: All systems are individually engineered and designed by considering the varied needs. A survey of heat loads, square and cubic footage, the manufacturing process, etc., determines the customer's individual humidification needs. The system is designed and manufactured to accurately meet the determined requirements. Field trained technicians install the systems and familiarise the customers' personnel with the various aspects and components. Installation supervisors provide assistance for customer-installed system. Field service technicians and in-house staff continue the after-sale support.

System components

- *Automatic drain valve:* On each cycling of the system allows any build-up of water or oil in the air line to be automatically vented into a plant's atmosphere or drain.
- *Air connection valve:* Rugged, solid brass air connector connects atomiser to clamp tee on air-line. Permits atomiser shut-off while system is in operation.
- *Water connection:* 8" x ¼" plastic tube assembly connects atomiser to the clamp teeon water line.
- *Silver mist atomiser:* Self-cleaning, no clog atomiser is made of rugged brass construction for long maintenance-free operation. Produces fine mist spray – quickly absorbed in plant's atmosphere.
- *Clamp tee:* Heavy-duty zinc-plated steel construction with ¼" brass connection and neoprene washer used to secure air and water connections to galvanize or PVC pipe.
- *Double ring pipe hanger and stay clamp:* Double ring pipe hanger used to hang system from ceiling, keeping water and air lines in proper distance from each other. Stay clamp adds rigidity and stability to system.
- *Air control unit:* Controls air to system has rugged construction and quick air relief for fast and accurate response. Solenoid receives electrical signal from control cabinet. Complete with air valve, regulator, filter, gauge and solenoid valve.
- *Fibre glass Water Supply Tank:* Automatically controls water to the proper level.

- *Humidity Controller:* Digital controller senses humidity to within ±2%. Automatically controls air control valve based upon humidity set point.
- *Humidity Sensor:* Rugged, yet extremely accurate. Humidistat senses humidity in the dirtiest environments.

4.2 Large space humidifiers

Covers up to 6,000 square feet for most applications. It is recommended for small areas. The unit is self-contained with self-cleaning atomisers, humidistat, fibreglass water tank and atomisers which are easy to install. The atomisers are of self-cleaning type and will not clog. It may be suspended from ceiling by use of 4 lengths of ¼" or 3/8" rod, placed on a flat surface or mounted on a column.

- 12, 24, 120, 230V electrical system
- 24, 36, 72 lbs of water per hour models available
- Pressure regulator valve included ½" water supply line to tank, ½" overflow drain
- *Air supply:* Minimum ¼ air supply line connected to the ¼ air regulator under the tank. The air is then reduced to 30 psi by adjusting the regulator knob on the front of the panel. The atomisers are designed to operate with 30 psi of air and 14.4 cfm of air.
- *Water supply:* A ½ pipe coupling is mounted on the right side of the water tank for water supply.
- *Drain:* A ½ pipe coupling is mounted in the bottom of the tank for connection of a drain pipe. This is an overflow safety drain.
- *Electrical supply:* 5' supply cord is provided with the unit.
- *Operation:* Set humidistat to desired setting. When the unit is on, the indicator light under the humidistat is on. The water level is properly maintained by a float valve in the tank.
- *Dimensions:* 20.5 inches wide x 15 inches high x 10.5 inches deep

5 Paasche airbrush company– Chicago, USA

5.1 Automatic humidifier units F810-2 andF830-4 – (2 or 4 Spray Heads)

These units provide an efficient method of maintaining uniform moisture in industry, at the same time helping to reduce possible fire hazards. The water is finely atomised and quickly absorbed into the room atmosphere. Each automatic sprayatomizer requires ½ CFM at 50 PSI air pressure for the finest uniform atomisation. Paasche humidifier units present an inexpensive means of reducing the danger of spontaneous combustion. Proper humidification limits dust diffusion and grounds static electricity.

Humidifier uses:

- *Printing and bindery plants:* They are of great value for keeping paper quality at its best to assure perfect registration when printing.
- *Tobacco Industries:* They require uniform RH control when aging and curing.
- *Textile mills, clothing factories and warehouses:* Fabrics tend to gradually deteriorate in dry atmosphere. Goods retain their original luster and sheen when RH is maintained. Dangers from lint and dust accumulations are greatly minimized with proper humidity.
- *Carton manufacturers:* Winter conditions can dry out materials used to make boxes, causing tears and improper folding during manufacture or damage while in use.
- *Dry and dusty factories:* Factories thatare dry and dusty due to insufficient humidity are constantly subject to the dangers of spontaneous combustion. One method of reducing this hazard is with a uniform, accurate and dependable automatic humidifier unit.

Humidifier units:

- F810-2 Automatic Humidifier (Two Spray Heads) Capacity – 30,000 Cubic Feet, maximum output 30 Gallons (227 l) in 24 hours.
- F830-4 Automatic Humidifier (Four Spray Heads) Capacity – 60,000 Cubic Feet, maximum output 60 Gallons (454 l) in 24 hours.

5.2 Paasche F800-Series automatic humidifier

The F800-Series Automatic Humidifier Unitsautomatically dispenses gallons of finely atomised water, providing controlled humidity when the dial on the humidistat is adjusted to the humidity required. The unit will automatically start and stop at the humidity setting. When properly connected to the water and electric systems of the building and to an air source that supplies constant air pressure (minimum 50 PSI) to the F800 series unit,these units will deliver finely atomized water for extended periods.

Humidifier installation: Mount humidifier unit in location using SM-42 mounting bracket. Connect air supply to HF-1/4" air inlet valve. Connect water supply to HF-1/4" water inlet valve. Open air inlet valve and water supply valve under air supply valveand check for leaks. Tighten connections. Turn humidistat control dial on item #1 counterclockwise to off position before plugging cord into grounded outlet. Turn U-3178 fluid adjusting knob on back of spray guns to closed position by turning clockwise – Do not over tighten. Set '0' on reference dial to line up with '0' on fluid adjusting knob and then open adjusting knob to number 30. Set regulator item #7, located behind gauge #2, at 50 PSI minimum air pressure and 60 PSI maximum. Plug units three wire electrical plug into properly grounded receptacle (115V, 60cycles). Turn humidistat control-dial to 'on' position and desired humidity. A 'click' sound is heard when activated. Water flow rate may be adjusted using the U-3178 Fluid adjusting knob on back of each spray gun. Mist should always be absorbed into the surrounding air without settling out onto the floor under the unit.

Do not aim directly at ceiling, wall, merchandise or machinery. Do not allow water vapour to condense or leaks to drip onto floor or unit, if this condition exists, check for water leaks from connections or spray gun nozzles. In case of insufficient atomisation correct by increasing air pressure and/or decreasing fluid flow. Lower humidistat setting if RH is high. Open drain valve on R-75AR Regulator/Condenser item #7 periodically to drain airline moisture buildup.

Guard face and eyes from escaping air and moisture. Do not use humidifier to create a fog condition as which water may collect on the unit and produce an electrical hazard.

6 Climaduct semi-central units – Ahmedabad, India

This is unique self-contained unit for complete industrial air conditioning. It comprises of axial flow fan, one or more centrifugal atomizers, air mixing chamber with interconnected fresh air and return air dampers to take fresh air as well as to re circulate departmental air. A set of large area 'V' shaped air filters and fabricated distribution duct with eliminator type grills. If required steam heating coils can also be mounted on R.A. damper of the mixing chambers. Different models with various fan capacities like12, 000 C.F.M., 15,000 C.F.M., 25,000 C.F.M are available.

KC – 4 Type Unitconsists of 34" Diaaxial flow fan, V belt driven by 5 H.P. electric motor. The air handling capacity is about 12,000 C.F.M. Unit is mounted with one double disc-type atomiser run by 1 H.P. electric motor.

KC – 5 Type Unit consists of 34" Diaaxial flow fan, V belt driven by 7.5 H.P. electric motor. The air handling capacity is about 15,000 C.F.M. Unit is mounted with one double disc-type atomiser run by 1 H.P. electric motor.

KC – 6 Type Unit consists of 41" Diaaxial flow fan, V belt driven by 10 H.P. electric motor. The air handling capacity is about 25,000 C.F.M. Unit is mounted with two double disc-type atomiser each run by 1 H.P. electric motor in a separate ring immediately after fan.

7 Faran Humidification –Loni, India

Pad-Type Air Washers and Spray-Type Air Washers produced by A.C. Humidifiers provide solutions for various ventilation problems for commercial and industrial establishments, which are widely used inTextile Mills, Paper, Tea Garden and Green House / Poultry Sheds. Available in modular designs and made out of pre-fabricated tank and panels that have long working life.

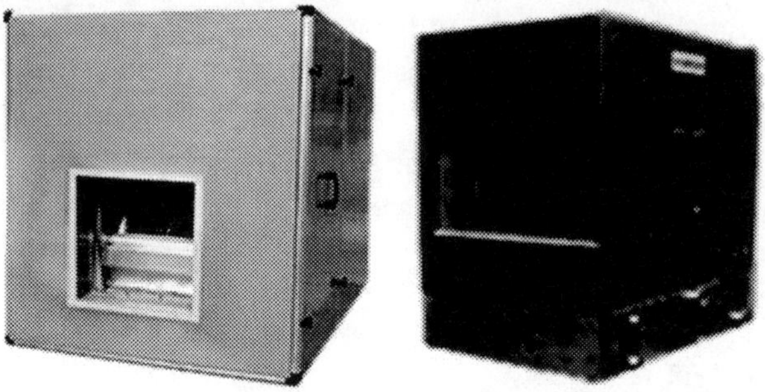

Pad type air washers spray type air washers

8 Luwa air engineering AG, Switzerland

Luwa's air washer unit with high speed drop eliminator is suitable for humidification and operation in connection with refrigeration. All build in parts are made of non-corrosive materials.

8.1 Uniluwa compact air-handling unit

The air treatment unit Uniluwa is ideal where fast installation is a must and only limited space is available. The unit is supplied as pre-fabricated and hence there is no time loss for the construction and installation. It is characterised by a self-supporting double panel construction.

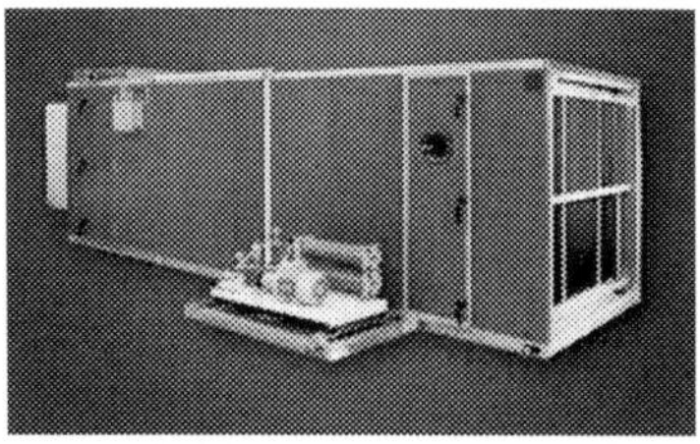

The unit is equipped with the TexFog technology, and no water tank is required. Time-consuming, expensive construction is superfluous. Uniluwa is a stand-alone unit and has been designed for small air volumes and high specific heat loads. Preassembled modules allow for short installation times. It is manufactured on a self-supporting base with double panel construction case. It is configured according to VDI 6022 hygienic requirements and provides air volumes from 20,000m³/h to 80,000m³/h

8.2 Luwair P – Insulated centralised plant

Luwair P is a centralised plant system designed for medium to large air volumes and high specific heat loads. Designed specifically to suit individual applications almost any shape variant can be achieved independent of building structure. Panels are available in galvanised and stainless steel in various finishes and colours for internal or external installation. Short installation times can be achieved through the use of an assembly-friendly supporting system and configured according to VDI 6022 hygienic requirements.

8.3 Texpac – Centralised plant

The plant is manufactured from 2 mm galvanized steel with a smooth internal finish. Suitable for medium to large air volumes and specifically to suit individual applications. Short installation times can be achieved and systems are re-usable; they can be dismantled and reassembled in a new location and configured according to VDI 6022 hygienic requirements. The Luwa Texpac centralised plant system has been designed for return air streams conveying high quantities of waste at high pressures.

All plants are available in multiple configurations incorporating air filtration either automatic or static filter banks, heating function either electric, gas, steam or LPHW, cooling either mechanical or evaporative, humidification with either high pressure systems or air washers and controls, either digital or pneumatic.

8.4 Luwa tex fog

Luwa tex fog is a high pressure cold water humidification system developed for installation within air conditioning plants. A compact pumping station pumps water via a two-stage filter to special nozzles where it nebulises under high pressure in aerosol mist and is mixed with air by means of turbulence elements. Due to this nebulisation and intensive air mixing, the greater part of the water evaporates. A droplet eliminator at the humidifier outlet ensures that no water is carried over and subsequent ductwork remains dry. Tex fog is a single pass system and sprays cold water at high pressure directly from the incoming mains supply. The system has integrated filters and UV disinfection can be added on request. The system does not store water and incorporates a control function which purges the system to prevent stagnation.

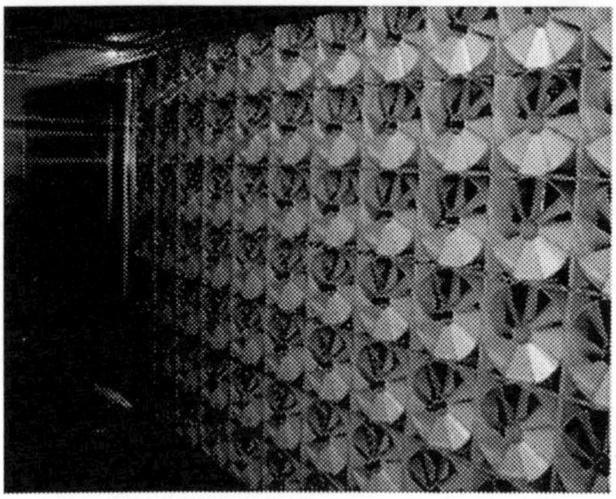

Advantages claimed are high degree of humidification and evaporation, hygienically safe (VDI 6022), no water tank or storage, small amounts of spray water used, minimum pump power requirement and maintenance and service friendly. The air washer can be applied either in centralised air handling systems or, as a stand-alone unit, in a central station of builders work. The housing stands clear of the ground on a base frame and consists of stainless steel panels. A pump draws the water directly from the tank. In case of water filtration or of chilled water operation, an additional reservoir is used. The air washer is also available, without the stainless steel housing, for installation within existing central stations.

8.5 Air washer luwa

The main features are constant and optimum humidification for dew point operation, variable humidification of the supply air for partial saturation, possibility of cooling, humidification and dehumidification with chilled water, housing or other elements which are in contact with humidity are made of plastics or of stainless steel, energy efficient centrifugal pumps and atomiser nozzle of plastics. Advantages claimed are modular construction, corrosion-resistant design and high efficiency with small pressure loss, adjustable humidification performance, high air throughput, low maintenance design and atomiser nozzle resistant to clogging.

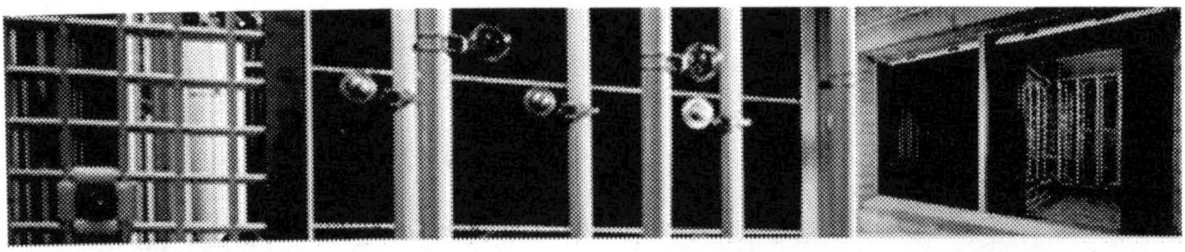

Air washer FRP construction Air damper Nozzle bank

9 Air washer-Pawc by abbot air systems – Faridabad, India

Air-handling units (AHUs) with large range of fans, designed and manufactured by Abbott Air Systems, have following features.

Air capacity: 6,000 m^3/h to 60,000 m^3/h in single unit. Equipped for central air cooling system installation. Connection to GI ducts for even distribution of cool and clean air over a wide area. One or multiple blowers to increase air capacity. Modular construction of high air capacity air washers for easy relocation and reuse. Efficient reduction of humidity through evaporation of water by cellulose-based corrugated paper pads with large surface areas in small units.

10 Air cleaning equipment by manvi textile air engineers – Mumbai, India

Manvi Textile Air Engineers produce various types of air washer plants that are simple to operate and cost effective and aredesigned to handle all types of spinning operations in a cotton textile mill. The spares are readily available.

10.1 Eliminators

Eliminator TA-100 –
Profile width 170 mm
For use at a pitch of 25 mm

Eliminator TA-W –
Profile width 170 mm
For use at a pitch of 29.5 mm

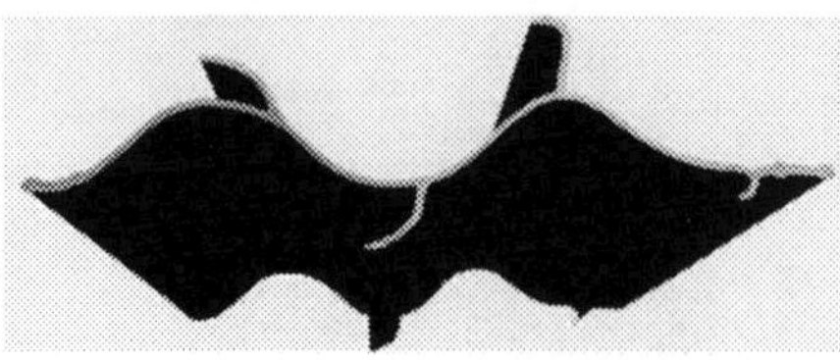

Eliminator TA-200 –
Profile width 150 mm
For use at a pitch of 30 mm

Eliminator TA-2W
Profile width 255 mm
For use at a pitch of 30 mm

Eliminator TA-K –
Profile width 250 mm
For use at a pitch of 30.5 mm

Eliminator TA-LW –
Profile width 220 mm
For use at a pitch of 30 mm

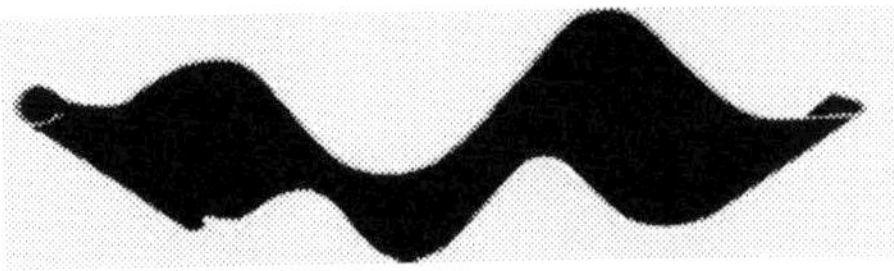

Eliminator TA-LT –
Profile width 215 mm
For use at a pitch of 25 mm

Eliminator TA-LH –
Profile width 280 mm
For use at a pitch of 25 mm

Eliminator TA-S100 –
Profile width 100 mm
For use at a pitch of 20

Eliminator TA-MH –
Profile width 270 mm
For use at a pitch of 25 mm

Eliminator TA-2LW –
Profile width 305 mm
For use at a pitch of 30 mm

Damper seal

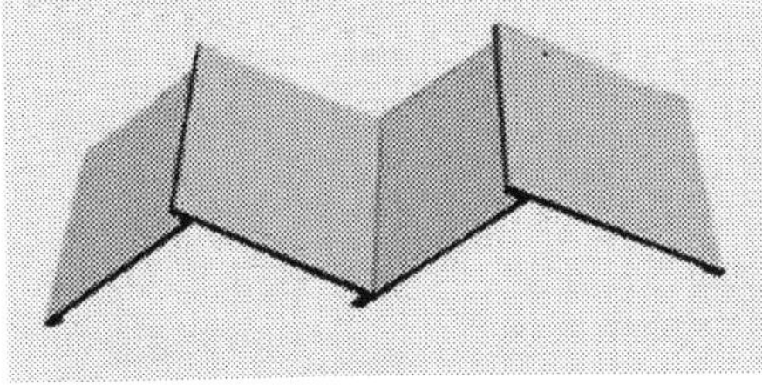

Eliminator TA-BAW –
Profile width 203 mm
For use at a pitch of 30 mm

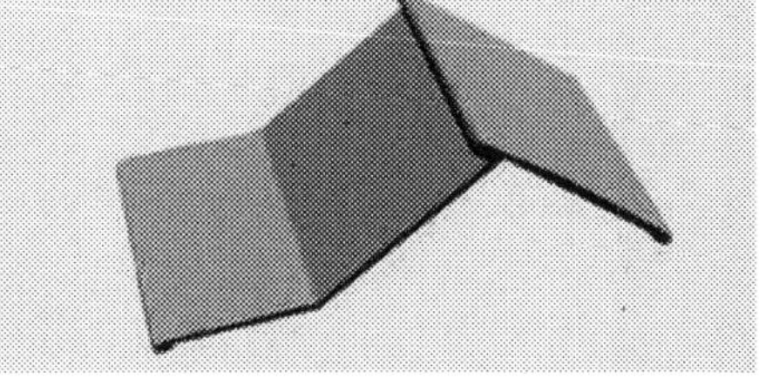

Eliminator TA-BAJ –
Profile width 160 mm
For use at a pitch of 30 mm

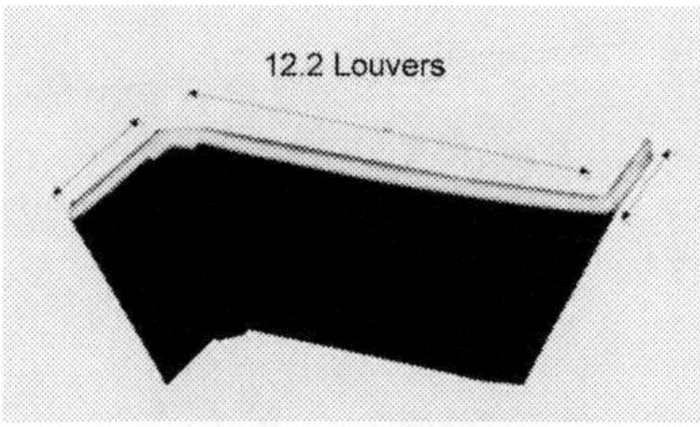

Louver – TA- plain Z type –
Profile width 25 + 84 + 13 mm
For use at a pitch of 60 or 75 mm

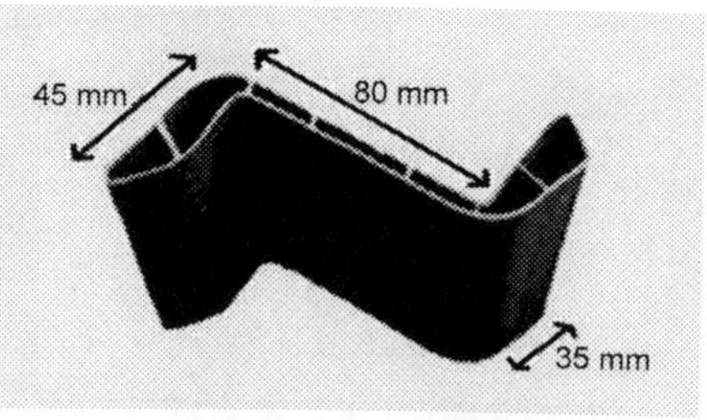

Louver – TA-HZ –
Profile width 45 + 80 + 35 mm
For use at a pitch of 75 mm

11 Air washer by excel airtechnique– India

Mist eliminators: Mist Eliminators are available in different shapes in extruded PVC and offer a low pressure drop. They precipitate the suspended water particles and allow only a homogeneous mixture of air with moisture. Aluminium fixtures with PVC notch plates in modular form are provided for quick removal and replacement.

Spray sets: It is an assembly (bank) of two horizontal header pipes on which number of vertical branch pipes / riser pipes are fitted. The spray nozzles are fitted on to the riser pipes. The spray header is connected to the delivery of the pump. The pipes are evenly distributed in turn to distribute nozzles to create a uniform mist.

Air diffusers: They distribute the humidified air in the department at various locations. They are available in PVC powder coated aluminium or stainless steel construction.

Air grilles: They distribute the humidified air at various locations. They are of single deflection and double deflection type. They are available in GI / SS and extruded aluminium construction.

Floor Grilles: They are the exhaust grilles fitted at different locations over the trench. The spent air along with floating fibrepasses through this grille into the trench.

Ducting: They carry the humidified air from the plant and a common masonry plenum and run generally above the false ceiling. They are fabricated at site with 20 / 22/ 24 G GI sheets and MS angles. They are generally supported from the roof purlins. The diffusers are fitted on the ducting.

Insulation: In order to prevent the heat transfer from the attic space (generally between 50° and 60°C) to the duct carrying the humidified air (around 25–28°C) the ducts are insulated with mineral wool. Insulation is also provided over the false ceiling /under the roof to arrest the roof heat entering into the department which in turn reduces the humidification plant capacity.

12 Roots air aystems, Noida, India

12.1 Air washers

The Roots CELdek air washer's surface contact material is specifically engineered for direct evaporative cooling and humidification. Manufactured from special cellulose paper impregnated with insoluble rot prevention compounds, it contains stiffening saturates and wetting agents. CEL dek is made of corrugated sheets assembled in self-supporting pads, which have a patented angle arrangement to direct water flow towards the air inlet side, where most of the evaporation occurs. It is also designed to flush away atmospheric dust. The pad has a contact surface between air and water of 440 m^2/m^3. The water distribution is specially designed for the Rootsdek air cooling machines. The water distribution cover is made of fibre glass-reinforced plastic for high temperature conditions and anti-corrosive properties. Metal parts are of 16 gauge galvanised sheet steel. The water tank is made of 3 mm fibre glass reinforced plastic which is steel reinforced. Salient features of roots air washers are as follows.

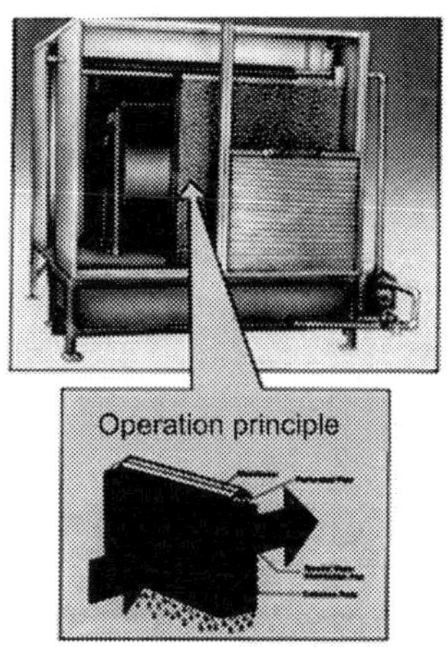

- Fil-type system can attain very high saturation efficiencies as against conventional system.
- Compact pre-fabricated design to take less space, consuming only 40% of conventional systems.
- No masonry work required and can be located inside or outside a building.
- Power consumption negligible as compared to spray machine due to low pump power and low pressure drop.

- Maintenance negligible as no nozzles are involved and hence trouble free service. Noise level only 65 db near machines as there are no high pressure nozzles.
- Quick installation as machine is factory fabricated and tested and has flexibility in regard to re-location at a later date.
- Custom Built ROOTS-DEK machines are available in stainless steel and/or FRP casings.

13 Mushroom industrial humidifiers

In Mushroom industrial humidifier, the humidity can reach 30% to 98.5%. It is automatic and can work 24 hours without manual work.

Model No sys 07-A

Technical parameter:

(a) Size of main engine: 880*1, 400*580 cm
(b) Weight of main engine: 118 kg
(c) Specified 0–3 kw (Automatic)
(d) Power: 220–440 V 50–60 Hz
(e) Feeding water quality: tap water, purified water or similar water
(f) Tank volume: 0.3 m^3

The industrial plunger pump transfers the Grade-3 purified water, which is pressed to 7MPa, to the spray nozzle through the high-pressure cooler pipes. The high-pressure water twists out from the specialised spray holes. After atomisation, the 3–15μm drops are sprayed into the air and to lower the air temperature. The spraying of the pressed water forms plenty of anion, making people comfortable. The features of high-pressure spray humidifier are good atomisation effect and low energy costs. Good atomisingeffect is obtained as the spray nozzle produces 5 billion atomised particles of 3–15μm diameter that is easily transformed into a gas, so textile machines won't be affected. Low energy loss achieved by direct use of water power, without supplementary equipment like air compressor. Energy loss for atomising 1kg water is 0.005kw average with the 0–3kw hydraulic power automatic frequency converter. Energy loss is only 1% as much of traditional ionic pump and 10% as that of centrifugal and gas-water mixed pumps. The cost is greatly reduced.

13.1 The composition of the high-pressure spray humidifiers

- *The composition of the converter power:* High-pressure ceramic pump combined with saline converter techniques is adopted. Automatic adjustment of the rotational speed and steady piping pressure make the humidifier run steadily and save energy.
- *Living monitor:* According to the demands for different RH in the area and process the monitors measure and display on-line and control the humidifier in different areas with the humidness sensor and controller. The system put the water-supply protection and backwater control into one system, which are monitored by computer online.
- *High-pressure transporter:* System adopts brass pipes with pressure-proof 15MPa above, acid-proof, alkaline-proof and good temperature resistance, keeps high hydrostatic pressure in the pipes and do not easily form dirty.
- *Waterway disposer:* The system adopts large water tank. The in-out opens of the tank are equipped with large-size filters to stop the blocks in the spray nozzle. Temperature can be lowered in the tank by adding ice, which reduces cost in summer.
- *Spray splasher:* Different super thin spray nozzles in different apertures and materials are accommodated in a set according to the requirement. A set of the system is equipped by 20-320 nozzles made of stainless steel Ti nozzle or ceramal supper-hard nozzle.

14 Neptronic SKE –Resistive Steam Humidifier – Canada

A compact unit controls humidity to within ±1% RH and delivers up to 80kg of steam per hour from a single unit. Made with 98% metal construction,

dramatically reducing the volume of spare parts required. Does not require replacement boiling cylinders. Due to its self-cleaning elements requires low maintenance. It also incorporates a scale inhibitor system to reduce the effects of lime-scale build-up and an AFEC system to prolong humidifier life. The unit's patented AFEC® (Anti-Foaming Energy Conservation) System protects the heating elements from burning out, enabling them to last much longer.

The system combines three safety features:

(a) *Internal Electronic Water Level Sensor:* Regular float switches can fail to give realistic water level readings due to the action of boiling and foam formation. If too little water is added to the tank, the level switch floats on the foam instead of on the water, thus resulting in the elements burning out in the foam. Therefore an internal solid state sensor is provided which gives realistic readings.

(b) *Internal and External Temperature Sensors*: Other units only have external temperature sensors to indicate when the elements are overheating. The addition of an internal sensor gives added protection from element burn-out.

(c) *Internal Foam Sensor:* The internal foam sensor conserves energy by responding to foaming only when necessary rather than having preset, regular drain cycles to avoid foam damage.

These features combine to provide a system that maximise the life of the elements.

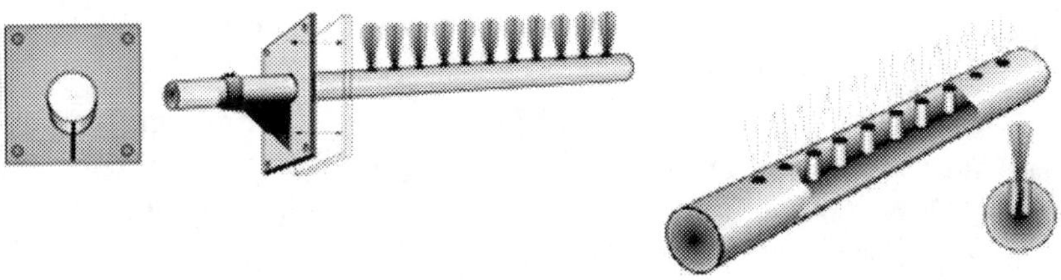

Super-dry stem pipes

The Neptronic SKE steam humidifier come with super-dry steam pipes, which provide an evaporation distance 40% less than standard pipe work. Each pipe has tiny internal tubes that take the steam from the centre of the pipe, where it is at its hottest and driest, guaranteeing condensation-free, dry steam delivery.

Main features are 580 kg steam output, self-cleaning super alloy, scale management system reduces scale build-up, close control within ±1%RH with RO water, fully CE compliant andWRAS approved, micro-processor control with LCD and fault diagnostic linked to BMS, 0–100% modulation as standard, duct pressures 1,250Pa–10,000Pa (with duct kit), rust-proof aluminium casing with baked enamel finish, stainless steel evaporation chamber with lifetime guarantee and wide, lockable hinged doors. Neptronic SKE Ultra process version is useful with ultra-pure water (2–18Megohms) and in extreme conditions. This version of the Neptronic SKE incorporates a stainless steel enclosure low smoke zero halogen wiring, no plastic components, special voltages and double nickel plated elements.

14.1 Neptronic SKE Dimensions, Output and Power

Dimensions and weights	SKE05M	SKE10M, SKE20M, SKE30M and SKE40M	SKE50M, SKE60M and SKE80M
Dimensions A (mm) / (in)	597 / 23½	724 / 28½	794 / 31¼
Dimensions B (mm) / (in)	470 / 18½	533 / 21	813 / 32
Dimensions C (mm) / (in)	292 / 11½	318 / 12½	318 / 12½
Weight (kg) / (lbs)	20 / 44	30 / 66	50 / 110
Room Distribution Unit	RDU –1	RDU –2	RDU –3
Dimensions D+A (mm)/(in)	737 / 29	890 / 35	890 / 35

Output and Power Consumption

Model	Power (kW)	Capacity (kg/h)	Current (A)		No of Outlets
			230 V/1ph	400 V/3ph	
SKE05	3.7	5	16		1
SKE10	7.5	10	11	1	
SKE20	15	20	22	1	
SKE30	22	30	33	2	
SKE40	30	40	44	2	
SKE50	36	50	53	2	
SKE60	44	60	64	2	
SKE80	60	80	87	3	

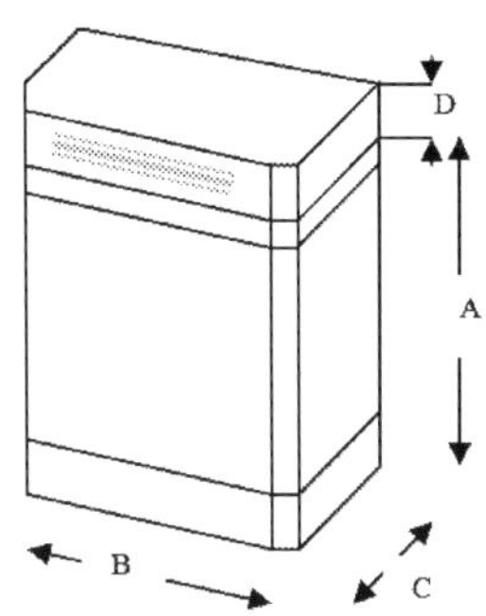

15 SAIVER A1 Air-Handling Unit– Italy

SAIVER Series A1 AHUs incorporate the finely tuned, value engineered cost-effective design aided by computer coupled with human ingenuity. SAIVER team is committed to produce one of the finest double-skinned AHUs range in the world to meet the requirements of most demanding cost and quality conscious customer.

The Frame

SAIVER unique frame design has inherent strength stability. The modular framework utilises a corrosion resistant, extruded marine, aluminium alloy, patented twin box section with True Thermal Break Construction. The entire module is subsequently mounted on a heavy sectional aluminium alloy or galvanised steel channel base.

Infill Panels

Standard 30-mm or 60-mm-thick infill panels are of double skinned construction from pressure injected polyurethane foam insulation with 'K' value of 0.02W/m°C and density 40kg/m3, sandwiched between galvanized steel with optional pre-plasticised or pre-painted finish, Pre-Aluman and stainless steel sheet is also available.

Filter, coils, air washers and fan sections requiring regular maintenance and inspection, have hinged or fully removable access panels. These are fitted to the frame with easy release, half-turn nylon handles and cam locks.

Handles can be operated internally for additional safety. Hinges are of heavy duty, load-bearing design with stainless steel pivot. Other panels can be detached, if necessary for access by removing screws with simple hand tools.

Inlet Section / Mixing Box

Plenum completed with dampers is specifically designed to minimise the stratification of entering air streams for maximum efficiency. Dampers are assembled within a rigid extruded aluminium frame, flanged and pre-drilled for easy fitting to connecting ductwork. Dampers are opposed blade type and available in both flat and double skinned aerofoil sections. Blades are formed from extruded aluminium with edge interlocks. Gaskets are provided to minimize leakage of air.

Coils are computer selected to obtain optimum psychometric efficiency with low air and water pressure drops. Chilled water, dire expansion, hot water and steam coils are constructed from copper tubes, mechanically bonded to aluminium fins as standard. Other fin materials are available including vinyl coated aluminium, copper, tinned copper and galvanised steel.

Coil Section

For corrosive flow media, stainless steel tubes and fins are available as an option. The coil assembly completed with carbon steel, copper or stainless steel headers is located within the coil section on aluminium support for easy withdrawal from either side.

On Site Assembly: The lightweight construction material and modular nature of the units make them suitable for lifting and maneuvering in difficult or confined locations. Modules can be easily aligned on site and locked together by sturdy stainless steel bolts, located in factory pre-drilled assembly holes. Continuous gaskets between each section ensure an airtight seal and thermal insulated. All fixings and gaskets are concealed within the unit

Filter Sections: Fully sealed filter sections are designed for easy withdrawal and renewal of filter cells and, are constructed to house any type of primary or secondary filters of different media with varies efficiencies. In areas of particular importance, such as hospitals and clean rooms, absolute filters can be provided to ensure safe human and machine environments.

Fan and Motor Section

SAIVER manufactured fans form the heart of all systems. Forward curved or backward curved non-overloading aerofoil centrifugal fans are available with various outlet configurations. All fan wheels and pulleys are individually tested and precision balanced, statically and dynamically, and keyed to the shaft.

Motors, mounted on slide rails with provision for easy belt tensioning, drive the fan with heavy duty V-belts. Combination spring and rubber vibration isolators are selected to match the power/weight ration of each fan for maximum isolation.

Intelligent Package

Intelligent Motor Control Centre

SAIVER Intelligent Package Air Handler equips with various operative and control devices to optimize unit running conditions. Motor Control Panel (MCP module) and Direct Digital Controller (DDC module) can be integrated into SAIVER intelligent air handler. All-in-one modular control centre results a fast and simple installation as well as a flexible and reliable operations. A unit mounted feature means space and cost saving.

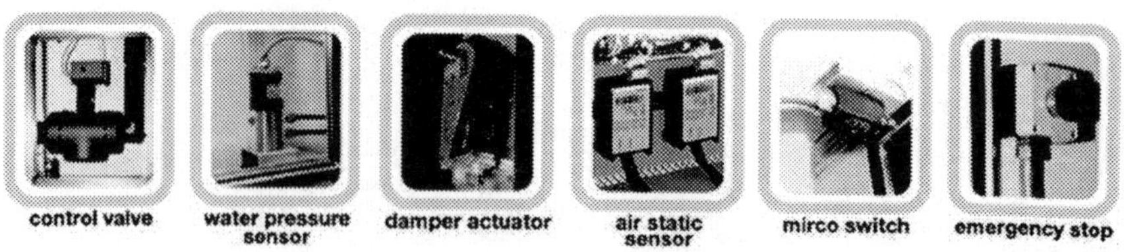

MCP *modules* comprise ofinverter completed with EMC filters to comply with EN regulations, Auto-bypass starting in case of inverter failure and Marshalling box for other services interfacing/connection, e.g., Fire Services/ BMS/M&E.

DDC *modules* comprise of DDC controller for local/remote controlling/ monitoring, chilled water valve completed with electronic control actuator, water/air differential pressure sensor, water/air temperature sensor, micro switch adjacent to access door, damper actuator at supply section, probe-type smoke detector at return air section, carbon dioxide sensor at return air section and filter differential pressure sensor.

Super Quiet Operation: SAIVER Acoustic AHU works with much lower noise level. For VAV application, with the combination of plug fan and SAIVER Acoustic Panel are able to meet NC39 at 9.0m3/s and 1,000Pa without supply silencer.

IAQ Package

Heat Recovery Unit: To improve indoor air quality (IAQ), one of the best solutions is to increase the fresh air quantity which is expensive. A rotary heat recovery unit allows energy exchange between supply and exhaust air streams. This high efficient heat exchanger can reduce the annual energy consumption in AHU by as much as 90%. (Latent and Sensible Heat Recovery). Alternatively, Heat Plate is also one of the best heat recovery devices which totally eliminate the potential problem of cross contamination.

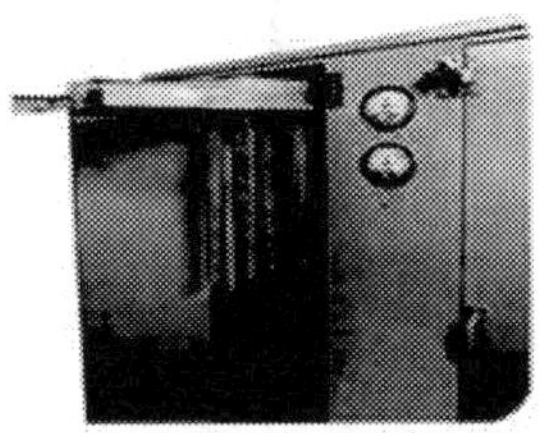

Ioniser: Ioniser destroys airborne and living micro-organisms by electrolysis process. The generator produces both positive and negative ions as they would occur under natural condition and the microbial control is performed by electrolysis (corona discharge) inside the Bi-polar unit. Single cell organisms are shocked/killed by the polar difference as negatively charged organisms collide with a positively charged particle.

UV Sterilizing Light: An UV system intends to 'capture and kill' airborne pathogens, improve IAQ and worker safety. The germicidal UV lamps in SAIVER air handler disinfect the air by irradiation and provide full coverage of the target surfaces. Installation sights include coils, drain pans, filters, exhaust systems, or anywhere mold, bacteria and pathogens can breed.

Desiccant Package: SAIVER provide all-in-one dehumidification control system (able to reach below 10% RH). Desiccant dehumidification ensures a hygienic and healthy environment by preventing the formation of molds and fungi inside air stream.

Heat Pipes: Besides the heat recovery application, heat pipes are used in dehumidification also. Heat pipes can increase an air handler moisture removal capability by 50% to 100%, reduce the chiller load by free pre-cooling but also provide free re-heating to lower the relative humidity of supply air. As most today's primary indoor air quality concerns are humidity related, the health benefits of heat pipes are noticeable. Run-around-coil is also an alternative system to improve dehumidification load on HVAC application.

Computer Selection Program: Saiver use own software program to make optimum equipment selection and submit quotation together with full technical information and drawings. Any variables such as local climatic conditions, unusual psychometric and physical parameters, are taken into account automatically. Clients are presented with computer generated, certified drawings for approval prior to equipment manufacturing.

Testing and Inspection: Saiver,IS09001 certified, works for consistent high standard maintained by a strict quality control program monitoring at all the manufacturing stages. Also variable speed dynamic fan test, sampling digital pressure test, sound performance test and coil performance test are conducted as needed.

16 Wet-Pad Humidifiers by Ningbo Yinzhou Chenwu Humidifying Equipment Factory–Zhejiang, China

Features:

(a) Enhances humidity and lowers temperature
(b) Components: special pads, water circulation system, fan, buoy, water supply system
(c) Full range of quality air pads (standard and customer-designed systems)
(d) Large airflow, operates smoothly, low noise, saves energy, easy cleaning
(e) Low invest and installation cost, high efficiency

Specifications:

- Humidification amount: 6–8kg / h. Humidification range: ≤750m3
- Voltage: 1) 380V, 50Hz, 2) 220V, 50Hz, 3) customisation available
- Fan power: 110W, Pump power: 85W
- Weight: 40kg, dimensions: 820 mm × 820 mm × 1,400 mm

- Ideally used for telecommunications rooms, computer rooms, electronic product roomsand precision instrument rooms (doors), Suitable for textiles and other industrial and commercial industries.
- Conditions for use: Environment temperature: 0–45°C, Water: city process water
- Humidity controller is needed for automatic humidification

17 GaSteam Gas Fired Steam Humidifier by Carel, USA

Carel USA supplies industrial and commercial humidifiers and controls for the HVAC industry, which includes steam, air/water atomising, high pressure water atomising, electrode, electric element, gas fired and direct steam humidifiers for printing, static control, museums, libraries, hospitals and many other applications. The GaSteam humidifier accomplished by the unique combination of modulating gas valve and modulating blower is tested at an astounding 89% efficiency by ETL without condensing, which is so efficient that Category IV venting is required.

- Type: Gas fired steam humidifier (Isothermic)
- Water: All source including potable, R/O and DI
- Capacity: 100, 200 and 400 lbs/h.
- Modulation: 25% to 100%
- Control type: On/off, modulating, proportional DDC, serial
- Tracking to set point: +/- 2%
- Efficiency: 89%
- Steam distribution: Duct, AHU, room, remote/room
- Small enough to fit in air handler vestibule

Patented Anti Foaming System (AFS)

Multiple communication capabilities (ModBus, BacNet, WebGate, etc.), integrated conductivity reading with control algorithms, triple external electronic water level float, aluminium cast teflon coated heat exchanger for ease of maintenance, Fully modulating gas valve (field convertible to LP), full top and bottom access for maintenance and Humivisor remote monitoring capability.

18 Armstrong International GFH Gas-Fired HumidiClean Humidifier

Gas fired humidifier reduces operating costs through significant energy savings, compared to electric humidifiers. The GFH Gas-Fired HumidiClean Humidifier from Armstrong with innovative Ionic Bed Technology reduces operating costs by eliminating the labour required by a surface skimmer. The GFH uses natural gas or propane for economical operation.

19 EvapoMist Humidifier by Field Controls Home Comfort Products, NC, USA

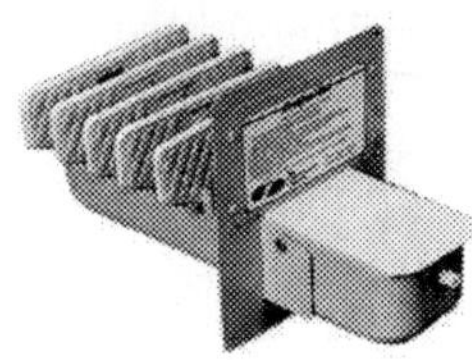

The EvapoMist™ humidifier is the solution to problems caused by dry winter air. They are easy to install and are durable and reliable with minimal maintenance.

Installation Instructions: Do not use the EvapoMist™ in a down-flow furnace (the type where the heated air output is from the bottom of the furnace).Select a location for your EvapoMist™ humidifier so that it will derive the most benefit from the normal air flow of the heating system. Care should be taken to install the unit in the area of highest air velocity and away from any turns in the heating system where turbulence can reduce evaporation. In installations where more than one humidifier is used, do not place them in a position where one obstructs the air flow to the other. Do not turn handle before or while installing the 'saddle tapping valve'. Be sure the piercing lance does not protrude beyond the rubber gasket, failure to do this may result in damage to the piercing needle. Do not install saddle valve on any water pipe which is downstream from a water softener. Assemble the 'saddle tapping valve' on copper tube with enclosed bolt and nuts. Tighten bolts evenly and firmly. Brackets should be parallel. Make the water connection to the outlet of the 'saddle tapping valve' and to the inlet of the EvapoMist™ humidifier. Use an insert, compression sleeve, and compression nut at both locations. Turn the handle clockwise until you feel it is firmly seated. You have now pierced the copper tube and the valve is closed. Turn on the water and open the saddle valve to check the entire system for leaks. Some tightening may be required but care should be taken not to over tighten small fittings. When the humidifier installation is satisfactorily completed, install the inspection cover. If the humidifier is installed on a slanting surface the inspection cover must be bent to the angle which the pan makes with the mounting bracket.

20 Air – O – Swiss Humidifiers.Roswell, GA 30076

Ultrasonic Technology uses high-frequency vibrations to turn water into a micro-fine mist where it is blown into the room to evaporate immediately. Includes the patented Ionic Silver Stick® that forms aqueous silver ions to prevent and control microbial growth for up to one year of continuous use. The ultrasonic system, with its high-frequency vibrations, destroys most bacteria and viruses. It includes a large, replaceable demineralisation cartridge with silver treated housing to keep the surface of the cartridge germ and bacteria free. No white dust is produced. The powerful demineralisation cartridge removes minerals and lime scale from the water. Variable low to high mist output control and rotating outlet for directing the mist. Built in hygrostat allows setting the desired humidity level. Humidifier switches off automatically when the selected humidity level is reached. It is easy to maintain. Just checking for deposit build up every fortnight depending on use is sufficient. A simple solution of white vinegar and water will remove the deposits. A bristle brush is used to clean around the vibrating membrane. It can be operated in continuous mode for nonstop humidification. Non wear titanium ultrasonic membrane gives long life. Demineralisation cartridge can be replaced or refilled with resin for more economical operating costs. Water reservoir empty indicator light and detachable transparent water tank are other features.

All models of Air-O-Swiss have pre-heating function heats water to 176°F before entering the nebuliser chamber. This kills microorganisms and allows the mist to exit the humidifier at a comfortable 104°F, preventing a drop in room temperature. All models are known for ultra-quiet operation of less than 25 dBA. All Air-O-Swiss equipments are UL listed

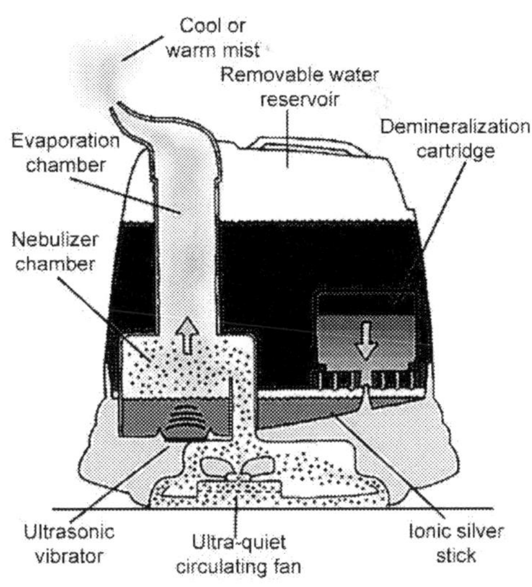

	20a Air-O-Swiss 7131	20b Air-O-Swiss 7133	20c Air-O-Swiss 7135
Power cord length	79"	79"	79"
Power consumption	45 W	45 W, 140 Wpre-heating mode	45 W, 140 Wpre-heating mode
Water capacity	1.4 gallons	2 gallons	2 gallons
Output	Up to 2.5 gallons per day	Up to 3.5 gallons per day	Up to 3.5 gallons per day
Weight empty	8.8 Lbs	7.7 Lbs	7.7 Lbs
Dimensions (L x W x H)	15.7" x 7.3" x 11.4"	15.7" x 8.7" x 13.8"	15.7" x 8.7" x 13.8"
Recommended room area	Up to 540 Sq foot	Up to 600 Sq Ft	Up to 600 Sq Ft
Optional items		Cool or warm humidification, detachable transparent water tank	

In model 7135 Electronic digital display and controls has several unique functions:

- Timer function for 1 to 9 hours of operation or continuous mode.
- Cleaning Indicator provides an illuminated light to remind when to check the cleanliness of the unit.
- Automatic Mode provides the ideal air humidity relative to the room temperature by regulating the output based on the selected humidity level.
- Sleep Mode sets the RH level to 60% and the operating time to 8 hours and turns on the pre-heating function. The output quantity is regulated based on the difference between the actual and selected humidity values.

20.1 Air-O-Swiss U7142 Ultrasonic Humidifier

- Electronic digital display and controls has several unique functions:
  - Timer function for 1 to 9 hours of operation or continuous mode.
  - Cleaning Indicator provides an illuminated light to remind when to check the cleanliness of the unit.
  - Automatic Mode provides the ideal air humidity relative to the room temperature by regulating the output based on the selected humidity level.
- Sleep Mode sets the humidity level to 60% and the operating time to 8 hours and turns on the pre-heating function. The output quantity is

also regulated based on the difference between the actual and selected humidity values

Optional cool or warm humidification.

Recommended for large spaces up to 860 square feet.

Power Cord Length: 79"

Power Consumption: 45 W, 140 Wpre-heating mode.

Water Capacity: 2.0 gallons.

Output: up to 3.5 gallons per day.

Weight Empty: 8.8 lbs.

Dimensions: 9.8"l × 11.0"w × 17.3"h.

Ionic Silver Stick: Maintains water purity without dangerous chemical additives. Patented technology uses the anti-bacterial effect of silver to eliminate over 650 bacteria and viruses. The anti-microbial protection works constantly when the device is in contact with water. Silver ions are released from the surface of the ionic stick to reach the required concentration in water and automatically maintains this safe concentration level. As the surface ions are released from the device, they are replaced by the internal concentration providing effectiveness for one year.

No maintenance is required. Place the ionic stick in the humidifier base and replace once in a year. Unlike other anti-microbial treatment methods, there is no hassle with additives needed every time the water reservoir is replaced, which can be 1 to 2 times a day. Safe enough to use with drinking water to maintain purity. Releases only the quantity of silver ions needed to kill microorganisms and meet standards of clean water maintenance. Economical operation as no expensive products is needed to buy continuously.

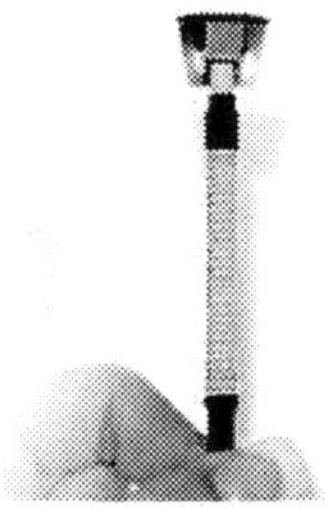

Air-O-Swiss Demineralisation Cartridge: Minerals present in all water such as calcium and magnesium can be released into the air by a humidifier with the water vapour. These minerals precipitate out of the water vapour as it evaporates and can leave a white dust on room furnishings. Minerals in water also have an impact on humidifier performance, cleaning and maintenance. Humidifier units that produce steam will lose their effectiveness as the heating elements get covered with scale build-up. These units develop heavy scaling over the heating element due to water minerals and require frequent and difficult cleaning.

Some humidifier manufacturers use mineral absorption pads that sit in the water base, which only have a limited capacity and effectiveness to capture minerals and require frequent changing to work. The Air-O-Swiss Demineralisation Cartridge is a powerful device that uses a large amount of granular adsorption media to capture and retain minerals from water.

All water traveling from the water storage tank into the humidifier base goes through the demineralisation cartridge, so the effectiveness of capturing minerals is maximized. The recommended cartridge replacement/refilling period is 2-3 months. Depending on the quality of water and mineral content, the cartridge may last longer or less.

Replacement of the Air-O-Swiss Demineralisation Cartridge

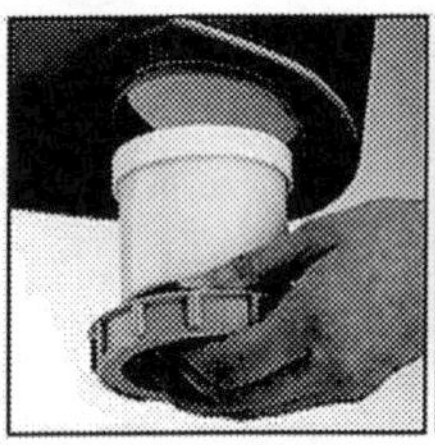

Remove the water tank cap. The cartridge is attached to the water tank cap.

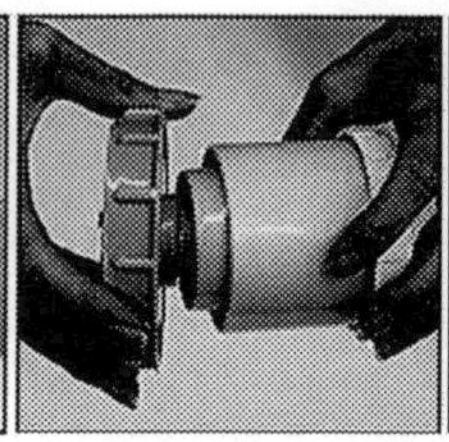

Unscrew the cartridge from the tank cap and dispose.

Unpack the new cartridge from the sealed bag.

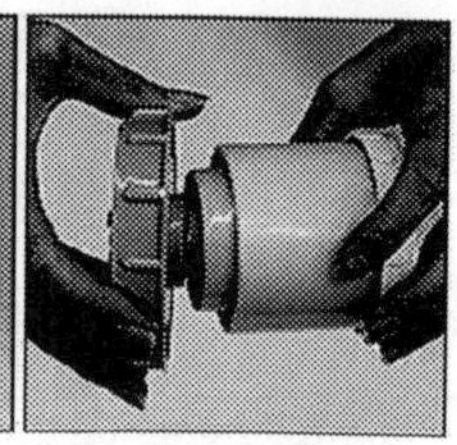

Screw the new cartridge to the water tank cap and replace cap on water tank.

Refilling the Air-O-Swiss Demineralisation Cartridge

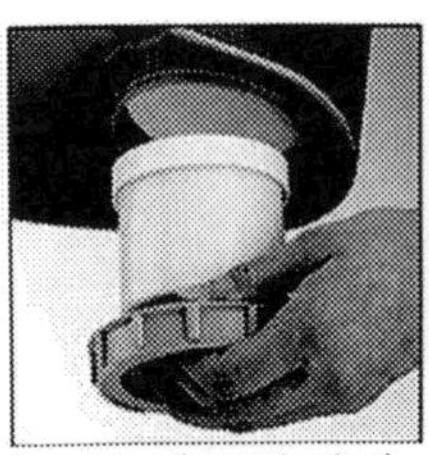

Remove the water tank cap. The cartridge is attached to the water tank cap.

Unscrew the cartridge from the tank cap.

Open the cartridge by unscrewing the top.

Dispose of the contents of the cartridge.

Features: Removes minerals from water. Helps prevent scale build-up inside the humidifier. Prevents 'White Dust' release from humidifier by removing lime scale and water minerals. Two options are available; complete self-contained replacement cartridge or mineral adsorption granule refill kits. Replacement is easy, unscrew the top lid, dispose of the used granules, pour in new replacement granules and screw on top. They are available as an economical refill 3 pack. Odorless resin. Replace every 2–3 months.

Open the refill bag of new resin. Note: granules must not enter centre area of cartridge. Pack centre with cloth, paper, or cotton prior to filling.

Pour the new resin into the cartridge. For easier filling, slip bag over cartridge and wrap an elastic band around the bag and top edge of cartridge.

Flip cartridge and bag over so new granules flow into cartridge. Remove any packing material in centre of cartridge. Screw top lid on.

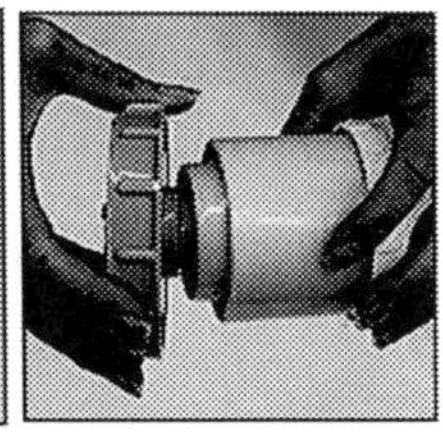

Screw the refilled cartridge to the water tank cap and replace cap on water tank.

20.2 Air-O-Swiss AOS1355 Cool Mist Air Washer

The AOS 1355 Air Washer washes and humidifies the air without the use of filters. Special humidifier discs turn constantly through water. This method removes many impurities (e.g. dust, pollen and particle-bound odors) from the air –as in nature, where the rain washes the air clean. The air washer works on the self-regulating evaporation principle and thus it requires no additional control systems to ensure optimal air humidity. The AOS 1355 is equipped with a special anti-bacterial protection system.

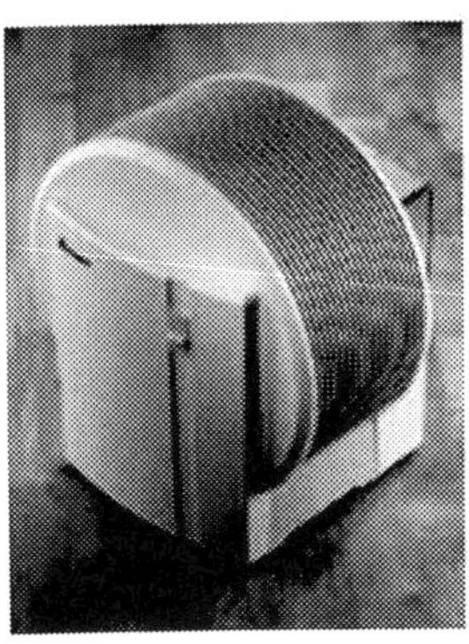

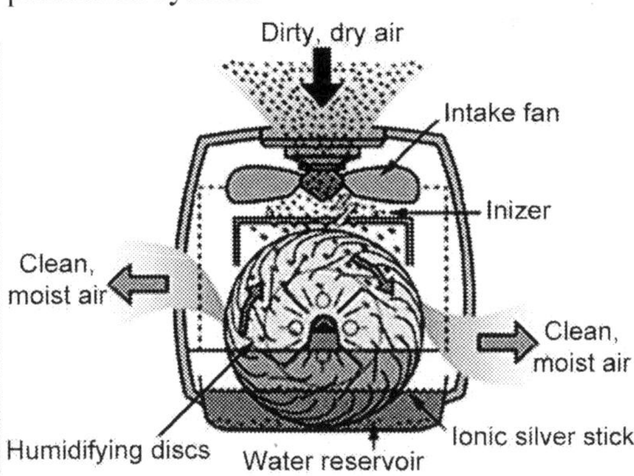

By drawing the air through water, the air washer naturally cleanses the air of smallest air borne particles, just like outside air is washed by rain. The air is humidified by special humidifying disks which render traditional replacement filters unnecessary. The self-regulating principle of evaporation guarantees an optimal level of humidity without additional control systems.

Features: Washes and humidifies the air in rooms up to 600 square feet, purifies the air as naturally as rain, captures particles down to 0.8 microns, removes impurities such as dust, pollen and particle-bound odours from the air. Totally, 22 humidifying disks spin to wash the air and provide constant humidity. Self-regulating humidification, no separate control systems are necessary. Simultaneously cleans the air of impurities and humidifies. Two ultra-quiet modes are available with transparent and removable water tank with handle. Easy handling and cleaning, disks are dishwasher safe up to 131 F. No replacement filters required. Patented Ionic Silver Stick Microbial Inhibitor prohibits bacteria growth, ensures safe and healthy air.

Specifications

Power: 20 W

Humidification Method: Cool Mist

Daily Output: -

Area Covered: 600 sq. ft.

Tank Capacity: 1.9 gal.

Speeds: 2

Dimensions and Weight:

Size: 12.6”D × 16.7”H × 14.9”W

Weight: 13 lbs.

20.3 Air-O-Swiss Combi –Humidifier and HEPA Air Filter

This combined air purifier and humidifier guarantees a healthy room climate. Not only does it ensure pure air by removing even the smallest particles (100,000 times finer than a human hair), it also ensures optimal air humidity.

By producing a healthy level of humidity, static electricity in the atmosphere is reduced and dust is suppressed. This makes this unit ideal for allergy sufferers as it not only reduces pollutants in the air with two HEPA filters but also creates a healthy atmosphere in a room. This is something that simple air filters cannot do.

Features include 3 extremely quiet performance settings, 2 HEPA particle filters to protect against allergens, such as pollen, dust, animal hairs and mites, static and dust suppression due to humidifier, odour/fume filter (active carbon filter) with increased efficiency due to a new filter arrangement, top performance air circulation, easy filter replacement, ergonomically designed handles, 2 removable transparent tanks with 2 handles for easy handling, tanks are easy to fill in even the smallest washbasin, fitted with an anti-bacterial evaporator cartridge, fragrance capsule for use with essential oils, clean, easy handling of the fragrance capsule without spillage or stickiness, operating light, easy-to-clean removable water tray and tidy cable.

The Air-O-Swiss Combi efficiently reduces the levels of air pollutants such as pollen, dust, animal hairs, mites, smoke and odours. The humidification facility of this unit is self-regulating, providing an optimal, hygienically humidified atmosphere. Italso has an air freshening system. By placing natural oils in the small fragrance capsule, aromas will be released into the room. With three performance settings, the unit is ready for immediate use and provides noticeably better air and reduces the amount of dust on the floors and furniture in rooms of up to 540sq ft. (50m2). It circulates the air in a room of 230 sq. ft. up to five times every hour and after a short period 99.97% of the harmful particles are removed from the air. Further advantages of this attractively styled unit are its extremely quiet operation, thanks to its enormous filter size and high quality fan, easy handling and exclusive functional design.

How Does the Combi Work?

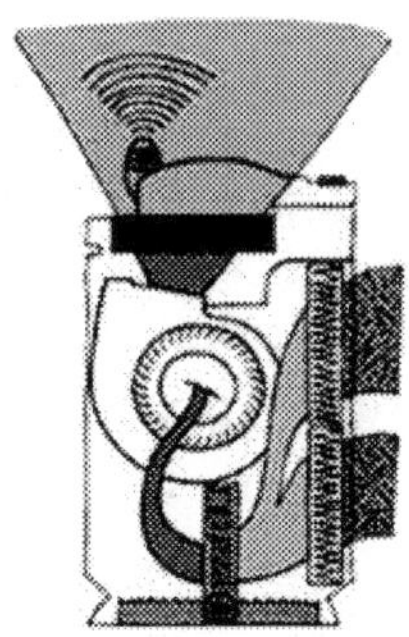

A high quality fan sucks in the polluted air and circulates it through two HEPA particle filters, which protect against harmful particles such as pollen, dust and spores that are up to 100,000 times finer than a human hair. The unit then draws the air through a wet mat and returns it to the room, optimally humidified. During the humidification process, the air is also washed, in other words, even more particles and odours are removed. Depending on its temperature, air can only absorb a certain amount of water. The process of evaporation thus provides self-regulating humidification, and as a result, additional control systems are not necessary. If desired the unit can also inject essential oils into the room making you feel relaxed or invigorated depending on the type of oil selected.

21 GS Industrial Humidifier & Atomizer, CA, USA

Introduction: The ultrasonic humidifier equipment features high output in water mist/fog; available selection from 6 lbs/h to 26 lbs/h, high efficiency in spraying fine mist/fog droplet 1–10um with average 5 micron, it is silent, low trouble rate, a relative humidity 99% can be achieved. It is equipped with standard water inlet, water outlet cock and overflow outlet, stainless steel body with painting treatment on surface. The ultrasonic humidifier can be classified into three control ways as following:

1. Manual control system. According to different situations, customers can use single equipment or use multiple equipments working independently. Multiple equipments can be controlled either independently or collectively.
2. Digital cycle work control system. The automatic cycle control period is between one minute and ninety-nine hours. Working time and time period can be set up according to user's need. Once it is set up, no worker is needed with its continual operation.
3. Digital humidity automatic control system.

The humidity control range is among 1–100% with error rate of ±5% and it is automatically controlled. Outlet pipe can be reasonably arranged according to user's requirement. Pipe connection should be airtight.

Humidifier Series: GS series industrial humidifiers have four types of industrial humidifier based on different fog/mist output and three types based on different control ways and there are two connection ways, single PVC pipe outlet connection (available for 6 lbs/h and 12 lbs/h) and double PVC pipe connection (available for 18 lbs/h and 26 lbs/h).

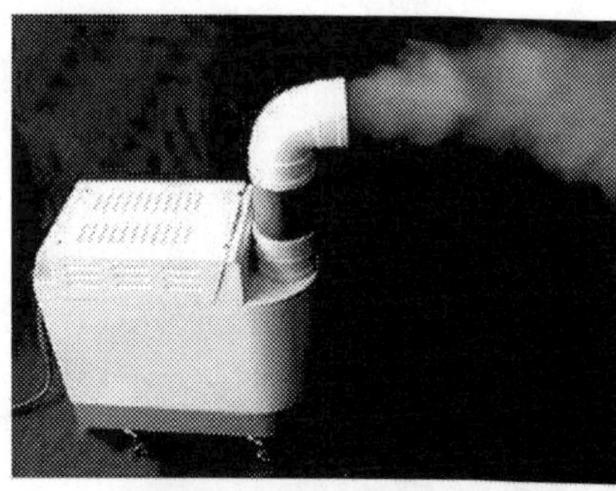

Double PVC pipe outlet

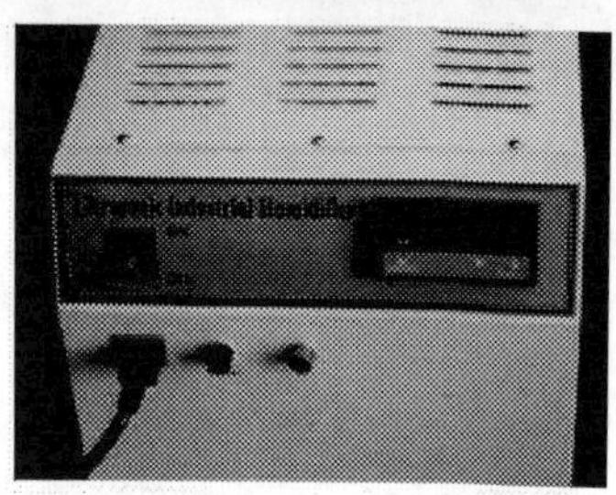

Humidifier front control panel

Technical Specifications

Humidifier Model	Mist/Fog lbs/h	Wind m³/H	Control	Power VAC/W	Unit Size(inch) –Weight(lbs)
GS-10	6	170	manual	110/300	21x19x10 –40
GS-20	13	170	manual	110/600	21x19x10 – 48
GS-30	20	350	manual	110/900	25x20x13 – 66
GS-40	26	350	manual	110/1200	25x20x13 – 84
GS-10S	6	170	schedule	110/300	21x19x10 – 40
GS-20S	13	170	schedule	110/600	21x19x10 – 48
GS-30S	20	350	schedule	110/900	25x20x13 – 66
GS-40S	26	350	schedule	110/1200	25x20x13 – 84
GS-10Z	6	170	humidity	110/300	21x19x10 – 40
GS-20Z	13	170	humidity	110/600	21x19x10 – 48
GS-30Z	20	350	humidity	110/900	25x20x13 – 66
GS-40Z	26	350	humidity	110/1200	25x20x13 – 84

Environmental temperature	1–40°C	Water pressure	14–55 lbs/inch²
Environmental humidity	less than 80% RH	Water temperature	1–30°C
Water quality	Tap water	Surrounding environment	No high temperature, radiation source, no high magnetic interference

KJS-3.0, KJS-6.0 & KJS-15.0

Industrial Ultrasonic Humidifier

Features: Fully automatic humidity control with auto self-cleaning function. Different from the GS series, KJS ultrasonic humidifier will clean itself when it is turned on or its continuous running time reaching every 4 hours, stale water will be flushed out and 85% of mineral deposits on surface of the ultrasonic oscillator will be cleaned and drained out from the water outlet; this keeps unit remaining in effective performance.

Easy in installation and operation. Just connecting a water inlet hose to the inlet connector and a water outlet hose to the outlet connector and placing humidity sensor to the location where humidity is needed and setting the required humidity is required. Then turn on the green power, humidifier starts cleaning itself and spraying fine mist within minutes through the sprayer mouth or PVC pipe.

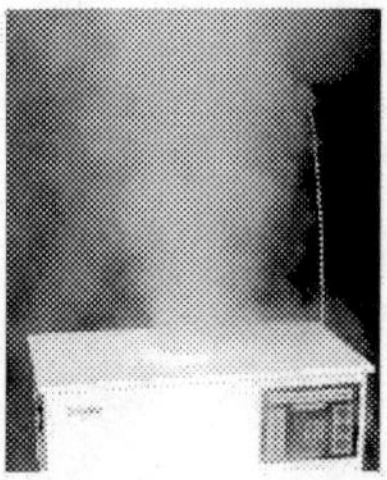

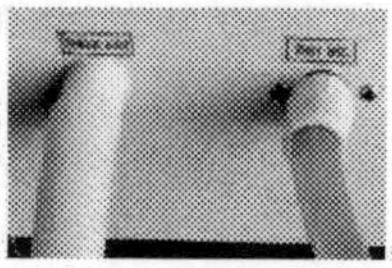

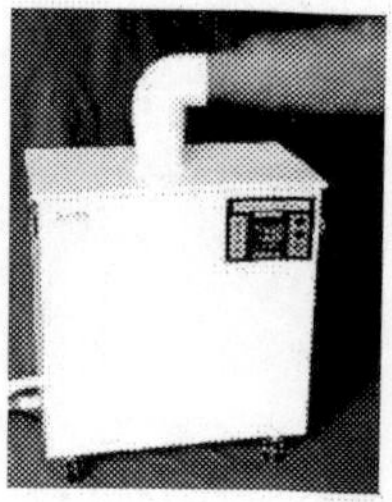

High efficiency in spraying fine mist/fog (droplet size 1–10um with average 5 micron), silent, low trouble rate and easy in maintenance, a relative humidity 99% can be achieved. KJS series is very good for using to increase humidity in room, warehouse, work shop, greenhouse, hydroponics application or using with air conditioning.

KJS Series Technical Specifications

Humidifier Model	**KJS-3.0kg/h**	**KJS-6.0kg/h**	**KJS-15.0kg/h**
Voltage (V)	AC110V	AC110V	AC110V
Power(W)	Less than 380W	Less than 750W	Less than 1,800W
Humidified space(m^3)	200m^3	390m^3	976m^3
Mist(fog) generated	3kg (6 lbs)/hour	6kg (13 lbs)/hour	15kg (30 lbs)/hour
Water volume atomized	3/4 gallon/hour	1.5 gallon/hour	3.75 gallon/hour
Control	Automatic humidity	Automatic humidity	Automatic humidity
Daily maintenance	Self.Auto-cleaning	Self. Auto-cleaning	Self. Auto-cleaning
Unit weight (lbs)	77 lbs	98 lbs	200 lbs
Unit dimension (inch)	14x20x27	14x24x27	18x33x37

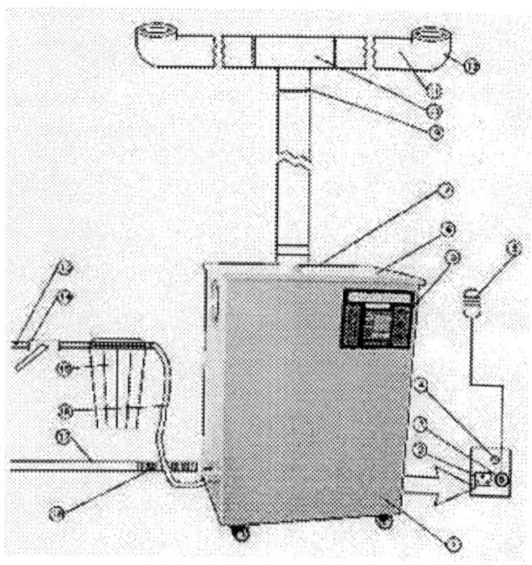

KJS series industrial ultrasonic humidifier is equipped with humidity sensor, which can be set within 30–100% (RH). The machine automatically switches on to humidify when humidity is less than 3% from set parameter. Automatic control precision of the sensor is ± 3% RH. Control panel indicator light will display 'low level water, add water automatically, humidify, auto-clean'. It tells its working state at the current time. JS humidifier will automatically switch the power off when water is at low level. It uses regular tap water and purified water is preferred. We should not use water which contains acid and alkali substance, or sewage.

Installing:

(1) Put humidifier on ground or workbench steadily, install 110v/50Hz single-phase three-wire power socket on the wall near the machine.

(2) Installing sprayer pipe (if necessary): Please refer to figure to install. (a) KJS-3.0kg/h humidifier, air pipe (9) (11), tee (10), elbow (12) uses 3 inches(outside diameter) PVC water outlet pipe; but KJS-6.0kg/h and KJS-15.0kg/h use 4 1/4 inches(outside diameter) water outlet pipe. (b) PVC pipe needs relative immobility. (c) To prevent that water drop down, PVC pipe (11) ought to tilt up slightly.

(3) Connect humidity sensor (8) to socket of humidity sensor (4 pins) on right side of the humidifier. Fix humidity sensor on the wall of humidifying room, and position and altitude of humidity sensor is determined by customers.

(4) You may connect water inlet hose (16)(3/4" prepared by customer) directly to tap water pipe, but if necessary, add water filter (15) (optional prepared by customer) between water inlet hose and the tap water pipe to get cleaner water for the machine.

(5) Connect hose (18)(1" prepared by customer) to 'water outlet' of humidifier, and the other end connects to your water outlet pipe (17). Make sure the level of water outlet pipe must be less than level of 'discharging water mouth'. Note: Avoid installing 'water inlet'or 'water outlet' mistakenly.

Usage and maintenance:

(a) After power on, insert power plug of the humidifier into power socket, switch the power button on, indicator on the power button will be lighted. If water tank has no water or water level is low, 'low level water' red lamp will be lighted, and 'auto-adding water' green lamp also will be lighted. The machine cannot spray fog; if water level in water tank is high level, 'low level water' red lamp will black out, but if 'auto-adding water' green lamp and 'spraying fog' blue lamp is lighted, the machine can spray fog. If water level in water tank reaches the highest setting level, and 'auto-adding water' green lamp is blacked out, 'spraying fog' blue lamp will be lighted, the machine can spray fog. The above states indicate that the humidifier runs normally.

(b) The machine has 'auto-cleaning' function. The humidifier will clean itself when it is switched on or its continuous running time reaches 4 hours, at same time, 'auto-cleaning' red lamp will be lighted. Cleaning time is about 4–6 minutes, and then sewage water will discharge from 'water outlet', but these operations do not influence on running of the humidifier. The machine also has 'manual cleaning' function. You can press 'cleaning button' down, and sewage water will discharge from 'water outlet'. Note: After discharging sewage, you must restore 'cleaning button' to 'auto' position; otherwise, the water will discharge from the machine continuously.

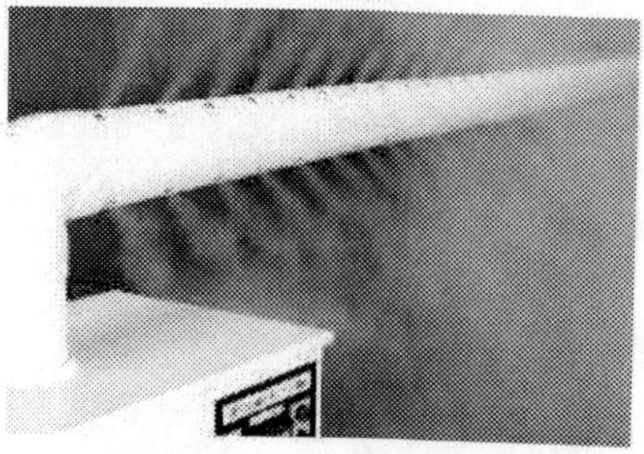

(c) The humidifier can control humidity in 'manual'; put 'humidity control' switch on 'manual' position, and then the machine will spray fog beyond control of humidity sensor.

Humidity Display and Setting: the machine is equipped with 'AI humidity controller' according to humidity demand of humidifying space, which has two lines digital display, upper line displays current room humidity, lower line displays setting humidity.

Setting Humidity: Press ▲ button to increase humidity and ▼ button to decrease humidity. To avoid influence of electricity failures, after finishing humidity setting, press 'memory' button, which saves the set parameter.

Maintenance: According to standard demand, it would be better to use a water filter to get purified water for the machine. It is suggested to run the machine once a week at least 30 minutes even if you donot use it, so it can be cleaned and flush out stale water, otherwise water mineral deposit could stay hard and become difficult to be cleaned out, and that could cause the misting discs not vibrating.

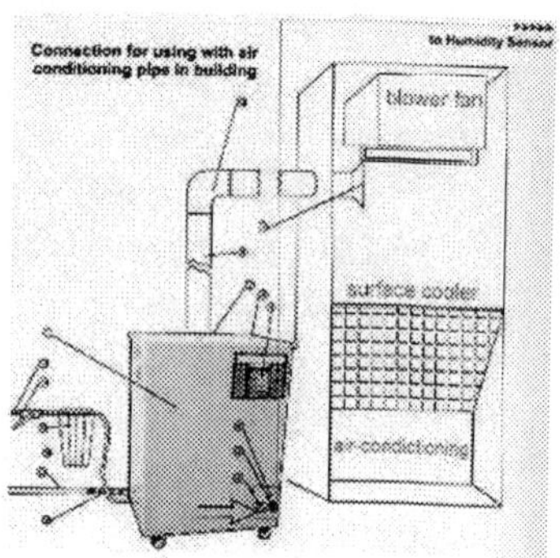

Service: The life of ultrasonic oscillator is about 5,000 hours. As the machine runs discontinuously according to humidity of workplace, generally, it is needed to replace a new ultrasonic oscillator after three years. The replacement oscillator disc (item K001) and humidifier part can be purchased on line at fogger, mist maker webpage.

General troubleshooting table is provided with each unit to the user.

Note:

(1) Sprayer mouth can be connected directly to air condition pipe by connecting air pipe to air returned end of air condition, in front of exhaust fan and in back of surface cooling. Owing to high air pressure in the pipe humidifier will run in high negative pressure condition. Under the condition, regulate wind inlet on the side of humidifier to avoid that water beads are sucked in air pipe of air-condition.

(2) Adding water from sprayer mouth is not feasible as manual control is not possible while adding water from sprayer mouth. If water level is too high, it will submerge air pipe and air cannot circulate and fog cannot spray out.

(3) When the machine is spraying fog the cover of the humidifier should not be taken out as it results in water beads sprinkle into the machine and cause humidifier failure.

22 Draft Air –Ahmedabad, India

22.1 Textile Air Conditioning: Air Washer System

The systems are designed and developed considering all heat sources in the department and required RH can be maintained constant at desired level with only air washers system. However to control inside temperature during hot and humid weather, chilled water spray is required and during winter, heating is recommended. Supply and return air system is designed considering concentration of heat load in the department to maintain uniform conditions in the department and to avoid any hot zones.

Spray Chamber: Corrosion free, low resistance, heavy duty rigid PVC piping, short polycarbonate spray nozzles with S.S. orifice and clamps for quick fixing, specially arranged to ensure fine spray spectrum for highest efficiency.

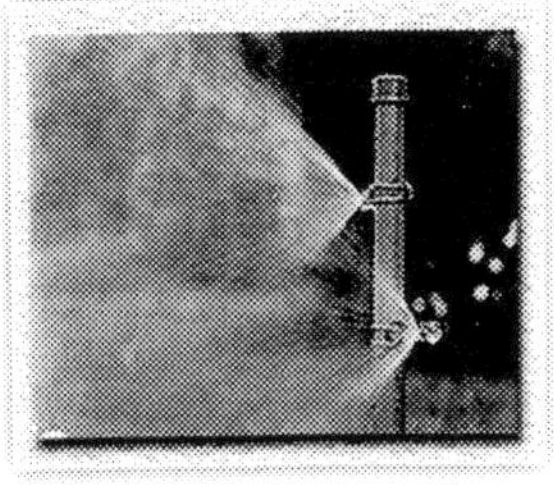

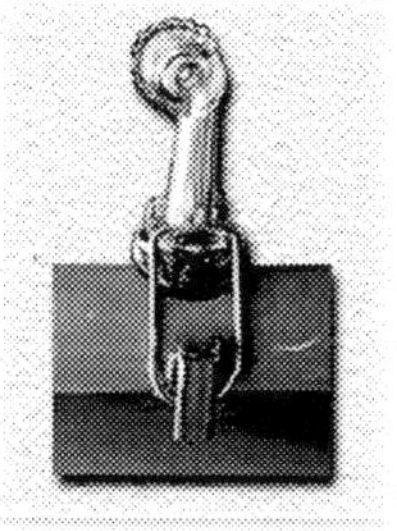

Louvers: Horizontal inlet distribution PVC louvers which evenly guide entry of air in the washer and also prevents back splashing.

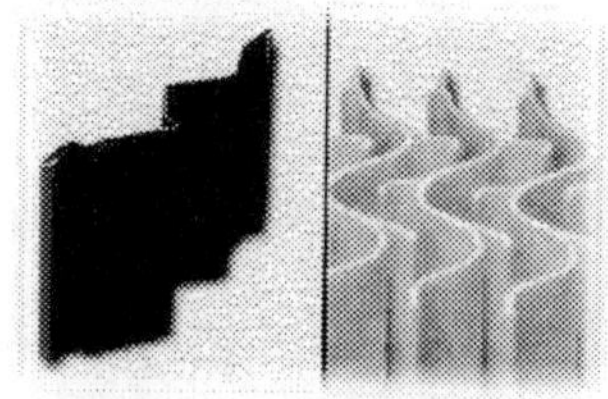

Dampers: Heavy gauge galvanised steel casing with airfoil aluminium extruded opposed blades Suitable for automatic operation.

Eliminators: 5 bend 6 pass PVC eliminators completely remove entrained free water particles and provide highest saturation efficiency.

Rotary Water Filter: Heavy gauge brass perforated drum fixed on hit-dipped galvanised structure. It is more efficient cleaning device than stationery screen where all water is returned to a central location. Water is filtered continuously. Drum is periodically flushed to remove accumulated waster matter into the G.I. Tray.

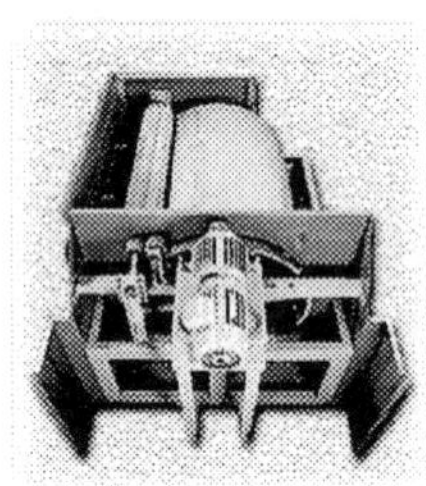

Rotary Air Filter: Automatic rotary air filter is provided for continuous cleaning of the return/exhaust air. A variety of filter media is selected for varying applications and fitted on galvanised perforated rotating drum of ample-free area to catch maximum fluff and dust from the air. Rotating action keeps the media clean and maintains uniform air flow. Fluff and dust deposited on the filter is sucked off by sliding nozzles connected to a powerful suction blower, which then is collected in the nylon bags. Clean air is passed on for either recirculation or exhaust.

Ventilation and Air Pressurisation System: The system is designed to maintain the desired purity, temperature and over-pressure in the department. The supply air quantities are estimated based on heat load, air leakages and ambient air quality. Ventilation systems can be with or without air washer. The exhaust system can be either mechanical or through gravity release dampers.

Centrifugal Ventilation Fans

Size: 200 mm to 2,500 mm impeller Dia

Capacity: 1,000 to 400,000 CMH.

Pressure: Up to 150 mm Wg.

Low noise high efficiency non-overloading impeller with backward inclined blades. Impellers statically and dynamically balanced as per ISO 1940. Casing fabricated from heavy gauge steel with heavy duty pedestals. Fan orientation is provided to suit the site requirement.

Beside axial flow and centrifugal ventilation fans, Draft-Air offers special purpose centrifugal fans for high temperature, fume/vapour exhaust, dust handling, material, conveying, induced draft and forced draft applications.

Axial Flow Fans

Size: 450 mm to 1,980 mm sweep. Capacity: 2,000 to 200,000 CMH. Pressure: Up to 65mm WG.

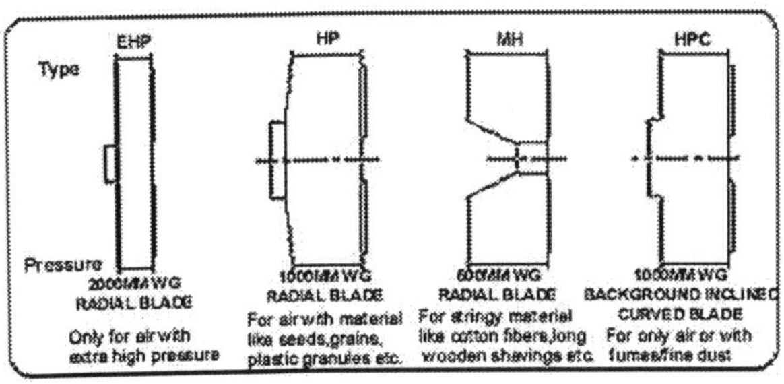

Hosing fabricated from heavy gauge steel sheet. Impeller is having high-efficiency aero-foil section adjustable pitch blades. Impellers are statically and dynamically balanced as per ISO 1940. Inlet cone and outlet diffuser are provided to further improve the efficiency.

23 Large Air Handler – By United Metal Products Textile facility, AZ, USA

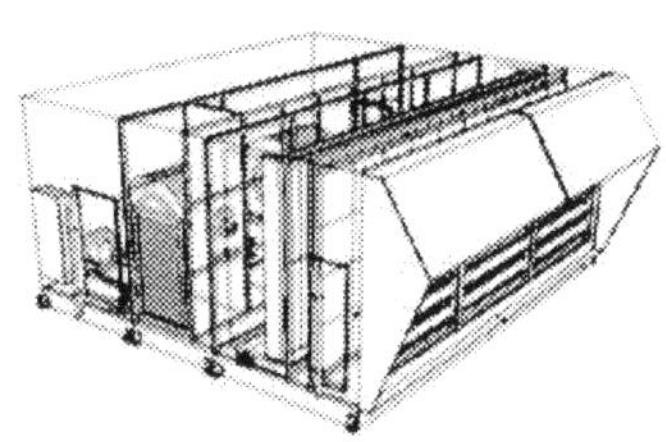

80,000 cfm @ 3.13 tsp

FC DWDI supply fan (2) 40hp

Inlet hood w/bird screen

OA dampers

CELdek media

Gas control vestibule

Hinged access louver

Operating weight:16,600 lbs.

24 Air Conditioning Projects – Standard and Specialized by Coolair (NI) Ltd– Northern Ireland

Coolair specialises in the air conditioning, ranges from the simplest to the most complex projects. They can be tailor made to the requirement of the user. Factors in which one could benefit from air conditioning are high density of occupants / customers, etc. high lighting levels, heat emissions from computers / displays / refrigeration equipment and other apparatus and solar gain from large areas of glass.

Where the room / area has a suspended ceiling or other ceiling void, ceiling cassette systems represent an effective and unobtrusive method of controlling temperature. Where no ceiling void is available (or for smaller rooms where cassette systems may be considered to be too large), the choice of equipment will normally be a high level wall-mounted system.Dependent on the number of rooms in the area to be treated and their location, multiple outdoor units or a single multi-split or VRV outdoor unit may be installed.

25 Ecoflair, Innovation, USA

When planning new buildings – especially when they have been earmarked for the tertiary sector, commerce or industry – creating a comfortable environment for the well-being of people who are going to work there, installing a flexible HVAC system and, above all, keeping energy consumption down should be recognised as elements of paramount importance. Ecoflair airconditioning system is developed by Uniflair to meet these three requirements in one fell swoop: based on the use of modular components, connected in a flexible manner offering limitless changeable setup options, Ecoflair provides a comfortable environment for each single individual while seeking to create the most energy-efficient conditions. The normal demands are individual well-being, flexibility and reduced energy consumption. The Ecoflair solution, the innovative air-conditioning and distribution system, can meet local comfort needs whilst assuring the building total flexibility, without foregoing the undeniable advantage of reducing energy bill outlay.

Each individual user can tailor environmental conditions to meet his/her comfort needs regardless of the conditions chosen by others. The user is free to choose either cooling or heating at any time with customisable temperature and ventilation levels, creating his/her own microclimate, even in an open-plan space, which may differ from that chosen by the colleague at the next desk.

A modular solution featuring connections of the plug and play kind, integrates the HVAC system with the access floor, or possibly the false ceiling, offering airconditioning and heating solution that suits the architectural needs of the rooms. With no HVAC system terminals to plan around, the designer finds drawing the layout of the rooms much easier with this system, the terminals are actually fitted under the floor or in the false ceiling. Moreover, the system can be installed with one configuration and changed to a new setup at any later date –at no cost and without interrupting service to accommodate the rearrangement of space in the rooms. With the exclusive Ecoflair control system, all the system's operating parameters can be measured and processed to define load profiles and the best control strategies, optimising the system's energy usage. To maintain thermal equilibrium, i.e. load balancing, traditional HVAC systems feature two different systems, one to generate heat and the other for cooling, which often have to run at the same time to cope with different load conditions in various parts of the building, with a negative effect on energy costs. Ecoflair, on the other hand, exploits heat transfer between various interior areas with a different thermal profile to achieve and maintain the climate wanted in the individual zones using as little energy as possible. That way, the system tends to strike an energy balance and only call on the services of the cold and heat production unit –consisting of a single generator for the whole system –for the portions of energy that fail to be balanced within the actual system.

The Ecoflair system, completely different from classic four-pipe systems with simultaneous cold and heat generation, is made up of a handful of modular elements. All parts are interchangeable and can be installed in the void under the access floor or behind the false ceiling. Furthermore, their plug andplay-type assembly procedure means they are easy to reposition.

The system's main components are as follows:

- The water distribution system consisting of a heating loop and a cooling loop which puts the hot and cold source at the disposal of all the various zones in the building;

- The Unitile zone terminal units, used for local air treatment (to control air temperature, humidity, ventilation, filtration and quality);
- Units transferring heat from one loop to the other to compensate heat loads of opposite sign inside the building; the energy provider effectively the generator producing heat and cooling required to cover the portion that has not been balanced, i.e. the amount of heat with a positive or negative sign for the portion not compensated by the transfer of heat between the two loops;
- The fresh air distribution and treatment units
- The humidity control units
- The main control system for combined management of all the system's units.

Unitile: Connected to the heating loop and cooling loop, the individual Unitile module, which can be installed under the floor or in the false ceiling, features an air/water coil and a fan. Based on the parameters set by the user – temperature and fan speed –which will determine his/her personal workstation microclimate, the module provides heating or cooling at the preferred flow rate. The terminal operates, cooling or heating each individual zone in the room with specific control capacity, thus providing the ideal conditions requested by the user.

Heat Transfer Units: Heat can be transferred from one loop to the other with a normal reverse-cycle unit, whose operation benefits from highly favourable heat levels, both at the condenser and evaporator. The unit maintains the temperature difference between the two loops in simultaneous hot and cold request mode, achieving a dual heating/cooling effect with a single compressor. A single unit can be used to handle all the system's needs, or it can consist of a number of small units to join the other equipment in the void under the floor or in the ceiling.

Air Treatment Units: Fresh air is delivered into the rooms to be airconditioned by means of various independent units serving individual zones or zone groups. The fresh air delivery rate is controlled with sensors, which adjust fan rotation speed. With the cross-flow recovery unit, air introduced can also be pre-cooled in summer and preheated in winter. Optimal recovery is a guaranteed trait of the unit's operation: the flow of fresh air is initially pre-cooled using expelled air and only then is the condenser used for cooling.

Energy Provider: The job of the energy provider is to ensure the heating loop and cooling loop temperatures always match those determined by the main control system as the best values at that particular point in time. It starts working when heating and cooling is requested at the same time and the heat transfer unit cannot maintain the temperature difference, resulting in loss of load balance.

Control System: The individual Unitile modules feature a control device located near the user. This device can maintain the conditions set by each user by adjusting the hot and cold water demand and fan speed accordingly. The transfer unit is controlled by monitoring the temperatures of the two loops and producing the necessary action when the temperature difference between them starts to dwindle. Fresh air control is adjusted by monitoring the area of the environment in terms of moisture, carbon dioxide and volatile organic content (VOC).In the end, the whole system is managed by a central device. Based on inside and outside temperatures, the temperature of the loops and quality of the air, the device can minimise the electricity demand, making for considerable energy savings whilst maintaining the highest levels of comfort.

26 Windchaser CA, USA

26.1 AAT2003 Ionic Humidifying Air Washer

The AAT2003 Ionic Humidifying Air Washer washes and humidifies the air by drawing air through water. This natural process eliminates small airborne particles just like the outdoor air is cleaned by rain.

Features

Purifies and humidifies the air at the same time, No traditional filters to replace; uses dishwasher-safe plastic discs. Twospeeds for added control; whisper-quiet operation, detachable water tank with handle; anti-bacterial safety system, 1.9-gallon capacity; suitable for a 600-square-foot room.

Equipped with an ion generator and two whisper quiet operating modes, the AAT2003 is suitable for operation during the pollen and heating seasons. Its self-regulating evaporation system guarantees an optimal level of humidity without additional control systems. The modern design and ease of operation and cleaning make the AAT2003 perfect for home or office. Ion generator cleans and refresh the air.

- Water Output – 3 Gallons/24h
- Water Capacity – 1.6 Gallons
- Applicable Area: Up to 600 sq. ft.

27 Atlas Sales and Rentals Inc.CA, USA

27.1 Air Cooled Portable Air Conditioners

Air-cooled portable air conditionersare most often the system of choice for end-users and industry professionals looking for effective spot cooling systems. The fact that air cooled portable air conditioning units can be installed almost anywhere within minutes explains their popularity as server room air conditioners, data centre air conditioners and cooling systems for a wide range of other applications. They can be used as a temporary air conditioning system as needed.

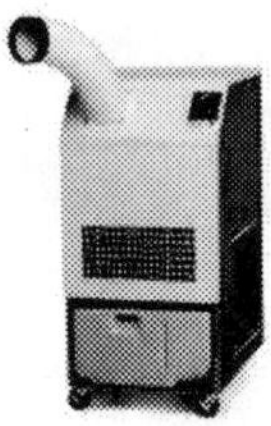
Classic Series

Office Pro Series

Cool Cube

Arctic Air Series

28 Skil-Aire, MD, USA

28.1 Roving Cooler – Spot Humidifier

The Roving Cooler is designed to be readily mobile to meet emergency needs for primary cooling or for transient supplementary needs. These units boast features normally found only in traditional central AC systems such as independent evaporator and condenser air streams and integrated digital

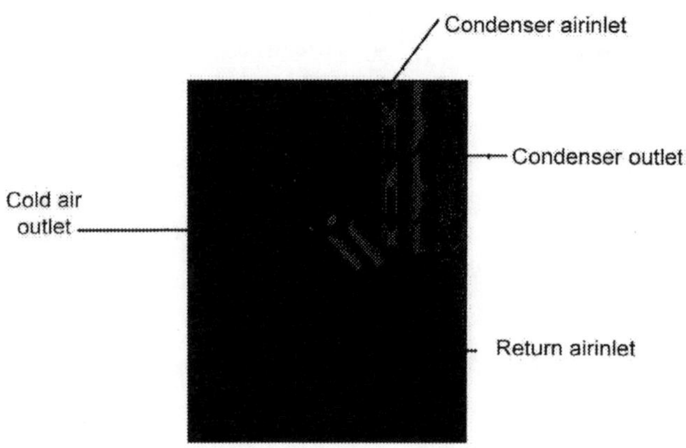

controls. Roving Coolers are designed to operate in conditions from 60 to 110°F which makes them perfect for many otherwise impossible applications both in and out of doors.

The Skil-aire Roving Cooler packaged air conditioners are available in air or water cooled models. All models have high external static capabilities which allow for the maximum airflow even with long flexible ducting up to 25′. These units fit through standard 30″doors and are mounted on four or six heavy duty locking casters for convenient location and storage. The durable 20-guage steel cabinet with special e-z clean, scratch-proof coating affords corrosion protection and provides an attractive appearance with virtually no maintenance. Evaporator air is continuously circulated through the front of the unit and condenser air through the top of the unit by draw through, forward curved, direct drive centrifugal blowers. Large face area coils eliminate condensate carryover while optimising heat transfer. Coils are constructed of heavy wall seamless copper tubes mechanically expanded to aluminium fins. A full condensate tank, with overflow safety cut out and alarms insures all condensate is collected and properly protected. Standard 1″ disposable filters with an arrestance value exceeding 90% are standard.

The factory installed water cooled condenser is a co-axial counter flow design selected to maintain low refrigerant operating pressures using a minimum quantity of water. Maximum operating water pressure with standard factory installed 2-way pressure regulating valve is 150 PSIG.

Heavy duty rotary or reciprocating compressors are standard, as are automatic restart and short-cycle protection. These quiet compressors provide dependable, proven, long-term service.

Each Roving Cooler includes an adjustable thermal expansion valve control with external equaliser, sight glass/moisture indicator, filter drier and liquid and suction Schrader fittings for charging. All controls, condensate system and major components are conveniently accessible from a single locking access door. Sight glass and service ports are located outside the unit.

All electrical components are conveniently located in the side of the unit, permitting fast inspection or servicing, even when the unit is operating. Factory installed electrical components include colour coded wiring, inherent protection on all motor legs, motor contractors, hi-lo pressure switches, oversized 24-volt control transformer and control circuit field terminals for interconnection with the controls.

Each Roving Cooler is controlled by a digital thermostat and comes complete with condensate pump and adjustable cool air plenums for maximum ease of use with minimum concern.

29 Jiangsu Jingya Environment Technology Co. Ltd. China

It is widely used for the textile, light industry, tobacco, electronics and various kinds of civil air conditioner systems.

Spray pipe banks

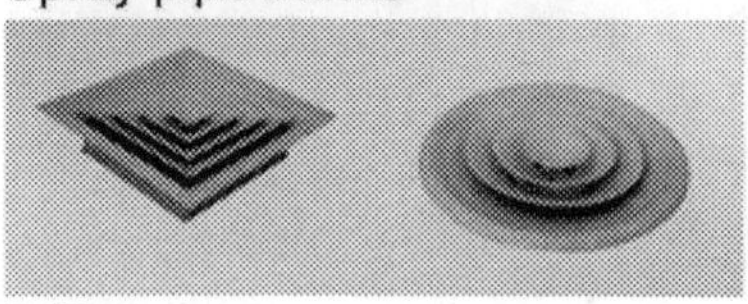

Diffusers

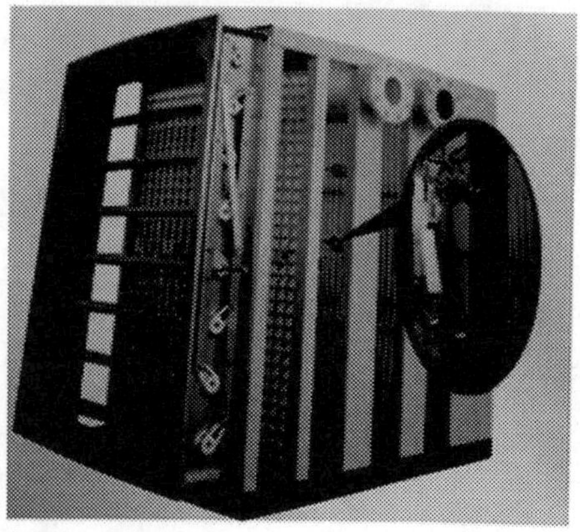

Main unit

- *Air Damper:* The made-to-order square-hole aluminium profile blade is smooth and straight with small air leakage rate. It can be flexibly and stably regulated with excellent linear regulation features and outstanding cooperation with the automatically controlled air valve actuator.
- *Fixed Louver:* Z-shaped aluminium alloy framework and the special aluminium blade are light weighted, firmed, anti-corrosive and durable. The blade is the streamline form with features of smooth, small ventilation resistance, energy saving and consumption reduction.
- *Spray Pipe Banks:* The plug-in type molding pipeline is excellently manufactured. The plug-in vertical pipe and clamp-type nozzle are conveniently disassembled, cleaned and maintained. The knock-on nozzle is equipped with the following features, including large amount of spraying water, good atomisation effect, high efficiency, block-up protection and corrosive resistance.
- *Water Eliminator:* The stream-line waveform water fender has small resistance and good moisture separation effect.
- *Air Straightener:* The stream-line grille has good stable and uniform airflow effect and less resistance.

29.1 ZKE Series Package Air-Handling Unit

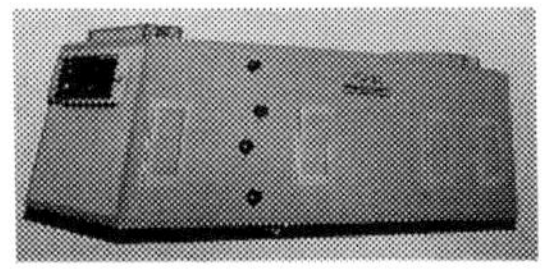

It is widely used for air conditioning in textile mills and number of other industries. It has beautiful appearance, fashion structure and captious materials. It has good air tight and thermal insulation performance, which is stable and reliable. It is simple to maintain and overhaul.

Rated air volume: 20,000m^3/h to 200,000m^3/h

Rated cooling capacity: 123kW to 1.258kW

Rated heating capacity: 229kW to 1,926kW

30 Amoto Engineering Pvt Ltd, Mumbai, India

30.1 Humidex 2000 Humidifier

Principle

HUMIDEX & SUSPENDEX series of humidifiers/ atomisers are unique devices or models of their own kinds and are scientifically designed for continuous working. They work on the principle of centrifugal atomisation and produce droplets free aerosol/mist with no wastage of water and maintain the desired humidity level (RH %) in the department without stoppage.

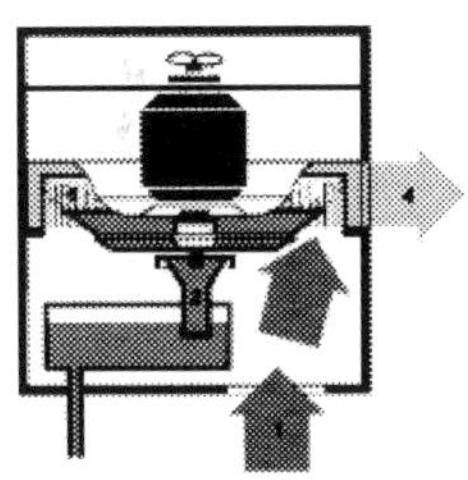

Accessories

The water is lifted upwards on the revolving PVC discs by a centrifuge, rotating on high speed (2,850 rpm).

Due to high centrifugal force, the water strikes the SS atomiser grid with great impact, where the water layer is broken into ultra-fine mist.

The flow of air carries the droplet-free aerosol/mist into the department.

Digital Humidistat: The Digital Humidistat controls the humidifier automatically and maintains the desired R.H. % in the department by simply selecting the desired humidity level with the help of the push button provided on the device. R.H. % can be set within the range from 10% to 95% to maintain the desired R.H. % in the department.

Automatic Flushing Device –Humidity Controller

The Automatic Flushing Device drains the water automatically at every 4 hr. from the water tray of humidifier. By this arrangement, the stagnant and dirty water is drained out from the water tray and new fresh water is available for the operation of humidifier. The automatic flushing device reduces the frequent maintenance of the humidifier almost negligible and makes the unit completely automatically monitored.

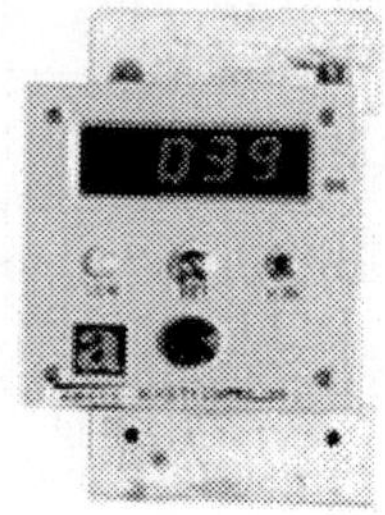

Technical Specifications

Dimension (W × H × D) in mm– 460 mm × 690 mm × 510 mm

Voltage Required–230VAC +/- 10%, 50HZ.

Air Circulation – 800 CMH (m^3/h)

Water Connection – By 3/8" Flexible Pipe

Evaporation – 6–7 L/h, and 9 L/h

Coverage – 6,000 CFT (Ft.3) Approx

Power Consumption –165 W

Water Particales Size – 5–10 microns

Weight (Dry) – 28 kg

Salient Features

No Corrosion problems, No shocking electricity bills, No wastage of waters, No water droplets carryover (water particle size 5–10 microns), No rusting of machineries, No water pump required, No plant room construction/structural modifications required, No special wiring and control panel required, No installation, operation and maintenance cost, No over flow of water (In-Built Float Valve).

Unique Features

Products of Hi-Tech R & D: The HUMIDEX & SUSPENDEX Series of Humidifiers/Atomizers are the products of intensive research and development and specially designed to solve the heat and dry air problems faced by various industries like Textiles (Spinning and Weaving), Paper, Tea, Tobacco, Nurseries, Warehousing, Food Storage Halls, Cold Storage and Cement Tiles and Pavers Block Industries.

Atomisation of Water: The Supplied water is fully converted into droplet free ultra-fine water particles of size 5–10 microns called aerosol. Hence, no need for drain and overflow connections of water in the departments.

Highly Energy Efficient: HUMIDEX ¨C 2000 Humidifier/Atomizer consumes very low power i.e. 165 W only, which is almost negligible as compared to the conventional humidification systems that too in single phase AC supply.

Accessories

Digital Humidistat: The Digital Humidist at controls the humidifier automatically and maintains the desired R.H. % in the department by simply selecting the desired humidity level with the help of the push button provided on the device. R.H. % can be set within the range from 10% to 95% to maintain the desired R.H. % in the department.

Automatic Flushing Device –Humidity Controller

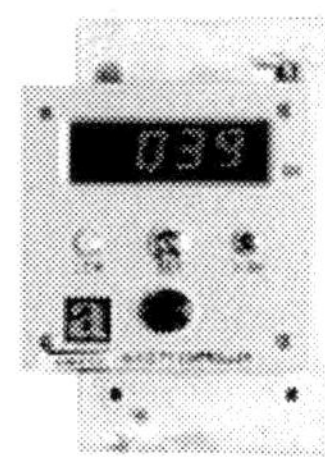

The Automatic Flushing Device drains the water automatically at every 4 Hrs. from the water tray of humidifier. By this arrangement the stagnant and dirty water is drained out from the water tray and new fresh water is available for the operation of humidifier. The Automatic Flushing Device reduces the frequent maintenance of the humidifier almost negligible and makes the unit completely automatically monitored.

Technical Specifications

Dimension (W × H × D) in mm	1020 mm × 690 mm × 510 mm
Voltage Required	230VAC +/- 10%, 50HZ.
Air Circulation	1,600 CMH (m^3/h)
Water Connection	By 3/8” Flexible Pipe
Evaporation	13–14 L/h and 18 L/h.
Coverage	12,000 CFT ($Ft.^3$) Approx
Power	330 W
Weight (Dry)	56 kg

Salient Features

- No Corrosion problems
- No shocking electricity bills
- No wastage of waters
- No water droplets carry over (water particle size 5–10 microns)

- No rusting of machineries
- No water pump required
- No plant room construction/structural modifications required
- No special wiring and control panel required
- No installation, operation and maintenance cost
- No over flow of water (In-Built Float Valve)

Unique Features

Products of Hi-Tech R & D: The HUMIDEX & SUSPENDEX series of humidifiers/atomisers are the products of intensive research and development and specially designed to solve the heat and dry air problems faced by various industries like textiles (Spinning and Weaving), paper, tea, tobacco, nurseries, warehousing, food storage halls, cold storage and cement tiles and pavers block industries.

Atomisation of Water: The supplied water is fully converted into droplet free ultra-fine water particles of size 5–10 microns called aerosol. Hence, no need for drain and overflow connections of water in the departments.

Highly Energy Efficient: HUMIDEX ¨C 2000 Humidifier/Atomizer consumes very low power i.e. 165 Wonly, which is almost negligible as compared to the conventional humidification systems that too in single phase AC supply.

30.2 Humidex – 40 Humidifier

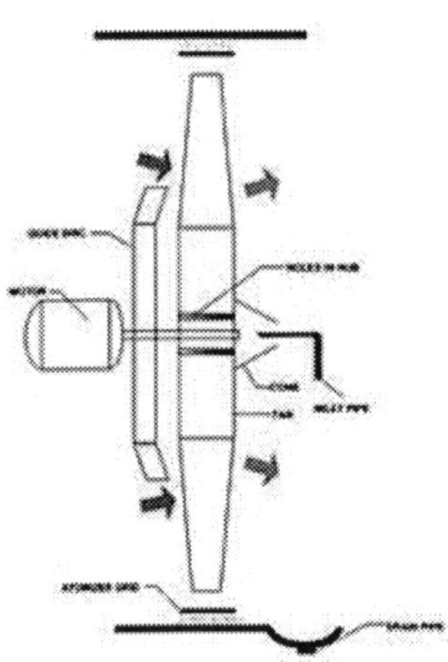

Hot and dusty air enters the unit through perforated filter plate. Fan assembly performs the dual function of blowing air and atomising water. Water at constant pressure is admitted at the centre of the fan. Water travels on guide disc through the holes in the fan hub. Due to centrifugal force, the water on the guide disc strikes the atomizer grid with great impact where the water layer is broken into the fine mist. The mist mingles with the air produced by the fan and spreads out uniformly into the room. Surplus unatomized water is drained out.

Technical Specifications

Dimensions	425mm × 422mm 5 515mm
Weight	15 kg
Air Circulation	1,700 cu.m/h.
Water Consumption	40 lit/Hr. max.
Power Consumption	350 W
Electrical Connections	230 V 50 Hz AC

Salient Features

- Corrosion Proof Construction.
- Low Power Consumption (350 W).
- Compact Heavy Duty Design.
- Light Weight and Portable.
- Easy To Install and Maintain.
- Low Initial Cost.
- 100% Factory Assembled.

30.3 Humidex – Mini, Table-top Humidifier

A sleek and compact Humidifier can be placed anywhere in your home, office, showroom and Laboratory. It is totally noiseless and maintenance free (except occasional cleaning). It cools in summer and maintains humidity evenly in winter. In short, it is highly efficient for both the seasons. It is a spot humidifier and can be used at any place.

Technical Specifications

Dimensions	360 Dia. × 230 mm Ht.
Voltage	230 V AC, 50 Hz.
Power Consumption	40 Wonly
Evaporation Capacity	0.75 to 1 L/h
Water Particle Size	5–10 microns
Water Reservoir capacity	5 litres Approx.
Air Circulation	80 CMH (m^3/Hr.)
Coverage	1,500 Ft^3 Approx.

Water Connection	By 3/8” flexible pipe
Dry Weight	3.5 kg

Salient Features

- Elegant and compact body
- Very low power consumption (40 W only)
- Totally noiseless
- Fine atomisation (5–10 microns)
- Corrosion proof, as all parts are made from engineered plastic
- Maintenance free
- Easy to operate and handle
- No nuts andbolts
- Portable
- External rotor motor
- No wastage of water
- No over flow of water (In-Built Float Valve)

Maintenance

How to Use

1. Connect the water supply to the inlet of Float Valve through 3/8” flexible pipe.
2. Let the water basin fill up to the blue level indicator.
3. Connect the 3 pin cord in 5 Amps 230 V AC supply.
4. Switch on the supply after the unit is filled with water up to the blue level indicator.
5. Ensure proper ventilation for better cooling.
6. The motor cover should never be removed.

How to Clean

To clean the plastic parts, use lukewarm water with some dish detergent. To clean the unit, proceed as follows:

Pull out the 3 pin cord from the 230 V AC supply. Press the Spinning Disc downward andunscrew the Suction Tube. Lift out the Spinning Disc and Guide Disc. Use a small brush to clean them. Rub the free shaft-end, dry and lightly oil it. Put back the Spinning Disc and Guide Disc and screw back the Suction Tube. Place the motor body assembly properly.

Do's and Don'ts

- Try to keep the unit near a window in summer (all windows should be closed during winter). The top portion should be at least 2 feet away from furniture and at least 5 feet below the ceiling.
- While cleaning, do not remove the plastic Atomizer Grid from the slots. Do not dip the unit into water or hold it directly under the running water. Make sure that water should not enter the motor at the cord outlet.
- The motor cover should never be removed.
- You can add perfumes to the water for a fragrant atmosphere. Or disinfectant for a hygienic atmosphere.

30.4 Suspendex – 360 Humidifier

'SUSPENDED'type atomisers are designed taking into consideration space constrains in the production and storage halls. The atomiser can be quickly suspended from overhead by means of chains. 'SUSPENDED'atomisers will deliver a fine mist in full 3600 pattern, 'SUSPENDED'atomisers require no pumps, no steam, no compressed air even NO FLOOR SPACE.

Salient Features

- No floor space required
- Delivers fine mist in full 3600 pattern.
- Low power consumption
- Corrosion proof construction
- Elegant andcompact design
- Portable andmaintenance free
- No wastage of water

Technical Specifications

Model	SUSPENDEX –3600
Dimensions	Dia 750 × 650
Dry Weight	20 kg
Air Circulation	800 m^3/ h
Evaporation Capacity	9 Litr. Hr.
Power Consumption	240 W
Electrical Connections	230 V + 10%, 50 Hz, A. C.

30.5 Suspendex – Mini Humidifier

SUSPENDED type atomisers are designed taking into consideration space constrains in the production and storage halls. The atomiser can be quickly suspended from overhead by means of chains. 'SUSPENDED'atomisers will deliver a fine mist in full 3600 pattern, 'SUSPENDED'atomisers require no pumps, no steam, no compressed air even NO FLOOR SPACE.

Salient Features

- No floor space required
- Delivers fine mist in full 3600 pattern.
- Low power consumption
- Corrosion proof construction
- Elegant and compact design
- Portable and maintenance free
- No wastage of water

Technical specifications

Model	SUSPENDEX –MINI
Dimensions	Dia 360 × 230
Dry Weight	4 kg
Air Circulation	80 m^3 / h
Evaporation Capacity	2 Litr. Hr.
Power Consumption	40 W
Electrical Connections	230 V + 10%, 50 Hz, A.C.

31 Coimbatore Air Controls Pvt Ltd. Coimbatore, India

31.1 Air Washer

Air washer will has single / multiple bank spray systems to suit high saturation efficiency. Also available with high-pressure pump to save the power of upto 50% of conventional type.

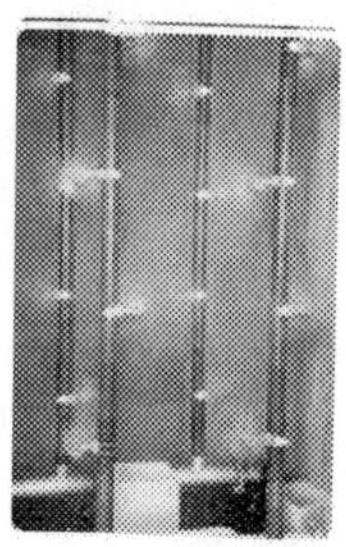

Heavy Duty Rigid PVC Piping to avoid corrosion. The spray nozzles are made of ABS / poly carbonite materials with S.S. orifice and S.S. Clamp for easy maintenance. They are specially designed for higher efficiency.

Airtight doors are made out of GI double-skinned insulated with glass wool to avoid the thermal radiation.

32 J. S. Humidifiers –Condair AG, Switzerland

32.1 HumiPac –Ceiling Mounted Humidifier

The HumiPac ceiling mounted humidifier with its patented design discretely maintains the relative humidity needed to create a refreshing and safe working environment. It can be used whether one already has an air conditioning or ventilating system or not. Hidden above the suspended ceiling, or mounted directly from the ceiling, quiet, energy efficient (only 300 W) and effective the JS HumiPac is the answer to office and workroom humidification. There are two versions available. The JS HumiPac CM20T is used above suspended ceilings and circulates air from suitable ceiling mounted intake and outlet grilles and flexible ducting. The CM20PT is for areas where no false ceiling is installed and has direct air intake and outlet louvers. HumiPac uses BS316 stainless steel, silver ion water sterilisation and flush and drain cycles to maintain absolute hygiene. Features suspended ceilings and circulates air from suitable ceiling mounted intake and outlet grilles and flexible ducting.

Sizing Data

One CM20 will provide humidification for 400 m^2 assuming outside conditions: −1oC at 100% RH, inside conditions: 21°C at 45% RH, 0.5 air changes per hour and ceiling height of 2.7 m.

To maintain a pleasant relative humidity of 45%, the moisture addition requirements in winter is approximately 3.2 g of water per m^3 of room volume.

Features

- WRAS approved for direct mains water connection
- Ceiling mounted –saves valuable floor and wall space
- Hidden –maintains original office design
- Unobtrusive –no visible sprays or steam
- Internal air filter removes up to 98% of air borne particles

- Adiabatic cooling effect
- Low energy –keeps costs down
- Easily installed and maintained
- Silver ion water sterilisation
- Safe 24 volt control connections
- Humidification section in 316 stainless steel
- Satisfies H&S (Display Screen Equipment) regulations

CM20T HumiPac –installation above suspended ceiling: Inlet air can be ducted or taken from the ceiling void. Outlet air is discharged at the point of need through adjustable ceiling mounted vents.

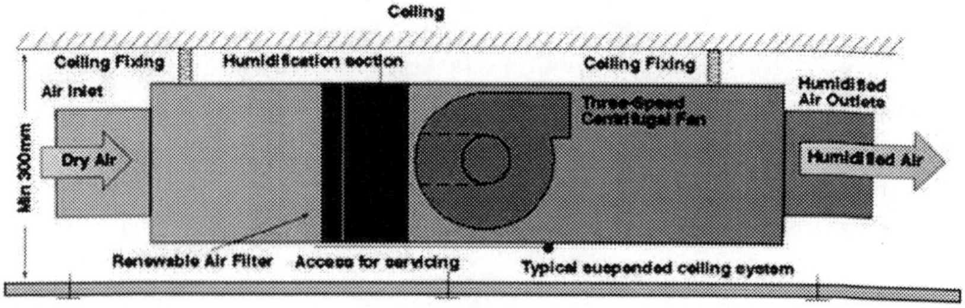

CM20PT HumiPac –exposed unit: Inlet air is drawn in through a grill and outlet air can be directed downwards through adjustable louvers.

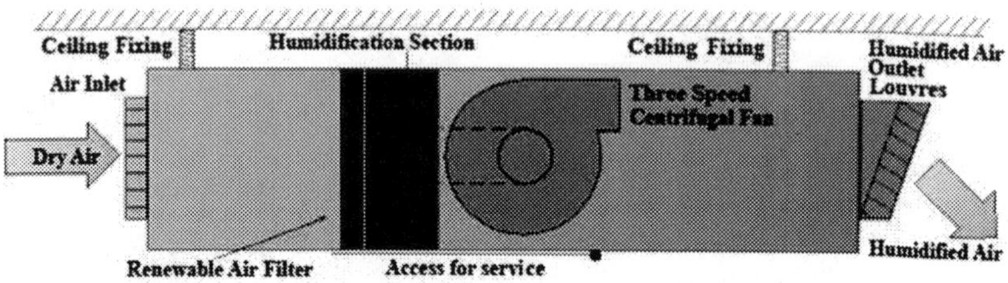

32.2 HumEvap MC2–Evaporative Humidifier/Cooler

The HumEvap MC2 sits inside a duct and raises the humidity by passing water through an evaporative matrix. The air passes through this moist matrix picking up water vapour from its wetted surface. This process consumes very little energy, far less than steam or even cold water atomisers. As well as humidification, the unit also offers up to 12°C of free evaporative cooling to an AHU system, making it a very economical alternative or supplement to mechanical cooling.It has just three 'plug and play' connections –water, drain and power – making it easy and simple to install. Therefore, this unit not only saves money in the long run through low energy use but also saves installation time and cost straight away.

Evaporative module

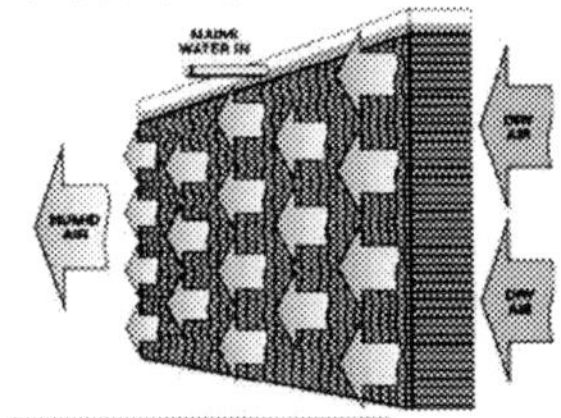

At the heart of the HumEvap MC2 is the evaporative module which is installed in an AHU or duct. Dry air passes through an absorbent glass fibre and ceramic composite matrix, picking up water vapour from its wetted surface. This process consumes very little energy, far less than steam or even cold water atomizers.

Water distribution across the top of the matrix is via durable stainless steel spray bars and uniform wetting is assured by distribution cassettes at the top of each vertical bank. As each vertical matrix bank has its own water supply, multi-stage control is an easy option where closer control is desired. For added hygiene, the evaporative media is impregnated with an exclusive anti-bacterial agent.

Features include extremely low energy use, exclusive anti-microbial impregnated matrix, quick and simple to install, high-grade stainless steel which do not rust, modular, compact space saving matrix, significant water saving technology, single phase evaporative humidifier/cooler and up to 12°C of evaporative cooling

Evaporative Cooling

The HumEvap MC2 can offer up to 12°C of free cooling and an economical alternative or supplement to mechanical chillers. As the hot outside air passes over the cool, wet surface of the evaporative matrix, there is a transfer of energy as the water evaporates. This action has two effects; it cools and cleans the air. Evaporative cooling is not a substitute for air conditioning as there are still some instances where direct mechanical cooling is required. However, as well as humidification the HumEvap MC2 provides low cost cooling without the costs associated with DX chiller units that cannot humidify in winter when the air is cool and dry.

Extremely low energy use

Humidifier	Energy consumption for 2000 operating hours
HumEvap MC2	0.5 kW
Air Atomizer	6.5 kW
Gas-fired steam humidifier	80 kW
Electric steam humidifier	76 kW

Compact Control Panel

The HumEvap MC2 Management System regulates operation of the evaporative module. This robustly constructed, micro-processor-driven unit controls the relative humidity level, flush, drain and other water quality maintenance features completely automatically. The operational interface is easy to set up and navigate and has multi-lingual options. The panel is compact at 475mm × 350mm × 155mm and can be situated up to 10m away from the evaporative module

Comprehensive real-time feedback shows RH, demand, hours run, water quality, component operating status and whether maintenance is required. Operation is programmable to suit the needs of the building, including occupancy, humidity level, proportional band and multi-stage operation.

All this is security-code protected to prevent unauthorised alteration of settings. Manual override of some functions, such as draining the unit, is available for maintenance purposes.

33 Condair Humidification and Evaporative Cooling AG,Switzerland

33.1 Draabe NanoFog Evolution High-Pressure In-room Humidifier

This quiet and discrete in-room spray humidifier is ideal for commercial buildings that do not have central AHUs. The nozzles are barely audible, just 10cm high and each can deliver up to 3kg/hr. of humidification directly to a room with rapid evaporation.

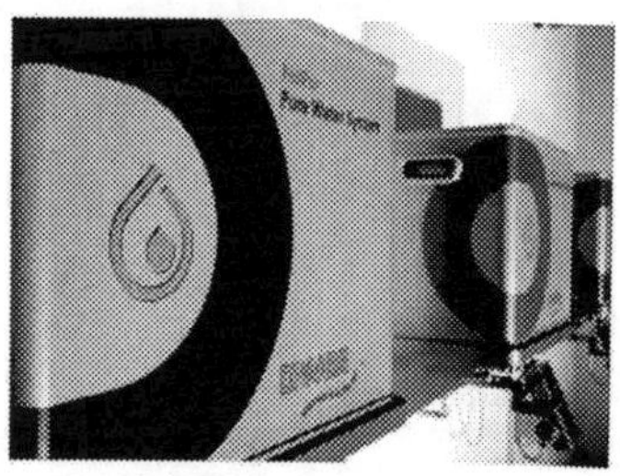

Multiple nozzles can be combined and controlled from a central panel to deliver up to 200 kg/hr, with zone control allowing humidity levels to be controlled independently in different rooms.

The system consists of a reverse osmosis (RO) water treatment system that removes the minerals from the supply water, a high-pressure pump, the in-room nozzles, hygrostats and a central control panel.Both the RO water filter and the pump are housed in self-contained modules that are exchanged with new ones delivered to site at appropriate service intervals. Used modules are returned to Condair for servicing with the new ones being easily fitted in minutes. By operating this service module exchange system, the user benefits from a very low on-site service requirement, minimal system downtime, unlimited warranty on these components and a humidification system that is reliable, efficient and hygienic.

Features and Benefits

- Ideal for commercial buildings without a central AHU
- Very quiet operation
- Small, attractive and discrete in-room nozzles
- Zone control for independent room humidity levels
- Low on-site service requirement

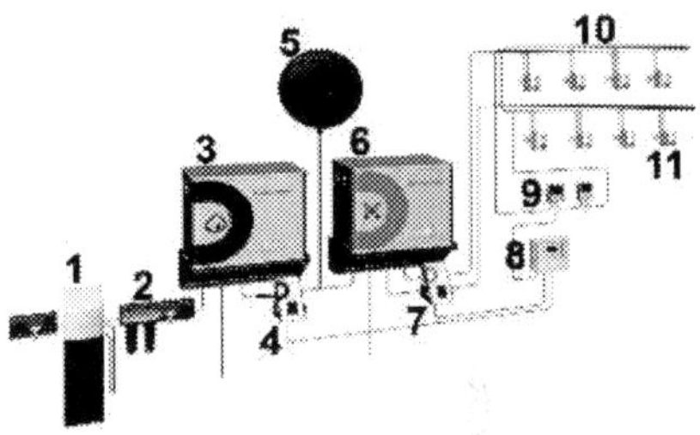

1. Water softener
2. Water filter to remove sediment from the supply water
3. RO water treatment module to purify supply water
4. RO touch screen control panel
5. Water expansion tank
6. High-pressure pump module supplying water at 85bar to the nozzles
7. High-pressure pump touch screen control panel
8. HumCenter central control panel
9. HumSpot zone control hygrostats
10. Zone A NanoFog Evolution nozzle line
11. Zone B NanoFog Evolution nozzle line

33.2 ML Princess High-Pressure Direct Air Humidifier

The ML Princess is a ceiling mounted humidifier delivering between 12 and 54l/h per distribution unit. It is ideal for use in production areas and doesnot require a compressed air supply.

An integrated fan draws air from below and distributes the moisture particles for rapid evaporation in the atmosphere and uniform humidification in a room.

The system has very low-energy consumption with a single distribution unit providing up to 54kg/h while operating on just 120W.

It has a very quiet operation at just 59dBa. Precision-engineered nozzles, water lines and valves combine with the fan unit to ensure drip-free performance. It is ideal for maintaining a humidity in industrial manufacturing environments.The humidifier operates on demineralised water to ensure minerals arenot introduced to an atmosphere.

Features and Benefits

- Low energy direct air humidifier
- Very quiet operation as no compressed air required
- Ideal for humidification of industrial manufacturing areas
- Low cost operation
- Operates on a demineralised water so does not introduce dust to the air

33.3 ML Solo High-Pressure Direct Air Humidifier

The ML Solo high-pressure humidifier provides lowenergy humidity control and evaporative cooling directly to a room's atmosphere.A series of spray units are strategically located in the area being humidified and are connected to a high-pressure pump and water treatment system. Supply water is filtered to remove any minerals and then supplied to the spray units at between 50 and 70bar. When released through the high-pressure nozzles, the water atomises and is absorbed by the air.

The ML Solo high-pressure humidifier provides low-energy humidity control and evaporative cooling directly to a room's atmosphere.A series of spray units are strategically located in the area being humidified and are connected to a high pressure pump and water treatment system. Supply water is filtered to remove any minerals and then supplied to the spray units at between 50 and 70bar. When released through the high-pressure nozzles, the water atomises and is absorbed by the air.

The nozzles can be suspended from a ceiling or mounted on a wall. Adjustable joints allow their sprays to be directed. A fan unit under the nozzles encourages evaporation and assist in directional control.Each spray unit can provide between 1.5 and 5 litres per hour and are available in white, grey or black.

Features and Benefits

- Very quiet operation
- Low running costs andenergy consumption
- Available in white, grey or black
- No compressed air needed
- Ceiling or wall mounted
- Demineralised water so no dust introduction

33.4 ML Flex High-Pressure Direct Air Humidifier

The ML Flex is a high-pressure spray humidifier for use in industrial manufacturing, horticulture, dust suppression and evaporative cooling. A series of high pressure nozzles are located on a custom made line for each application and are connected to a high-pressure pump station and water treatment system.

Minerals are removed from the supply water and delivered to the nozzles at between 60 and 70bar. On release the water atomises and is absorbed by the air raising its humidity.

The ML Flex is manufactured in high-quality stainless steel. As there is no fan unit to disperse the spray, no electrical connection is required at the point of distribution so the system can be employed in explosion proof environments. It also allows it to be used in very dusty atmospheres, such as those experienced in textile manufacturing.The nozzle line can be suspended from a ceiling or mounted on a wall.As the humidifier operates with a demineralised water supply no dust is introduced to the atmosphere.

Features and Benefits

- Very quiet operation at just 32dBa
- No compressed air required
- Very low running costs
- Stainless steel construction
- No integrated fans so ideal for dusty or explosion-proof environments

33.5 JetSpray Compressed Air and Water Spray Humidifier

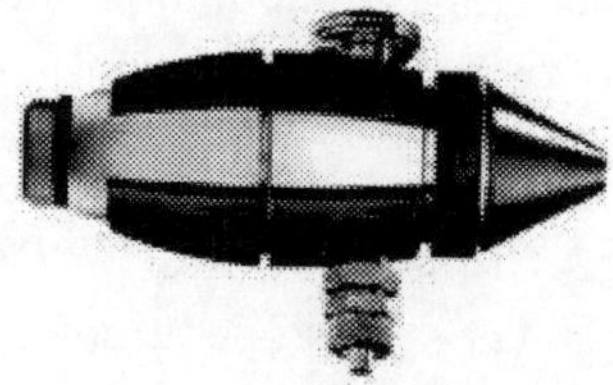

Robust operation without the need for RO water. Compressed air ensures very rapid evaporation without the risk of drips. The JetSpray provides very accurate humidity control directly to a room's atmosphere using a combination of compressed air and water. A series of strategically located nozzles provide consistent humidity throughout the space.

The JetSpray is available in 60 and 600L/h capacity models. The 600L/h model is available with on/off control or fully modulating 0–100% output for highly accurate +/-2%RH humidity control.

The use of compressed air ensures very rapid moisture evaporation, without the risk of wetting or drips, and highly directional spray aerosols.Self-cleaning nozzles enable the humidifier to work with any potable water quality without the need for a demineralised supply. Its fan-free design ensures robust operation in dirty, dusty industrial environments with minimal maintenance. The JetSpray nozzle contains a self-cleaning mechanism that prevents blockages. Installation is easy and can be carried out by any competent HVAC contractor without the need for special tools. It incorporates standard plumbing fittings and can be supplied with stainless steel, copper or plastic pipework.

Features and Benefits

- Robust self-cleaning operation ideal for industrial use
- Provides accurate and consistent humidity across an area
- Very low energy use and operating costs
- Can run on mains, softened or demineralised water
- Easy to install and operate
- 0–100% modulation for close control

33.5 Condair EL Electrode Boiler Steam Humidifier

The Condair EL electrode boiler steam humidifier provides reliable steam humidification and is easy to install, use and service. The steam cylinders last up to three times longer than similar models due to advanced water quality management. The mineral content of the water is precisely monitored and

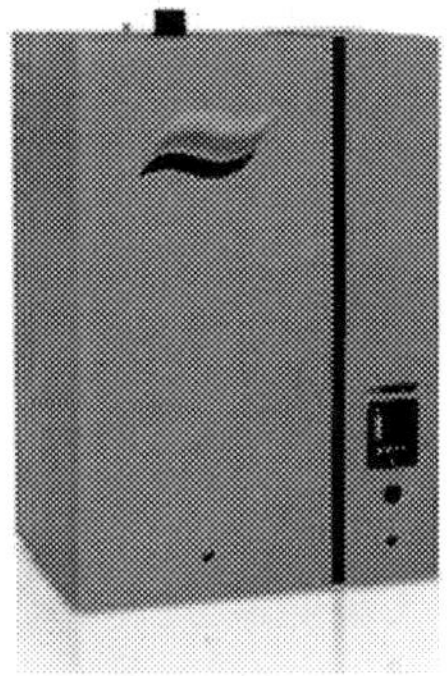

managed. Water with high mineral content is drained and fresh water added as required, thus minimising scale build-up and reducing the energy needed to heat fresh water.

When servicing is needed, the Condair EL's design makes it easy to drain, remove and replace the disposable steam cylinders, keeping downtime to a minimum. Cleanable cylinders are also available. An advanced touch screen control panel makes operation simple while providing detailed diagnostic reporting. A USB connection allows performance data to be downloaded into Excel and the humidifier's software to be easily updated.

Features and Benefits

- Reliable steam humidifier that's easy to install, use and maintain
- Long lasting steam cylinders due to innovative water mineral management
- Advanced touch screen control for easy operation and detailed reporting
- Convenient and rapid servicing through cylinder replacement
- BMS connectivity as standard
- Operates on standard mains water

33.6 Condair RS Resistive Steam Humidifier

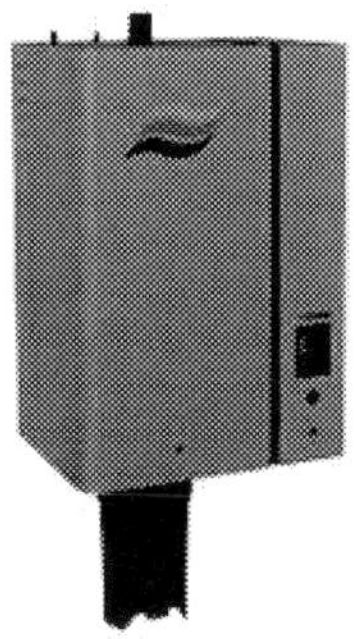

This electric steam humidifier delivers accurate humidity control without the expense of plastic disposable boiling cylinders and simplifies maintenance with its patented scale management system.Scale that forms on the heating elements breaks off under normal operation and falls into the externally located scale collector tank. Scale removal is a simple process of draining the scale collector tank with the push of a button, allowing the unit to cool then disconnecting the tank and emptying it.

The Condair RS can operate with mains water to provide ±3%RH control or on RO water to offer ±1%RH control. When operating on RO water, scale build-up is virtually eliminated, significantly reducing the unit's required maintenance.

By locating the water inlet and drain between the twin walls of the outer cylinder wall and an inner cylinder liner, the temperature is kept at a level that inhibits scale formation in this area. This 'cold water pool' prevents blockages of the inlet and drain by scale to further enable extended periods between servicing.An advanced touch screen controller provides intuitive control over operation and very detailed reporting, including downloading performance, fault and service history into Excel for advanced diagnostics. BMS connectivity is standard with BACnet and Modbus protocols.

Features and Benefits

- Patented scale management system for easy removing of scale
- Cleanable stainless steel cylinder eliminates the need for disposable plastic cylinders
- Close humidity control of ±3% with mains water and ±1% with RO water
- Can be used with RO water to significantly reduce service requirements
- 'Cold water pool' prevents water inlet and drain blockages due to scale

33.7 Condair GS Gas-Fired Steam Humidifier

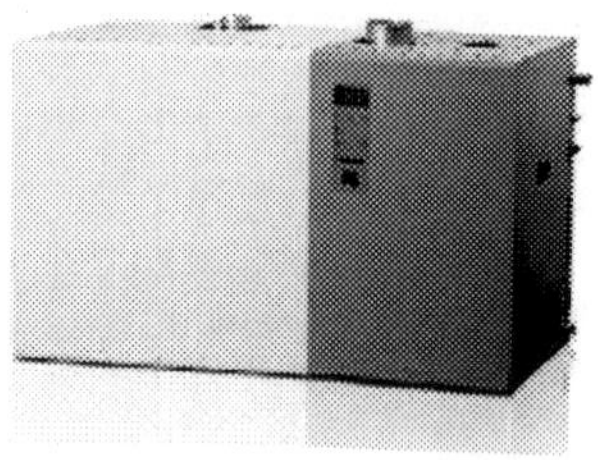

The Condair GS offers high capacity steam humidification with low operating costs. The economy of using gas makes this humidifier around 60% cheaper to power than an equivalent electric steam humidifier. This can provide a fast payback when replacing electric steam, especially on high capacity systems.

This gas-fired humidifier incorporates a 360° flame encircling the burner, which provides a high thermal efficiency of over 90%. This results in a faster response to humidification requirements, closer humidity control and minimal gas consumption.

The Condair GS has been designed to incorporate large flat heat exchangers that shed their lime-scale during operation, helping to reduce maintenance and maintain efficient heat transfer.

The humidifier can be connected to a BMS or operated via its control panel with backlit LCD display. It has a continuous self-diagnostic function with self-correction and fail-safe operation. An on-screen trouble shooting feature walks the user through corrective processes that can avoid the expense of an engineer's callout.

Features and Benefits

- 60% lower energy costs than electric steam humidification
- Large steam outputs

- Can be used with mains, demineralised or softened water
- Very efficient 360° gas burner (90% efficient)
- Certified by British Gas

33.8 Condair LS Live Steam Humidifier

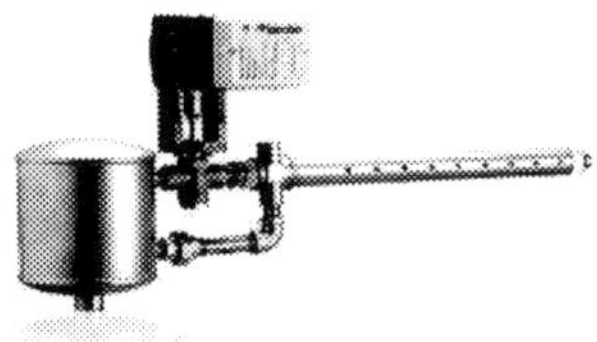

Condair LS live steam humidifiers are designed to control and distribute steam under pressure from a central steam boiler for direct introduction into a duct or AHU.Condair LS steam lances are constructed of high quality 304 stainless steel and are available in single or multiple configurations. There is a wide variety of lengths ranging from 12" to 144" to suit any duct size and give the best possible steam dispersion.

The steam lance is pre-heated with a 97% jacketed exterior. When steam is first introduced to the humidifier it flows around this outer jacket. This heats the inner lance so that when steam is introduced into it, the walls of the lance are hot and therefore condensation is prevented. This ensures only hot, dry steam is released into the duct without any 'spitting'.

An optional stainless steel insulation jacket encompasses 1/2" fibreglass insulation to improve efficiency and minimize heat transfer to the air stream.

Features and Benefits

- Economic steam humidification using a building's existing steam network
- High capacity humidification up to 1,452kg/hr.
- Lance encased in 97% pre-heated jacket to avoid condensation and 'spitting' into the duct
- Electric or pneumatic actuator
- Optional fibre-glass insulation to avoid heat transfer to air stream.

33.9 Condair CP3 Mini Low Capacity Steam Humidifier

The Condair CP3 Mini will provide up to 4kg/hr. of steam either directly to a room or into an AHU. It has a very attractive yet inconspicuous design with concealed rear connections for power and water.This low capacity steam humidifier can be serviced without needing to call a technician. All electronics and current carrying parts are protected against water splashes and inadvertent contact allowing cylinders to be changed safely. A large LCD display provides information on all operating parameters and the unit can be controlled with the integrated keypad. An optional battery-powered humidity sensor can be located in the room and transmits a signal back to the humidifier to control output to the desired level.

The Condair CP3 Mini operates on mains water and is ideal for both homes and offices.

Features and Benefits

- Discreet, low capacity steam humidifier
- Ideal for domestic or light commercial applications
- Concealed connections for an attractive appearance
- Easy to service without needing a technician
- Can be used in-room or in-duct

33.10 Condair EL Steam Humidifier with Fan Unit

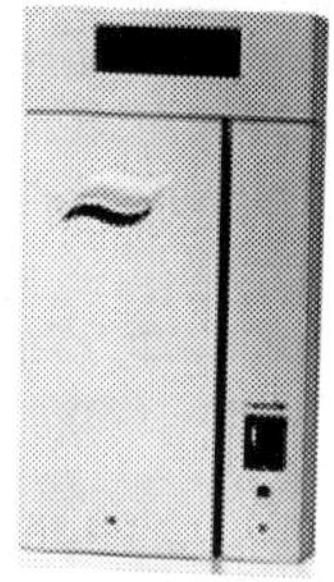

This electrode boiler steam humidifier provides humidification directly to a room's atmosphere. Its fan unit can provide up to 45kg of steam per hour making it ideal for direct room applications such as laboratories, printers, bikram yoga studios or manufacturing facilities.The fan unit can be mounted directly on top of the humidifier or remotely and connected to the humidifier with an insulated steam hose.The unit is easy to install, simple to use and very reliable. It operates on regular mains water and incorporates many innovative features to ensure long operational periods between service requirements. When lime scale build-up reaches a high level, the disposable steam cylinders are easy to remove and replace, keeping downtime to a minimum.

Automatic water conductivity monitoring and staged electrode use improves the efficiency of the cylinders while a pumped drain removes limescale from the system minimising service requirements.The Condair EL incorporates the latest touch screen technology to provide intuitive control and advanced reporting features.

Features and benefits

- Provides pure hygienic steam humidification directly to a room's atmosphere

- Easy to install, use and service
- Long last steam cylinders due to innovative water mineral management
- Advanced touch screen control for easy use and detailed reporting
- Operate on standard mains drinking water

33.11 Condair OptiSorp Short Evaporation Steam Distribution Manifold

The OptiSorp steam distribution manifold shortens the evaporation distance of steam released inside a duct or AHU. This prevents steam from coming into contact with components or corners within the AHU and ensures reliable and hygienic steam humidification.Tiny patented inner nozzles on the manifold draw steam from the centre of the steam pipe where it is at its hottest and driest. This guarantees that steam is introduced to the AHU without 'spitting' that can occur when condensation forms around the cooler outer surface of the pipe.A uniform spread of these nozzled outlets across the whole manifold provides a consistent release of dry steam across the air stream and reduces the evaporation distance compared to standard steam pipes.

The OptiSorp is made of robust stainless steel for a long operational life.

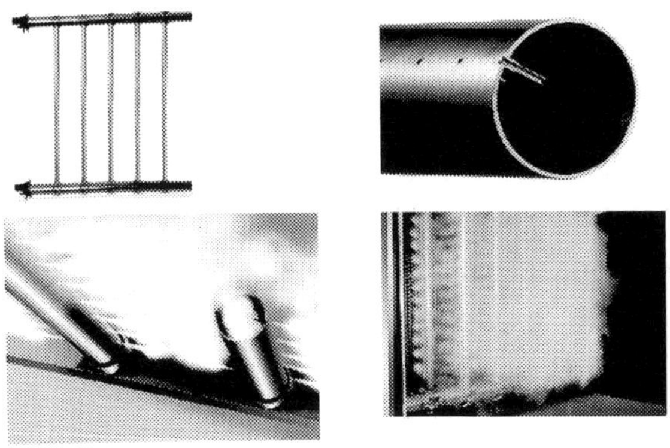

Hot dry steam is distributed uniformly across the duct

Features and Benefits

- Up to four times short evaporation than standard steam pipes
- Patented inner nozzles guarantee dry steam deliver
- Uniform steam distribution in duct for accurate humidity control
- Stainless steel construction for long operational life

Appendix – 2

Cooling and heating systems

1 NewTech cooling towers – Mumbai, India

NewTech Cooling Towers are made from corrosion proof-engineered plastics, and hence the cooling tower does not rust, flake, chip, peel and does not need painting or protective coating. These are totally seamless cooling towers, i.e. there are no seems panel revert or hundreds of fasteners to fail or compromise the performance of the product. Known as a heat-transfer unit, it is used to remove heat from any water-cooled system. Assembled with high efficiency fan it features a wide-chord design for maximum efficiency at low fan tip speeds. It is also featured with gravity flow distribution system that allows easy and non-restrictive maintenance. Provided with 15 years of warranty on the cooling towers structural shell and 18 months warranty on Cooling Tower motors.

1.1 Air cooled chilling plants

Made from best available compressor and constructed with quality components. Durable, dependable and compatible in nature, it is fully-integrated, turnkey cooling systems. Fully wired and charged and tested under load conditions to ensure easy installation and successful start-up. The plant is manufactured with non-ferrous material to ensure a corrosion-free environment. Being compact in size it is ideal for environment where physical space is at prime concern, or for small-scale process cooling applications in the plastics, pharmaceutical, chemical processing and metal finishing industries.

Other salient features

Imported / Indigenous Compressor, Imported Refrigeration Controls, Air Cooled Condenser, S.S. Tank insulated with PUF, Electric Control Panel, Pipe Section Body, Mobile and Compact

1.2 Water cooled chilling plant

Manufactured for indoor location, Water Cooled Chilling Plant comes complete with an integrated pump/tank. Suitable for process cooling applications. It is totally assembled with premium quality components and test-run before dispatch. Fitted with highly efficient stainless steel water pumps and stainless steel tanks, it is rated for ambient and cooling water temperatures.

Featured with Refrigeration Controls, Electric Panel Board it is placed on a common base which is compact and movable on wheels.

Other salient features

Imported / Indigenous Compressor, S.S. Tank insulated with PUF, Imported Refrigeration Controls, Electric Control Panel, Water Cooled Condenser, Pipe Section Body, Mobile & Compact and FRP Cooling Tower.

2 Airtech Faridabad, India

2.1 Chilling plant

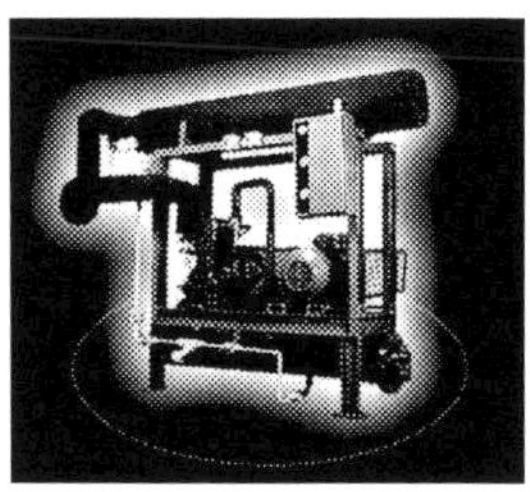

Airtech Chilling Plant is being manufactured from 3 TR to 150 TR capacities in compact design covering minimum floor area.

Reciprocating compressor of reputed indigenous make.

Condenser is Compact design shell and tube type. It is being tested hydrostatically at 20 kg/cm^2

Condensed liquid refrigerant is being expanded in DX Type Chiller through expansion valve. Water is being circulated through Chiller shell side having baffles arrangement; chiller is also being manufactured and tested hydrostatically 20 kg/cm^2 pressure.

Low pressure, high pressure cutout is being provided for the safety of plant. Thermostat is provided for automatic tripping, in case when temperature gets maintained.

Model	ACHW-3	ACHW-5	ACHW-7.5	ACHW-10	ACHW-15	ACHW-20	ACHW-25	ACHW-30
Temperature Range	5°–15°C	5°–15°C	5°–15°C	5°–15°C	5°–15°C	5°–15°C	5°–15°C	5°–15°C
Refrigeration Capacity	9,000 Kcal/hr	15,000 Kcal/hr	22,500 Kcal/hr	30,000 Kcal/hr	45,000 Kcal/hr	60,000 Kcal/hr	75,000 Kcal/hr	90,000 Kcal/hr
Refrigerant	R-22/R-12	R-22/R-12	R-22/R-12	R-22/R-12	R-22/R-12	R-22/R-12	R-22/R-12	R-22/R-12
Tonnage (TR)	3	5	7.5	10	15	20	25	30
Power Supply	415V, 50 Hz, AC, 3 PH.	415V, 50 Hz, AC, 3 PH.	415V, 50 Hz, AC, 3 PH.	415V, 50 Hz, AC, 3 PH.	415V, 50 Hz, AC, 3 PH.	415V, 50 Hz, AC, 3 PH.	415V, 50 Hz, AC, 3 PH.	415V, 50 Hz, AC, 3 PH.
Compressor	Open Type	Open Type	Open Type	Open Type	Open Type	Open Type	Open Type	Open Type
Chiller	Dx-Type	Dx-Type	Dx-Type	Dx-Type	Dx-Type	Dx-Type	Dx-Type	Dx-Type
Condenser	Water Cooled	Water Cooled	Water Cooled	Water Cooled	Water Cooled	Water Cooled	Water Cooled	Water Cooled
Chilled Water Flow (LPH)	2,250	3,750	5,625	7,500	11,250	15,000	18,750	22,500

- Due to continuous product development and improvement specifications and technical details are subject to change without notice.
- Capacities are based on 11°C temperature of water.
- ΔT for chiller shall be 4°C.
- Condensed, water inlet temperature should be 32°C Max.
- Higher capacity chilling plants are also available on demand and fabricated accordingly to temp requirements

Airtech pressurisation and air cooling plant are used to supply the filtered cool air in the industrial shed to avoid outside dust and foreign material entry into the working zone. This also helps in increasing work efficiency, especially in the dry and hot season of May and June. These types of plants involve low initial investment as compare to air conditioning plants.

Salient features include Compact Blower Modular Design, Rugged and Precision Engineering, Economical Construction, Filtration up to maximum peak efficiency and available from 10,000 CFM to 200,000 CFM.

2.2 Ammonia refrigeration plant

Airtech Ammonia Refrigeration plants are designed and manufactured with minute quality control installation with an aim of trouble-free operation of

plant. All major parts such as Oil Separator, Condensers, Ice Bank Tank System & IBT Coil, Receiver Accumulator, Air Cooling Unit are manufactured in-house. They have Non CFC Ammonia Gas, hence low running cost and easy maintenance. Normally used for Water Chilling Plant, Brine Chilling Plant, Cold Storage, Deep Freezing Chamber, Gas Liquefaction Plants, Ice Plant and Freezing Complex.

Scientifically designed cooling system combined with high-efficiency condenser, chiller, compressor, expansion valve provide more efficient water cooling with consequent reduction in operating cost. Quality minimises the maintenance but motor requires lubrication very often.

2.3 FRP Induced draft counter flow cooling towers

Salient Features

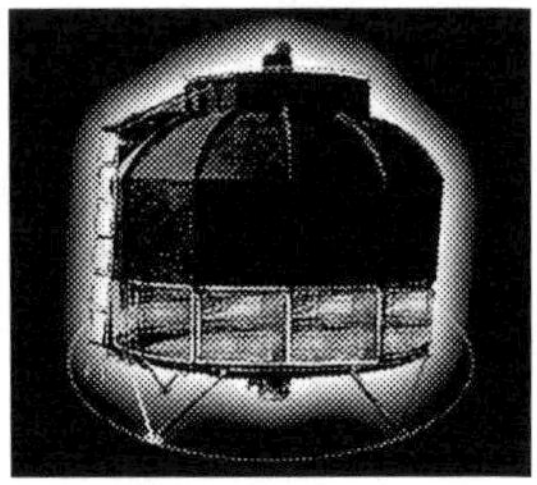

In_FRP Cooling Tower, metallic parts are fully galvanised/powder coated/epoxy coated and hence completely corrosion resistant. Axial fans are direct driven. Bottle-type design to withstand high wind velocity, sudden shocks and vibrations. Unit can absorb air from all directions and is absolutely leak proof.

Honey comb PVC fill – increases surface area, splits incoming air and water into several streams spreading equally across the tower cross section. Can take water temperature up to 80°C. The unit is aesthetically designed, compact and low weight.

Specification

Model	Nominal Capacity y TR	Heat Rejection Kcal/Hr x 1000	Dimensions Height mm	Dimensions Diameter mm	Fan Dia mm	Motor HP	Operating Weight (kg)	Pump Head mm H_2O
ACT - 7.5	7.5	28	2,000	1,000	600	1.0	250	2,000
ACT - 15	15	56	2,000	1,000	700	1.0	320	2,000
ACT - 25	25	94	2,200	1,800	900	2.0	450	2,400
ACT - 30	30	113	2,200	1,800	900	2.0	500	2,400
ACT - 40	40	150	2,800	2,100	1,200	2.0	700	3,000
ACT - 50	50	189	2,800	2,100	1,200	2.0	800	3,000
ACT - 60	60	226	2,800	2,100	1,200	3.0	1,300	3,000

Contd...

Model	Nominal Capacity y TR	Heat Rejection Kcal/Hr x 1000	Dimensions		Fan Dia mm	Motor HP	Operating Weight (kg)	Pump Head mm H_2O
			Height mm	Diameter mm				
ACT - 80	80	300	3,000	2,800	1,500	3.0	1,500	3,000
ACT - 100	100	378	3,000	2,800	1,500	5.0	1,500	3,000
ACT - 125	125	469	3,000	2,800	1,500	5.0	1,700	3,200
ACT - 150	150	563	3,000	2,800	1,500	7.5	1,800	3,200
ACT - 175	175	656	3,200	3,600	1,800	7.5	2,100	3,400
ACT - 200	200	750	3,200	3,600	1,800	10.0	2300	3,400
ACT - 250	250	938	3,500	4,000	1,800	10.0	3000	3,700
ACT - 300	300	1126	3,500	4,000	1,800	10.0	3500	3,700
ACT - 400	400	1500	3,700	5,800	3,000	15.0	5500	4,000

***Note:-** Nominal data based on 15 L.P.M. (4 USGPM per TR) Water Inlet 36.4°C (97.5°F), Water outlet 32.2°C (90.0°F) and Wet Bulb 28.3°C (83.0°F). Bigger capacities cooling towers are available on demand

3 Air filt Technologies P Limited, Delhi

These evaporative coolers use the principle of simple evaporation of water in air. These cooling devices and are quite effective in moderate humidity locations and are cost-effective finding use in many fields. In highly humid climates, evaporative coolers provide thermal comfort benefit apart from the increased ventilation and air movement. The units are available in single-skinned and double-skinned PUF insulation. It is odourless and provides a dust-free environment.

The Evaporative Coolers are made out of GI sheet sections and mainly consists of DIDW blowers / axial flow fans, pre-filter, water circulation pumps and cooling pads. Evaporative Coolers can be supplied in single skinned and double skinned (PUF insulation).

Features include high rate of cooling without excessive relative humidity, providing dust-free environment

Unlike the conventional coolers this system is odorless and has low running cost.

Specifications:

- Available in a wide range from 1,200 CFM to 40,000 CFM.
- Hollow extruded aluminIum profile structure with sandwich type P.U.F. injection panels.
- Construction out of pre-coated plasticised sheet.
- Density of P.U.F. insulation.
- Corners, joints and handles are made of Nylon Die Cast.

- Vibration isolators mounted under the blower.
- Fireproof talon canvas connection between the blower outlet and the main body.
- Choice of top, bottom and side discharge.
- Dynamically balanced DIDW blowers.
- Self-priming/monoblock pump.

3.1 Single skin evaporative coolers

Airfilt Technologies is one of the prominent manufacturers and suppliers of Single Skin Evaporation Coolers that can be availed at industry leading prices. To cater to the varied clients' demands, these high functionality coolers have an open-able design to allow field assembly. Equipped with dynamically balanced DIDW blowers and self-priming monoblock pump, the entire assemblage is tested for its performance, before transmission.

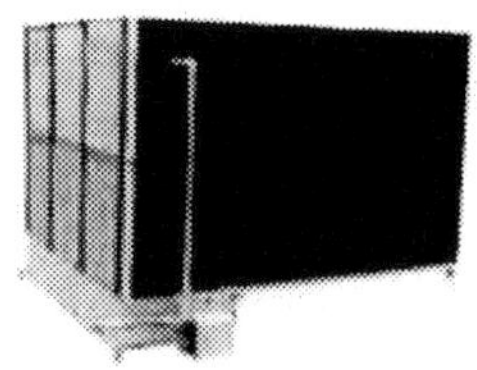

Features:

- Energy-efficient
- Longer service life
- Rugged in construction

Specifications:

Available in a wide range from: 3,000 CFM to 60,000 CFM.

Sheet metal bolted construction with angle iron base SS tank /18 SWG stainless steel tank.

Openable design to allow field assembly.

Varoon isolators mounted under the blower.

Fire proof Tatrlon canvas connection between the blower outlet and the main body.

Choice of top, bottom and side discharge.

Dynamically balanced blowers.

Self-priming / monoblock pump.

3.2 Double-skin evaporative coolers

Our company is considered a reputed name to manufacture and supply Double Skin Evaporative Coolers that come in hollow extruded aluminium profile structure, with sandwich-type P.U.F. injection panels. Handed over in different CFM ranges, these efficient coolers have vibration isolators mounted under the blower.

Features:

- Excellent performance
- Tough in construction
- Easy to maintain

Specifications:

- Available in a wide range from: 1,200 CFM to 40,000 CFM
- Hollow extruded aluminium profile structure with sandwich type P.U.F. injection panels
- Construction out of pre-cooled plasticised sheet
- Density of 0-U.F. Insulation is 40 kg/rns.
- Corners, joints 8 handles are made of Nylon pre cast.
- Vibration Isolators mounted under the blower.
- Fire proof TatrIon canvas connection between the blower outlet and the main body.
- Choice of top, bottom and side discharge.
- Dynamically balanced D1OW blowers.

4 Nuberg Engineering Pvt. Ltd. Noid

4.1 Air washer and textile air conditioning plants

This system has unique feature of close control of temperature as well as humidity. The main features include refrigeration equipment, machine working in conjunction with cooling tower and pumps, using of chilled water to control inside design conditions on by means of cooling coils placed within the air flow to be cooled. Cooling coils will be chilled water type. This system has the advantage to maintain close control of temperature and humidity uniformly in all seasons and with varied equipment load and help in long life of textile machinery.

5 Farnam Custom Products – Arden, USA

5.1 Custom frame duct heaters

Farnam's open frame duct heaters are very rugged and efficient air heaters. They are designed to heat large volumes of air in ducts. The open designs provide very low pressure drops across these heaters. Their low static pressures allow customers to use smaller and more economical blowers since there is

little restriction of air flow. Systems can be more efficient. These heaters themselves are inherently very efficient since the circuits are exposed directly to the air flow. So, heat transfer from the wire to air is direct and immediate. As a result, these heaters give their heat readily to the air and remain relatively cool themselves which affords long life. Fast warm-up and cool-down times are inherent advantages.

Heaters are designed for easy installation into a duct. There are several standard sizes available which can be tailored for particular needs. Voltages up to 480 V single or three phase are available and have a long history of trouble-free service in the field. The inherent efficiency of the design assures that excessive temperatures don't build up in the heater to shorten its life. And, since the heater is so open, it does not congest or tax the overall air delivery system.

5.2 Heat Torch™ Process Heater

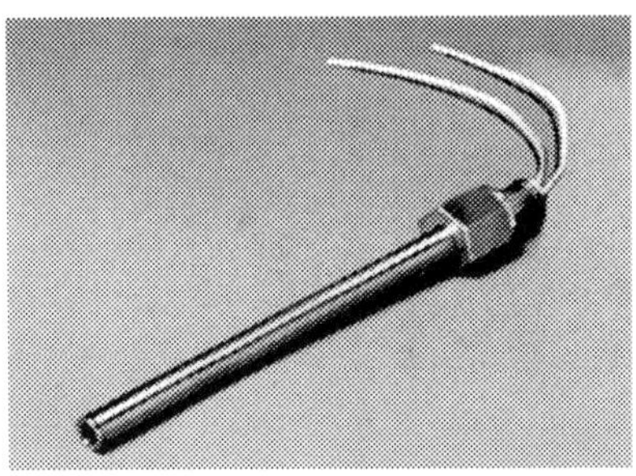

The Heat Torch™ 030 is the smallest of Farnam Custom Products' standard process heaters. These 8.6 mm diameter air heaters are ideal for concentrating hot air streams into small areas. It is a highly efficient tool with relatively negligible heat losses. The small size and focused air stream support great precision. These air heaters operate best with air flow in the range of 0.3 to 6 Standard Cubic Feet per Minute. Power options are available up to 300 W in either 120 V or 240 V supply. Maximum recommended output temperature for our smallest Heat Torch process heaters is 700°C (1,292°F). There are many applications for these hot air tools such as: melting, cutting, adhesive activation, de-soldering, heat staking and sterilisation

Wattage: 50–300 Ws,

Voltage: 120 V, 50–300 W, 240 V, 200–300 W

Inlet Fitting: Inlet fitting, #10-32 UNF thread,

Exhaust Fitting: NF – no fitting

5.3 Flow torches

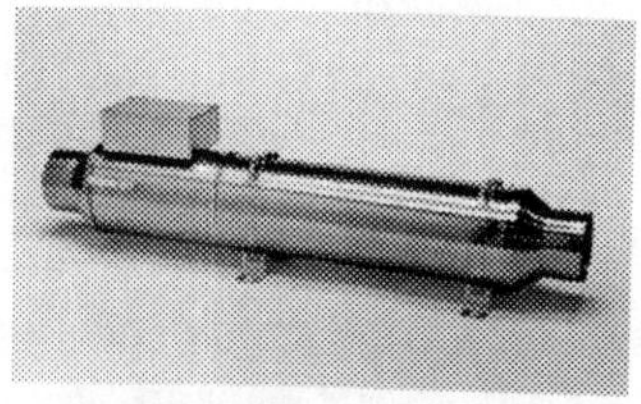

ARDEN, NC, 1/1/07 – Farnam Custom Products announced their new line of high-performance Flow Torches today. These open coil heaters are designed for high flow rates and low-pressure drop due to their efficient design and minimal flow restrictions. The robust construction of these stainless steel heaters offers reliable, long-life performance and reduced operational cost.

Standard power from 1 kW to 75 kW is available for a wide range of applications, including hot air curtains, air drying, dehumidification, paint baking/drying, plastic curing and exhaust gas heating. Airflow of up to 2,000 SCFM and up to 3 PSIG can be handled by the 8" Flow Torch. Maximum output temperature is 932°F, depending upon airflow. Farnam works closely with OEMs, plant engineers and end product designers to provide custom air heaters for a broad array of applications. Additional specifications and contact information are available at: www.farnam-custom.com.

5.4 Axial fan heaters

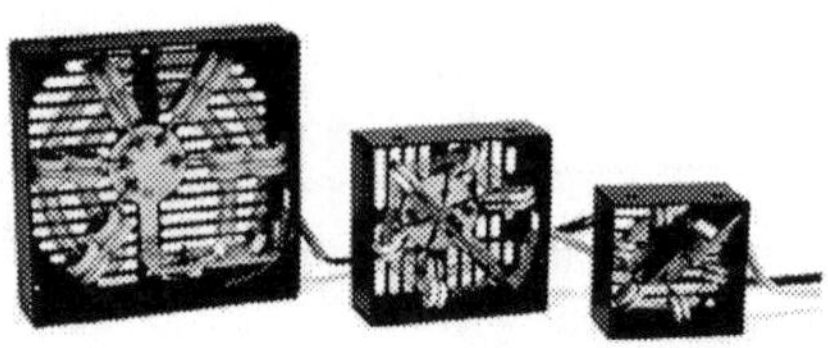

Axial Fan Heaters (Series AF-20) mount directly to square axial fans. The AF-20 heater housing uses the same mounting holes and functions as a guard for both heaters and fans. (Fans available separately.) They are available in several sizes with a broad range of wattages and custom features, which are ideal for drying, curing and warming applications. All Axial Fan Heaters are UL recognized components under UL file number E154979.

6 GEM Equipments (P) Ltd, Coimbatore, India

6.1 Refrigerated air dryer

Compressed refrigerated air dryer

Designed to have better turbulence and maximised heat transfer rate. Salient features include compact design, low-pressure drop, consistent dew point,

power saving, high-quality finishing, non-cyclic system, more reliability, ease of installation, environment friendly and reduced maintenance.

Unique 7x Heat Exchanger with moisture separator used is a combination of pre-cooler, evaporator and built in stainless steel demister moisture separator. Pre-cooler (air to air heat exchanger) and evaporator (air to refrigerant heat exchanger) are cross flow type with more cross-sectional area for flow to make it non-clogging.

Heat Exchanger is designed to have better turbulence and maximised heat transfer rate. Lesser difference in temperature between inlet air and outlet air ensures better effectiveness. Piping is completely eliminated and the air passage is non-ferrous. The heat exchanger assembly is encapsulated in PUF Insulation (Eco Friendly) to prevent the loss of cooling effect.

Special feature of 7x tower includes high efficiency, corrosion resistance, lesser weight, compact in size, ease of access to inlet and outlet, user friendly and less maintenance.

Dependable automatic condensate drain with the following features:

- Adjustable electronic timer controlled, pilot operated / direct, compressed air powered, auto drain valve.
- Positive discharge of even heavily contaminated condensate.

Non-Cyclic Refrigeration System: Hot gas by-pass value automatically maintains dew point temperature across a wide range and ambient conditions without the need any adjustments. High pressure, high temperature refrigerant vapour is introduced after the expansion valve to ensure temperature control. Direct expansion, non-cycling allows rapid response to changes in operating conditions.

Installation and Service

- *Ease of Installation*: All dryers are shipped pre-piped and wired, ready to install and operate, installation is made easy with conveniently located air and drain connections.
- *Electrical*: In accordance with applicable codes with professional harnessing and wires as per IS: 694-1990. Compressors for larger capacity dryers are protected with overloads and safety trips.
- *Environmental*: GEM Dryers are designed to have low-energy usage, helping to conserve the earth's resources and minimise pollution. Refrigerants are with zero ozone depletion factor, and thereby making

GEM dryers' 'Ozone Friendly'.

- *Service*: GEM Dryers are designed to require little maintenance. In case a service is necessary, a team of trained technicians is available to answer your questions about installation, operation and maintenance or repair. A complete inventory of spare parts is maintained at the factory and channel partners and local service providers located all over India.

Specifications

Base model	Model variance		Nominal capacity		Maximum pressure#		Electrical connection		Air connection	Rated power *kW	
	B	E,F	cfm	m^3/h	bar g	psi g	220V/ 1Ø/ 50Hz	415V/ 3Ø/ 50Hz	In/out	Air cooled	Water cooled
2KD+ 002	✓		20	34	16	232	✓		½"BSP(F)	0.15	
2KD+ 004	✓		40	68	16	232	✓		½"BSP(F)	0.2	
2KD+ 006	✓		60	102	16	232	✓		1"BSP(F)	0.4	
2KD+ 008	✓		80	136	16	232	✓		1"BSP(F)	0.5	
2KD+ 010	✓		100	170	16	232	✓		1"BSP(F)	0.6	
2KD+ 015	✓		150	255	16	232		✓	1½"BSP(F)	0.9	
2KD+ 020	✓		200	340	16	232		✓	1½"BSP(F)	1.2	
2KD+ 025	✓		250	425	16	232		✓	2"BSP(F)	1.4	
2KD+ 030	✓		300	510	16	232		✓	2"BSP(F)	1.6	
2KD+ 040	✓		400	680	16	232		✓	2"BSP(F)	1.9	
2KD+ 050	✓		500	850	16	232		✓	2"BSP(F)	2.3	
2KD+ 060	✓		600	1020	16	232		✓	2½"BSP(F)	2.8	
2KD+ 075		✓	750	1275	16	232		✓	3"NB ASME Flg	3.8	3
2KD+ 100		✓	1000	1700	16	232		✓	3"NB ASME Flg	5	4.1
2KD+ 125		✓	1250	2125	16	232		✓	4"NB ASME Flg	5.7	4.8
2KD+ 150		✓	1500	2550	12.5	180		✓	5"NB ASME Flg	6.8	6.2
2KD+ 200		✓	2000	3400	12.5	180		✓	5"NB ASME Flg	8.7	8
2KD+ 250		✓	2500	4100	12.5	180		✓	6"NB ASME Flg	11	9.6

Wall-Mounted refrigerated air dryer

Compact, lightweight and easy to operate. Salient features include compact design, wall mounting type, ecofriendly system, consistent dew point, low-pressure drop and power saving.

General Features

- Copper 'Tube in Tube' non-corrosive heat exchanger.
- Inbuilt Non-Corrosive Centrifugal specially treated Aluminium Moisture Separator.
- PUF Insulation for Heat Exchanger Assembly.

- Timer based Zero air loss Auto Drain Valve.
- Filled with 'Ozone Friendly' R134a refrigerant.
- Wall Mounting type – can be mounted on any machine tool.
- Noiseless operation.

Digital Dew Point Indicator: Digital Dew Point indicator which illustrates the pressure dew point accurately. Inbuilt adjustable timer for Auto Drain Valve.

Specifications

Model	Air flow		Work pressure	Power consumption	Dimensions (mm)			Weight
	CFM	m³/hr	barg(g)	KW	A	B	C	Kg
2KW010	10	17	16	0.15	520	230	450	35
2KW020	20	34	16	0.21	520	230	450	38
2KW040	40	68	16	0.375	870	230	500	45
Electricty connection						230V, 1Ø,50HZ		
Refrigrant						R134A		
Inlet/Outlet						½"BSP		

Working Conditions

Pressure	7 to 16 bar
Inlet Temperature	45°C
Ambient Temperature	35°C (40°C Max)
Pressure Dew point	3°C
Atmosphere	Dust/Dirt Free

High-pressure refrigerated air dryer

Designed to have low pressure drop and higher efficiency. Salient Features include complete range with ozone friendly refrigerant, compact in size, wide range, lower power consumption, reliable components and consistent dew point.

General Features

- Copper 'Tube in Tube' non-corrosive heat exchanger, designed to have lower pressure drop and higher efficiency contains inbuilt centrifugal moisture separator having dead zone to avoid carryover of condensate.
- PUF insulation for heat exchanger assembly.
- Electronic timer operated auto drain valve.

- Universally available refrigeration components such as compressor, fan motor, filter dryer and expansion valve.
- Non cyclic refrigeration systems, with the help of hot gas bye pass solenoid valve, for varying load condition and seasonal adjustments.
- GEM PET series dryers are filled with 'Ozone Friendly' R134a refrigerant.

Installation and Service

- Installation: GEM PET dryers are pre-wired/pre-piped and are ready to install and operate.
- Electrical: Electrical is in accordance with application code and uses international components.
- Serviceability: A 'Fit and Forget' unit and requires least maintenance (trouble shooting and installation guidelines are given in manual that will be supplied along with dryer).

Specifications

Base Model	Air flow		Ref. Power	Electrical connection		Inlet/Outlet	Dimension (mm)			Weight
	CFM	m³/hr	kW*	220V/ 1Ø/50Hz	415V/ 3Ø/50Hz	BSB	A	B	C	Kg
2KD0027	20	34	0.21	✓	–	½"	360	475	550	45
2KD0047	40	68	0.36	✓	–	½"	360	475	550	47
2KD0067	60	100	0.36	✓	–	½"	360	475	550	47
2KD0087	80	136	0.48	✓	–	1"	500	600	730	78
2KD0107	100	170	0.6	✓	–	1"	500	600	730	90
2KD0157	150	255	0.68	✓	–	1"	500	600	730	90
2KD0207	200	340	1	✓	–	1½"	700	700	830	120
2KD0257	250	425	1.4	✓	–	1½"	700	700	830	135
2KD0307	300	510	1.4	✓	–	1½"	700	700	830	135
2KD0407	400	680	1.8	✓	–	2"	900	900	1230	170
2KD0507	500	850	2.2	✓	–	2" NB flg	900	900	1230	180
2KD0757	750	1275	3.3	✓	–	2" NB flg	750	100	1400	250
2KD1007	1000	1700	4.2	–	✓	2" NB flg	900	1200	1475	350

Nexgen Refrigerated Air Dryer

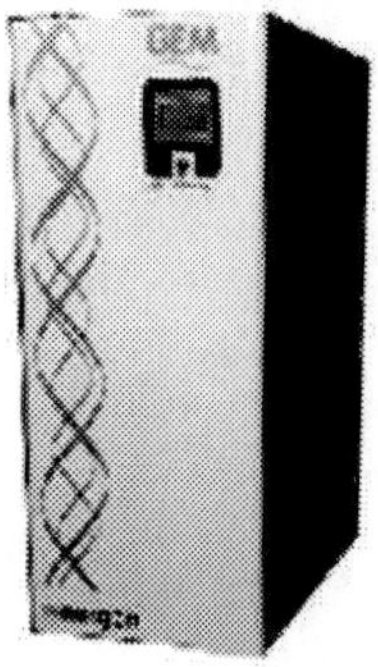

Salient Features:

Compact, lightweight & easy to operate

Low Pressure Drop

Consistent Dew Point

Power Saving

High Quality Finishing

More Reliability

Ease of Installation
Environment Friendly
Reduced Maintenance

Specifications

Tower

- Non-corrosive aluminium body
- Amply sized to save heat of adsorption.
- Large desiccant candle, ensures sufficient contact time to produce −40°C dew point (−60°C with molecular sieves).
- Each tower is provided with pressure gauge.

Filters

- Each dryer is equipped with a pre-filter to protect desiccant from free water contamination, dust and scales, etc. and an after filter to avoid desiccant dust carry over.

Pressure Gauge

- Pressure gauges on each tower for easier illustration of working of dryer.

Valves

- Highly reliable solenoid valves, to international standards, which act as main flow and exhaust valves. Precision orifice for highly accurate purge flow.

Why super pack instead of other dryers?

- Occupies no floor space as it is wall mounted.
- Easy to operate.
- No moving parts and hence high reliability.
- Least maintenance required.
- Electronic timer enables accurate cycle timing.

General Features

- Well-designed mounting brackets
- Purge exhaust muffler for noise control.
- Highly accurate electronic timer.
- Dried air as per requirements of ISO 8573-1 class 3-3-4.
- Designed to meet the requirements of ISO 7183.
- Quick exhaust facilitates sudden de-pressurisation and thereby better regeneration.

- Re-pressurisation cycle to avoid desiccant dusting.
- Designed for efficient working with least pressure drop.
- Powder coated for aesthetic look and additional protection.

Operating Conditions

	Ideal	Maximum
Working Pressure	4 bar to 10 bar	12 bar
Inlet Temperature	40°C	50°C
Atmospheric Dew Point	-40°C	-60°C
Cycle Time	4 minutes	4 minutes
Electrical	220V / 50 Hz / Single Phase	220V / 50 Hz / Single Phase

6.2 Heatless air dryer

SPD Series heatless air dryer

Dryer equipped with a pre-filter to protect desiccant from free water contamination, dust and scales.

Salient Features:

Compact in size

Ease of operation

Light weight

Non-corrosive Aluminium body

Aesthetic look

Silent - low purge noise

Specifications

Tower:

- Non-corrosive aluminium body
- Amply sized to save heat of adsorption.
- Large desiccant candle, ensures sufficient contact time to produce −40°C dew point (−60°C with molecular sieves).
- Each tower is provided with pressure gauge.

Filters:

- Each dryer is equipped with a pre-filter to protect desiccant from free water contamination, dust, scales, etc. and an after filter to avoid desiccant dust carry over.

Pressure Gauge:

- Pressure gauges on each tower for easier illustration of working of dryer.

Valves:

- Highly reliable solenoid valves, to international standards, which act as main flow and exhaust valves. Precision orifice for highly accurate purge flow.

Why Super Pack instead of other dryers?

- Occupies no floor space as it is wall mounted.
- Easy to operate.
- No moving parts and hence high reliability.
- Least maintenance required.
- Electronic timer enables accurate cycle timing.

General Features

- Well-designed mounting brackets
- Purge exhaust muffler for noise control.
- Highly accurate electronic timer.
- Dried air as per requirements of ISO 8573-1 class 3-3-4.
- Designed to meet the requirements of ISO 7183.
- Quick exhaust facilitates sudden de-pressurisation and thereby better regeneration.
- Re-pressurisation cycle to avoid desiccant dusting.
- Designed for efficient working with least pressure drop.
- Powder coated for aesthetic look and additional protection.

Operating Conditions

	Ideal	**Maximum**
Working Pressure	4 bar to 10 bar	12 bar
Inlet Temperature	40°C	50°C
Atmospheric Dew Point	−40°C	−60°C
Cycle Time	4 minutes	4 minutes
Electrical	220 V / 50 Hz / Single Phase	220 V / 50 Hz / Single Phase

Heatless desiccant air dryer

Salient features include Power On light, On-Off switch, Tower pressure gauges, highly accurate motorised cam timer, Differential pressure gauge

for filters, PLC, Cycle failure indication, Sized to save heat of adsorption – minimises purge air usage and Limited velocities through tower prevent bed fluidisation – stops desiccant dusting.

- Large desiccant bed – ensures sufficient contact time to produce −40° C Dew point(−80°C Dew point with Molecular sieves)
- Uses bulk desiccant – no special cartridges required.

Can be offered for higher pressure and lower dew points also.

Optional Fill and Drain Ports:

- Separate fill and drain ports for ease of desiccant replacement.
- Heavy duty purge exhaust mufflers for quite operation.
- Non-lubricated inlet control valves.
- Purge flow valve for adjusting purge rate.
- Pressure relief valves on both towers.

Stainless Steel Support Screens and Air Diffusers:

- Easy removable for cleaning.
- Filters out gross contaminants protect valves.
- Prevents channeling.

Structural Steel Frame with Complete Floor Stand:

- Easy installation.
- Dryer is completely assembled, piped and wired before shipping.
- Optional factory mounting of pre and after filters.
- Shipped with full charge of desiccant. Choice between alumina or molecular sieve.
- Only hookup utilities need to operate.
- Lifting lugs for easy handling.

Filters

- Pre-Filter:
  - Every dryer should be equipped with pre-filter to protect the desiccant from free water contamination, dust, scales, etc., from compressed air supply. This assures top efficiency and longer desiccant life.

- Zero Oil Filter:
  - Oil from the compressor's crank case will be passed on to the line with the compressed air. This will severely affect the life of the desiccant. To enhance, oil particles are removed in this filter up to 1 micron rating.
- After Filter:
  - This is provided after the dryer to remove the desiccant dust carried over to the equipment.

Technical specifications

MODEL	Air Flow		Electrical Connection			Inlet / Outlet	Dimensions (mm)			Desiccant	Weight
	SCFM	Nm³ / hr	V	Ph	Hz	BSP	L	W	H	Qty/tower	kg
HLD-030	300	510	230	1	50	2" NB	1,400	1,000	2,400	110	350
HLD-040	400	680	230	1	50	3" NB	1,400	1,200	2,480	140	400
HLD-050	500	850	230	1	50	3" NB	1,600	1,300	2,230	180	500
HLD-060	600	1020	230	1	50	3" NB	1,600	1,300	2,450	215	650
HLD-075	750	1275	230	1	50	4" NB	2,000	1,300	2,200	260	800
HLD-100	1000	1700	230	1	50	4" NB	2,000	1,300	2,500	360	950
HLD-125	1250	2125	230	1	50	6" NB	2,000	1,300	2,750	450	1,150
HLD-150	1500	2550	230	1	50	6" NB	2,200	1,400	2,200	520	1,400
HLD-200	2000	3400	230	1	50	6" NB	2,200	1,400	2,860	720	1,750

Operating conditions

	Min	Max
Working pressure	7 bar to 12.5 bar	70 bar
Inlet temperature	45°C	60°C
Atmospheric dew point	−40°C	−80°C
Pressure drop	0.2 bar	
Code of construction	IS2825/ ASME Sec VIII Div 1 *	
All flanges	As per ANSI B 16.5 *	
Electric	230 V AC, 50 Hz	

6.3 Evaporative cooling tower

Bottle cooling tower

The tower casing is corrosion-free and lightweight.

Design: Evaporative (FRP) Cooling towers are of vertical induced draft counter flow design with uniform water distribution and optimal heat transfer. Towers can be installed independent of wind direction. The tower casing is made from tough fiberglass and has sufficient strength to

withstand high velocities and vibrations. The water collection sump is leak proof and avoids water spillage. The performance of cooling tower greatly depends upon the water distribution over the fills. SCSP nozzles distribute water evenly through a wide spray angle without any dry pockets. The fill is of rigid poly vinyl chloride (PVC) and is of honeycomb design with very large contact surface area.

- *Casing*: The tower casing is made of tough fiberglass-reinforced plastic (FRP) and has sufficient structural strength to withstand high wind velocities and vibrations. UV-stabilised resin is used along with gel coat for longer life. It is resistant to local impacts and even if slight damage occurs, local repairs can easily be done. The portion of casing housing fill and eliminator has a round cross section. The water collection sump, also of FRP, is leak proof and avoids water spillage.
- *Fill*: The fill is of rigid PVC and is of honeycomb design with very large contact surface area. The fill splits the air and water into several streams, increasing the time of contact and also heat transfer between water and air. The fill is in modules and is packed in the tower casing without any cutting for curves. The air pressure drops through the fill is negligible. The fills are available with flute height of 6 mm, 12 mm and 19 mm with sheet thickness of 1 mm and 1.2 mm.
- *Sprinkler*: Automatic rotary sprinkler system made of Nylon 66 material, with rotary head and sprinkler pipe distributes the hot water over the entire space of the filler. Sprinkler pipes are non-clogging, require low pressure to operate and assures uniform water flow with minimal operating pump head. The FRP eliminators attached to sprinkler pipe are specifically designed for low pressure drop and minimises the drift loss of water.
- *Drift Eliminator*: Reduces carry over losses of water. The eliminator is of rigid PVC (Applicable for square type cooling tower). The individual drift eliminators of S-shaped corrugated sheets are bonded with subsequent layers to create the structure. The entire area is thus divided into several fine S shaped mini zones each removing water droplets on the entire surface of the cell.
- *Axial Fan*: Specially designed energy efficient fans are of induced-draft axial type with adjustable pitch. Material chosen are non-corrosive of plastic, FRP or aluminium alloy. The high efficiency

design ensures low running cost and the lowest possible noise level. Fan blade pitch is factory set and dynamically balanced.

- *Motors*: The motors are totally enclosed (IP55), flange type, 415 V, 3 phase, 50 Hz, induction weather proof with SS304 extended shaft and are specially designed for cooling tower application.
- *Corrosion Free*: The tower casing is of FRP, fill and eliminator are of PVC, and SS304 fasteners are used, thus eliminating corrosion, the biggest enemy of cooling tower. All steel components such as motor support, water distribution pipes and hardware are hot dip or spray galvanised.
- *Lightweight*: The towers are compact and light weight resulting in easy delivery to site and installation. Light weight also saves on structural and masonry. Roof installation can also be done without any special reinforcements.
- *Installation, Service and Maintenance*: Towers of lower capacity are completely factory assembled before dispatch. At site the tower has just to be bolted, on the RCC / brick masonry foundation, thus saving a lot of installation time. Towers of higher capacity can easily be assembled at site and then installed in a manner similar to small towers. Maintenance is considerably reduced because fan, sprinkler fill and eliminator can easily be approached from the top without disturbing the cooling tower casing.

Technical Specification

Sl. No.	Motor	Dia of Fan	In / Out Size	Overall Dimension	Shipping Weight	Operating Weight
	HP / RPM	mm	NB	D x H in mm	kg	kg
GCT-7.5	0.5 HP / 1440	450 (PVC)	1½" / 1½"	930 x 1500	90	225
GCT-10	0.5 HP / 1440	450 (PVC)	1½" / 1½"	930 x 1690	90	340
GCT-15	0.75 HP / 1440	700 (FRP)	2" / 2"	1180 x 1650	90	440
GCT-20	1 HP / 950	750 (NL)	2" / 2½"	1380 x 1800	110	470
GCT-30	1 HP / 950	750 (NL)	2½" / 3"	1660 x 1860	180	550
GCT-40	1 HP / 950	850 (NL)	3" x 3"	1780 x 1950	210	710
GCT-50	2 HP / 950	900 (NL)	3" x 3"	1990 x 2120	275	775
GCT-60	3 HP / 950	1160 (AL)	3" x 3"	1920 x 2720	340	810
GCT-80	3 HP / 950	1160 (AL)	4" x 4"	2150 x 2670	385	900
GCT-100	5 HP / 950	1400 (AL)	5" x 5"	2895 x 3050	625	1900
GCT-125	5 HP / 950	1400 (AL)	5" x 5"	2950 x 3250	810	2310
GCT-150	5 HP / 950	1600 (AL)	5" x 5"	3000 x 3950	1010	2700
GCT-200	7.5 HP / 710	1800 (AL)	6" x 6"	3300 x 4230	1100	3060
GCT-300	12.5 HP / 600	2400 (AL)	8" x 8"	4450 x 3900	2200	4070

Square cooling tower

Sufficient strength to withstand high velocities and vibrations

- *Design*: Evaporative (FRP) Cooling towers are of vertical induced draft counter flow design with uniform water distribution and optimal heat transfer. Towers can be installed independent of wind direction.
- *Casing*: The tower casing is made of tough FRP and has sufficient structural strength to withstand high wind velocities and vibrations. UV-stabilised resin is used along with gel coat for longer life. It is resistant to local impacts and even if slight damage occurs, local repairs can easily be done. The portion of casing housing fill and eliminator has a round cross section. The water collection sump, also of FRP, is leak proof and avoids water spillage
- *Fill*: The fill is of rigid PVC and is of honeycomb design with very large contact surface area. The fill splits the air and water into several streams, increasing the time of contact and also heat transfer between water and air. The fill is in modules and is packed in the tower casing without any cutting for curves. The air pressure drop through the fill is negligible. The fills are available with flute height of 6 mm, 12 mm and 19 mm with sheet thickness of 1 mm and 1.2 mm.
- *SCSP (Solid Cone Square Pattern) nozzles*: The performance of cooling tower greatly depends upon the water distribution over the fills. The nozzles distribute water evenly through a wide spray angle without any dry pockets. They are lightweight and reduce the frequency of clogging. The nozzles produce a solid cone spray of water that is distributed in a square pattern onto the fills.
- *Drift Eliminator*: Reduces carry over losses of water. The eliminator is of rigid PVC (Applicable for square type cooling tower). The individual drift eliminators of S-shaped corrugated sheets are bonded with subsequent layers to create the structure. The entire area is thus divided into several fine S shaped mini zones each removing water droplets on the entire surface of the cell.

- *Axial Fan*: Specially designed energy efficient fans are of induced-draft axial type with adjustable pitch. Material chosen are non-corrosive plastic, FRP or aluminium alloy. The high efficiency design ensures low running cost and the lowest possible noise level. Fan blade pitch is factory set and dynamically balanced.
- *Motors*: The motors are totally enclosed (IP55), flange type, 415 V, 3 phase, 50 Hz, induction weather proof with SS304 extended shaft and are specially designed for cooling tower application.
- *Corrosion Free*: The tower casing is of FRP, fill and eliminator are of PVC, and SS304 fasteners are used, thus eliminating corrosion, the biggest enemy of cooling tower. All steel components such as motor support, water distribution pipes, hardware etc., are hot dip or spray galvanised.
- *Lightweight*: The towers are compact and light weight resulting in easy delivery to site and installation. Light weight also saves on structural and masonry. Roof installation can also be done without any special reinforcements.
- *Installation, Service and Maintenance*: Towers of lower capacity are completely factory assembled before dispatch. At site the tower has just to be bolted, on the RCC / brick masonry foundation, thus saving a lot of installation time. Towers of higher capacity can easily be assembled at site and then installed in a manner similar to small towers. Maintenance is considerably reduced because fan, sprinkler fill and eliminator can easily be approached from the top without disturbing the cooling tower casing.

Technical specifications

Sl. No.	Motor	Dia of Fan	In / Out Size	Overall Dimension	Shipping Weight	Operating Weight
	HP / RPM		NB	LxWxH in mm	kg	kg
SCB-10	0.5 HP / 1440	450 (PVC)	2" / 2½"	815 x 815 x 1600	80	220
SCB-15	0.5 HP / 1440	450 (PVC)	2" / 2½"	815 x 815 x 1950	80	220
SCB-30	1 HP / 950	750 (NL)	2" / 2½"	1130 x 1130 x 2235	235	790
SCB-40	2 HP / 950	1000 (AL)	3" / 3"	1440 x 1440 x 2590	255	805
SCB-60	3 HP / 950	1000 (AL)	3" / 3"	1740 x 1740 x 2810	310	1070
SCB-100	5 HP / 950	1400 (AL)	5" / 5"	2130 x 2130 x 3100	575	1675
SCB-150	5 HP / 950	1400 (AL)	6" / 6"	2650 x 2150 x 3200	690	2075
SCB-200	7.5 HP / 710	1800 (AL)	8" / 8"	2650 x 2650 x 3225	950	2200
SCB-250	10 HP / 710	1800 (AL)	8" / 8"	3150 x 2650 x 3350	1075	2340
SCB-300	12.5 HP / 600	2200 (AL)	8" / 8"	3150 x 3150 x 3500	1270	2605

Cross flow cooling tower

The tower casing is corrosion-free and lightweight.

The GEM Cooling Towers is a pioneering product in Design & manufacture of FRP Cooling towers in South India. We are continuously striving to offer better Cooling solutions to customers, which are both, innovative and also state of the art.

ECT is the Evaporative Coil Cooling tower which is an induced draft counter flow design Closed Circuit Cooler.

Salient Features include design, casing, fill, sprinkler, SCSP nozzles, drift eliminator, axial fan, motors, corrosion free, light weight, installation, service and maintenance

Principle of Operation: Heat from the Process Fluid is circulated through the coil of the closed circuit cooler gets dissipated through the coil tubes to the water cascading downward over the tubes. Simultaneously, air is drawn from air inlet louvers from the base of the cooler and travels upward over the coil. A small portion of the water gets evaporated and the warm moist air is discharged to the atmosphere which removes the heat from the process fluid.

Functions and features:

- *Structure*: Frame and all supporting structures are hot-dip galvanised steel to minimise rusting and corrosion ensuring long life. Casing made of fiberglass-reinforced polyester (FRP), it is lightweight, easy to assemble, no painting required and thus reducing maintenance cost. Inspection doors are furnished to provide convenient access to the interior for inspection, maintenance, adjustment of float valve, cleaning of lift-out strainer and flushing out the sump.
- *Air Distribution*: The fan motors are totally enclosed weather proof type. The motors are mounted on easily adjustable base located inside the fan stack ensuring lower noise level. Specially designed energy-efficient fans are of induced-draft axial type with adjustable pitch. Materials chosen are non-corrosive plastic, FRP or aluminium alloy. The high-efficiency design ensures low running cost and the lowest possible noise level. Fan blade pitch is factory set and dynamically balanced.
- *Water Distribution Fills*: The fill is rigid PVC and is honeycomb design with very large contact surface area. The fills split the air and water into several streams, increasing the time of contact and also heat

transfer between water and air. The fill is in modules and is packed in the tower casing without any cutting for curves. The air pressure drop through the fill is negotiable. The fills are available with flute height of 6 mm, 12 mm, and 19 mm with sheet thickness of 1 mm and 1.2 mm.

Evaporative coil cooling tower

The tower casing is corrosion-free and lightweight.

- *Design*: The GEM Cooling Towers is a pioneering product in Design & manufacture of FRP Cooling towers. We are continuously striving to offer better Cooling solutions to customers, which are both, innovative and also state of the art. ECT is the Evaporative Coil Cooling tower which is an induced draft counter flow design Closed Circuit Cooler.

- *Principle of Operation*: Heat from the Process Fluid is circulated through the coil of the closed circuit cooler gets dissipated through the coil tubes to the water cascading downward over the tubes. Simultaneously, air is drawn from air inlet louvers from the base of the cooler and travels upward over the coil. A small portion of the water gets evaporated and the warm moist air is discharged to the atmosphere which removes the heat from the process fluid.

Functions and Features:

- *Structure*: All supporting structures are hot-dip galvanised steel to minimise rusting and corrosion ensuring long life. Casing made of fiberglass-reinforced polyester (FRP), it is lightweight, easy to assemble, no painting required and thus reducing maintenance cost. Inspection doors are furnished to provide convenient access to the interior for inspection, maintenance, adjustment of float valve, cleaning of lift-out strainer and flushing out the sump.
- *Air Distribution*: The fan motors are totally enclosed weather proof type. The motors are mounted on easily adjustable base located inside the fan stack ensuring lower noise level. Fan specially

designed energy-efficient fans are of induced-draft axial type with adjustable pitch. Materials chosen are non-corrosive of plastic, FRP or aluminium alloy. The high-efficiency design ensures low running cost and the lowest possible noise level. Fan blade pitch is factory set and dynamically balanced.

- *Water Distribution Fills*: The fill is rigid PVC and is honeycomb design with very large contact surface area. The fills split the air and water into several streams, increasing the time of contact and also heat transfer between water and air. The fill is in modules and is packed in the tower casing without any cutting for curves. The air pressure drop through the fill is negotiable. The fills are available with flute height of 6 mm, 12 mm, and 19 mm with sheet thickness of 1 mm and 1.2 mm.

Modular cooling tower

The tower casing is corrosion-free and lightweight.

- *Design*: Evaporative (FRP) Cooling towers are of vertical induced draft counter flow design with uniform water distribution and optimal heat transfer. Towers can be installed independent of wind direction.

- *Casing*: The tower casing is made of tough FRP and has sufficient structural strength to withstand high wind velocities and vibrations. UV-stabilised resin is used along with gel coat for longer life. It is resistant to local impacts and even if slight damage occurs, local repairs can easily be done. The portion of casing housing fill and eliminator has a round cross section. The water collection sump, also of FRP, is leak proof and avoids water spillage.
- *Fill*: The fill is of rigid PVC and is of honeycomb design with very large contact surface area. The fill splits the air and water into several streams, increasing the time of contact and also heat transfer between water and air. The fill is in modules and is packed in the tower casing without any cutting for curves. The air pressure drop through the fill

is negligible. The fills are available with flute height of 6mm, 12 mm and 19 mm with sheet thickness of 1 mm and 1.2 mm.

- *Sprinkler*: Automatic rotary sprinkler system made of Nylon 66 material, with rotary head and sprinkler pipe distributes the hot water over the entire space of the filler. Sprinkler pipes are non-clogging, require low pressure to operate and assures uniform water flow with minimal operating pump head. The FRP eliminators attached to sprinkler pipe are specifically designed for low pressure drop and minimises the drift loss of water.
- *SCSP Nozzles*: The performance of cooling tower greatly depends upon the water distribution over the fills. SCSP nozzles distribute water evenly through a wide spray angle without any dry pockets. They are lightweight and reduce the frequency of clogging. The Solid Cone Square Pattern (SCSP) nozzles produce a solid cone spray of water that is distributed in a square pattern onto the fills.
- *Drift Eliminator*: Reduces carry over losses of water. The eliminator is of rigid PVC (Applicable for square type cooling tower). The individual drift eliminators of S-shaped corrugated sheets are bonded with subsequent layers to create the structure. The entire area is thus divided into several fine S-shaped mini zones each removing water droplets on the entire surface of the cell.
- *Axial Fan*: Specially designed energy efficient fans are of induced-draft axial type with adjustable pitch. Material chosen are non-corrosive of plastic, FRP or aluminium alloy. The high-efficiency design ensures low running cost and the lowest possible noise level. Fan blade pitch is factory set and dynamically balanced.
- *Motors*: The motors are totally enclosed (IP55), flange type, 415 V, 3 phase, 50 Hz, induction weather proof with SS304 extended shaft and are specially designed for cooling tower application.
- *Corrosion Free*: The tower casing is of FRP, fill and eliminator are of PVC and SS304 fasteners are used, thus eliminating corrosion, the biggest enemy of cooling tower. All steel components such as motor support, water distribution pipes and hardware are hot dip or spray galvanised.
- *Lightweight*: The towers are compact and light weight resulting in easy delivery to site and installation. Light weight also saves on structural and masonry. Roof installation can also be done without any special reinforcements.

- *Installation, Service & Maintenance*: Towers of lower capacity are completely factory assembled before dispatch. At site the tower has just to be bolted, on the RCC / brick masonry foundation, thus saving a lot of installation time. Towers of higher capacity can easily be assembled at site and then installed in a manner similar to small towers. Maintenance is considerably reduced because fan, sprinkler fill and eliminator can easily be approached from the top without disturbing the cooling tower casing.

Sl. No.	Motor	Dia of Fan	In / Out Size	Overall Dimension	Shipping Weight	Operating Weight
	HP / RPM		NB	L x W x H in mm	kg	kg
SCT-400TR	15 / 360	2400 (FRP)	10" / 10"	4000 x 4000 x 3965	2,150	4,300
SCT-500TR	20 / 360	3000 (FRP)	10" / 10"	4500 x 4500 x 3965	2,400	4,700
SCT-600TR	20 / 270	3600 (FRP)	10" / 10"	5200 x 5200 x 4465	2,700	5,200
SCT-800TR	25 / 270	4270 (FRP)	12" / 12"	6000 x 6000 x 4465	3,200	5,600
SCT-1000TR	25 / 200	4800 (FRP)	14" / 14"	6700 x 6700 x 5020	3,800	5,00

Double cross flow cooling tower

The tower casing is corrosion-free and lightweight.

Design: The double cross flow cooling tower operates according to the cross flow principle. The cooling tower is manufactured using FRP with a stainless steel supporting structure. The water is collected in a flash box for equal distribution and then it is recirculated once it collects at the bottom in the basin. The casing of the tower is extremely strong, and it can withstand velocities and strong vibrations. The motor has a totally weather proof IP-55 degree of protection, suitable for hot and humid atmosphere and it is covered by FRP motor cover

Readily adaptable to any water cooling applications, these cooling towers are specially designed for hard water, ceramic industries, steel industries, oil refineries and dusty atmospheres. It is highly acclaimed by clients for chemical and corrosion resistance, maximum service life and high performance.

6.4 Dry cooling towers

Dry cooling tower

The main function of the dry cooling tower is to cool and maintain the temperature of process hot water at a particular level. The tower works on a principle of heat transfer and is driven by an electric motor. The wound copper tubes eliminate any unnecessary leakage after installation. The tubes also allow better heat transfer efficiency.

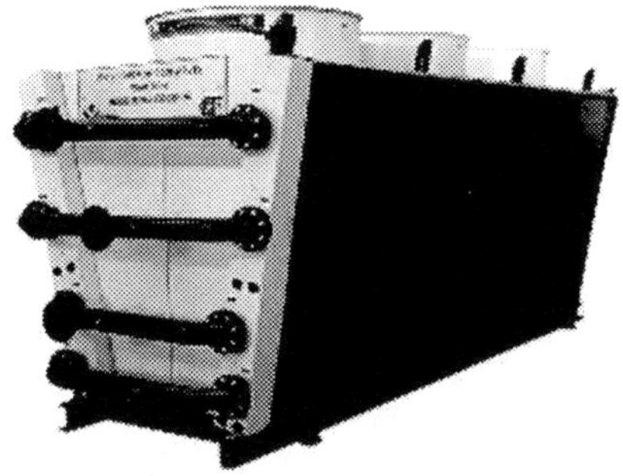

The positive and controlled expansion between the tube and fin provides a clean, smooth inner tube surface for water pressure drop and guarantees uniform heat transfer.

The rigid galvanised steel frame provides protection against tube damage during expansion and installation. Also, it is mounted on a heavy duty channel base frame. The axial flow fans are noise-free. The low speed of the motor minimises noise and increases efficiency. Motors are specially designed to withstand moisture, rain and dust.

Aqua saver

The aqua saver is built with corrosion-resistant aluminium sheets. The casing is divided into baffles and fan sections, to ensure proper air distribution through the coils and maximise efficiency at part loads. The copper tubes and aluminium fins facilitate maximum performance.

Applications:

- Water Cooled Compressor Cooling
- Hydraulic Power Pack Cooling
- Plastic Injection Moulding Machine Cooling
- Aluminium Die Casting Machine Cooling
- Induction Furnace Coil & Electrical Control Panel Cooling

7 AESA Air Engineering Private Limited – France, India, Singapore and China

7.1 Cooling washer

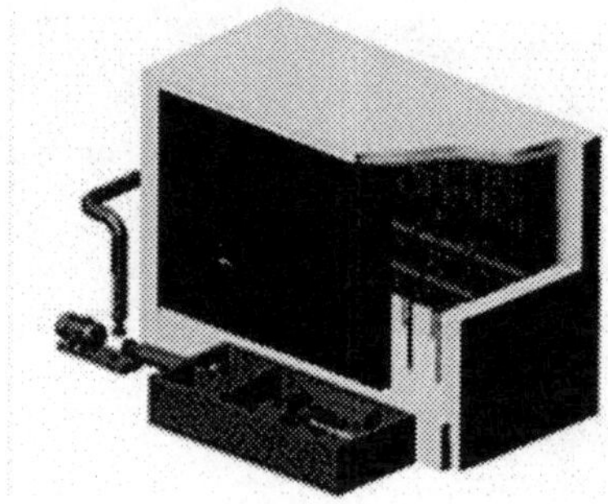

- Pre-fabricated or concrete housing
- For adiabatic cooling with automatic spray water control
- For chilled water cooling with automatic dew point temperature control
- PVC Baffler and Eliminator Blades
- 1 or 2 spray bank system
- Air volume: 57,600–558,500 m^3/h

Appendix – 3

Fogging fans and jets

1 US Global Resources USA

Humidifan: The Nozzle-Free, Horizontal Airflow Fogger

These are all-purpose units for industries for humidification and/or fogging. There are two series for humidification and fogging, namely, HRSM and GRSM. HRSM model is the same as the Turbo XE series, but is on wheels and has a tank. GRSM is used for odour control. CRSM is the series used for application of chemicals.

1.1 The aqua turbo xe serie

Aquafog's unique design simplifies principal moving parts to one motor and its integral fan. Water or other fluids are fed into the specialized hub of fan blade assembly where it is subdivided and channeled into passageways running the length of each blade. As liquid exits, it is atomised into ultra-fine fog particles. This does not require any permanent plumbing installation, high pressure pumps, complicated filtratio, and constant maintenance of clogged nozzles. It has streamlined one motor/one blade design producing and dispersing fog using a single nozzle-free blade assembly.

1.2 Aquafog turbo XE horizontal airflow fogger

Hanging Turbo XE units give the best versatility and dependability in the Aquafog line. They generate micron-fine fog particles with excellent forced circulation, providing uniform fog distribution and maximum coverage.

Additional Turbo XE features

When mounted high overhead, it allows more time for the cold fog particles to evaporate as they naturally descend. Exclusive nozzle-free design eliminates plugging problems and need for expensive filtering systems.

It uses ordinary water and power supply to produce high quality, high volume fog. Fans can be stationary or equipped with oscillation. It is excellent for safe, effective chemical fogging. Used for pesticide and fungicide application, insecticide use and foliar feedings, the Aquafog Turbo XE is indispensable for use in greenhouses. They are capable of continuous duty operation and come largely pre-assembled, ready to install. Convenient panel mount flow meter and in-line strainer provide control needed for required fog output. Dwyer flow metres provided are easy to clean, available in seven sizes from broad volume use to narrow ranges, providing better accuracy at low volume output. The high volume ranges from 1 to 11 Gallons Per Hour to 4–34 GPH and low volume range from 5 to 50 cc/min to 20–300 cc/min.

1.3 Turbo XE Oscillator

When increased fog coverage and circulation (especially when coverage area exceeds twenty feet) is needed, an oscillator helps doing the job right. Its rotational movement lets the operator fog at higher outputs while maintaining dryness underneath. In addition to improving circulation, sweeping the fog with an oscillator alleviates droplet collisions and fallout. Turbo XE Oscillators provide a continuous and consistent (dwell-free) movement. There is a choice of ninety six different settings, with sweeps ranging from 3.75 degrees to a two-minute, full rotation of 360 degrees.

Coverage range

A. 1,200 average, up to 2,000 square feet
B. 1,800 average, up to 2,400 square feet
C. 2,200 average, up to 3,000 square feet
D. 2,800 average up to 4,200 square feet
E. Contact your Odour Control Consultant
F. 1 unit per cooling chamber

Fan size	Max dry fog output	Cycle	CFM rating	Propulsion distance	Noise @ 10' distance	Energy consumption	Weight	Largest flow meter
1 HP	30 GPH	60hz	3,260	35 feet	76-79 dB(A)	10.6 amps @ 120V	50 lbs.	4-34 GPH
3/4 HPtt	24 GPH	50hz	3,180	30 feet	73-76 dB(A)	4.0 amps @ 240V	52 lbs.	2-24 GPH
1/2 HP	20 GPH	60hz	2,730	30 feet	74-77 dB(A)	7.4 amps @ 120V	39 lbs.	2-24 GPH
1/3 HPtt	14 GPH	50hz	2,580	25 feet	69-73 dB(A)	3.0 amps @ 240V	42 lbs.	1-11 GPH
1/4 HP	12 GPH	60hz	2,160	25 feet	68-72 dB(A)	7.2 amps @ 120V	40 lbs.	1-11 GPH

All Turbo XE Fans are standard equipped with dark green housing and a dual voltage, white epoxy painted wash down duty TM motor (WDD) 115/230 V 1 phase 60/50 hz. Suffixes are added to standard product number to describe special features. For e.g.

Available fan options

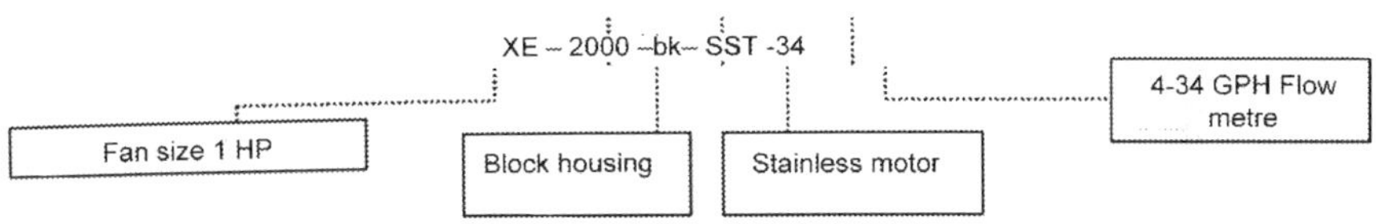

Fan Size	Manufacturer's Standard P.N.	Black Housing (bk)	WDD 3 phase 230/460 V (3ph)	Explosion Proof 1ph 115/230 V (Exp)	Explosion Proof 3ph 230/460 V (3Exp)	Stainless Steel motor 1ph 115/230 V (SST)
1 HP	XE 2000-	✓	✓		✓	✓
3/4 HP	XE 1500-	✓				✓
1/2 HP	XE 1000	✓		✓	✓	
1/3 HP	XE 660-	✓				
1/4 HP	XE 500t-	✓				

ᵗXE 500 comes equipped with a 1/2 HP blade assemble, this unit can easily be upgraded to an XE 1000

P. N.	Suffix	Volume Range
F-34	(34)	4–34 GPH
F-24	(24)	2–24 GPH
F-11	(11)	1–11 GPH
F-5	(5)	20–300 cc/min*
F-3	(3)	10–110 cc/min*
F-1	(1)	5–50 cc/min*

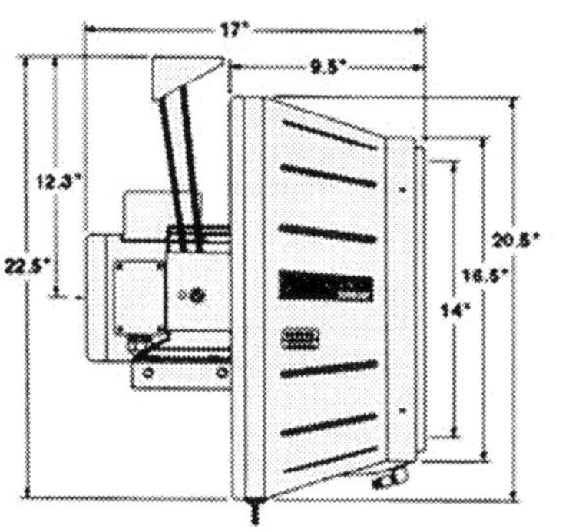

Flow meter sizes:

*60 cc/min = approximately 1 GPH

Choosing an appropriate size flow meter can sometimes be difficult. If a flow meter size is not indicated, the manufacturer will select the appropriate size flow meter considering the industry.

Accessories

When fog circulation at ground level or bench height is required floor stand provides convenient, rapid mobility and a safe, secure Aquafog mounting support.

Travelling floor stand

More travelling stand features:

- Semi-pneumatic tires for easy travel over rough surfaces.
- Tripod design with small 30" footprint for fitting through aisle ways and tight spaces.
- Generous 6' 8"-high stand dollies easily for clearance through doorways.
- Support boom is fully adjustable for pre-setting operational height.
- Compatible for use with a Turbo oscillator.
- Easily assembled; ships via UPS.

1.4 Aquafog Turbo XE RSM

Standard Features

- Large tank holds up to 18 US gallons (68 liters).
- Engineered for harsh environments, units have all stainless steel hardware and heavy wall thermoplastic construction.
- Equipped with flow control, automatic fill float assembly, overflow port, water hose connection and liquid level pointer gauge.
- Wide-mouth opening for chemicals.

- Draining plug for easier cleaning.
- Sturdy molded handles and 10" semi-pneumatic wheels for rugged mobility.

Available in two models, CRSM for pesticide application and HRSM for humidification, the RSMs are compact, mobile foggers. They combine the advantage for the no-clog, continuous duty atomising fan with a large-capacity tank for recirculation of all wastewater.

1.5 Aquafog Turbo XE CRSM

When the job calls for a low-volume method chemical application, the CRSM is top performer. Even when the tank is full, one can easily wheel this sturdy unit from one place to another. Here's more about the CRSM:

- CRSM's chemical pump operates at a maximum rate of 5 US GPH, producing particles that average 5 microns in size.
- A circulation pump vigorously agitates the solution, allowing use of heavy powders.
- Features a no-clog, nozzle-free atomization head and made of thermo-plastic that's highly corrosion-resistant.
- Fluid level gauge and large-mouth opening make it easy to mix and fill tank with chemicals.
- Units treat up to 28,000 square feet.
- Uses standard electrical supply; available in 1/2 HP, 115V, single phase or 3/4 HP, 240 V, 50 Hz, single phase.

1.6 Aquafog Turbo HRSM

The HRSM provides high fogging, maintaining the weight and appearance of produce. It is ideal for use in produce storage. Humidity management promotes wound healing and reduces weight loss and sprouting, enhancing the quality and appearance of all fruits and vegetables. The HRSM is designed to humidify large storage areas requiring a fogging machine that can recirculate its own wastewater, and/or a humidifier that can be easily transported from one location to another. Features same nozzle-free, no-clot atomszing head as the Turbo XE fan, but has large recirculation tank. Liquid chemicals can also be added to large recirculation tank for chemical treatments. It is designed to be compact, sturdy and portable which are constructed with watertight electrical components and all stainless steel hardware. It is equipped with water hose connector and automatic-fill float assemble for continuous duty operation.

Specifications:

- Voltages:1HP, 115 V, 60 Hz, h PH or 1 HP, 28-–230V, 60 Hz, h PH o/4¾ HP, 240 V, 50 Hz, PhPH
- CFM: 1HP = 3,260 o/4¾ HP = 3,180
- Energy consumption: 1 HP = 10.6 AMPS or 1Hh 3PH = 3.0 AMPS or 3/4HP = 4 AMPS
- Noise @ 10ft. distance: 1 HP = 6-–79 dB(A) or 3/4HP = 3-–76 dB(A)
- Weight: 1HP = 80 ls. or 1HP/3Ph = 76 lbs. or 3/4HP = 83 lbs.
- Tank Capacity18 U.S. Gal. /68 Liters
- Output capability up to 30 GPH

Standard Features:

- Large tank holds up to 1 US gallons.
- Engineered for harsh environments, units have all stainless steel hardware and heavy wall thermo-plastic construction.
- Equipped with flow control, automatic-fill float assembly, overflow port, water hose connection, and liquid level pointer gauge.
- Wide-mouth opening for adding chemicals.
- Draining plug for easier cleaning.
- Sturdy molded handles and 10" semi-pneumatic wheels for rugged mobility.
- Delivered fully assembled, ready for operation; shipped by common carrier.

1.7 Aquafog Turbo XE HRSM

The HRSM is ideal for high fogging output with mobility or applications where drainage is not available, such as vegetable and fruit storage. HRSM's powerful submersible pump is capable of pumping up to 30 gallons of fluid per hour to the fan'g atomising head. Unit automatically fills and can be operated for extended periods or on a continuous basis. Compact in size, the HRSM combines a powerful humidifier with a re-circulatory sump module. Available in I HP, single or three-phase and 3/4 HP, 240 V, 50 Hz, single phase.

1.8 Aquafog 4er Fogger fer SmalleasAreas:

Aquafog® 400's adjustable fogging output and extensive air circulation make it convenient for use in small places. It can be used for mixing conditioning in cotton mills, in loom sheds, etc.

- Ideal for cooling
- Lightweight; easy installation
- Consumes extremely low amounts of energy less than 1 amp
- Features fully adjustable fogging output range
- Propels fog and air up to 20 feet
- Operates very quietly
- Unit fills automatically and connects to ordinary garden hose

Specification:

- Fogging Capability: 0–2 GPH /0–7 Liters per hour
- CFM: 840 @ 60 Hz /710 @ 50 Hz
- Energy Consumption: .7 AMP @ 115 Volt / .35 AMP @ 230 Volt
- Noise @ 10 ft. distance: 62 dB(A) @ 60 Hz / 61 dB(A) @ 50 Hz
- Weight: 8.5 lbs. / 3.8 kg
- Dimensions: Width 15"/ 38 cm; Depth 12.5" / 32 cm; Height 20.5" / 52 cm
- Voltages: −115 V 60 Hz 1ph, −200–225 V 50 Hz 1ph, −225–250 V 50 Hz 1ph

Maximum Coverage:

A. 2,000 average, up to 4,000 cu. ft. per unit,
B. 3,000 average, up to 5,000 cu. ft. per unit and
C. 4,000 average, up to 6,000 cu. ft. per unit

1.9 Humidistat

Humidistat automatically controls Aquafog® 400 for applications requiring fewer than 85% humidity. It incorporates a man-made sensing element with large, easy access control dial and needs no wiring. Standard features are compact, lightweight made of thermo-plastic that›s highly corrosion resistant, hangs high overhead, out of the way. Features no-clog, nozzle-free atomisation head with adjustable fogging capabilities up to 2 US GPH, producing particles of average 35 microns in size. A small, built-in sump area that reuses all of its internal condensation and hence no drainage line is required. Unit is equipped with flow control, automatic fill float assembly and water hose connection. Comes as fully assembled and is quick to install.

Available Models

XE-400	2 GPH Mister Fan unit	115V 60 Hz
XE-400-H	Fan Unit /w Humidistat	115V 60 Hz unit w/ Humidistat control pkg.
XE-400-50L	0–7 liters hour Mister Fan unit	200–225 V 50 Hz
XE-400-50L-H	Fan unit /w Humidistat	200–225 V 50Hz unit w/ Humidistat control pkg.
XE-400-50H	0–7 liters hour Mister Fan unit	225–250 V 50 Hz
XE-400-50H-H	Fan unit /w Humidistat	225–250 V 50 Hz unit w/ Humidistat control pkg.

1.10 Industrial Humidifier

Aquafog fans are useful in a variety of large-scale manufacturing operations. Units are powerful and ideal for spot, evaporative cooling for workers in hot working environments. Manufactures of products that need specific RH levels can utilise the hanging Turbo XE units to humidify large environments. Aquafog units produce high-quality fog with fully adjustable output for versatility and control in keeping nearby equipment dry. In addition, this low-maintenance, nozzle-free units are self-flushing and will not clog. Aquafog fogging fans are successful in re-humidifying, suppressing dust and static electricity in many indoor manufacturing operations, such as wood, paper, textiles and leather goods applications, as well as providing for the proper relative humidity environments in storage facilities. With industrial-grade components and unique engineering, they provide the performance and dependability highly sought after in the manufacturing industry.

The key fogging equipments provided are

- Spot and Worker Evaporative Cooling - XE-1000, XE-2000, XE-1500 for 50 Hz, XE-HRSM,
- Mobile Humidification - XE-HRSM,
- Hanging Industrial Turbo XE Fans
- Cooling/Humidification - XE-1000, XE-2000, XE-1500 for 50 Hz
- Fumigation - XE-HRSM, XE-ORSM, XE-CRSM

2 Pacific Northwest Garden Supply, Surrey B.C. USA

2.1 Hydrofogger

The Hydrofogger is ideal for mid-size humidification. With coverage up to approximately 2,000 sq. ft. per unit, the Hydrofogger produces the

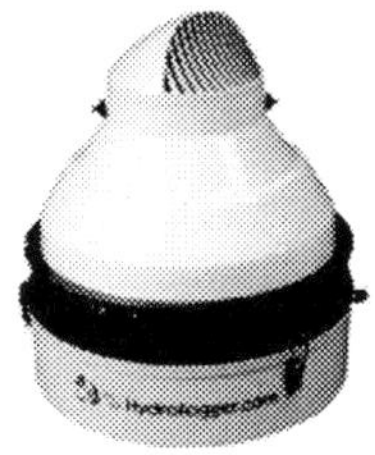

humidification needed simply, quietly, and efficiently. For applications requiring less humidification, Minifogger is available. These humidifiers have fewer moving parts and do not require steam or treated water. They both come with a smooth and powerful industrial-grade motor that will produce clean and ample humidification for years to come. The units are designed to install in minutes, and they are fed automatically by a simple connection to existing cold water supply. No need to ever fill the tank! Humidistat Controls humidity level and Hydrofogger. Units are also available in 230 V/50 Hz.

3 Keystone Ahmedabad, India

3.1 Centrifugal spot humidifier units

The Keystone unit runs by specially designed 3/4 H.P. high speed electric motor suitable for 400/440 V, 3 Phase, 50 cycles, A.C. Supply. The humidifier throws approximately 2,200/2,400 C.F.M. of air and evaporates up to 40 liters water per hour. Available in two models (Fixed Type & Revolving Type). Compact and easy to install. Independent and light weight. No ducting required. High evaporating capacity. Uniform distribution of humidified air. Low operating cost with minimum maintenance.

4 A. C. Humidification Loni, India

4.1 Spot humidifier

'Humidair' (Spot Humidifier) with high efficiency evaporation, effective mist dispersal, completely self-contained, water atomizer and air distributor are designed for minimum maintenance and very low operating cost.

Features

High-efficiency evaporation, designed for minimum maintenance, atomises water into a fine mist for rapid evaporation and distributes the humidified air uniformly throughout room.

5 Mist Magic, New Delhi, India

5.1 Industrial Humidifier / Cool Mist Humidifier F-SP-10002-HP

	Conventional air washer humidification systems	Mist magic free space cool mist humidifier
Energy Usage	25–50 HP	2–8 HP only
Water Consumption	All extra water is lost in evaporation or wasted	All extra water is recycled hence no wastage of water
Ducting	Conventional, expensive and cumbersome ducting	In-built low cost ducting
Plant & Machinery	Large area required for the bulky air compressor and air washer system	Small size machinery which takes the space of an air conditioner
Maintenance	Expensive and problematic involving constant breakage	Hassle free low cost maintenance with low cost spare parts
Power Back-UP	Large generators required which consume huge amounts of fuel	Very small domestic generator required which consumes less fuel, proving to be extremely cost effective for India
Humidity Regulation	No provision for scientific maintenance of humidity levels	Scientific control of humidity levels through a humidistat.

Energy Efficiency: The F-SP-10002- HP Industrial Humidifier consumes only 2 HP of electricity per 30,000 cubic feet.

Key Features are that they can be run by a very small generator, uses 50% water less than normal plants, plant and machinery requires the same space as a Window Air conditioner, easy maintenance, controls fluff & dust, easily replaceable spare parts and after sales service available by local representative

5.2 Cool Mist Fan MF-10003-HP

Mist Fan offers active cooling and is positioned directly at the intended cooling area. Cool Mist Fan provides enhanced performance compared to Mist Line as the high velocity air increases the evaporation rate which in turn improves the cooling effect. Cool mist fan can be mounted on walls or columns. It can also be in the form of mobile pedestal unit.

6 Luwa Air Engineering AG Switzerland

6.1 TexFog – Hygienic and Energy-Saving Humidification

The high-performance humidifying system TexFog was developed specially for use in the textile industry, in order to obtain the optimal dosage of moisture in the supply air. It is specifically useful in Nonwoven manufacturing. TexFog uses only fresh water and does not require a water tank. Using TexFog is safe from a hygiene point of view and fulfills the German hygiene regulation VDI 6022. It requires up to five times less energy than a conventional air washer and also allows a super saturation of the air.

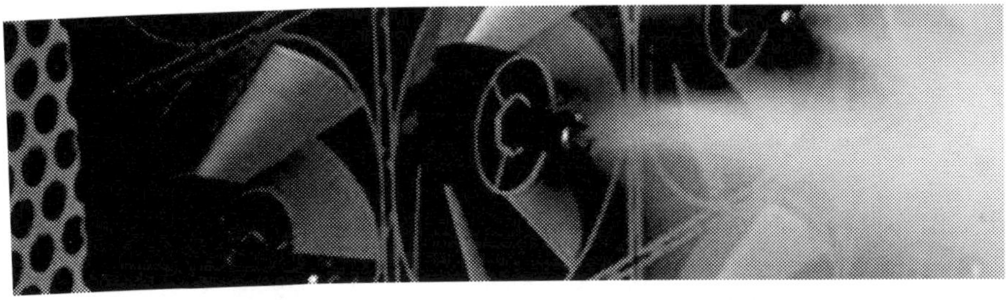

7 Condair – JS Humidifiers, West Sussex U.K

7.1 Jet spray air and water atomising humidifier

The Jet Spray humidifier provides economical and consistent humidification directly into work spaces and within air handling systems. These humidification systems have been humidifying industries and air handling systems around the world, which incorporates the latest in controls, hygiene and environmentally sound technology. Jet Spray's precision engineered self-cleaning nozzles are claimed having the lowest air consumption. The drip-free nozzles produce sprays of just 7.5 μm and are built to last carrying a 10-year warranty. The control panel uses remote digital interface, which can be mounted up to 100 m away from the main control unit. The whole system is easy to install and can run on mains, demineralised or softened water. Due to a self-cleaning mechanism, maintenance is minimal and often comprises of just an annual check, making the JetSpray ideal for use in busy industrial environments.

The Jet Spray induct humidifier is an energy efficient way to humidify air handling systems. Excellent moisture distribution is guaranteed by the choice of nozzle capacity, suitable spacing on the manifold, correct positioning of the manifold in the AHU and the capacity to change the orientation in relation to the airflow. Evaporation distance varies with air temperature, velocity and desired RH but, where space is limited the JetSpray Eliminator Matrix System (JEMS) can be installed downstream to guarantee evaporation within 1 m.

In-duct humidification

Features include close humidity control, low energy consumption, low running costs, low maintenance and significant cooling effect with 10-year warranty on all nozzles. Lowest compressed air consumption compared to any air/water atomising humidifier.

Environmentally friendly

The Jet Spray humidifier uses 90% less energy than electric steam humidifiers. Not only are there considerable cost savings but, for a system producing 100 kg of moisture per hour, this equates to a reduction in CO_2 emissions of around 260 tonnes over five years. As an ISO14001 company, environmental impacts and sustainability are important features of JetSpray design. Water and energy use are minimised during operation and at the end of JetSpray's long service life, a high proportion of materials can be reclaimed and recycled, reducing impacts on landfill.

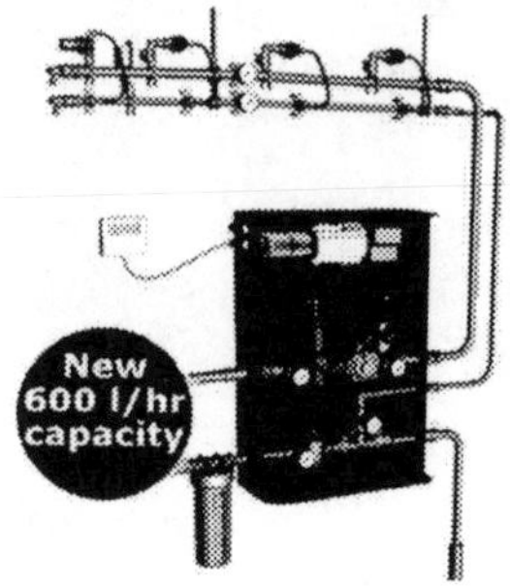

Remote interface

Microprocessor controlled units have an easy-to-read multi-lingual interface, which shows operational status and makes commissioning simple. Set point, ambient humidity and temperature, purge and flush cycle operation, hours run, service interval times, occupancy operation and fault diagnostics are shown. The display can be installed up to 100m away from the control panel. For improved control in large areas, up to three averaging humidity sensors can be connected to the microprocessor controller or it can be linked to a building management system. On/off and modulating control is available on high capacity systems allowing ±2% RH control. Modulating systems also incorporate information on water flow.

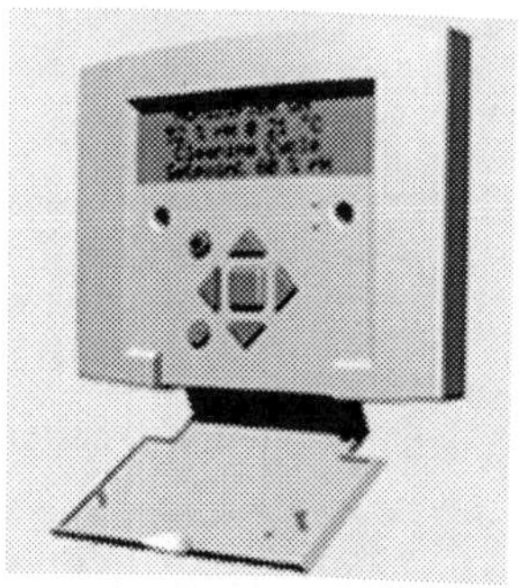

The nozzle jets are kept free from blockages with a self-cleaning pin

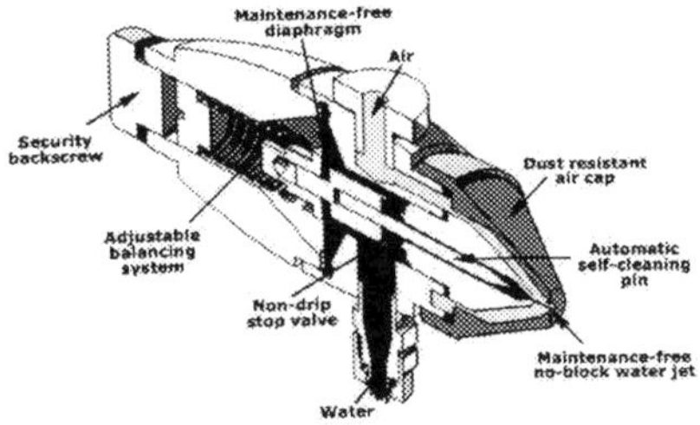

7.2 Armstrong live steam humidifier

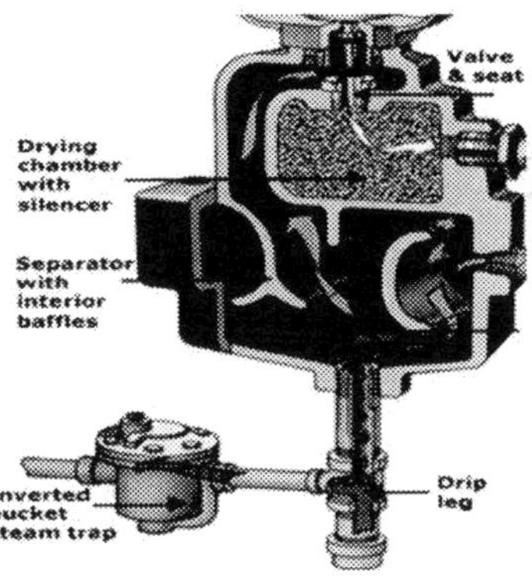

The Armstrong live steam humidifier is ideal for situations where there is a ready supply of steam, as it removes the expense of generating steam to humidify using electrode or resistive boiler-type humidifiers. Unlike a steam-to-steam humidifier, a live steam system will give a very fast response to humidification requirements and has minimal maintenance requirements. It also has the advantages of guaranteed dry steam injection and no bacteria risk as the steam itself is totally sterile.

The Armstrong live steam humidifier is capable of running at pressures up to 4 bar producing hotter, drier steam than other 2 bar systems. This high pressure also gives a huge output capacity of up to 1,823 kg/h from a single unit. By running at a higher pressure, the size of pipe work and pressure reducing valve required to install the humidifier are much smaller making it a cheaper system to set-up.

The steam is initially put through a Y-strainer to remove most of the stream's dirt and scale particles. It is then fed through a jacket that surrounds the steam lance situated in the duct. In so doing, the lance's temperature is maintained at a very high level to ensure the steam passing through it and into the duct at a later, more critical stage doesn't produce condensation on its surface. Once the steam has traveled through the lance's jacket, it is fed into the separator. This chamber contains two baffles, which reduce the speed of the steam and reverse its flow in order to separate the condensation from the steam.

Features include running up to 4 bar pressure delivering very dry steam, large capacity from a single unit - 1,823 kg/h, fast response and excellent control with up to 123:1 turndown ratio and ± 1%RH. Silencer provided reduces noise. Fully engineered and pre-set humidifier with actuator pre-stroked to control valve and excellent controllability with control valve integrated into separator to maintain its temperature and reduce condensation. Zero leakage on valve due to opening against steam pressure.

The humidifier's control valve is situated in the separator chamber and is controlled by an actuator mounted on the top of the unit. By situating the valve within the separator its temperature is kept at a very high level and condensation through the control valve is eliminated.

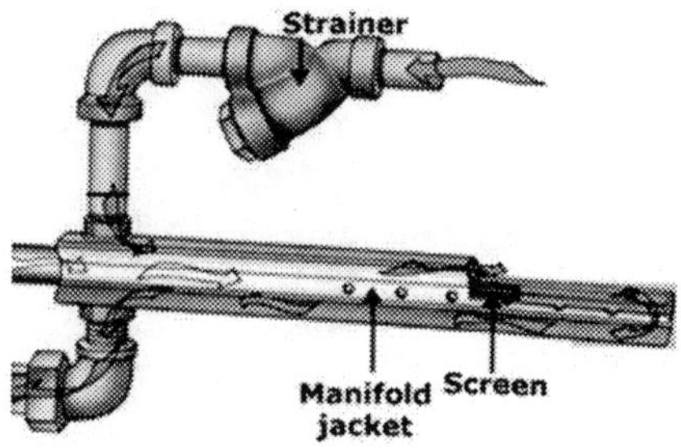

Condensation is removed through a condensate trap located under the separator chamber. A temperature switch is located before this trap, which will only allow the humidifier's control valve to open once it reaches a set temperature. This ensures that the humidifier will only start to inject steam into a duct once the system is fully heated with super-dry steam. In other live steam systems that don't incorporate a temperature switch as standard it is often recommend that the control valve is not opened until 10 minutes after the humidifier is switched on. This is often ignored in plant rooms, which leads to condensation being injected into a duct.

Stainless steel version

The Armstrong live steam humidifier is available in cast iron with stainless steel steam lance or a fully stainless steel model for critical applications.

Direct air version

For applications that require direct humidification of an atmosphere, a wall-mounted live steam system is available. It has a maximum capacity of up to 130 kg/h and ensures evaporation of the steam with a fan mounted behind the steam outlet.

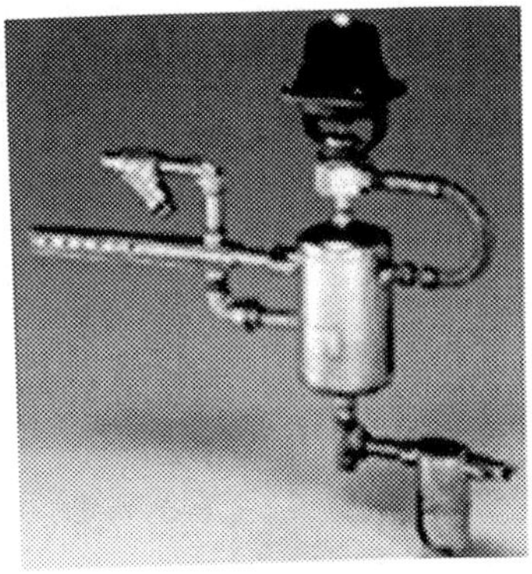

Stainless steel version

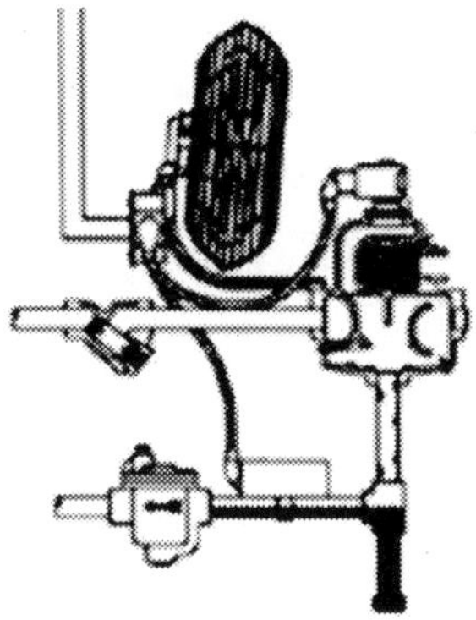

Direct air version

7.3 J.S. Mistifier – Cool Mist Humidifier

This humidifier is ideal for greenhouses, mushroom or orchid growing, storage areas, research chambers or industrial environments such as textile or tobacco processing. The JS Mistifier emits impressive clouds of cool, humidifying mist. It has three levels of output – 2.5 litres, 3.5 litres and 5 litres per hour that are achieved by removing the upper sections of the unit. This humidifier will maintain the optimum environment required for growing plants that need high humidity levels. The unit has a very quiet operation and requires no wiring as it plugs directly into a standard electrical plug. The Mistifier is plumbed into a mains outlet and an internal float valve automatically controls the level in the water reservoir. Alternatively, it can be used without a mains connection for short periods and manually topped-up.

Features

- Easy to install and use
- Durable trouble-free structure
- Good for chemical disinfection and deodorisation
- Maintains humidity from 10% to 80%RH
- Manual or automatic fill
- 2½ litre, 3½ litre or 5 litre output per hour
- Generates particles 5–20 microns in size
- Low energy

Three different outputs

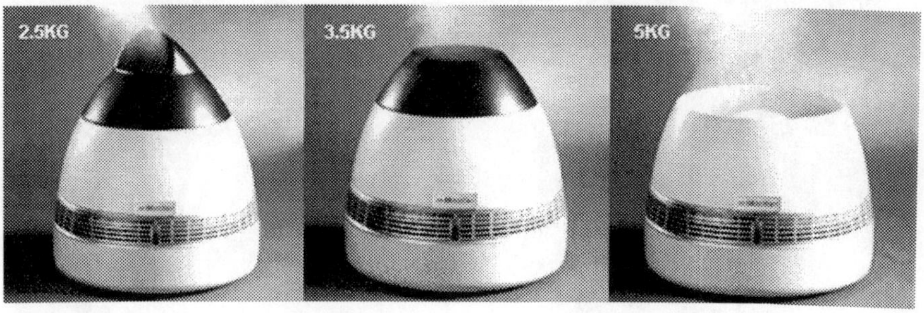

8 Pinnacle Climate Technologies/Schaefar – Sauk Rapids, MN 56379, USA

Misting Fan VersaFog Oscillating Portable

8. A. 36" Versa Fog ½ HP Misting	8. B - 30", 36" Misting Fan Pedestal

	8. A. 36" VersaFog ½ HP Misting Fan	8. B - 30", 36" Misting Fan Pedestal
Features	High Velocity Oscillating Portable Misting Fan. VersaFog: 36", 1/2 hp, 2 speed,12 nozzles Fan Misting Cart: 24", 1/4 hp, 2speed, 6 nozzles.	Pedestal Portable: 24" High Velocity 2 Speed Oscillating Misting Fan 24", 1/4 hp, 2 speed, 6nozzles
Functions	Increase air humidity, air recirculation, lower the air temperature	Lower the air temperature, air recirculation, increase air humidity

Technical Specifications		
Voltage	110v. (Specify if 220 V)	110 v. (specify if 220 V)
Ampere		3a, (VersaFog 6a)
Horse Power	1/4 H/P, (VersaFog 1/2 h/p)	
CFM Rating		5,100 CFM high/3,920 CFM low
Weight	160 lb.	107 lbs (VersaFog 240 lbs)
Oscillating Fan	24in. (VersaFog 36")	
Nozzles	6 nozzles (Versafog12)	4 nozzles, (VersaFog 12)
Speeds		Two speeds
Water Pressure	1000 pci.	
Primary Water Filtration	Built-in	
Primary Water Filtration	Built-in	
	Non Corrosive Materials	
	Continuous Operation	
Power Consumption	95 W (250w)	
Humidification Capacity	up to 10 w/gallon/hour	
Humidification Coverage	50,000 sq.ft (and more)	
Misting Ring	Stainless-Steel, powder-coated	
Water Source	Regular Cold Water System	
	Auto drain valve reduces nozzle clogging	
	Brass and stainless steel nozzles and fixtures	
	Mist Fan Humidifier easily integrated with THC-1" Humidistat (optional, not included) for automatic area humidification.	

9 Buyamag Inc. Carlsbad, CA, USA

9.1 Humidifier NP – 50

Two versions are available in NP – 50 type, viz Deluxe Industrial Humidifier and Centrifugal Humidifier.

Humidifier NP-50

This Deluxe Industrial Humidifiers delivers excellent quality performance, easy fast installation and low maintenance with dependability. It is designed on a centrifugal concept to produce fog mist. This concept design allows to bypass all of the burdens of complicated installations, need of compressors, pumps, nozzles, drainage pipes, high maintenance, waste of space, heavy power consumption. It is trouble-free having simple structure with high durable qualities. It has direct cold water system connection. Very quiet and continuous operation which has centrifugal principle applied. No special installation required. Aerosol mist is completely droplet free because of powerful and heavy duty atomizer. There is no need of high pressure pumps or water filters. The nozzle does not clog.

Humidifier Centrifugal NP-50

Centrifugal Humidifier is connected directly to a cold water supply, gets filled itself automatically with built in Automatic-Fill Float Assembly without interruptions. Centrifugal Humidifier quietly delivers up to 2 gal/hr of perfect Mist Fog. Each Humidifier can be integrated with HCH-3" Humidistat (optional, not included) for automatic area humidification. It does not require drainage line. It is reliable with low space requirement. Has low energy consumption with maintenance free Squirrel Cage Motor with built-in Thermal Overload Switch. The structure is simple and trouble free with high durability and easily portable.

Technical specifications

Humidification Coverage:	5000 sq.ft (and more)
Humidification Capacity:	Up to 2.5 gallon/w/hour (2,000–4,000 cc/hour)
Power Supply:	110 V/60 Hz (or 220 V/50 Hz European system available)
Power Consumption:	90 W
Particle Size:	2–20 microns
Water Source:	Regular Cold Water System
Water Connection Fitting:	1/4"
Water Pressure:	No special requirements
Air Humidity Control:	Automatically Controlled by 'HCH-3' (humidity control humidistat, optional item)
UL, CE:	Approved

10 H. Ikeuchi & Co. Ltd. Osaka, Japan

10.1 Air AKI® DRY Fog Industrial Humidification System

Total humidification system for humidity control in large space

Features

- *Humidity control in large space*: The high-precision-made Dry Fog spraying unit, with its sprayed fog reaching a distance of 4 m ahead, produces the stable humidity in large space (within the whole factory).
- *For solution of troubles caused by electrostatic*: Solving the troubles caused by electrostatic, the system drastically improves productivity of machinery.
- *High performance energy saving*: No steam by boilers required. Energy saving of over 30% expected.
- *The least maintenance necessary*: With supply of clean air and purified water, the system requires the least maintenance.

10.2 AKIMist® "D" Dry Fog Humidifier

Dry Fog humidifier in ultra-compact body with its high spraying performance has following features.

1. Ultra-compact design: This is claimed as the most compact humidifier in the world with high performance.
2. High spraying performance: World-wide patented AKIJet® nozzle sprays fine quality fog over 4 m to provide effective humidification.

3. Easy-to-maintain mechanism: Ease of maintenance by one-touch disassembly. Connection to the compact RO water purifier AKIMist® keeps it in almost maintenance-free condition.
4. The least clogging nozzle: 2 types of the least clogging nozzle, based on the new technology, are available according to the application.
5. Good extensibility of the system: Up to 4 pieces of the nozzle mountable to a humidifier.
6. Flexible operation ability: Spray direction is adjustable within 10 to 50% with the Ball Adaptor (option) attached. Automatic humidity control is exercised by the timer or the humidity controller.

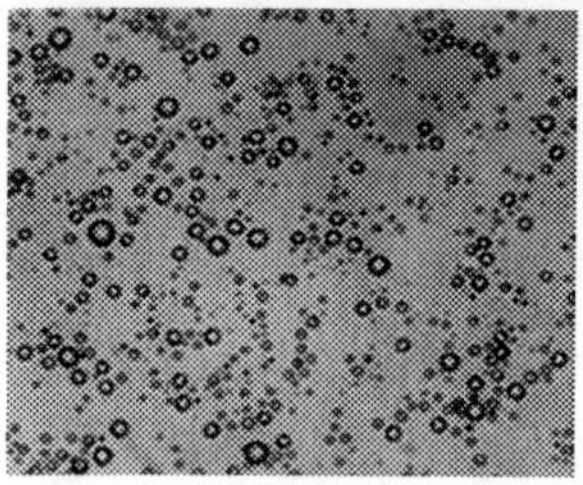

High-Quality Fog (S-type)

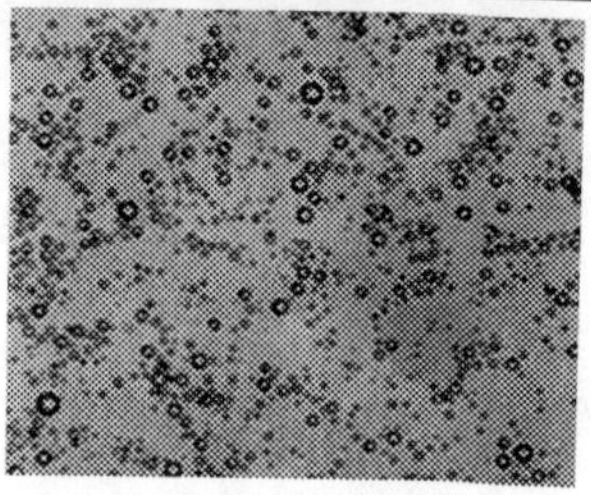

Large Volume Fog (D-type)

	S. Type	D. Type
Fog size	Ultra-fine Dry Fog with a silky touch.	Large volume cotton-like fine fog
Sauter Mean Droplet – dia:	7.5 microns	12 microns
Spray volume per each nozzle	2.4 L/hr (0.63 GPH) at Air Pressure 0.3 MPa (42 PSI)	3.0 L/hr (0.79 GPH) at Air Pressure 0.3 MPa (42 PSI)

S-Type Performance

Air Pressure Mpa (PSI)	Spray Volume L/hr (GPH)	Air Consumption NL/min (SCFM)
0.3 (42)	2.4 (0.63)	36 (1.34)
0.35 (50)	2.7 (0.71)	41 (1.53)

Model No	A pressure 0.3 Mpa (42 PSI)		Number of Nozzle
	Spray Volume L/Hr (GPH)	Air consumption NL/min(SCFM)	
AD-1S	2.4	36	1
AD-2S	4.8	72	2
AD-3S	7.2	108	3
AD-4S	9.6	144	4

This is more suitable for humidification without wetting the wall and floor even within a room of limited space.

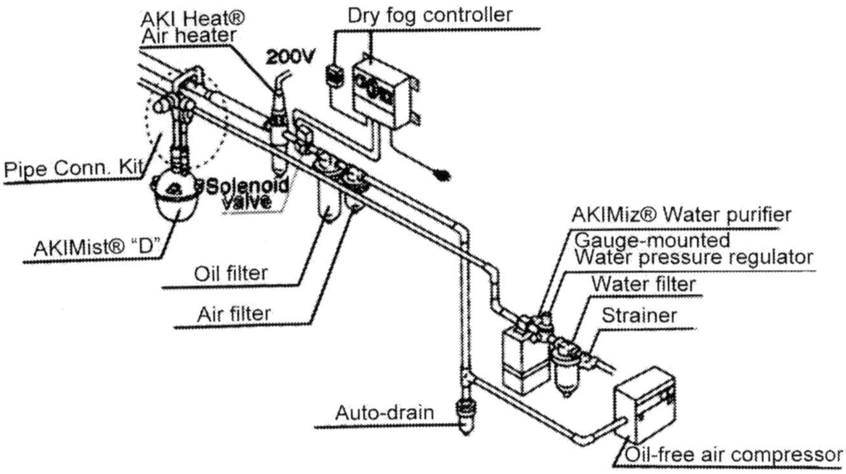

D – Type Performance

Air Pressure Mpa (PSI)	Spray Volume L/hr (GPH)		Air Consumption NL/min (SCFM)
0.3(42)	3.0(0.79)		36(1.34)
0.35(50)	3.5(0.92)		41(1.53)
Model No. D. Type	at Air Pressure 0.3 Mpa (42 PSI)		Number of Nozzle
	Spray Volume L/hr	Air Consumption NL/ min (SCFM)	
AD-1D	3.0	36	1
AD-2D	6.0	72	2
AD-3D	9.0	108	3
AD-4D	12.0	144	4

The humidifier produces large volume of fog with less air. It is suitable for humidification in large volume in a spacious room. For minimal maintenance and long life, clean air and purified water must be supplied. IKEUCHI provides wide range of ancillary devices for purifying water/air and automatic humidity control from years of experience.

10.3 A - AD – T Dry fog portable humidifier set

Features

- *Portable Humidifier Set*: No piping work necessary. Portable humidifier set for immediate use at any place just with air supply.

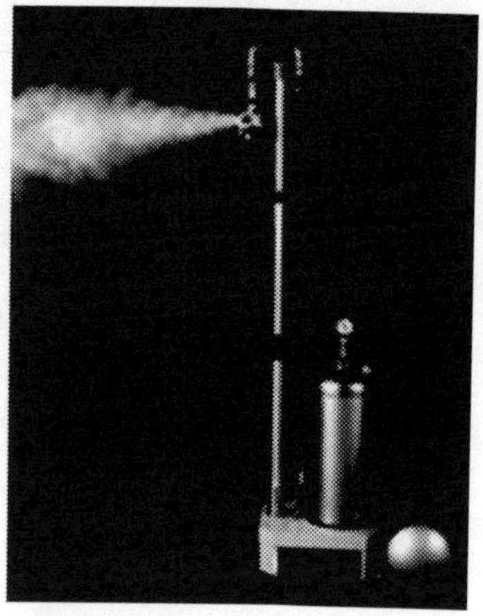

- *Choose your Fog*: Two types of spraying nozzle are available with the spraying unit; one for high quality fog (2.4 L/h or 0.63 GPH) with a silky touch, and another for large volume cotton-like fog (3.0 L/h or 0.79 GPH), both per one nozzle at air pressure 0.3 MPa.
- *High Extensibility*: Up to 4 nozzles can be mounted. The humidifier with 4 nozzles has large spraying capacity, maximum 12 L/h or 3.16 GPH at air pressure 0.3 MPa or 42 PSI.
- *Flexible Operation*: Easy automatic control by Humidity Controller (option) or its timer.
- *Portable*: Easily portable by attaching casters.

Main applications are in various factories such as of printing, textile and paper fabrication for spot humidification around machinery and electrostatic prevention, for electrostatic prevention and cooling by dry fog in factories of plastic, film fabrication, etc., for humidification within a horticultural facility, for humidification within a storage of vegetable, fruit, and fresh produce, for spraying of deodorant in sewage dump, for easy installation of a humidifier and when a small volume of dry fog is required

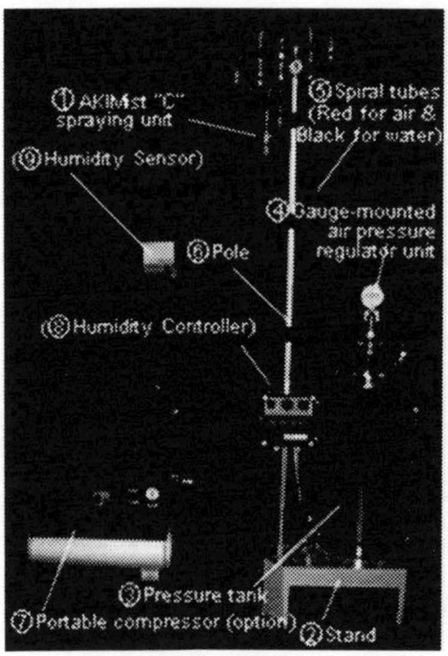

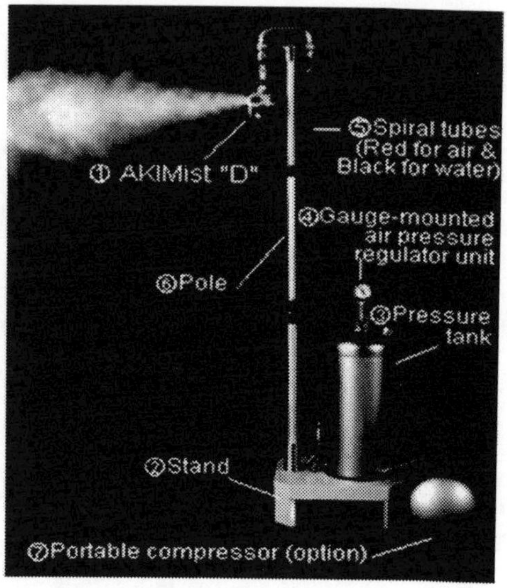

The standard contents include along with AKI Mist units are a stand, pressure tank, gauge-mounted air pressure regulator unit, spiral tubes for air and water, pole, portable compressor, humidity controller and humidity sensor are optional.

Portable Compressor: The oil-free type of compressor with its piston ring made of Teflon for the longer life and clean air

Model No.	Motor		Max. Pressure	Output	Tank Capacity	Weight	Dimensions
TO5P5S17	100 V 0.55 kW	50 Hz	0.96 MPa (9.8 kg/ cm^2)	60NL/min at 0.68 MPa (7 kg/cm^2)	12 L	20 kg	L: 418 mm W: 495 mm H: 230 mm

Note: These portable compressors are to operate the humidifier with one nozzle only.

Pressure Tank

Tank Capacity	Material	Max. Pressure	Weight
18 L	SUS304	0.68 MPa (7 kg/cm^2)	5 kg (23 kg in full)

Note: Relief valve is already set at 0.4 Mpa.

Specification

Portable humidifier sets are suitable for small space and spot humidification.

(a) *Portable HumidifiertSet*: No piping work necessary. Portable humidifier set for immediate use at any place just with power and air supplies.

(b) *Flexible SprayingnDirection*: Spraying direction 360 degrees rotatable. Best for spot humidification at any place.

(c) *Best for SmalleSpace*: For spot humidification around machinery and humidification in such small room up to 100 m3 volumes as an inspection room.

(d) *Flexiblen Operation*: Easy automatic control by Humidity Controller (option) or its timer.

(e) *Portable*: Easily portable by attaching casters.

Required supply air: 36 46 NL/min (at air pressure 0.29 0.49 MPa). Portable humidifiers are used to prevent electrostatic destruction and particle adhesion in electronics processes, to prevent paper jam, and paper expansion and contraction in printing factories, to prevent dust adhesion and ignition, when a small volume of dry fog is required. When air supply is available from utilities, the portable compressor is not necessary.

11 S.K.B Systems, Coimbatore, India

11.1 Industrial humidifiers

Range: Bahnson Spot type, Oscillation, Fogaire, E-fog, humilab, Con-Air, Con Air Max, Custom-built Humidifiers with Automatic Controls all to suit specific requirements.

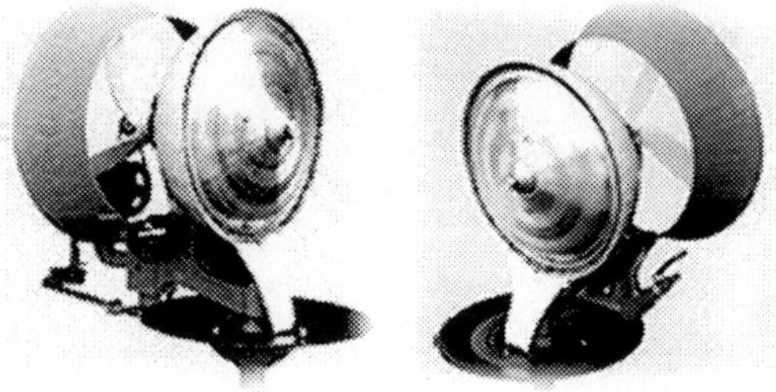

Application industries

Textile Industry, Tea and Tobacco Manufacturing Units, Pharmaceutical Industries, Wood Processing Industries, Cold Storage, Horticulture Industries and Green Houses.

11.2 "B" Spot humidifier

The features are low operating and recurring cost with efficient fractional HP Motor, better humidity control at lower cost, correct humidity maintained, self-contained, requiring no duct work or auxiliary equipment, minimum and easy maintenance, quicn atomisation of water into a fine mist for rapid evaporation, effective mist dispersal due to positive directional air flow, high capacity evaporation. Installed room shall get uniform distribution of humidified air. Easy to install, flexible and self-contained unit

11.3 "H" Oscillation humidifier

Features: Total area covered by single system shall be 2,200 sq. ft. One Oscillation Humidifier shall replace 3 Spot type humidifiers because of its sweep area of 170 degrees oscillation arrangement. No condensation of water in front of humidifiers. They are easy to install and smooth in operation, simple and silent due to the genuine grade bearing for the energy efficient motors. Evaporating capacity of rs litres per hour, but can be manipulated to suit customer's requirements using imported gate valve.

12 Air humidifier by ningbo chenwu humidifying equipment facto – China

12.1 Description

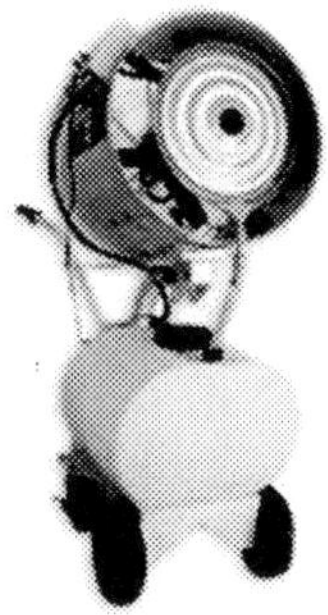

Mainly used in all kinds of clothing factories, all kinds of textile factories, high temperature and workshop with polluted, ironware factories which need to lower temperature, waiting room, supermarkets, restaurants and other public places, and it is also used in the hothouse and the farm.

The technical parameter is as follows:

- Mist Output: 0–35 L/H.,
- Air Capacity:7,200 m^3/L.,
- Humidified Space: 300m^2.,
- Spray Direction: Mobile.
- Power : 600 W., Power Supply: 220 V, 50 HZ
- Weight: 60 kg,
- Dimension: 550*450*1,400 mm

13 Faran Industrial Co. Ltd., Loni, India

13.1 Model : HR-15 (Industrial Humidifier)

Specifications

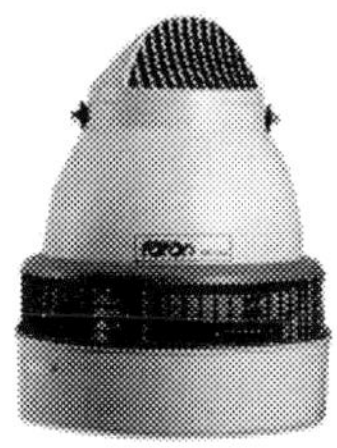

Coverage : 15–30 Sq. Mtr

Power supply : 110 V or 220 V / 50 Hz or 60 Hz

Power consumption : 90 W

Humidification capacity : 1,500 cc/hour (0.4 gallon/hour)

Dimensions : 310 mm × 310 mm X 470 mm(12.20" × 12.20" × 18.50")

Features

Easy to install(connect to water tap or water tank)

Various space usage (on-wall)

Easy to carry by handle

Trouble free simple structure (high durability)

Good for chemical disinfection and deodorisation

Good for electronics, textile, tobacco factories, low temperature warehouse and mushroom cultivation.

Applications

Greenhouse, propagation room, research chamber, storage room, nursery, mushroom and orchid cultivation, chicken breeding, textile, tobacco and timber industries.

Notes

Set it up horizontally to control the water level

Make sure the connecting wire is long enough

Check and clean the water tank periodically

If there is no humidification, turn power off and check the water level. If water overflows, turn the water tap off and check the float valve.

15 Manvi Textile Air Engineers, Mumbai, India

15.1 Power mist humidification system

This system offers balanced humidification to the applied environment. A balanced amount of humidity is required in a manufacturing process, which is required to stabilise moisture in wood, paper or textiles. These are designed using quality raw material and tested on different parameters to ensure that the products are free from all defects. Features include design and manufacturing for masonry and sheet metal plant, overhead and underground return air system and necessary air control dampers, diffusers, nozzles, spray header.

Centralized fogging system for blow room, yarn conditioning area and weaving area consists of high pressure pump, high pressure piping and nozzles spread on bale plucker area, yarn condition area and weaving area. Uniform RH is maintained in every corner of the department as it is distributed as per heat load of the department. This is connected to automatic control system which will control humidity up to set percentage with 1.5 % accuracy. Desired RH for blow room, yarn conditioning and weaving can be maintained separately as needed.

15.2 Humidifying equipment

Manvi have introduced humidifying equipment manufactured in accordance with the approved industry norms. These devices are used for increasing the ambient humidity. For the functioning of these products, water pressure pumps or compressed air is not required. This range has gained huge popularity as capable of producing fine mist support, resulting in quick evaporation and finally, air humidification. Special features are light in weight, silent operations and low power consumption.

15.3 Ultra-Fog Mini Humidifier

Technical specifications

Design	Portable.
Dry Weight	6 kg
Particle Size	< 5 microns
Dimensions	310 mm x 310 mm x 470 mm. (14.56" x 14.56" x 18.90").
Power Input	Single phase, 220 – 240 V / 50 Hz
Power Consumption	120 W (Max.)
Area Coverage	200 to 300 ft2.
Humidifying Capacity	2 to 3 Ltrs. / Hr.
Water Connection	0.5 inch (< 5 kg / cm^2).

MODEL NO. : TA-MHT-003

Features

(a) Noiseless operation. Quiet and peaceful working.

(b) Independent spot humidification system.

(c) No air compressor or water pumps required.

(d) Lowest energy consumption in its category (only 120 W).

(e) No harm to electrical equipment due to very fine particle size.

(f) Easy to install. Shipped in ready-to-use condition.

(g) No corrosion effects and no rusting of machineries.

(h) Negligible maintenance, simplified even further due to portability of unit.

(i) No condensation and no water droplets occur.

(j) Good for chemical dis-infection and deodorisation.

15.4 Ultra – Fog Table Mounted

Design	Portable.
Dry Weight	9 kg
Particle Size	< 5 microns
Dimensions	370 mm x 370 mm x 480 mm. (14.56" x 14.56" x 18.90").
Power Input	Single phase, 220 – 240 V / 50 Hz.
Power Consumption	90 watts (Max.)
Area Coverage	500 to 800 ft^2.
Humidifying Capacity	6 to 8 Ltrs. / Hr.
Water Connection	0.5 inch (< 5 kg / cm^2).

MODEL NO. : TA-MHT-006

Features

(a) Independent spot humidification system.

(b) No air compressor or water pumps required.

(c) Lowest energy consumption in its category (only 90 W).

(d) No condensation due to very fine water particle size.

(e) Easy to install. Shipped in ready-to-use condition.

(f) No corrosion effects and no rusting of machineries.

(g) Negligible maintenance, simplified even further due to portability of unit.

(h) Good for chemical dis-infection and deodorisation.

15.5 Ultra-Fog 360°-Ceiling Mounted

MODEL NO. : TA-MHC-006

Particle Size:	< 5 microns
Power Consumption:	90 W only.
Area Coverage:	300–600 sq. fts.
Fog Output:	4–6 Ltrs./Hr.
Power Supply:	Single phase, 220 V / 50 Hz.
Water Connection:	0.5 inches (< 5 kg/cm2)
Packing Dimensions:	44 x 44 x 46 cms
Dry Weight:	9 kg

MODEL NO. : TA-MHC-006

Features / Advantages:

- When installed in the centre of the room, it will ensure uniform distribution of fog particles inside the entire room as fog is delivered at 360 degree.
- Independent spot humidification system.
- No compressed air/water pressure pumps required.
- Lowest energy consumption in its range.
- Negligible maintenance.
- Specially designed motor with inbuilt thermostat capable of generating 20,000 hours of non-stop power to ensure consistent humidification.
- No condensation due to very fine particle size.
- Easy to install. No structural modifications required to install the machine.
- Specially designed for Q.A. / Q.C. laboratories, research chambers and electronics industries.

15.6 High fog humidifier trolley mounted

The High Fog system consist of spinning disc and teeth surrounding it. The water is run onto the disc spinning at 3,500 rpm with certain pressure to produce atomisation down to 18 microns. Studies have shown that water droplets of this size tend to follow the air stream. Hence, high capacity fan is provided behind the disc, which can deliver humid air to long distance. Water can also be provided which tend to evaporate faster if air quantity is increased. Hence no corrosion effect will be there in surrounding area.

Model No	TA-SHT-050
Particle Size:	18–20 microns.
Power Consumption	550 W only
Area Coverage	1,600– 2,600 sq. fts
Fog Output	0– 50 Ltrs./Hr
Power Supply	Single phase, 220 V / 50 Hz
Effective Distance	10– 15 m
Air Flow	8,280 m3 /Hr
Tilt Angle	30°/–15°
Oscillation	120° (Auto)
Water Tank Capacity	60 ltrs.

Salient features and advantages

- Designed to deliver very fine droplets size of 18–20 microns and provided water and air control also.
- It has control of water output from 0 to 25 ltrs, if water quantity is reduced then particle size will be finer.
- Provided with 120 degree oscillation to cover large area.
- During working the heat will be absorbed by mist and the temperature will be decreased as well as increased air relative humidity, decreased bug dusts and purified air.
- Heat will be absorbed by mist and temperature will be decreased to 5–12 degree.
- Can achieve humidity 60–80% in the department.
- 304 stainless steel mist maker, anti-corrosion.
- Saving space, low noise, durable and thermal protection motor, strong structure, easy to operate and maintain, low power consumption.
- No blockage and leakage, portable design.
- Mounted on trolley for more convenient movement with built in tank of 60 ltrs capacity.

High-Fog is specially designed for cotton storage area, Blow room bale plucker, jute mills, ginning mills and cotton mixing areas. **STEM FOR CENTRALISED AIR WASHER WITH CHILL**

15.7 Fogging System for Centralised Air Washer with Chiller

Features

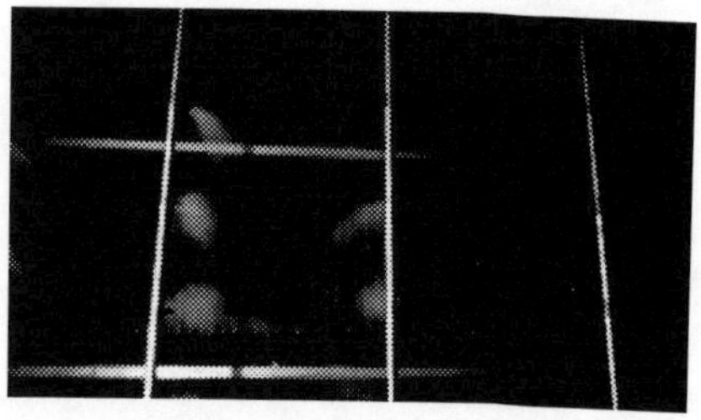

- Additional fogging bank in the air washer plant increases saturation efficiency
- Increase in saturation can achieve required RH with less CMH, which indirectly saves power.
- Fogging Bank can be connected to small chiller to achieve desired temperature in the department

Advantages

(a) Due to increased saturation desired RH can be maintained with lower CMH or with same CMH is can maintain better RH% in the department.

(b) Power saved in pump HP and supply air fan hence less running cost.

(c) In humidification, only RH can be achieved but we can connect small chiller of 3 to 10 TR depending on department heat load and can maintain department temp less than outside wet bulb.

(d) About 60 to 70% less power consumed compare conventional chilled water spray air condition.

15.8 Semi Central Fog Humidification Plant

It consists of one wall mounted fan with 5/7.5/10;15 HP motor and v filter grouted in masonry chamber behind the fan so that filtered air will be sucked by fan and fogging box with water eliminator will be mounted in department. From the mouth of the box we can mount duct with grills on the both sides of duct through which humid air will be delivered in the department.

It is designed on the based on the heat load of the department and can achieve relative humidity up to 85% in the department. It will consist of one wall mounted fan with 5/7.5/10;15 HP.

Advantage

(a) Power saved in pump HP and supply air fan hence less running cost.

(b) Very less construction required hence capital investment reduces.

(c) In humidification only RH can be achieved but we can connect small chiller of 3 to 10 TR depending on department heat load and can maintain department temp less than outside wet bulb.

(d) About 60 to 70% less power consumed compare conventional chilled water spray air condition.

Appendix – 4
Spray nozzles and mist eliminators

1 Nozzles

1.1 Ikeuchi USA

Hydraulic Nozzles (Following varieties are available)

 **Flat spray nozzle**	**Standard Flat Spray** 1. One-piece structure standard flat spray nozzles (VP, VVP) 2. One-piece structure standard flat spray nozzles in plastic (UVVP, VVP-U) 3. Three-piece structure standard flat spray nozzles (V, VV) 4. Coin-shaped standard flat spray nozzles(CVVP) **Even Flat Spray** 5. Even flat spray nozzles (VE, VEP) 6. High pressure cleaning Even Flat spray nozzles (VNP) 7. Descaling nozzles (DSP) 8. Wide-angle Flat Spray 9. Wide-angle Flat Spray Nozzles (YYP) 10. Wide-angle Flat Spray Nozzles for Low Pressure Operation (LYYP) Off-center Even Flat Spray Nozzles (OVVEP) Self-cleaning Flat Spray Nozzles (MOMO) Quick Installation Nozzles (QB) Quick-detachable Type Flat Spray Nozzles (ISVV) Air Mixing type Flat Spray Nozzles (EJVV) Air & Steam Spraying Flat Spray Nozzles (VZ)

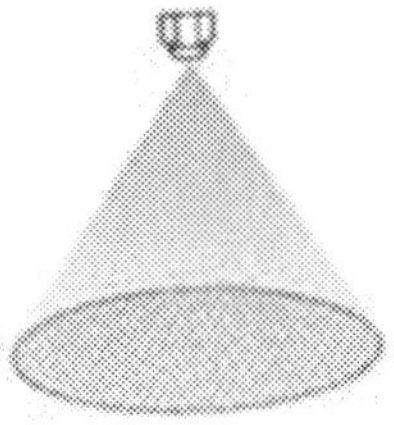 **Full cone spray nozzles**	1. Standard Full Cone Spray Nozzles (JJXP) 2. Ceramic Orifice and Whirler Inserted Full Cone Spray Nozzles (JUP) 3. Alumina Orifice and Whirler Inserted Full Cone Spray Nozzles (JUXP-AL92) 4. Flange-type Large Capacity Full Cone Spray Nozzles (TJJX) 5. Wide-angle Full Cone spray Nozzles (BBXP) 6. Narrow-angle Full Cone Spray Nozzles (NJJP) 7. Small Capacity Full Cone Spray 8. Small Capacity Full Cone Spray Nozzles in Plastic (JJRP) 9. Small Capacity Full Cone Spray Nozzles in Metal (J) 10. Minimal Clogging Full Cone Spray Nozzles (AJP) 11. Quick Detachable Type Full Cone Spray Nozzles (ISJJX) 12. Full Cone Spray Nozzles for High-efficient Etching (SNAPJet)
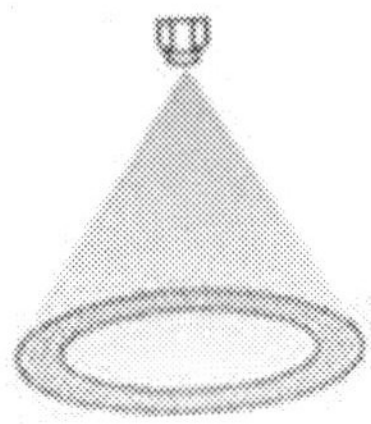 **Hollow cone spray nozzles**	1. Extremely Fine Mist and Ultra Small Capacity Hollow Cone Spray Nozzles (KB) 2. Spray header with KB-CERTIIM series nozzles 3. Fine Atomisation and Small Capacity Hollow Cone Spray Nozzles (K) 4. Medium Capacity Hollow Cone Spray 4.1. Medium Capacity Hollow Cone Spray Nozzles (AAP) 4.2. Alumina Ceramics and Medium Capacity Hollow Cone Spray Nozzles (AP-AL92) 5. Flange-type Large Capacity Hollow Cone Spray Nozzles (TAA) 6. Small Capacity Hollow Cone Spray Nozzles (KD)
Special cone spray nozzles	1. Square Full Cone Spray Nozzles (SSXP) 2. Multiple-nozzle head Full Cone Spray 2.1 Seven head Full Cone Spray Nozzles (7JJXP) 2.2 Extremely Fine Atomisation 7-head Full Cone Spray Nozzles (7KB) 2.4 Gas Cooling Nozzles SPILLBACK (SPB)

 **Solid stream jet nozzles**	1. Standard Solid Stream Jet Nozzles (CP, CCP) 2. Round Inlet Solid Stream Jet Nozzles (CRP, CCRP) 3. Trimming Nozzle (CM, CTM) 4. Multiple-orifice Solid Stream Jet Nozzles 2CCP / 7CCP, 2CP / 7CP series 5. Self-cleaning Solid Stream Jet Nozzles (MOMOC)
Special Solid Stream Jet Nozzles	Ejector Nozzles for Solution Agitation (EJX)
Special Purpose Nozzles and Accessory	1. Air Nozzles (TAIFUJet®) 2. Water/Air Curtain Nozzles (SLNH / SLNHA) 3. Accessory - UT Ball Joints

Pneumatic Air Nozzles TAIFU Jet®

 **Pneumatic Spray Nozzles**	**Features** • Taking in surrounding air through holes around nozzle edge, TAIFUJet nozzles spout out double volume of air supplied. • Uniform distribution resulting from unique design achieves an efficient air blow and saves air consumption. • Designed to reduce a noise level for improving working environments. **Materials** ABS (Acrylonitrile butadiene styrene) / Stainless steel **Applications** • Cooling • Blowing off water droplets, drying • Blowing off dust • Transporting • Air curtain

<table>
<tr><td colspan="2">Comparing with hydraulic spray nozzles applying only water pumps, pneumatic spray nozzles have the following features.
1. Excellent atomizing performance
2. Large turndown ratio (large flow control coverage)
3. Large free passage diameter</td></tr>
<tr><td>Fine Mist Nozzles
BIM / GBIM / GSIM_s series</td><td>Pneumatic spray nozzles developed to satisfy crucial requirement on spray nozzles in continuous casting process in steel making</td></tr>
<tr><td>Minimal Clogging Fine Mist Nozzles
SETOJet series</td><td>Pneumatic spray nozzles developed especially for spraying / atomisation of viscous liquid</td></tr>
<tr><td>Impinging-type Fine Mist Nozzles
AKIJet® series</td><td>Impinging-atomisation type pneumatic spray nozzles featuring uniform distribution of droplet size</td></tr>
<tr><td>Semi-fine Mist Nozzles
LSIM series</td><td>Pneumatic spray nozzles producing semi-fine atomisation, applying very low pressurised air from conventional blowers</td></tr>
<tr><td>Steam Driving Nozzles
JOKIJet® series</td><td>Innovative pneumatic nozzles using steam instead of compressed air to generate fine mist of liquid components</td></tr>
<tr><td>Water/Air Curtain Nozzles SLNH / SLNHA series
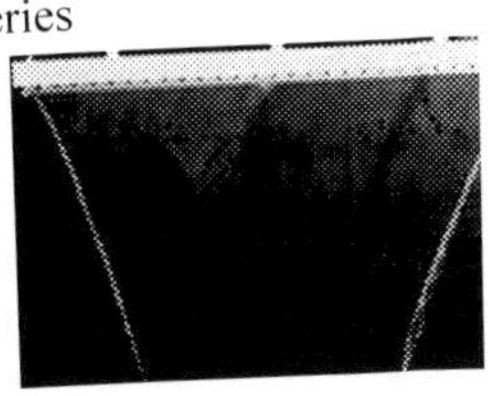</td><td>Features
• Spray uniformly in width direction with soft spray impact for minimal damage to the products.
• Thinner liquid film spray saves cost of chemicals and water.
• Available in liquid spraying type (SLNH series) and air spraying type (SLNHA series).
Materials
• Stainless steel 304 / PVC (Polyvinyl chloride)
Applications
• Air Blow-off
• Etching
• Cleaning</td></tr>
</table>

1.2 Condair West Sussex-UK

JetSpray AHU Cold water spray humidifier

The JetSpray AHU system can provide close humidity control to +/-2% relative humidity and is the energy efficient way to humidify air-handling systems. It can be retro-fitted to existing AHUs or specified for new systems. Either way the savings in running costs and low maintenance routines mean that this remarkable system will more than pay for itself within one year.

JetSpray AHU is hygienic in use, self-cleaning and self-purging, uses mains, demineralised or RO water and gives excellent control of the relative humidity.

A JetSpray AHU humidifier comprises a suitable number of atomising nozzles mounted onto one or more custom-designed manifolds, installed across the airflow. Excellent moisture distribution is assured by the choice of nozzle capacity, suitable spacing on the manifold, correct positioning in the AHU and flexible orientation of the nozzles.

Evaporation distance varies according to air temperature, velocity and RH but, where space is at a premium, a JetSpray Evaporation Matrix can be installed to guarantee evaporation in just one metre.

JS Pure Tec ensures hygienic operation, with UV water sterilisation, active drain to empty the nozzle line when not in use and regular purging of pipe work to the JetSpray control so water does not remain static in the system.

JetSpray AHU – Low energy, Hygienic Humidification for Air-Handling Units

Features

- ±2% relative humidity control
- WRAS approved for direct mains water connection
- Saves energy and money
- Low maintenance costs
- Hygienic and clean:
  - UV sterilisation
  - Self-purging
  - Self-cleaning

– No water recirculation

- Mains and RO water
- Complete evaporation
- Excellent control of RH
- Easy installation
- Better for the environment

Precision engineered nozzles

Precision engineered nozzles combine compressed air and water to achieve maximum evaporation at minimum distance.

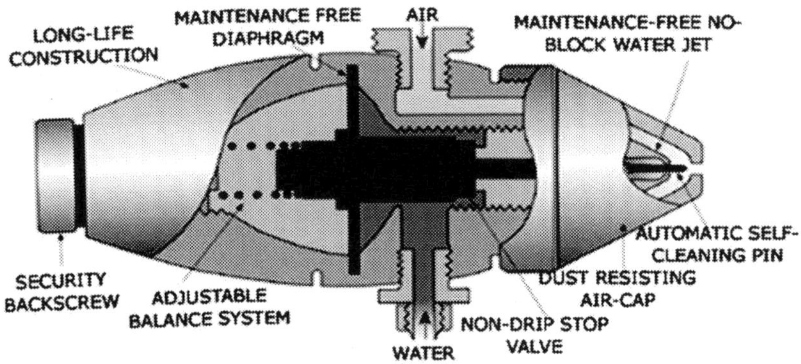

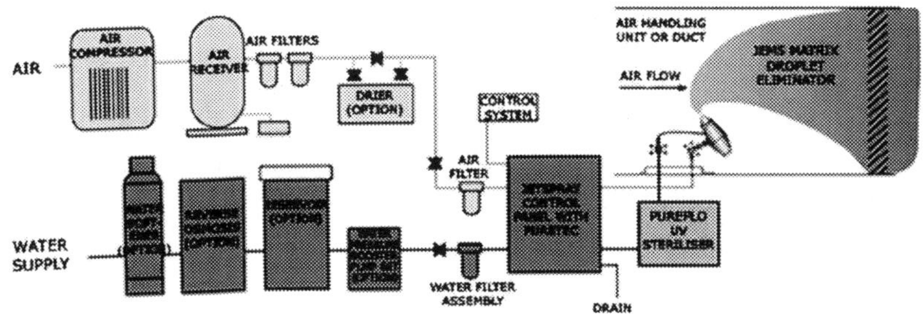

Jet Spray Complete Package: A clean, dry compressed air supply of 0.56 m^3/h (0.033 cfm) is required for each litre of atomised water, at a minimum of 4.5 bar (67 psi). A final filter with auto drain is recommended immediately before each JS control panel. Mains, softened or demineralised water supply should be available at a minimum pressure of 4.0 bar (58 psi), with no significant fluctuation, max pressure 7.5 bar (109 psi). Even though the JetSpray can run with mains water, demineralised water should be used for a dust-free environment, where the application demands it.

1.3 Excel Air Technique – India

Spray Nozzles

Spray Nozzles produce a fine mist (fog) and allow the passing air stream to absorb the same, thereby lowering its dry-bulb temperature closer to its wet-bulb temperature. They are available in PVC, PP and PC with S.S. caps and clamping provision.

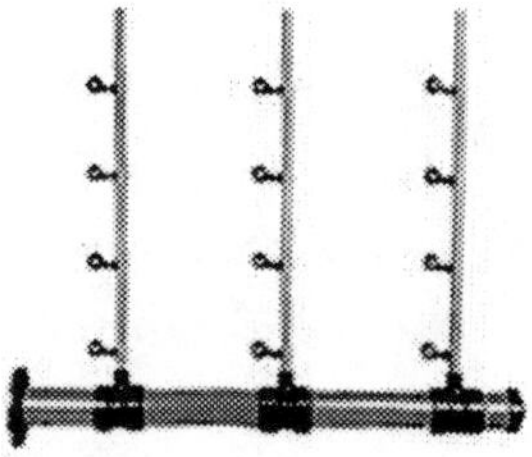

2 Eliminators

2.1 M.M. Aqua Mist Eliminator – Gurgaon, India

Mist eliminator produced by MM Aqua is U V stabilised made out of virgin PVC in 3 different designs to meet engineering applications in textile units, air washers / air-handling units, paint booth gas turbine air intake systems, etc. Mist eliminator is an impingement separator designed for horizontal gas flow. It is composed of sine curve shaped flow contours assembled with phase separating chambers our three types of mist eliminators are T-100, T-200 and T-400

Profile depth	170 mm.
Maximum length	5,000 mm.
Material	P V C.
Uses	Mist eliminator as well as air straightener.

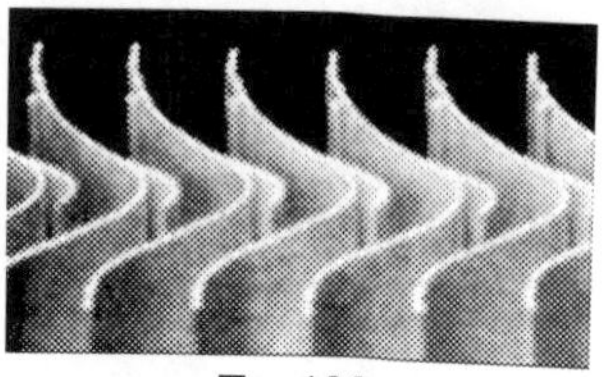

T – 100

Profile depth	150 mm.
Maximum length	5,000 mm.
Material	P V C.
Uses	Mist eliminator

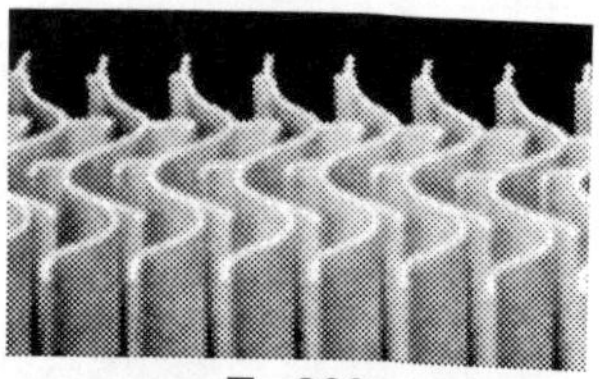

T - 200

Profile depth	105 mm
Maximum length	5,000 mm
Material	P V C.
Uses	Mist eliminator

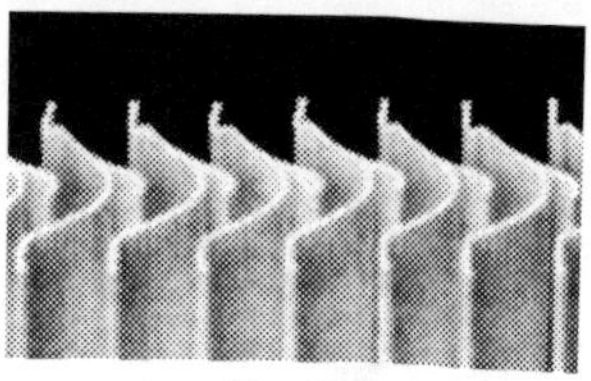

T - 400

Advantages of M.M Aqua Mist Eliminators

Manufactured through continuous extrusion process gives uniform thickness to eliminator profile. Lower pressure drop due to special aerodynamic contour and smooth surface. Longer life due to corrosion resistant plastic material. Light weight so easy to handle. Up to 3 m height no intermediate support is required.

2.2 ACS Mist Eliminators – Texas, USA

ACS is pioneers in mesh pad eliminators which have a very high efficiency of vapour elimination.

ACS Mesh Pad Mist Eliminators

ACS offers a large variety of mesh materials, densities and configurations to offer exact replacement-in-kind for virtually any existing mesh pad within days or to optimise your mist elimination application, most often with no modifications being required to your equipment. In most applications, the ACS mist eliminator can be sized, designed and rated based on the given process data for up to 99% removal of liquid droplets 3 microns and larger.

Chevron vane mist eliminators

ACS offers exact replacement-in-kind for most chevron-shaped vane mist eliminators and two specialty designs that offer improved high performance mist elimination at very high flow capacities and velocities. Our patented MultiPocket® vane is truly revolutionary in the vane industry of low fouling, durable, high capacity products. In most applications, the ACS MultiPocket® product can be sized, designed, and rated based on the given process data for up to 99% removal of liquid droplets or small particulates 8 micron and larger.

Mesh-Vane assembly mist eliminators

After years of testing and field verification experience, ACS offers the industry the very best combination of capacity and fine liquid droplet micron removal performance with the mesh-vane assembly product. Let ACS show you how to handle 200% greater flow rates at up to 99% removal of liquid droplets 3 microns and larger with no modifications to your vessel.

Appendix – 5

Fans and blower

1. Aerotech Ghaziabad India

1.1 Centrifugal Blower

AeroTech manufacture a wide range of centrifugal fans to suit number of industrial applications. Centrifugal Blowers are well engineered and having high efficiency incorporating all the latest technologies. All types of centrifugal fan that manufactured are designed to meet customer's specific requirements.

Type of Fans

High Volume Limit Load Blowers*(SISW Type)

High Volume Limit Load Blowers*(DIDW Type)

Low Pressure Blowers

Plug Fans

Medium Pressure Blowers (Backward Curved Blades)

Medium Pressure Blowers (Self Cleaning Blades)

Medium Pressure Blowers (Straight Radial Blades)

Induced and Forced Draft Fans

High Pressure Blowers

Two Stage Blowers

*Available in class 1, 2, 3 and 4 constructions.

Technical Specification

Air volume 2,50,000 C.M.H

Air pressure up to 1,200 mm w.g in Single stage

Air pressure above 1,200 mm w.g in Two Stage

Temperature up to 380°C with cooling disc

Salient Features

Easy maintainability

Low power consumption

Low noise

High efficiency

Customize drive arrangements

Customize discharge directions

Extra Features Available

Available in M.S /S.S /aluminium /F.R.P

High temperature fan with cooling disc

Drain pug and inspection window

Horizontally split housing for easy handling

F.R.P / rubber lining

Inlet / outlet damper

Special paint finish

1.2 Axial flow fan

AeroTech Axial Flow Fans are specially designed and manufactured for optimum relation between air quantity and power consumption and to meet the exact demands of commercial and industrial ventilation in hot, humid, dusty and corrosive environments. Cylindrical casing of fans is fabricated out of mild steel /stainless steel sheet with adequate size for fine clearances as is practicable for better efficiency. Impeller will be of cast aluminium alloy and aerofoil of blade is designed to have maximum efficiency at lesser H.P.

Type of Fans

Tube Axial Fan (Belt Driven/Direct Driven)

Heavy Duty Wall Mounting Fan/Ring Fan

Mine Ventilation Fan/Contra-Rotating Fan

Vane Axial Fan (Belt Driven/Direct Driven)

Bifurcated Axial Fan

Power Roof Ventilator/Extractor

Technical Specification

Air volume up to 130,000 C.M.H

Sizes from 300 mm to 1,600 mm.*

Available 4, 6, 7, 8 & 11 blades impeller

Available with fixed and variable pitches impeller

Larger sizes also available on demand.

Salient Features

Cost effective

Large air volumes

Low power consumption

Easy to install

Low maintenance

Can be connected to duct

Extra Features Available: Available in direct drive and V-Belt drive, Long and short casing, F.R.P or rubber lined, Gravity Louver or Bird Guard (Front/Back) and special paint finish.

The axial flow fans are widely used for providing the required airflow for the heat and mass transfer operations in various industrial equipment and processes. These includes cooling tower for air conditioning and ventilation, humidifiers in textile mills, air-heat exchangers for various chemical processes, ventilation and exhaust as in mining industry. All the major industries use large number of axial flow fans operation, such as Humidifiers fans, Ventilation fans, Cooling towers, Heat exchangers Industrial air circulator and Man cooler

The FRP fans offer certain critical advantages as optimal aerodynamic design of fan impellers to provide higher efficiency for any specific application, reduction in overall weight of the fan, thereby extending the life of mechanical drive system. It requires lower drive motor rating and light duty bearing system. Lower power consumption is resulting in appreciable energy savings. FRP fan fabricated compression moulding/resin transfer moulding technique would have uniform dimensions and consistent quality. Lower flow noise and mechanical noise level compared to conventional metallic fans. Longer life of the fan is achieved due to improved mechanical strength.

1.3 Man coolers

'Aerotech' man coolers are meant for industrial use where forced draft circulation of air is needed. These blasters discharge high velocity powerful air offering effective cooling at a particular spot or long distance. They are unique for the comfort for the man, machine and materials. Built in robust construction and fitted with TEFC motors for 3 phases, 50 cycles, A.C. supply operation, these air blasters employ high efficiency aluminium alloy impellers with aerofoil section blades and are designed for optimum relation between air quantity, pressure and power consumption. The air blasters consist of strong wire guards, tilting and swiveling arrangement.

Air Filters are designed to provide better filter efficiency, and ideal air flow characteristics, constant performance greater fire retardency. It is easy maintenance and simple handling provide it a longer life thus eliminates numerous operational constraints. Available in standard as well as tailor made sized.

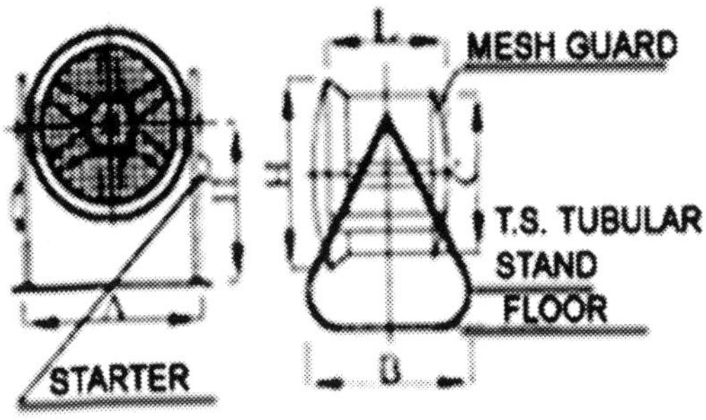

TUBULAR MOUNTED									
TYPE	SIZE	A	B	C	D	H	L	MOTOR	
								H.P	RPM
AETM	500	745	55	510	650	500	460	2	2850
AETM	600	845	655	610	750	600	520	1.0	1440
AETM	700	1000	655	710	850	700	620	2	1440

COLUMN MOUNTED								
TYPE	SIZE	A	B	C	D	e	MOTOR	
							H.P	RPM
AECM	500	255	205	600	485	325	2	2850
AECM	600	255	205	700	485	325	1	1420
AECM	700	255	205	800	585	325	2	1440
AECM	800						3	1440

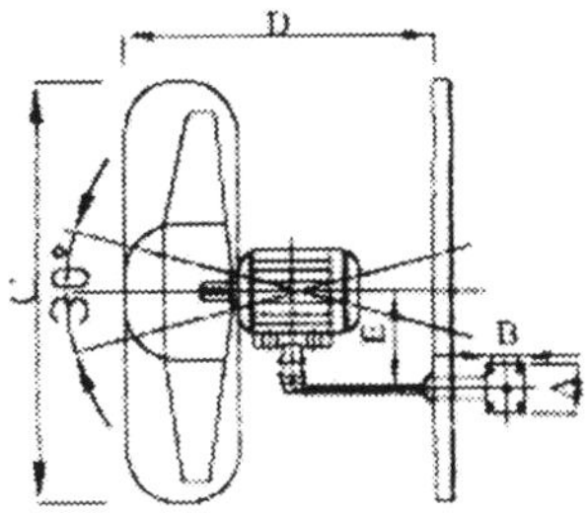

PEDESTAL TYPE							
TYPE	SIZE	A	B	C	D	MOTOR	
						H.P	RPM
AEPM	500	555	1155	600	430	2	2850
AEPM	600	555	1155	700	415	1	1440
AEPM	700	555	1155	800	485	2	1440

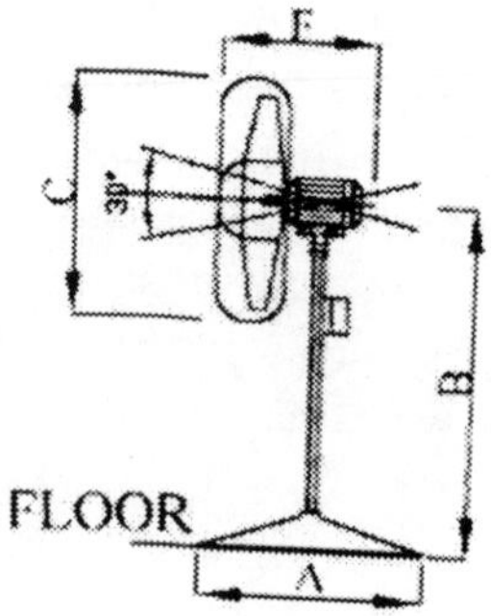

ACM	A	B	ØC	ØD	E	F
50	160	6	750	500	540	350
60	160	6	900	600	640	430
70	200	6	1050	700	746	440
80	210	6	1200	800	846	560
90	210	6	1350	900	950	575
100	225	6	1500	1000	1046	590
120	290	6	1800	1200	1265	680
140	300	8	2100	1400	1410	760
160	325	8	2400	1600	1525	800

1.4 Air Curtain

Air Curtain is the environmental separation which prevents outside air, insects or hot air entry to the conditioned area through the open door ways or the door being opened at frequent intervals. The principal of air curtain is that it blows the atmospheric air at constant velocity at the door ways and forms an air stream for the entire width and height of the door openings. The velocity will vary depending on the effective trough of the air stream. Moreover, it can reduce energy consumption considerably by holding the uniform cold air conditioned area. It also eliminates the flies and other small insects' infiltration through an open door, and helps maintaining the required hygienic standard in the specified areas.

Applications: Aerotech Air Curtains are ideal for Clean Rooms, Computer centers, Air conditioned Buildings and offices, Research laboratories, Medical institutions, Hospitals, Kitchens, Departmental Stores Restaurants, Electronic Component Manufacturing, Pharmaceutical Processing Optical instruments Assembly units

Features

Easy installation

Engineered to provide many years of maintenance free service

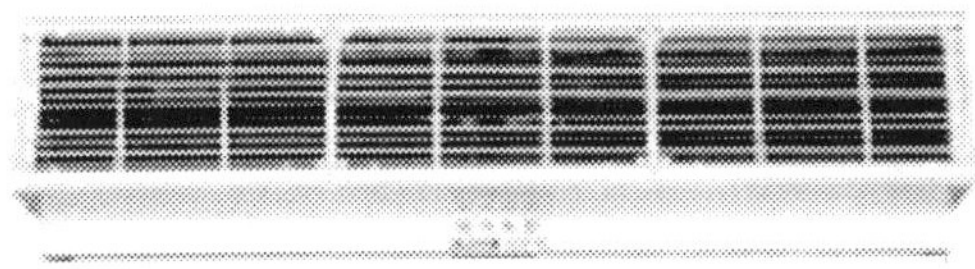

Aesthetic design

Constructed from high-quality materials.

1.5 Roof Extractors

Aerotech roof extractor is a motor-driven fan unit and is provide with durable weather proof exhaust hood. This is ideal for extraction or air supply through the roof, usually preferable in Factories, D.G. rooms, Steel plant and Foundries.

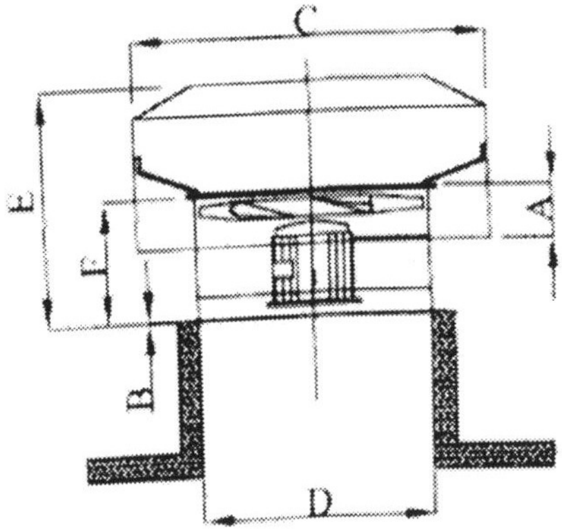

Roof extractor on flat roof

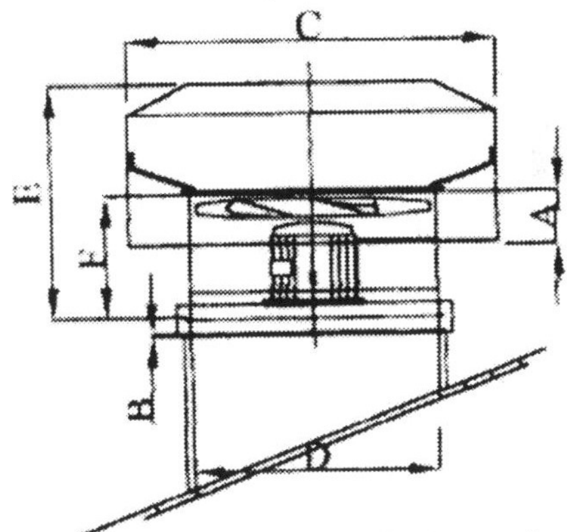

Roof extractor on sloping roof

ACM	A	B	ØC	ØD	E	F
50	160	6	750	500	540	350
60	160	6	900	600	640	430

Contd.

Contd.

ACM	A	B	ØC	ØD	E	F
70	200	6	1050	700	746	440
80	210	6	1200	800	846	560
90	210	6	1350	900	950	575
100	225	6	1500	1000	1046	590
120	290	6	1800	1200	1265	680
140	300	8	2100	1400	1410	760
160	325	8	2400	1600	1525	800

2 Keystone Air Systems Ahmadabad, India

Centrifugal blower

Axial flow fans

- To achieve high efficiency, fans have a defined hub ratio 71 to 25 for given capacity, pressure and speed.
- Impeller blades are adjustable type and can rest at designed angle to suit the system resistance.
- Fan can run in parallel to suit capacity and in series for higher pressure. Impeller is made of aerofoil profile from cast aluminium alloys.
- Bifurcated-type fans are available for handing corrosive gases.
- Capacity: Up to 300,000 m^3 /hr.
- Pressure: Up to 100 mm of WG

3 Humidin – Ioni – UP, India

3.1 Backward curved centrifugal blowers

Humidin centrifugal blowers are multi-bladed and designed to deliver large volume of air when operating at the lowest possible peripheral speeds. Ideal for heating, ventilating, air conditioning and industrial process duties, they can be used for all type of air-moving applications where relatively clean air is moved. Ruggedly built, quiet operating and dependable, these blowers would perform year after year with a minimum of attention.

Limit Load	Backward Curved		Backward Curved with Radial Tip		Straight Blade		High Pressure	
Type	Type	Total Pressure mm WG	Type	Total Pressure mm WG	Type	Total Pressure mm WG	Type	Total Pressure mm WG
ACLL-S & ACLL-D Class 1	ACBL	100	AMRL	300	AMSL	250	ACHS	700 and 1000
Class II	AMBM	450	AMRM	600	AMSM	400	2 stage fan which develops up to 2500mm WG (On request)	
Class III	AHBM	800	AHRM	900	AHSM	500		
Class IV	AHBM	1200			AHSH	1100		

3.2 Forward curved centrifugal blowers

Humidin is manufacturing Forward Curved Centrifugal Blowers in DIDW as well as SISW designs. As mentioned below:

- These fans have forward curved impellers with forward curved blades having special profile.
- Gives maximum efficiency at low noise level.

The fan impeller is statically and dynamically balanced.

Humidin forward curved centrifugal blowers are suitable for various applications, where high volume air displacement is required like – Air Cooling, General Ventilation Pressurisation, Air Conditioning, Heating and

Commercial Processes. They are available in various capacities from 1,000 m3 /hr., 100,000 m^3 /hr. and static pressure is ranging from −10 mm WG – 100 mm WG.

3.3 Centrifugal blowers (ID & FD)

The ID & FD Centrifugal Blowers are carefully designed to meet ideal requirement for various industrial applications. Humidin makes 3 basic types of blowers in 23 sizes; are made in heavy duty construction with statically and dynamically balanced impellers which give trouble free performance.

Available air delivery 500 cum/hr., to 1, 50,000 cum/hr. These models consist of completely backward curved, backward curved with self-cleaning and straight blades.

3.4 Axial flow fans

Humidin also offers axial flow fan through which the air or gas flows parallel to the impeller axis. Available in various types, covers a wide range of quantities and pressures. They are widely used in industry for providing general ventilation, fume exhaust and removal of hot spots and as roof extractor. Designed for an optimum relation between air quantity, pressure and power consumption, these ACA Axial Flow Fans have MS Casing/ Bracket with cast aluminium alloy impeller along with aerofoil section blades suitable for various mountings.

They offer the most economical and efficient installation for industry such as thermal power stations, fertilisers, heavy engineering, chemical, textile, food general, paper, sugar and pharmaceuticals etc.

3.5 Green house fan

Increase demand for vegetables, flowers, fruits has resulted for cultivation in green houses. Humidin systems ensure proper climatic condition with low pressure drop and high efficiency equipments with minimum energy requirements.

Environmental control for birds has become essential as they are put in enclosed place.

3.6 Roof Extractor

It is a motor-driven fan unit equipped with cast aluminium alloy impeller, aero foil section blades and is provided with a durable weather proof exhaust cowl. This is ideal for extraction or air supply through the roof and is usually preferable in factories, power plant, steel plant and foundries. The fan which is incorporated here is of our ACA series and can be selected accordingly.

4 Trane Climate Changer USA

4.1 Fans

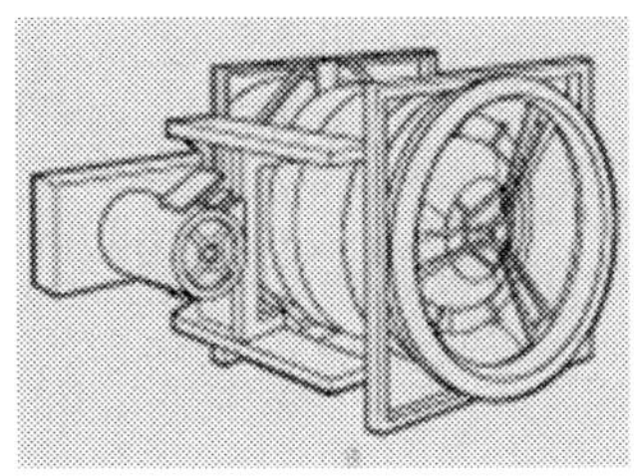

Trane Climate Changer™ air handlers offer an extensive array of fan types and options with different capacity. These fans can optimize not only the airflow and static pressure requirements, but also the acoustical, efficiency and discharge requirements. Variable-frequency drives for modulation in variable-air-volume systems gives even more precision.

Fan selection should be based on the fan airflow (CFM), total static pressure (TSP), efficiency (BHP), operating range and sound levels. Trane fan curves show fan CFM, TSP, fan RPM and fan wide-open CFM (WOCFM) performance relationships with the fan in the module. The effect of the air handler casing on fan performance is included in the fan curve. Fan curve catalogues are available from Trane sales engineer. Trane offers fan types that include forward-curved (FC) airfoil (AF), backward-curved (BC) plenum (unhoused AF, or plug) and direct-drive plenum fans.

5 Airfilt Technologies P Limited, Delhi

5.1 Centrifugal Blowers

They are specially designed for an efficient functioning and come in a variety of sizes. These centrifugal blowers comprise of housing, impeller and drive arrangements. The housing consists of a sheet iron with circular inlet and rectangular outlet. The impellers precision balanced for a smooth vibration less operation. These blowers are fabricated in M.S., stainless steel, aluminium, etc., depending upon the kind of usage. Apart from the usual material these can also be made from various other coatings like F.R.P, Hot dip, Galvanizing, Epoxy, etc.

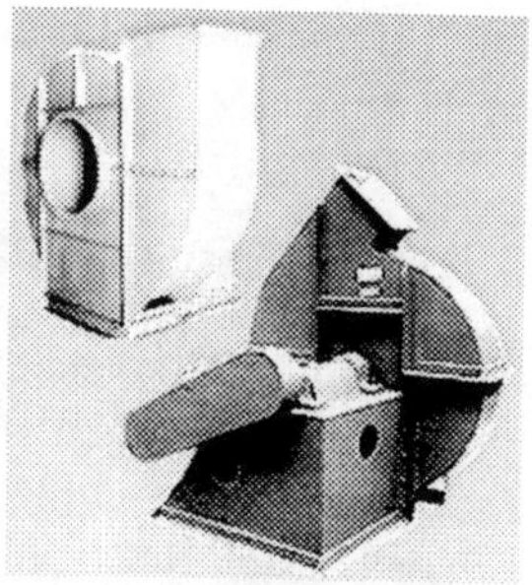

5.2 Axial Flow Fans

Airfilt brings a variety of axial fans that find their usage in all the major industries where a required air flow and mass transfer operations are involved. They are conventionally designed with impellers made of aluminium and mild steel. The new and improved aerodynamic fan designing, composite development, structural design combined with latest manufacturing process are aimed to bring out consistent quality and higher productivity. The fans have a capacity of 500 CMH to 100,000 CMH and an impeller size of 300 mm to 1,600 mm.

5.3 Roof Exhausters

Airfilt offers specially designed roof exhausters for efficient ventilation in the humid and hot workspaces. They are available in various capacities having high volume sweeping actions to assure removal of humid air. The body of the exhauster is made up of MS and has an impeller directly connected to the motor.

5.4 Tunnel Ventilation

Tunnel ventilation assures an efficient exhaust system. The unit consists of axial impellers that deliver large quantities of air at high pressures. The system consists of either a single impeller or two impellers having a varied blade width.

6 C.D. Blowers – New Delhi India

6.1 Axial Flow Fans

C.D. Blowers offer Axial Flow Fans, which are claimed to be highly efficient. The number of blades in the rotors could vary from 2 to 50. As the name suggests, the flow of air is occurred in axial direction and the casing is done in such a way that it recovers static pressure. As soon as the rotor rotates using an electric motor, a flow establishes through rotor causing an increase air stagnation pressure. There is a cylindrical casing, which encloses rotor. There are well shaped converging and diverging passage to receive and discharge the flow. The fan wheels or rotors are constructed using heavy gauge MS sheet with an aluminium casting. Widely applicable in ventilation process, which could be mines or tunnel where the large pressure is required along with compact dimensions.

7 DraftAir – Ahmadabad, India

7.1 Centrifugal Ventilation Fans

Size	:	200 mm. to 2,500 mm. impeller Dia
Capacity	:	1,000 to 400,000 CMH.
Pressure	:	Up to 150 mm. Wg.

Low noise high-efficiency non-overloading impeller with backward inclined blades. Impellers statically and dynamically balanced as per ISO 1940. Casing fabricated from heavy gauge steel with heavy duty pedestals. Fan orientation is provided to suit the site requirement.

7.2 Centrifugal Process Fans

Beside axial flow and centrifugal ventilation fans, Draft-Air offers special purpose centrifugal fans for high temperature, fume/vapour exhaust, dust handling, material, conveying, induced draft and forced draft applications.

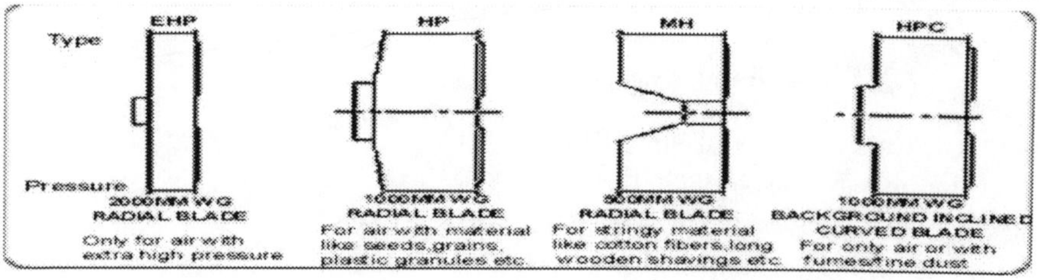

7.3 Axial Flow Fans

Size	:	450 mm to 1,980 mm sweep.
Capacity	:	2,000 to 200,000 CMH.
Pressure	:	Up to 65mm WG.

Hosing fabricated from heavy gauge steel sheet. Impeller is having high-efficiency aero-foil section adjustable pitch blades. Impellers are statically and dynamically balanced as per ISO 1940. Inlet cone and outlet diffuser are provided to further improve the efficiency.

8 Ningbo Yinzhou Chenwu Humidifying Equipment Factory – China

8.1 Automatic Centrifugal Atomisation Humidifier

Features:

(a) Centrifugal atomisation, directly atomises water, vents to space via fan

(b) Enhances air humidity

(c) Operates smoothly, immediately improves humidity

(d) Our proprietary designs significantly improve humidifier performance

(e) Suitable for use in textile factories manufacturing wool, hemp, cotton, silk, knitting, tie and non-woven textiles

(f) Also suitable for humidity adjustment

(g) Easy maintenance

Specifications:

(a) Atomisation amount: 10–12l / h. Atomisation direction: 360°C

(b) Humidification cubage: 1,200 m^3

(c) Power: 250 W. Voltage: 220V, 50 Hz

(d) Hydraulic pressure: 0.6–3 kg / cm^2

(e) Dimensions: 550 x 620 x 720 mm

(f) Conditions for use:

1. Environment temperature: 0–40°C
2. Water: city processed water
3. Place humidifier where working humidity is needed

8.2 Automatic Centrifugal Atomisation Humidifier

Features:

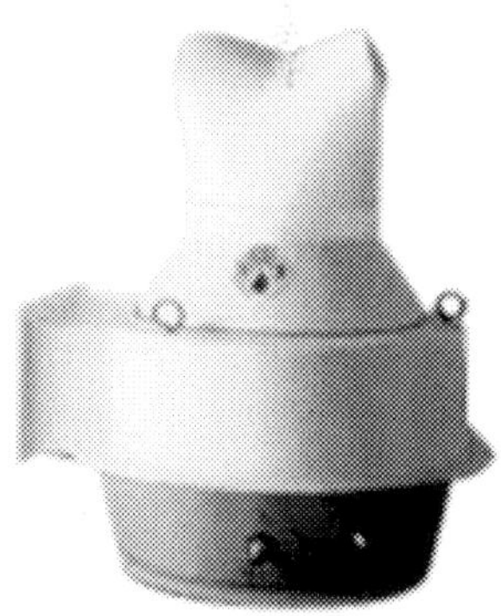

(a) Centrifugal atomisation, directly atomises water, vents to space via fan

(b) Enhances air humidity

(c) Operates smoothly, immediately improves humidity

(d) Our proprietary designs significantly improve humidifier performance

(e) Suitable for use in textile factories manufacturing wool, hemp, cotton, silk, knitting, tie and non-woven textiles

(f) Also suitable for humidity adjustment

(g) Easy maintenance

Specifications:

(a) Atomisation amount: 7–8 l / h, Atomisation direction: 2 nozzles

(b) Humidification cubage: 750 m^3

(c) Power: 200 W, Voltage: 220 V, 50 Hz

(d) Hydraulic pressure: 0.6–3 kg / cm^2

(e) Dimensions: 550 × 620 × 720 mm

(f) Conditions for use:

1. Environment temperature: 0–40°C
2. Water: city processed water
3. Place humidifier where working humidity is needed

9 Luwa Air Engineering AG Switzerland

9.1 Quench Air Outlet for Filament Spinning

Machine air conditioning is required in filament spinning. The quench air outlet produces a laminar air flow with which maximum product quality is achieved in the spun-bond process. Luwa guarantees a speed variation of max. ± 0.05 m/s in the range of 0.3–0.7 m/s with minimum pressure loss. Luwa offers a comprehensive system including extraction, depending on requirements.

10 Manvi Textile Air Engineers – Mumbai, India

10.1 Fans

Manvi Textile Air Engineers Pvt. Ltd have developed advance fan system which can save power and give you much cleaner department due to high pressure developed by the fans. Advance fan system consists of Heavy duty fan casing, high-efficiency energy efficient aerodynamic fan impeller, Inlet cone, Outlet cone and Nose fairing.

10.1.1 Heavy Duty Fan Casing

Fan casing is manufactured from heavy gauge with adjustable motor mounting base plate and support frame.

Advantages:

1. Heavy duty casing last long and does not get corroded completely.

2. Adjustable bolted type of motor base plate and support frame allows you to open completely and paint it easily after making it rust free.
3. In future, if there is any change in machinery load then without changing fan casing you can upgrade your motor to higher H.P by re drilling in its base plate. Fan impeller can be utilized to deliver higher CFM after setting blade angle to deliver desired CFM.

10.1.2 High Efficiency Energy Efficient Aerodynamic Fan Impeller

Fan blades are designed with latest aerodynamic software to achieve high efficiency.

Fan blades are manufactured from high grades of FRP resin to achieve desire shape twist along length and very smooth surface finish.

Advantages:

1. Due to aerodynamic shape its efficiency is achieved up to 85% hence saves power compared to any other manufacturer in the field of humidification.
2. Due to FRP it is light weight. It is easy to install and maintain. Due to light weight it also reduces starting torque and current requirement.
3. Life of bearings of motor increases leads to less down time.

10.1.3 Bell Mouth Inlet Cone

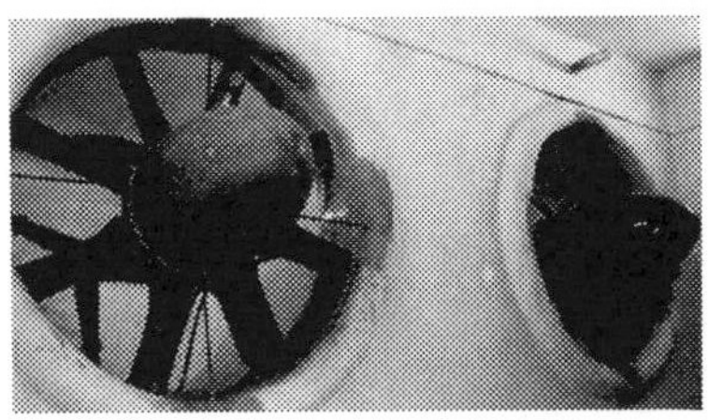

Bell mouth inlet cone allows air to enter freely leads to negligible losses of mean dynamic head.

Advantages:

Due to easy flow or reduced resistances to flow of air it saves power by 3 to 4%

10a.4 Outlet Cone

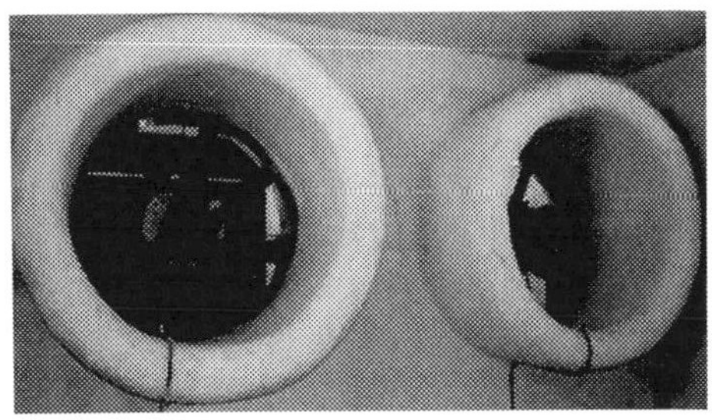

Outlet cone of fans helps to diffuse the air to reduce back pressure and recovers velocity pressure losses.

Advantages:

Due to pressure recovery it reduces power by 3 to 5%.

10a.5 NoseF or Spinners

Spinners are placed in front of the hub hence when fan is sucking air, it reduces the static loss, i.e. resistance to fan. High pressure fan hub ratios are often up to 35% to 50% hence large portion of fan is covered by hub if no spinner is used then there is substantial loss of pressure and efficiency.

Advantages:

1. Due to spinner department gets more CFM and better pressure.
2. Fan pressure can be better utilised to deliver air till last diffuser and maintain whole department under pressure. Hence with good RH and pressure from supply system, fly or floating fiber tends to settle down and then return air fan can suck the fly. This leads to cleaner department with spinners in return air fan pressure is utilized to suck fly from tail end of the department.

10.1.6 Advantages of Advance Fan System

Due to design of advance fan system in which we have incorporated bell mouth inlet cone and outlet cone with spinner, will save client 15% of power compared to any other manufacturer. Part of the same power can be utilized to increase pressure and keep cleaner department.

If you save even 10 HP power = 10 HP x 746 = 7.46 kw

Power saved per year in terms of units per year = 7.46 kW x 24 hrs x 355 days = 63,559 units saved per year.

Moreover you will have much better cleaner department than any other manufacture who don't uses bell mouth inlet cone, outlet cone and spinners and waste fan pressure to overcome these resistance.

11 Multi-Wing Group – Denmark

11.1 X-large Diameter Fans

The group of X-large axial fans covers a diameter range from 2,134 mm to 2,743 mm. You will find the Multi-Wing axial fan profiles Airfoil and Sickle represented within this diameter range. The X-large size fans are typically used in applications like cooling towers and large condenser units.

Airfoil profile

Multi-Wing's airfoil axial fan series can be used in almost any air moving application such as cooling towers, engine cooling.

Sickle Profile

Multi-Wing's sickle axial fan series is the answer for generating pressure with low noise axial fans.

12 HVLS Fan Co., Lexington, KY

When summer arrives and temperatures rise in industrial facilities, productivity can plummet. The most effective way to improve summer working conditions is to increase air circulation within its un-air conditioned facility. Rather than rely on personal air-circulation fans, one can chose large, 20-ft. diameter slow-moving fans from the HVLS Fan Co., Lexington, KY. The installation of these fans measurably improves the thermal comfort conditions within the entire plant area. The HVLS fans use 10 hollow-core extruded aluminium blades. They can be made in 4-ft. to 12-ft. lengths, creating fans that range in size from 8 ft. to 24 ft. in diameter.

Unlike a high-velocity fan, which moves a lot of air by rotating quickly and repeatedly stirring a small amount of air, an HVLS fan moves a lot of air by moving a very large column of air very slowly, at about 3 mph. It can move from 15,000 cfm (8-ft. diameter fan) to 122,000 cfm (24-ft. diameter fan) of air, and will cover an unobstructed area of up to 20,000 sq. ft. with no significant reduction in the air velocity at the outer perimeter of the column of air. The effect throughout the area of coverage is a gentle breeze.

13 Jiangsu Jingya Environment Technology Co. Ltd., China

13.1 SFF232 – 12 &JYC-1type Centrifugal Ventilator

It is mainly used for the textile dust removal system on various ventilation systems.

SFF232-12 fan is a kind of centrifugal ventilator which is redesigned on the basis of 11 type fan, which has been enhanced by a large margin in the performance and efficiency. It is more silent.

JYC-1 draught fan is a kind of centrifugal ventilator about new type energy saving weaving technology, which is improved and designed on the basis of SFF232-12 fan. It has increased efficiency.

- Full pressure of the fan: 460 Pa to 2,800 Pa
- Fan flow 12,500 m^3/h to 96,000 m^3/h
- Installed power – 4 kw to 75 kw

13.2 JYFZ-2 Energy Saving Textile Axial Flow Fan

It is mainly used for the air supply and exhaust system for air conditioner in the textile industry. This unique textile axial ventilator has 1-class energy rate; it can save more than 15% electricity under the same air volume comparing with the domestic traditional fan. Casted by applying the high strength aluminium alloy, the fan blade is equipped with features of high strength and excellent cleanliness on the surface. Considering that the false ogive applies the new-type structure, the entrance channel of fan will be smoother. The impeller applies the new-type wheel hub structure so that the fan blade can be adjusted conveniently. It is light, compact, durable, solid, and has good appearance.

Full pressure of the fan: 500 Pa to 850 Pa

Fan flow: 32,000 m^3/h to 260,000 m^3/h

Installed power: 7.5 kw to 55 kw

14 AESA Air Engineering Private Limited – France, India, Singapore and China

14.1 Industrial Fans

Return and supply air fan
Radial flow fan
6–12 blades
Air Volume: 30,000– 130,000 m3/h
Air Pressure:150–700 Pa

High-efficiency fan
Centrifugal Fan, suitable for the conveyance of dust laden air
Air Volume: 400– 7,000 m3/h
Air Pressure: 1,500 – 0,000 Pa

15 Blowtech Engineers Pvt Ltd. Thane, India

15.1 Motorized Roof Extractors

Blowtech developed new FRP base motorised roof exhaust fans as per customers' existing roof profile for zero leakage and perfect fitment. This new invention in compare of powerless wind driven roof fan (Turbo Fans) works excellently to maximise the cross ventilation by 10- time compared to powerless fans.

This fan helps in providing the natural ventilation with the help of induced draft pressure inside the plant by extracting fume, heat, humidity and stale. These fans most effectively used for fume/heat extraction and ventilation system by providing de-humidification and cross ventilation in production area. This invention by Blowtech Engineers Pvt. Ltd reduced the cost of big HVAC system of AHU, duct line and fresh air grill system which are 6-times higher in cost with less output in terms of efficiency and de-humidification for any working atmosphere. The fans are offered in diameter- 24" & 32" as per roof sheet width.

15.2 FRP Blowers

Blowtech provide FRP blowers that are fabricated using high-grade material that further make them more reliable and durable in nature. The FRP centrifugal blowers offered can be availed by spending very less from the pocket. Also, we are one of the well-established manufacturers and suppliers of FRP centrifugal blowers.

Capacity : From 500 CMH to 60,000 CMH

Static Heads : 50 mm to 750 mm of water column

Our centrifugal blowers are available in different capacities and pressure ratings and with belt driven and direct driven option.

MOC of DME blowers: PP, PVC, FRP, PP+FRP, PVC+FRP

15.3 Vane Axial Fans

Product Description:

Our range of Vane Axial Fans is compact in size and are easy to install. Reliable in operation, these fans are made for optimum realisation between air qualities, pressure, and power consumption.

Features:

- High efficiency
- Aerofoil design
- Robust construction to reduce vibration
- Low noise level
- Special coating and lining for casing and impeller (GRP, FRP, PP, rubber, epoxy)
- Low weight/high strength
- Identical air performance for all blade impellers
- Vibration free as VDI-2056
- Aerodynamic performance tested as per IS 3588 and reverse direction of air flow possibility by turning the blade through.

15.4 Industrial Exhaust Fans

Industrial Exhaust Fans are propeller type and are used for general ventilation. Exhaust fans are used to remove the heated air and fumes from ovens, workroom, kitchen, etc., and are also used for ventilating commercial buildings.

Exhaust fans are of two types:-

- Light Duty Fans: It consists of totally enclosed single phase motor and the impellers are dynamically balanced for trouble free operation. They are used in offices, cabins, kitchens, bathrooms, etc.
- Heavy Duty Fans (Ring Mounted) 12" to 48": The fans are specifically designed for industrial/Commercial use.

The body is constructed of a heavy mild steel, blades are broad. In order to discharge large volume of air for smooth operation, single phase/3 phase motors with condensers are fitted within the dome shape.

15.5 Tubular Man Coolers

We are manufacturing and supplying tubular man coolers. We supply two types of man coolers, i.e. pedestal and tubular types. Both the types are quality tested as per the specified standard code. Pedestal man cooler is fabricated of cast aluminium alloy. The impeller is fitted to cast iron base and has been mounted on steel columns and. coupled directly to the motor on iron base. Tubular coolers consist of impeller that is manufactured using cast aluminium alloy coupled on motor shaft. The motor with impeller assemble is mounted in a tubular cradle. Also, there is a provision to adjust the angle of air.

15.6 Fume Exhaust Hoods

We are leading manufacturers of fume exhaust hood. With its precisely engineered and simple design, the offered Fume Exhaust Hood is known to be one of the best options available. Manufactured in accordance with the set industrial guidelines, using the highest grade of raw materials, its functionality, sturdiness, durability and resistance to corrosion is quite commendable. In addition, this product is highly acclaimed for its user-friendly and economical nature.

Features:

- Superior functionality
- Commendable sturdiness and durability
- User friendly and economical nature

15.7 Exhaust Blower

We are manufactures a large range of Exhaust Blower and heavy duty Axial Flow Exhaust fans. The centrifugal fans can be providing SISW section. This fan can be operated at temperatures up to 72°C.

These blowers are installed with exhaust duct to extract kitchen fumes and fresh air ducts for fresh air input to provide cross ventilation inside the kitchen area. It is made of using superior quality stainless steel, mild steel with proper anti corrosive epoxy paints.

15.8 Axial Flow Fans

We are manufacturing Axial Flow Fans. Direct driven axial fan, where the impeller is mounted on the motor shaft and as another option our Axial Flow bifurcated fans construction, where the motor is completely isolated from the air stream. In addition to the standard fan designs, Blowtech Engineers Pvt. Ltd. provide axial fans to meet the clients specification with designed air changes required inside the kitchen for proper ventilation.

Products are available in five series that include axial flow fans in sizes like:

- 300 mm–1,600 mm with fixed pitch impeller of 6 Nos blade.
- 300 mm–1,200 mm with adjustable pitch impeller of 4/6/8/Nos blades.
- 300 mm–900 mm.
- 450, 600 and 900 mm with fixed and adjustable pitch impeller.

The different types of axial flow fans available are:

- Belt drive
- Bell mouth
- Duct type
- Roof fixing
- Bifurcated
- Bed type
- Split type (marine duty)
- Wall mounting.

16 Coimbatore Air Control Systems Private Limited, Coimbatore, India

16.1 Axial Flow Fan

Impellers are placed in MS casing and directly coupled with motors. These impellers are made out of aluminium alloy die-casting and specially designed for high static pressure and high efficiency.

The blades are adjustable and the impellers are dynamically and statically balanced

Appendix – 6

Dampers and diffusers

1 Aerotech Damper, Ghaziabad, India

Aerotech damper grill and register fabricated in high-quality extruded profile, adjustable horizontal and vertical single/double deflection louver with opposed blade aluminium extruded section damper. Horizontal fixed bar linear grill with 0 to 45 degree louver inclination in heavy and slim line sections.

AeroTech has been the leader in recognising and meeting the need of air management product and system. They are the manufacturer and supplier of comprehensive range of air terminal equipment to suit decor and interior. To meet with modern sophisticated architectural design AeroTech manufacture horizontal fixed bar grill in any curvature shape and mitered corner.

Diffuser Square/Rectangular from aluminium extrusion an ideal for most commercial project having high quality diffuser with 1,2,3,4 way directional flow. Multi-slot diffuser in 2,3,4,5 slots with hit and miss damper with special coating to give them a smooth surface. AeroTech has complete range of Air Terminal Equipment in extruded aluminium and steel construction.

2 Hiross Flexible Space System, Italy

2.1 Hiross C and Hiross V

Hiross C and Hiross V range cover cooling and heating needs by distributing the air under the floor: either chilled or hot water is supplied to a changeover coil or a double coil system which have each a 3 way valves incorporated. Conditioned air module Hiross C generate conditioned air to be delivered into the under floor supply plenum and draw air back through grilles into return under floor plenum thus reducing the need for a false ceiling, offering up to 10% reduction of floor-to-ceiling height and 5–7% of building construction costs. It is suitable for both open plans and cellular offices space and in height-limited space.

Hiross variable air flow unit Hiross V generate conditioned air to be delivered into the under floor plenum and draw air back through the space or, un-ducted, through the false ceiling, thus allowing full use of the under floor void as supply plenum with no restrictions as to positioning of air terminal units. This continuously monitors and controls the air volume and temperature both in the floor plenum and in the ambient for each assigned zone by means of the intelligent data link to each fan tile. Particularly suitable for open space plans and non-height-limited areas.

High-efficiency filters provide air cleanliness and an optional electrode steam humidifier (5 kg/h) provides humidity control.

They may be enclosed in partitions or left free standing in shafts, corridors, technical rooms or in the working space.

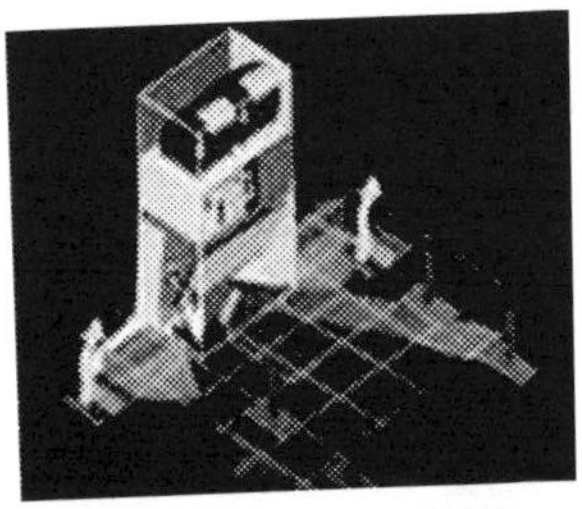

Liebert HIROSS

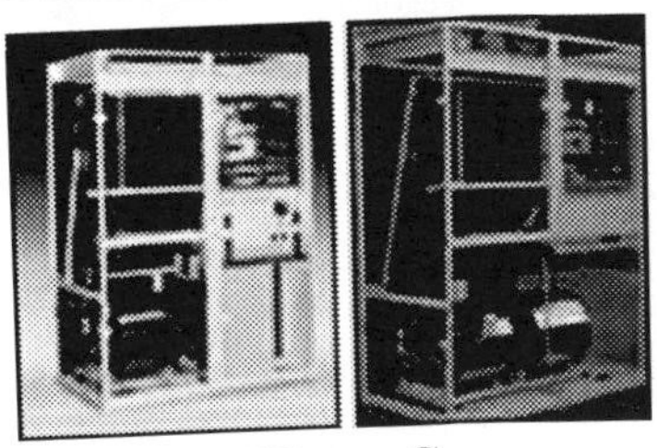

Hiross C

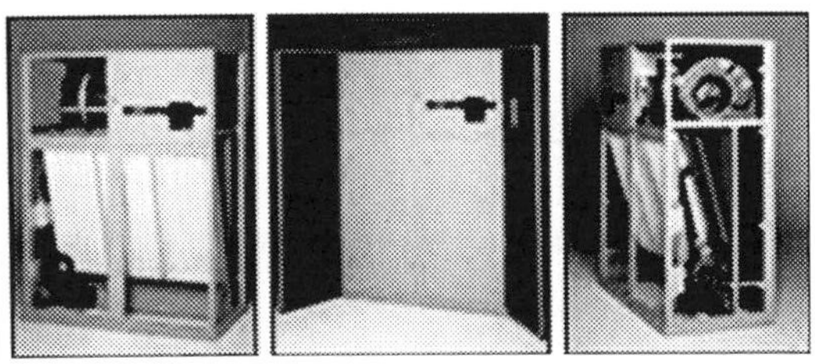

Hiross V

	Hiross C	Hiross V
Models	3 models: HirossC15-25-35 for under floor air supply and return.	4 models: 11-22-33-44 for under floor air supply with room or ceiling return
Cooling capacity	Up to 32 kW (26°C-50%R.H.,CW in/out 7/12°C)	Up to 40 kW (26°C-50%R.H.,CW in/out 7/12°C)
Heating capacity	Up to 50 kW (20°C-50%R.H., HW in/out 80/70°C)	Up to 58 kW (20°C-50%R.H.,HW in/out 80/70°C)

Contd.

Contd.

	Hiross C	Hiross V
Air flow	Up to 9500 m3/h	Up to 9500 m3/h // Automatic variable flow speed controller
SPL (Noise level)	40–45 dB(A) 2m free field (generally 35NR in room)	40–50 dB(A) 2m free field
Filter	- Filter from G3 to F7 grade	- from G3 to F5 grade
Electric heater	- up to 14.85 kW	- up to 9.1 kW
Length mm	1000–2000	850–1450
Height mm	1950	1950
Depth mm	750	750
Automatic control:	Zone AHUs have on-board microprocessor control, micro face, intelligent, user friendly interface for continuous HirossTU / HirossTUC network monitoring and optimisation of FSS system working parameters. The controller may interlink with higher level interfaces and supervision system	

3 Atlanta Supply Company – USA

3.1 Plastic T-Bar Ceiling Air Diffuser by Eger Products

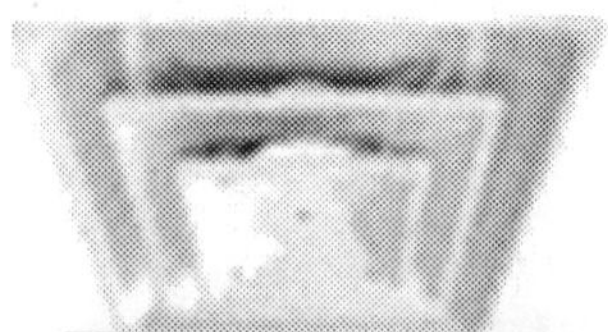

Eger Products have all plastic 24” T-Bar Ceiling Air Diffuser designed for drop in ceilings that will not rust, will not sweat, and it will not chip; color stays consistent throughout the part. Neck sizes — 6”, 8”, 10”, 12” & 14”.

Perfect for all applications.

3.2 Stepdown Supply Diffusers Model P470

ITEM #	DESCRIPTION	
P4700B06	24x24 Black T-Bar Supply	Overall Dimensions: 23.5”x23.5”
P4700B08	24x24 Black T-Bar Supply	Overall Dimensions: 23.5”x23.5”
P4700B10	24x24 Black T-Bar Supply	Overall Dimensions: 23.5”x23.5”
P4700B12	24x24 Black T-Bar Supply	Overall Dimensions: 23.5”x23.5”
P4700B14	24x24 Black T-Bar Supply	Overall Dimensions: 23.5”x23.5”
P4700W06	24x24 White T-Bar Supply	Overall Dimensions: 23.5”x23.5”
P4700W08	24x24 White T-Bar Supply	Overall Dimensions: 23.5”x23.5”
P4700W10	24x24 White T-Bar Supply	Overall Dimensions: 23.5”x23.5”

Contd.

Contd.

ITEM #	DESCRIPTION	
P4700W12	24x24 White T-Bar Supply	Overall Dimensions: 23.5"x23.5"
P4700W14	24x24 White T-Bar Supply	Overall Dimensions: 23.5"x23.5"
P4703B06	24x24 Black T-Bar Supply	Overall Dimensions: 23.5"x23.5"
P4703B08	24x24 Black T-Bar Supply	Overall Dimensions: 23.5"x23.5"
P4703W06	24x24 White T-Bar Supply	Overall Dimensions: 23.5"x23.5"
P4703W08	24x24 White T-Bar Supply	Overall Dimensions: 23.5"x23.5"

3.3 Lay in Air Diverters – Model #: PLEADIV*

ITEM #	DESCRIPTION	
PLEADIVB	24x12 Black Air Diverter	Overall Dimensions: 23.75"x12"
PLEADIVCB	Corner Blow Black Air Diverter	Overall Dimensions: 23.75"x23.75"
PLEADIVCW	Corner Blow White Air Diverter	Overall Dimensions: 23.75"x23.75"
PLEADIVW	24x12 White Air Diverter	Overall Dimensions: 23.75"x12"

3.4 Lay in Air Diffusers

ITEM #	DESCRIPTION
PLEADEFC	24x24 Clear Air Deflector
PLEADEFW	24x24 White Air Deflector

3.5 Step Down Return Grills

ITEM #	DESCRIPTION	
PLEARDW	24x24 White T-bar Return	Overall Dimensions: 23.875"x23.875"
PLEARDW06	24x24 White T-bar Diffuser	Overall Dimensions: 23.875"x23.875"
PLEARDW08	24x24 White T-bar Diffuser	Overall Dimensions: 23.875"x23.875"
PLEARDW10	24x24 White T-bar Diffuser	Overall Dimensions: 23.875"x23.875"
PLEARDW12	24x24 White T-bar Diffuser	Overall Dimensions: 23.875"x23.875"
PLEARDW14	24x24 White T-bar Diffuser	Overall Dimensions: 23.875"x23.875"

3.6 Perforated Return Grills

ITEM #	DESCRIPTION	
PLEAPERF	24x24 Perforated Return Grille	Overall Dimensions: 23.875”x23.875”

3.7 Filter Grills

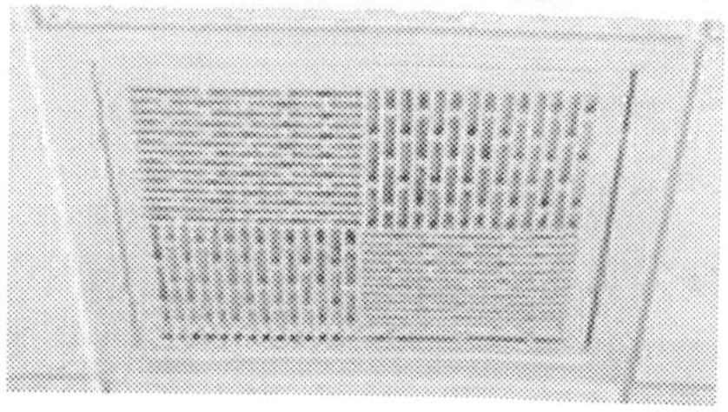

ITEM #	DESCRIPTION
PLEFAR06B	6” Black Perf. Filter Grille
PLEFAR06W	6” White Perf. Filter Grille
PLEFAR08B	8” Black Perf. Filter Grille
PLEFAR08W	8” White Perf. Filter Grille
PLEFAR10B	10” Black Perf. Filter Grille
PLEFAR10W	10” White Perf. Filter Grille
PLEFAR12B	12” Black Perf. Filter Grille
PLEFAR12W	12” White Perf. Filter Grille
PLEFAR14B	14” Black Perf. Filter Grille
PLEFAR14W	14” White Perf. Filter Grille
PLEFAR16B	16” Black Perf. Filter Grille
PLEFAR16W	16” White Perf. Filter Grille
PLEFAR18B	18” Black Perf. Filter Grille
PLEFAR18W	18” White Perf. Filter Grille
PLEFARB	24x24 Black Filter Grille
PLEFARW	24x24 White Filter Grille

3.8 Lay in Corner Blow Diverter

This is used to cover some area from not getting humidified.

3.9 Lay in Air Deflector

This is used to prevent humidified air for certain specific machines or area.

4 Luwa Air Engineering AG Switzerland

4.1 Baffle Plate Air Outlet

Luwa baffle plate air outlet guarantees an optimum air distribution of the supply air and creates a predominance of vertical air movement so that the humidity can reach the yarn optimally and the dust produced is pushed downwards. There, the dust is taken up by the exhaust air system where the air is re-conditioned in the filter and washer plant. After this re-conditioning process, the air is returned to the spinning hall.

4.2 RotorSphere Air Outlet

Since rotor spinning is a closed production process compared to ring spinning, Luwa have developed the RotorSphere system for this application. RotorSphere requires less air and can be run with a smaller air-conditioning plant. Thus, initial investment and operating costs are reduced. In addition, with RotorSphere a much improved homogeneity of the micro-climate is achieved along the rotors and the feed cans situated between them. RotorSphere also ensures a good working environment for the factory staff.

4.3 LoomSphere® Air Outlet

The patented air outlet LoomSphere® – allowing direct air conditioning of the weaving machines is the main feature of the air conditioning concept for weaving mills. It guarantees the desired conditions regarding humidity and temperature on the warp.

This spot air conditioning in combination with a reduced air volume for the room, called TAC® Combi air-conditioning, is the ideal and most economical solution for any weaving mill. The TAC® Combi air conditioning offers a high degree of efficiency and comfort with low investment and subsequent low running costs.

4.4 Air Bell Air Outlet for Circular Knitting

The AirBell air outlet is used for air conditioning of circular knitting machines. AirBell has the effect that the air flow above the machine forms a sort of bell around the machine. Cross contamination between machines is eliminated with the Air Bell system. Furthermore, with the AirBell air outlet, the work area is freed from fly and dust and fault frequency is reduced by up to 50%.

4.5 The FlowMaster Laminar Flow Outlet for Non-Woven Plants

The FlowMaster laminar flow outlet is suitable for local air conditioning in manufacturing halls with an irregular local heat load distribution. The FlowMaster is a fresh air outlet which distributes the incoming air at a low speed near the floor, with little induction and turbulence. The air that has been warmed by machines and people rises and is guided via the outlet openings out of the room or removed directly as machine exhaust air.

4.6 Quench Air Outlet for Filament Spinning

The quench air outlet produces a laminar air flow with which maximum product quality is achieved in the spun-bond process. Luwa guarantees a speed variation of max.± 0.05 m/s in the range of 0.3–0.7 m/s with minimum pressure loss and offer a comprehensive system including extraction, depending on requirements.

5 Excel Airtechnique – India

5.1 Air Distributors

They streamline the air flow after the fans entering the spray chamber so that the humidification process is uniform for every CFM of air passing through the spray. They are also available in extruded PVC. They are assembled with a set of PVC fixtures and can be quickly removed allowing an easy entry into the spray chamber.

5.2 Distribution Louvers

Air dampers are provided at fresh air entry, hot air exit and recirculation and by pass locations. They are used for controlling the air flow in the required proportion. They are made in M.S. steel construction as well as aluminium foil extruded construction. They could be connected with actuators for automation purpose.

5.3 Air Dampers -

5.1.1 Gravity Louver Damper:

Gravity Louver Damper also known as 'Auto shutter' is an arrangement provided at the delivery side of the axial fan and remains open when the fan is switched on and remains closed when the fan is switched off. This allows using less number of fans in a multiple fan system design without short circuit. Further when a motor trips and hence the fan is not able to run the short circuit is prevented.

5.4 Air Diffusers

They distribute the humidified air in the department at various locations. They are available in PVC powder coated aluminium / S.S construction.

6 China Powergrand HVAC Industry Co., ltd

6.1 Swirl Air Diffuser

- The vane is designed on the principle of fluid dynamics to provide the swirling air flow with a certain primary speed.
- The angle of the vane can be adjusted in accordance with the working conditions to provide rotating air in horizontal, vertical and diagonal directions.
- Vane adjusting device can be manually controlled, electrically controlled.

6.2 Swirl Air Diffuser Square

- The outside frame and the inside vane of the cone are of detachable structure, easy to install and adjust. Air volume regulating valve can be added so as to adjust airflow.
- Powder coated and anti-oxygenation can be used for surface treating.
- Aluminium and stainless steel is for option, size can be adjusted as per requirement

6.3 FKP – SAD Plastic Air Diffuser

Size: Out size 395x395mm, adapt to the size of ceiling mounted. Neck size dia 150, 200, 250, 300, 350 mm.

Material: Plastic, ABS and HIPS.

Features:

1. Less condensation than metal grills
2. Do not rust
3. Easy to install and clean
4. Match adjustable damper to control the air volume

6.4 FKP-RAD Air Diffuser

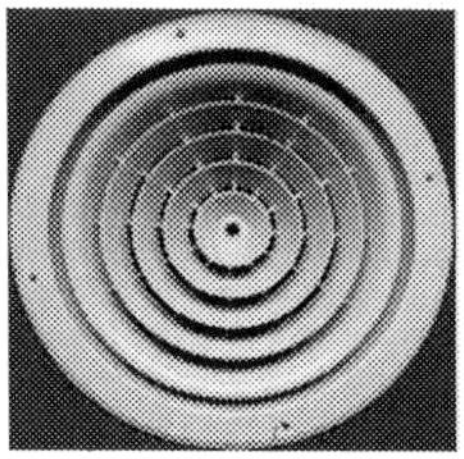

Size: Out size dia. 405mm, neck size dia. 150, 200, 250, 300 mm

Material: Plastic, ABS and HIPS

Colour: The standard size is white RAL9016, if the quantity is big enough, we can choose color according to customers' requirements.

Features:

1. Less condensation than metal diffuser
2. Light weight
3. With countersink in the out frame, easy to install

6.5 Round Air Diffuser

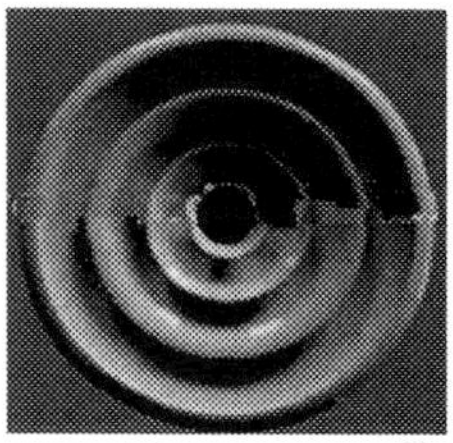

1. Used for upward airflow, suitable for ceiling mounted
2. Excellent appearance with molding vane and outside frame into integrity
3. Inside core can be dismantled, install easily
4. Neck size can be adjusted as per customers' requirements
5. Dampers can be installed to adjust the airflow

6.6 Jet Diffuser

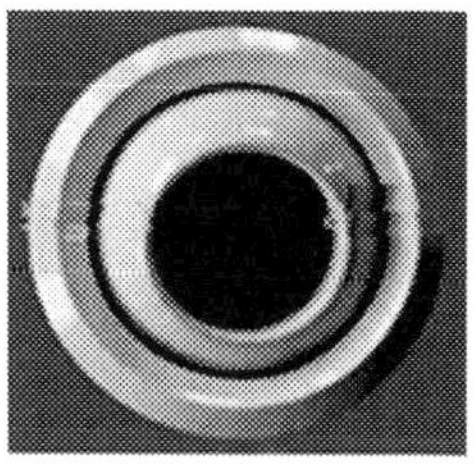

FK-BS ball spout is mainly used for the odium, airport, theater, museum, etc.

Excellent appearance, long space of airflow, low noise

Airflow direction can be adjusted manually or electrically

Plastic spraying, outside color can in harmony with architecture requirements. aluminium alloy

6.7 Square Air Diffuser

Development of the FK-SADE series comes from the FK-SAD series. Compared with FK-SAD series, the outside frame is extended, and size is normally at 595 x 595mm and 603 x 603mm so as to match the ceiling grid. Material can be aluminium alloy and ABS plastic.

6.8 Ceiling Diffuser

1. Outside frame and inside vane of the cone are of detachable structure, easy to install and adjust.
2. Air volume regulating valve can be added so as to adjust airflow.
3. Powder coated and anti-oxygenation can be used for surface treatment.
4. Aluminium and stainless steel material for option, size can be adjusted as per requirements.

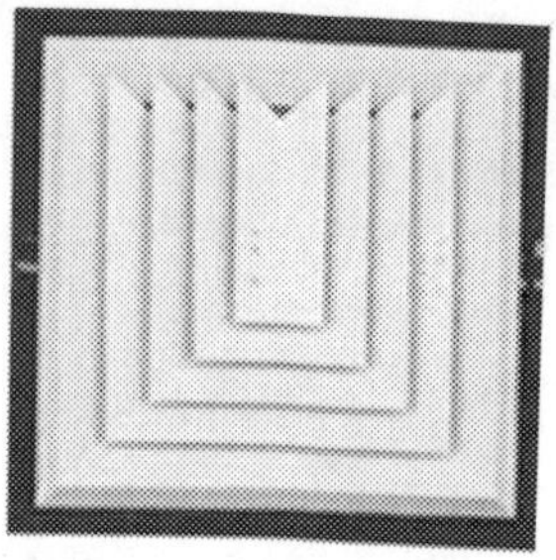

6.9 Ball Spout

1. Mainly used for stadiums, airports, theatres andmuseums
2. Excellent appearance, long space of airflow, low noise
3. Airflow direction can be adjusted manually or electrically
4. Dampers can be installed to adjust the airflow

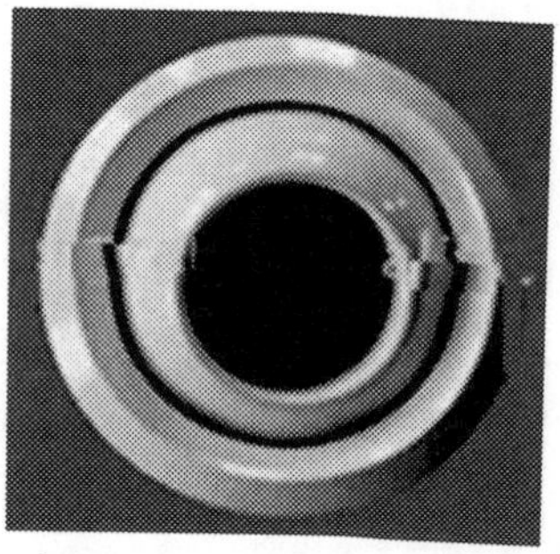

6.10 Air Intake

1. Used for exhaust and air intake of exterior walls
2. With excellent waterproof function owing to special designed blade
3. Air filter or insect filter can be inserted

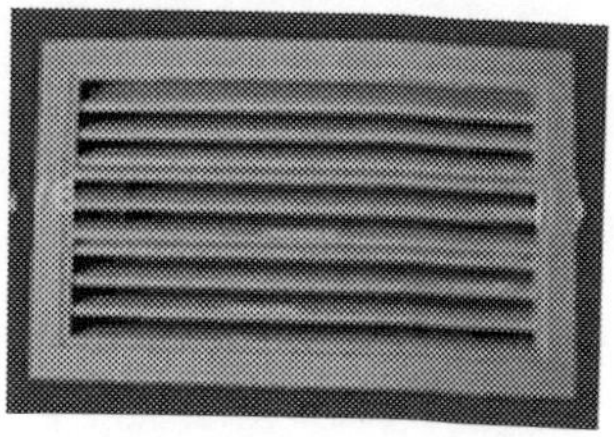

6.11 Air Grille

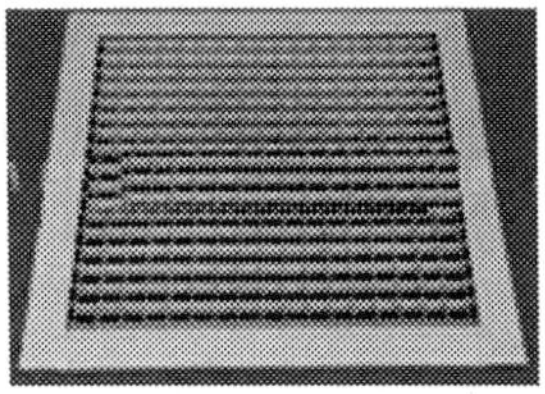

1. Excellent appearance, long space of airflow, used as return air of ceiling and side wall.
2. Egg crate design, with solid structure, easy for balancing airflow.
3. Dampers can be installed to adjust the airflow.

6.12 Air Louver

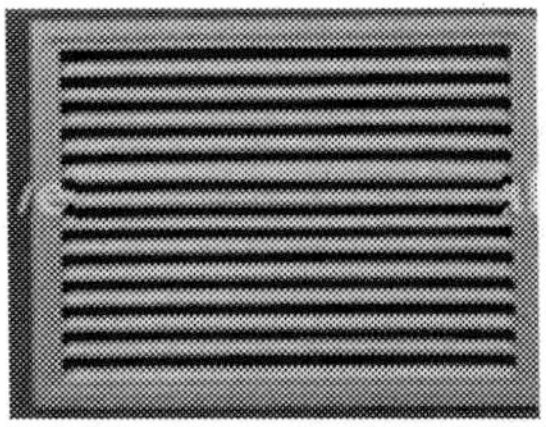

1. Excellent appearance, mainly used for return air of ceilings and side walls
2. Fixed blade with solid frame
3. Blade angle is 45°

6.13 Linear grille

1. Mainly used for supplying air in halls, lobbies.
2. Two adjustable vanes in each seam so as to control direction and volume of airflow.
3. Can be assorted with static pressure box.
4. Swiftly balances the temperature of airflow and shows velocity of airflow to avoid dust wind.
5. Maximum length is 2.5m, can be assembled with two or more parts.

6.14 Volume Control Damper Square

1. Opposite multi-blade regulating damper made by our plant has the features of good design, solid and practical, nice looking and easy to operate
2. Can be manually operated or motor operated
3. Galvanized sheet and stainless steel for option, size can be adjusted as per requirements

6.15 Volume Control Damper Round

1. Can be manually-operated or motor-operated
2. Galvanized sheet and stainless steel material for option, size can be adjusted as per requirements

6.16 Manual Actuator for Dampers

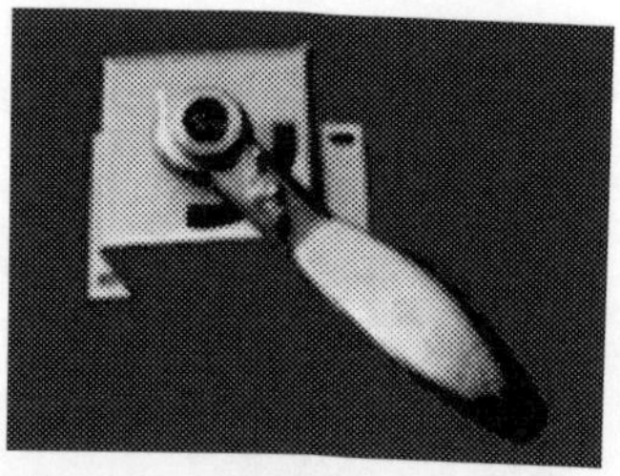

1. Handle design fit for somatology
2. Galvanized or chromium-nickel plating (optional)

6.17 Manual Actuator

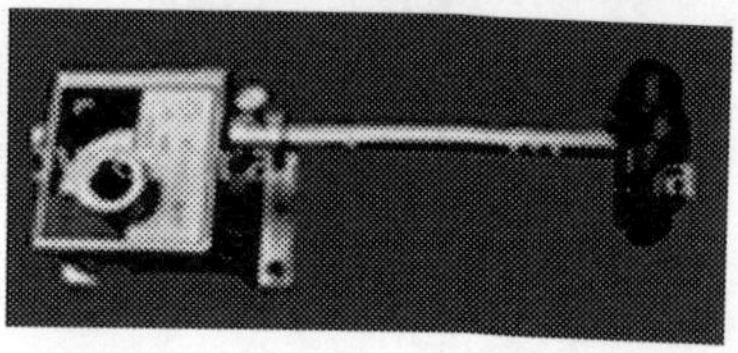

1. Transmission mechanically by worm wheel so as to control accurately and be convenient for operation
2. Galvanised or chromium-nickel plating (optional)

6.18 Flexible Duct

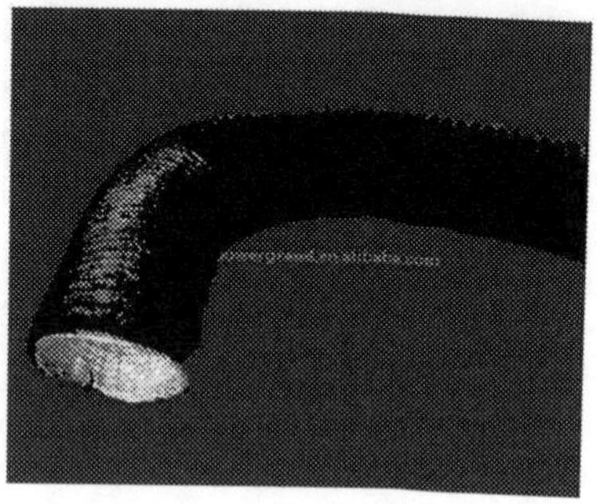

Constructed of 1-layer of aluminium film and 1-layer of polyethylene film 100% bonded together, encapsulating the steel wire.

Apply for low/medium/high pressure air ventilation and air condition system; easy to install with oval or round connection, nontoxic gas and can resist high temperature.

6.19 Flexible Air Duct

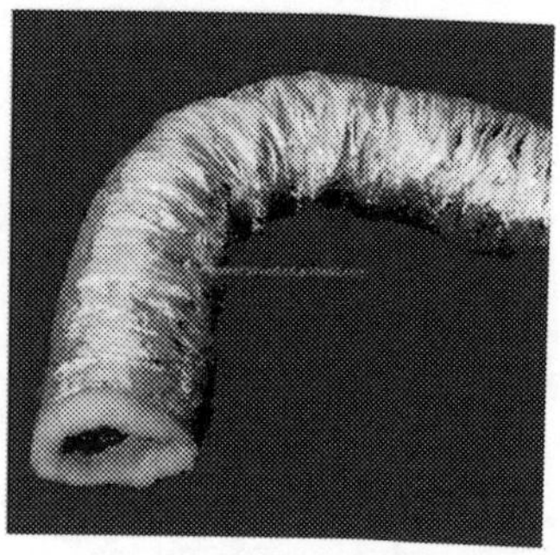

Constructed of 2-layers of aluminium film 100% bonded together, encapsulating the steel wire. With a thick blanket of fibreglass insulation and jacket with an aluminium sleeve. Micro-hole made in the aluminium foil inside can absorb the noise effectively.

Apply for low/medium/high pressure air ventilation and air condition system; easy to install with oval or round connection, no toxic gas and can resist high temperature.

6.20 Flexible Air Duct

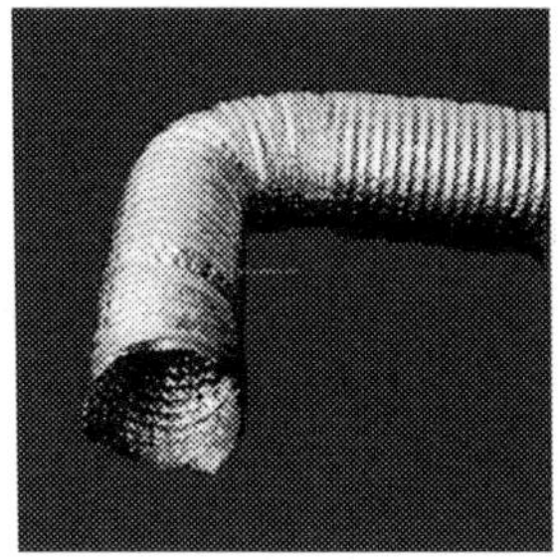

Constructed of 2-layers of aluminium film 100% bonded together, encapsulating the steel wire. Apply for low/medium/high pressure air ventilation and air condition system; easy to install with oval or round connection, no toxic gas and can resist high temperature.

Diameter Range: 75–1200 mm (3.5 inch–48 inch)

Temperature Range: −30 to 140

Max. Air Velocity: 30 m/s

Working Pressure: 2500 pa

Standard Length: 10 m

6.21 Flexible Air Duct

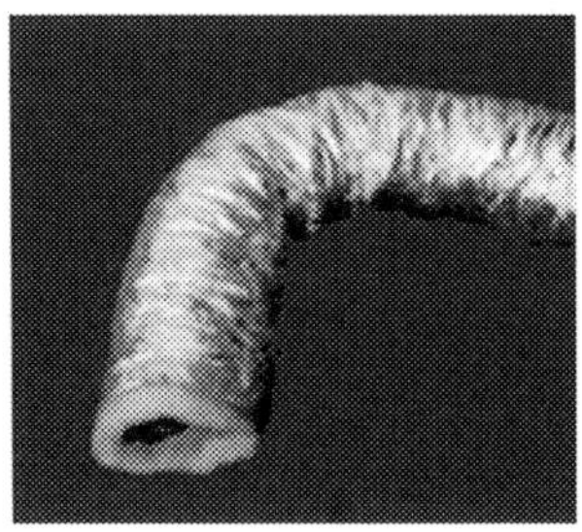

Constructed of 2-layers of aluminium film 100% bonded together, encapsulating the steel wire with a thick blanket of fibreglass insulation and jacket with an aluminium sleeve. Apply for low/medium/high pressure air ventilation and air condition system; easy to install with oval or round connection, no toxic gas and can resist high temperature.

Diameter Range: 75–1200 mm (3.5 inch–48 inch)

Temperature Range: −30 to 140

Max. Air Velocity: 30 m/s

Working Pressure: 2500 pa

Fiberglass Thickness: 25 mm

Fiberglass Density: 20 kg/m^3

Standard Length: 10 m

7 Dunham Rubber and Belting – Greenwood USA

7.1 Item # 290 0072, SILBRADE® Braid Reinforced Silicone Hose.

- Open mesh polyester braiding incorporated within the wall of silicone-silicone tubing.

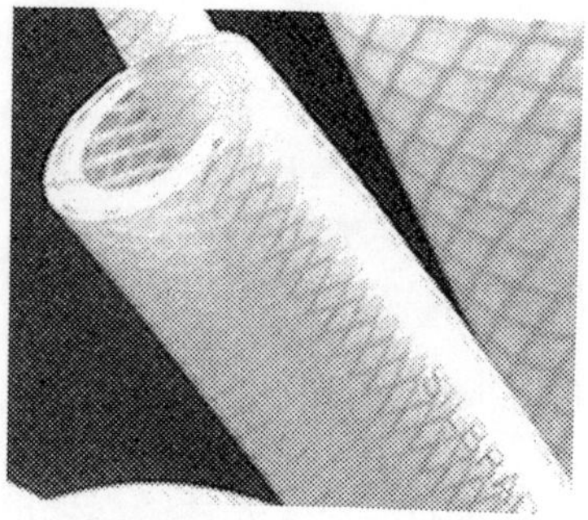

- Made from FDA-sanctioned ingredients.
- Translucent natural colour for visual contact with the flow.
- Able to resist extreme temperature variation: −80°F to 350°F.
- Odourless, tasteless and inert.
- Excellent weatherability properties – resists UV, ozone, gases, moisture and extreme temperatures.
- Offers far higher pressure capabilities than similarly sized unreinforced silicone tubing.
- Listed by the National Sanitation Foundation (NSF 51).

I.D.	1/8 inches
O.D.	0.365 inches
Wall	0.12 inches
Standard Length	100 ft.
Working Pressure at 70°F	233 psi
Burst PSI at 70°F	699 psi
Weight per 100 FT.	5 lbs.
Standard Coil Length	100 ft

7.2 Super Vac-U-Flex®

- Material: High strength fibre- reinforced PVC hose cover bonded to coated spring steel wire helix.
- Color: grey.
- Temperature: −20°F to 150°F.

Applications: Air, Chemical fumes, Dust, Exhaust and Ventilation, General purpose, Material handling, Moisture/Hydraulic, Oil mist, Special purpose, Vacuum and suction.

Features: Cuffed ends, Fume removal, Moisture resistant, Oil resistant and UV/ozone resistant.

- Offers flexibility for easy handling and installation
- Great for industrial and commercial vacuum cleaners and ventilating or cooling industrial machinery
- Ideal for bilge pumps and bilge ventilation

7.3 Flexible duct clamp, band and band lock

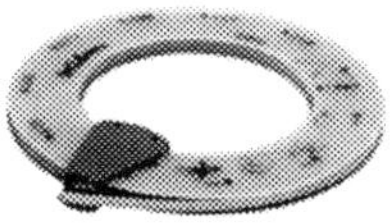

FDB

FDBL

FVA

MDC

7d Insulated Ducts

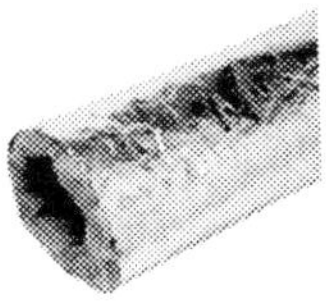

FDB

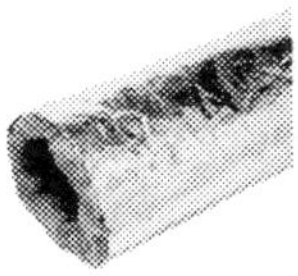

FDB

FDB

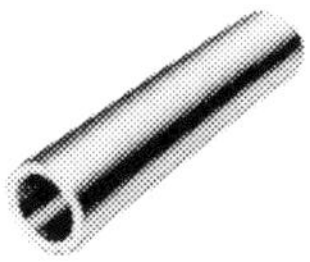

FDB

8 Climavent Fume Extraction Systems – England

Climavent Fume Extraction Systems have been built to handle the obnoxious fumes and odours in a range of industrial, commercial and educational environments. They are available in a variety of sizes and finishes, and each unit can be customised to suit specific applications. These powerful systems have been proven to provide maximum efficiency alongside minimal maintenance costs. Fume extractors can be wall or stanchion mounted. Portable fume extraction fans and self-supporting extraction arms are also supplied.

Depending on the application fume extraction accessories, carbon filtration units and pipe work can be manufactured in polypropylene, UPVC, stainless steel and powder coated mild steel. Units can be built to occupy single positions in a laboratory or to offer multi point extraction along production lines.

9 AESA Air Engineering Private Limited – France, India, Singapore and China

9a Pneumatic and Volumetric Dampers

Pneumatic Damper type PKA

- Specification: & 150, & 200, & 250, & 315
- Automatic Shut-Off Dampers for intermittent Suction systems
- Air Speed: 20 m/s
- Pressure loss max: 5 Pa

Volumetric dampers

- Centrifugal Fan, suitable for the conveyance of dust laden air
- Aluminum blades
- Electronic Actuator
- Size from 600x620 to 2000x2820

10 Coimbatore Air Control Systems Pvt Ltd. Coimbatore, India

10.1 Dampers

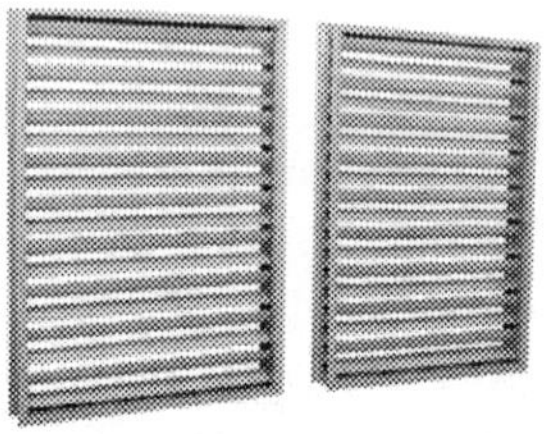

Aerofoil aluminium extruded leaves hosted in sturdy powder coated MS Box with ABS linking arrangements for smooth operation to suit auto control actuator. The linking arrangements are designed for opposite blades operation to precise control of air flow.

10.2 Diffusers

Supply Air Diffusers

The Diffuser diffuse the air uniformly from the supply air duct. Diffusers are made out of GI / MS / SS / PVC duly powder coated for better appearance and long life.

Linear Diffusers

The Linear Diffusers will be made out of galvanized iron duly powder coated.

Appendix – 7
Sensors and data loggers

Majority of modern air handling and humidification systems are having programmable logic controls and automatic monitoring. Following are examples of some commercially available systems.

1 CiK Solutions GmbH · Karlsruhe Germany

1.1 Vaisala Veriteq – Data Loggers

High-end 'Vaisala Veriteq' data loggers are ideal for precise measurement of temperature and humidity, yet can also be used for current loops, CO_2, door contacts, differential pressure and many more. They can be deployed as autonomous data loggers in validation or mapping tasks and can also be integrated into the fail-safe 24/7 monitoring system Vaisala Veriteq 'viewLinc' for high-precision temperature monitoring.

The Vaisala Veriteq data loggers and the Vaisala Veriteq 'viewLinc' monitoring system are ideal for use in all fields where temperature, humidity and other parameters are critical, such as pharmaceutical, biotech, health sector, aerospace, calibration laboratories and many more.

Highlights:

- Hybrid technology allows both validation and continuous monitoring with the same unit.
- Up to 10 year's typical battery operating life.
- Available for GxP and FDA-regulated operation.
- Calibration certificates traceable to international standards.

1.1.1 Multi-Function Data Loggers

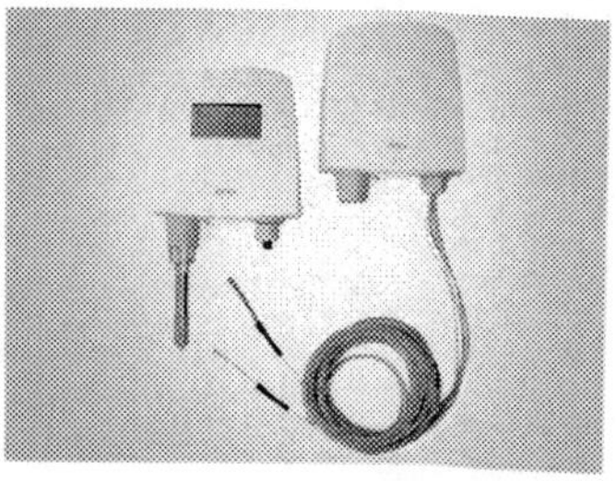

Wireless Data Logger HMT140 – Temperature, Humidity, Analog Signals and Door Contact

High-end data logger for temperature and humidity monitoring as well as analogue signals that can be validated.

Due to its high precision and wide range of features, the HMT140 is predestined for monitoring tasks in almost any application, including cleanroom and life sciences applications

Can be integrated into the central 'viewLinc' monitoring system via WLAN

- Fully autonomous power with 18-month battery life
- Pre-calibrated and field replaceable humidity/temperature sensor
- Inputs for humidity/temperature, temperature, voltage, current or door contact
- Optional LCD display
- Select from integrated or external sensors
- Traceable and accredited calibration including certificate

1.1.2 Data Loggers for Temperature and Humidity

Validatable high-end data logger for temperature and humidity monitoring

- Internal sensor for temperature and humidity
- Temperature measurement range: −35°C to +85°C
- Accredited calibration including certificate

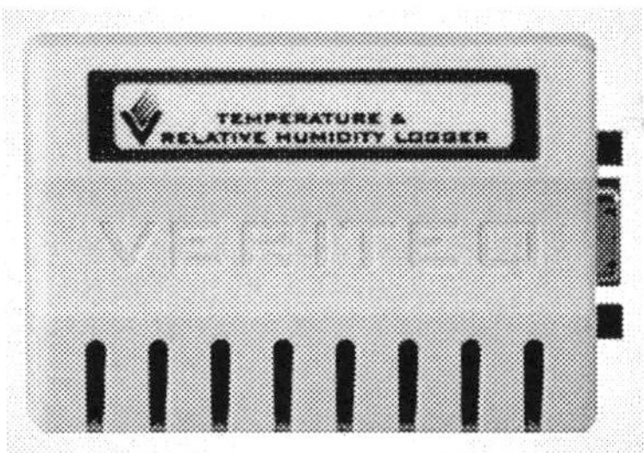

Data Logger DL2000 – Temperature and Relative Humidity

Choice of 3 temperature and 3 humidity calibration points, or 5 temperature and 6 humidity calibration points to match ICH requirements.

- Humidity accuracy up to ±1%RH at 0.05%RH resolution
- Temperature accuracy up to ±0.1°C at 0.02°C resolution
- Accuracy of ±2%RH and ±0.15°C guaranteed (!) after first year of deployment
- 10-year's battery operating life

1.1.3 Data Loggers for Temperature

Data logger DL1000/1400 – internal and external temperature

- Validatable high-end data logger for temperature monitoring
- Internal and/or up to 4 external probes (3 m or 7.5 m)

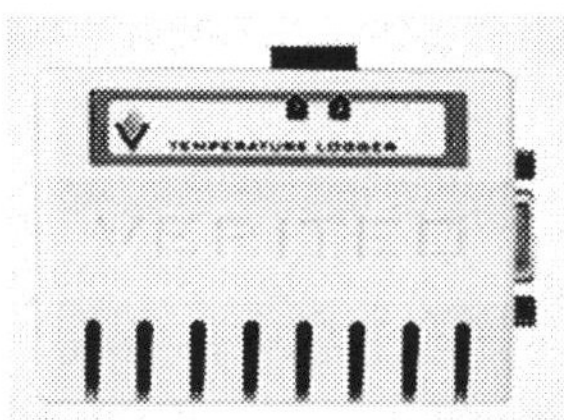

- Temperature measurement range: −90°C to +85°C (with 3 probe types)
- Temperature accuracy up to ±0.1°C at 0.02°C resolution
- Accuracy of ±0.15°C guaranteed (!) after first year of deployment
- 10-year (!) battery operating life
- Accredited calibration including certificate

1.1.4 Data Logger DL1016/1416 – Multi-Application Logger

This high end, multi-application data logger can simultaneously monitor different temperature ranges in up to four chambers, from −80°C ultra-low temperature freezers, freezer/refrigerator units to temperature controlled warehouses and incubators.

- Up to four external probes (3 m and 7.5 m) possible
- Temperature measurement range: −95°C to +70°C with only one type of logger
- Temperature accuracy: ±0.25°C at 0.01°C resolution
- Accuracy of ±0.35°C guaranteed (!) after first year of deployment
- 10-year (!) battery operating life
- Accredited calibration including certificate

1.a.5 Data Logger DL1200 – Low Temperatures

High-end data logger for monitoring low temperatures

- Optionally with one internal, one external or two external temperature sensors.
- Temperature measurement range: −55°C to +40°C.
- 10-years battery operating life.
- Accredited calibration including certificate.

Data Logger DL1700 – Thermocouple Logger

High-end data logger ideal for extreme temperatures from −240°C to +1760°C

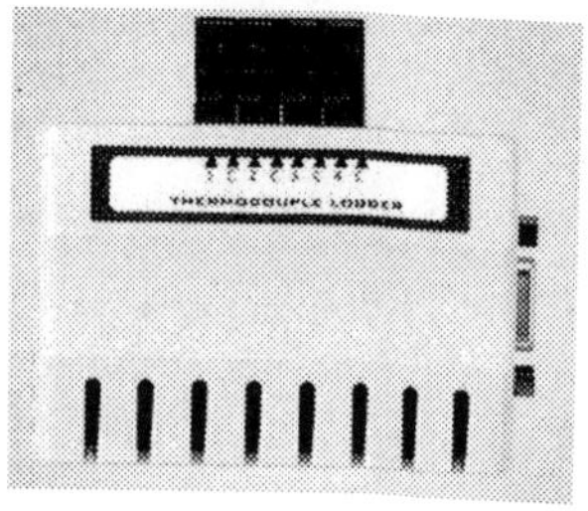

- Accepts type J, K, T, E, R and S thermocouples
- No programming or complicated equations required
- Highly accurate replacement for bulky data acquisition systems
- Traceable calibration including certificate

1.1.6 Data Loggers for Special Applications

Data Logger 4000 – Voltage and Current Loop

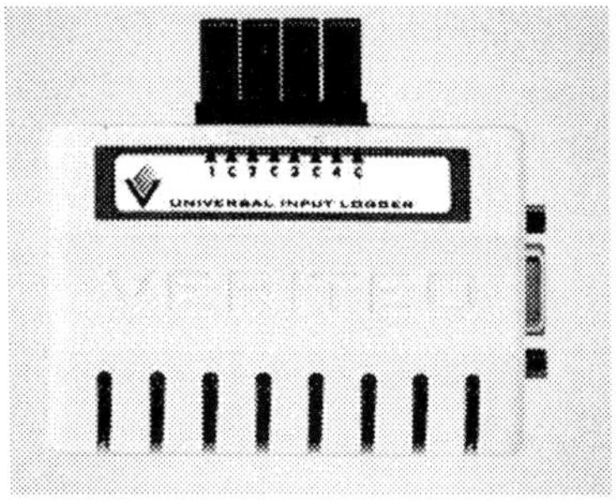

- High-end data logger for current loop and voltage monitoring
- Measurement ranges: 0…1 V, 0…10 V, 4…20 mA
- Single and multi-channel versions available
- Logging of industry-standard sensors, transmitters and transducers
- 10-year (!) battery operating life
- Accredited calibration including certificate

1.1.7 Mid-Range Data Logger

Data Logger Mid-Range DL1000MR / DL1016/1416MR / DL2000MR - Temperature, Humidity and Contact Switch

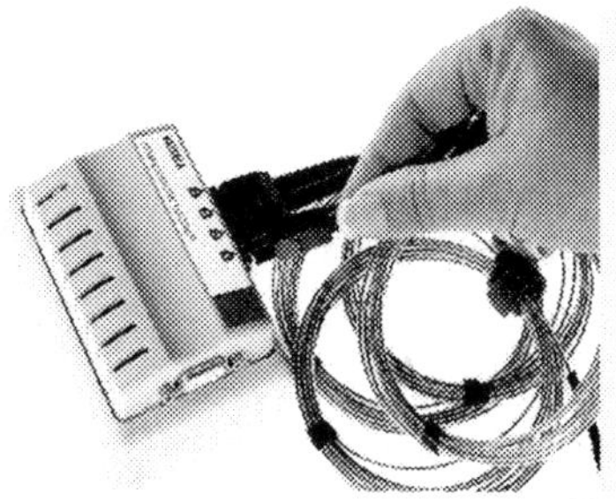

The Mid-Range data loggers are designed for early phase drug and device development applications where speed and economy are critical. The MR loggers are used with the Vaisala Continuous Monitoring System to monitor and analyse environmental data and provide presentation-quality records that are easily exported to PDF and spreadsheets. The loggers can also be used with Vaisala's vLog Mapping software for validation studies.

1.1.8 Spectrum 2000 Data loggers

Spectrum 2000 data loggers incorporate data logging and RH-sensing technologies with integral precision RH and temperature sensors with long-term high-accuracy measurement performance, sensitivity and responsiveness. Accurate to ± 2% RH and 0.15°C, the data loggers can sense changes as minute as 0.05% RH and 0.02°C with a time-base accuracy of better than 5 sec per day. The rugged thermoset polymer-based capacitive RH sensor offers outstanding resistance to air-borne contaminants and condensation, delivering exceptional in-calibration service and reliability. Data loggers interface to any PC (through any serial or USB port) with Spectrum Software, a Windows-based package for configuring, downloading, displaying, analysing and reporting collected humidity and temperature information. Humidity data are displayed in graphical or tabular form or can export results for use in other programs. The logger can also be used with the Spectrum Excel add-in, a utility program designed to work directly within Microsoft Excel.

1.2 Transmitters

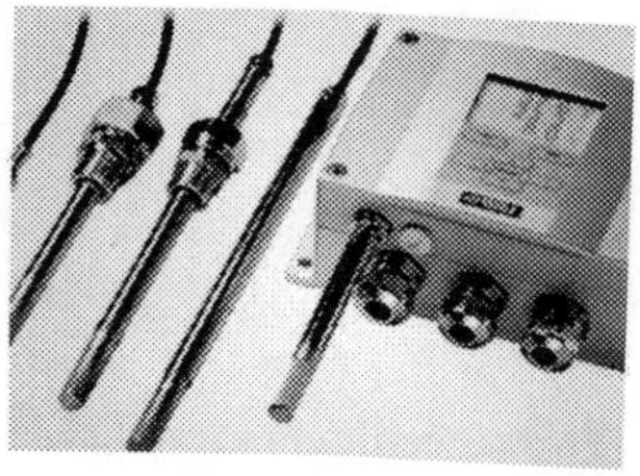

Transmitter Overview

The following details a variety of high quality transmitters (sensors) with Current Lop/ Voltage outputs, which can be used as stand-alone units or adapted to the Vaisala Veriteq viewLinc monitoring system via the DL4000.

1.2.1 Humidity and Temperature

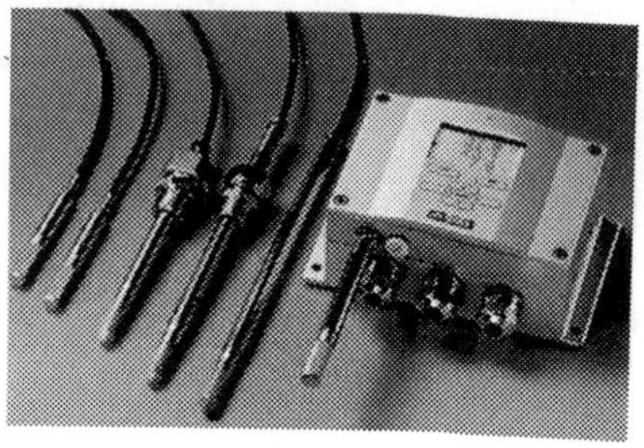

Humidity and Temperature Transmitter for Wall Mounting. Vaisala HUMICAP® Series HMT330

Top-of-the-Line models for Demanding Industrial Applications.

Choose among six models for almost every application:

- Fixed sensor
- Duct mounting and tight spaces (pipes, channels, etc.)
- High pressures up to 100 bar and vacuum conditions
- High temperatures up to 180°C
- High humidity applications
- Pressurised pipelines up to 40 bar

Features:

- Full 0 ... 100% RH measurement
- Temperature range up to +180 °C (depending on model)
- Pressure up to 100 bar (depending on model)
- Next generation Vaisala HUMICAP® Sensor for excellent accuracy and stability
- Graphical display of measurement trends and over four-year history
- Multilingual user interface
- Excellent performance in harsh chemical concentrations

- Heated probe for superior performance in condensing environments
- Shows measurement trends and history graphically
- Corrosion resistant IP65 housing
- NIST traceable (certificate included)
- Optional integrated data logging, with over four years of measured history
- LAN and WLAN communication options

Humidity and Temperature Transmitter for Wall Mounting. Vaisala HMT360 Series

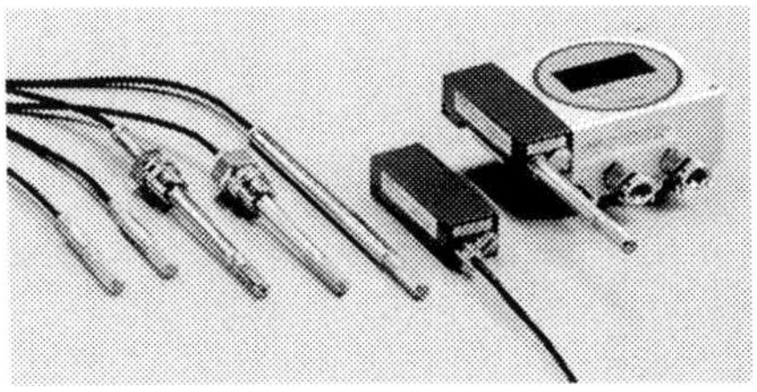

Intrinsically Safe Humidity and Temperature Transmitters for hazardous areas

Features:

- Measures humidity and temperature, outputs also dew point, mixing ratio, absolute humidity and wet bulb temperature
- Safe operation with the entire transmitter in hazardous areas: Divisions 1 and 2 (USA, Canada), Categories 1G/ Zone 0 and 1D / Zone 20 with protection cover (EU)
- Intrinsically safe
- Designed for harsh conditions
- Vaisala HUMICAP Sensor features high accuracy, excellent long-term stability, and negligible hysteresis
- Six probe options
- Temperature range between −40 and +180°C depending on the probe option
- NIST traceable (certificate included)

1.2.2 *Pressure Transmitters*

Pressure Transmitter PTB110

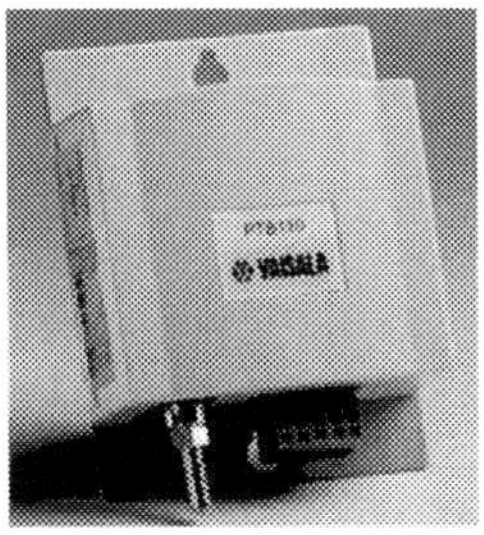

The pressure transmitter PTB110 has been designed to measure exact air pressure at ambient temperature, as well as for the general monitoring of the ambient pressure in a broad temperature range. Additional pressure transmitters on request.

Differential Pressure Transmitter PDT101

The differential pressure transmitter PDT101 has been specifically designed to meet the demands of challenging applications in life sciences, as well as for high-tech clean room applications. The PDT101 can be attached to the wall, a DIN rail or a panel and is ideally suited to be integrated into the continuous monitoring system Veriteq 'viewLinc', as is required in controlled environments for measuring and monitoring of critical environmental parameters.

- Includes NIST traceable calibration
- 4 different models available

Differential Pressure Transmitter PDT102

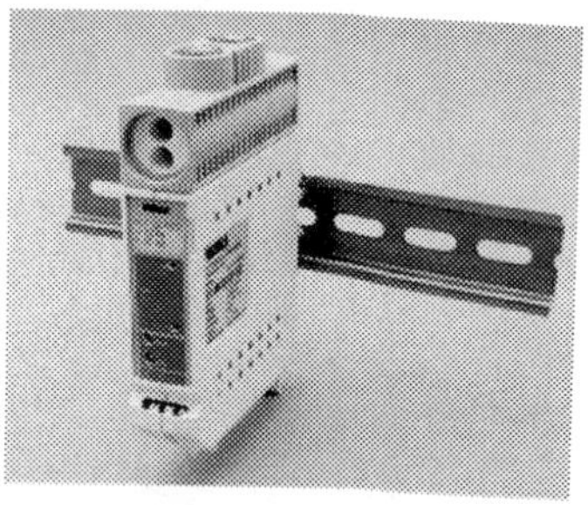

The differential pressure transmitter PDT102 stands out in particular through its ultra-slim design and clips directly to a DIN rail. Like the PDT101 it is designed for demanding applications in life sciences as well as for high-tech clean room applications.

- Includes NIST traceable calibration16 different models available

1.2.3 DewtPoint

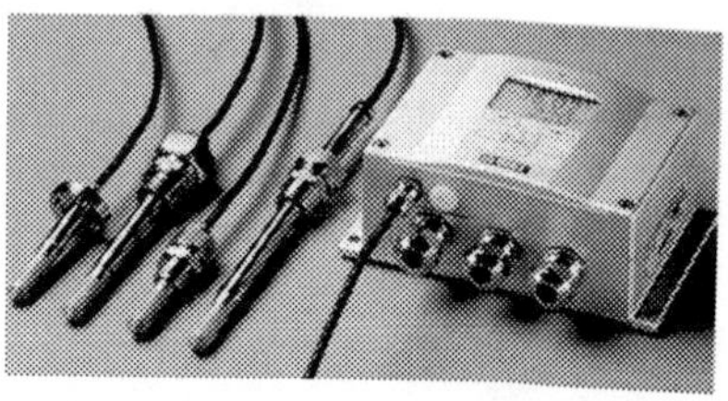

Dew point and temperature transmitter for wall mounting Vaisala DRYCAP ® Series DMT340

The first choice for demanding industrial applications with low dew points

Features:

- Measures dew points from −60 °C to +80 °C with the accuracy of ±2 °C
- Vaisala DRYCAP® Sensor provides accurate, reliable measurement with excellent long-term stability and fast response
- Immune to condensation
- Unique auto-calibration feature
- Optional alarm relays, local display and mains power supply module
- Compatible with the hand-held dew point meter DM70 (see below)
- NIST traceable (certificate included)
- 3 analog outputs and a serial interface, WLAN/LAN

1.2.4 Carbon Dioxide (CO2)

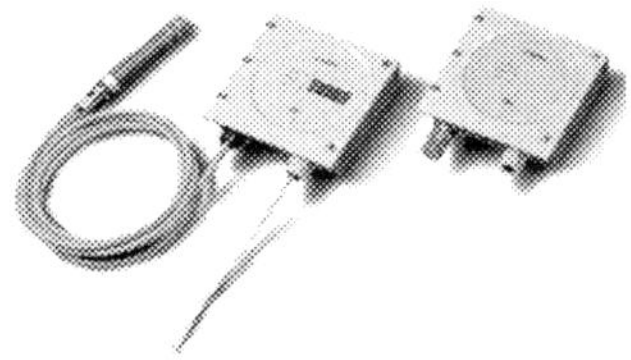

Carbon Dioxide Transmitter for wall mounting

Vaisala CARBOCAP® Series GMT220

Carbon Dioxide Transmitter (CO_2) – designed for harsh and humid environments.

Applications include:

- Horticulture and fruit storage
- Greenhouses and mushroom farming
- Safety alarming and leakage monitoring
- Demand controlled ventilation in harsh environments

Features:

- Incorporates Vaisala CARBOCAP® – the silicon-based NDIR sensor
- Choice of several measurement ranges
- Excellent long time stability
- Easy installation
- IP65 protected against dust and spray water

The Veriteq SP 2000 RH and data loggers are portable, battery operated self-contained RH and temperature recorder with internal sensors and memory. It is an easy-to-use alternative for RH chart recorders and bulky data acquisition systems. The battery provided can work for ten-years. Data loggers record RH and temperature without wires, power cords or connections. Simply place them where you need to measure humidity and temperature and they will record accurate time-based readings up to six times per minute - covering periods of up to several years in duration. No charts, wires, connections or power cords are required.

2 Sensirion AG – Switzerland

2.1 CMOS Humidity Sensors

Humidity sensors are gaining more significance in diverse areas of measurement and control technology. Manufacturers are not only improving the accuracy and long-term drift of their sensors they are also improving their durability for use in different environments, and simultaneously reducing the component size and the price. Following this trend, Swiss-based Sensirion AG has introduced a new generation of integrated, digital, and calibrated humidity and temperature sensors using CMOS 'micro-machined' chip technology. The new product, SHT11, is a single chip RH and temperature multi sensor module with a calibrated digital output which allows for simple and quick system integration.

Conventional sensors determine RH using capacitive measurement technology. For this principle, the sensor element is built out of a film capacitor on different substrates (glass, ceramic, etc.). The dielectric is a polymer which absorbs or releases water proportional to the relative environmental humidity, and thus changes the capacitance of the capacitor, which is measured by an onboard electronic circuit.

Weaknesses: Stefan Christian, product manager at Sensirion, lists three primary weaknesses of conventional analogue humidity sensors.

- Poor Long-Term Stability: Due to the relatively large dimensions of the sensor elements (10–20 mm^2), as well as the aging of the polymer layer, current capacitive sensors exhibit varying degrees of sensitivity to the same external influences. Therefore, the drift per year, i.e. the yearly change in error tolerance of the sensor, is becoming an important criterion for quality. The aging of the metallic layer electrodes can also lead to errors in the humidity signal.
- Complicated Calibration: Before use, capacitive humidity sensors must undergo a complicated calibration process. For this purpose, the end user must have complex and expensive calibration and reference systems, as well as external electronic components, such as memory components.
- Analogue Technology: Additional problems arise directly from the analogue measurement principle, which links the stability of the operating voltage inseparably to the sensor accuracy. This problem can only be counteracted by increased spending on electronics and inevitably leads to higher integration costs.

By combining CMOS and sensor technologies, Sensirion has released this new standard a highly integrated and extremely small humidity sensor, the SHT11. The device includes two calibrated micro sensors for relative humidity and temperature which are coupled to an amplification, analogue-to-digital (ND) conversion and serial interface circuit on the same chip.

A micro-machined finger electrode system with different protective and polymer cover layers forms the capacitance for the sensor chip, and, in addition to providing the sensor property, simultaneously protects the sensor from interference. Total coverage with condensation or even immersion in liquid presents no problems. The temperature sensor and the humidity sensor together form a single unit, which enables a precise determination of the dew point without incurring errors due to temperature gradients between the two sensor elements.

3 Titan Products Ltd, UK

3.1 Duct Humidity Sensor: Product Code: TPDH, TPVDH, TPVDHT and TPDHT

The Titan Duct Humidity Sensor measures humidity and temperature conditions in ventilation ducts. The measurement element is based on a monolithic integrated circuit combining a capacitance measurement for humidity and PT 100 element for temperature giving 0–10 V or 4–20 mA outputs.

TPDH 4–20 mA Humidity Only

TPVDH 0–10 Volt Humidity Only

TPDHT 4–20 mA and 0–10 V Temperature and Humidity

TPVDHT 0–10 V Temperature and Humidity

TPVDH-RT 0–10 V Humidity and Thermistor Temperature

TPDH-RT 4–20 mA Humidity and Thermistor Temperature

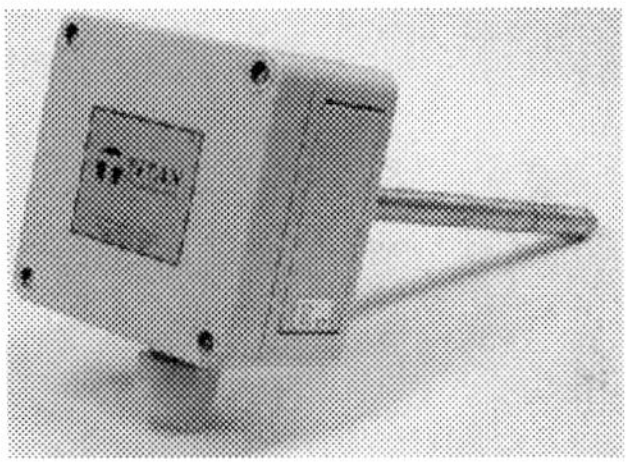

Specifications

Material: Body: Polycarbonate, Probe: 20 mm PVC Tube

Sensing Elements: Monolithic integrated circuit (combined) Alternative Thermistor for temperature

Supply: 24V DC± 15%mA Loop Powered. 24 V ± 15% AC/DC Combined Voltage

Output: 4–20 mA, 0–10 V, Optional Thermistor Temperature

Range: Humidity 0–100%. Temperature 0–50°C

Accuracy: Humidity ±2%. Temperature ± 0.2°C

Operating Temperature: 0–50°C

Terminals: 1 mm recommended, 2.5 mm max

4 Honeywell USA

4.1 Humidity Sensors

Relative humidity/temperature and relative humidity sensors are configured with integrated circuitry to provide on-chip signal conditioning. Absorption-based humidity sensors provide both temperature and %RH outputs. On-chip signal processing ensures linear voltage output versus %RH. Sensor laser trimming offers 5% RH accuracy and achieves 2%RH accuracy with calibration. Packages are chemically resistant and operate in ranges of −40°C to 85°C to accommodate harsh environments. Relative humidity sensors contain a capacitive sensing die set in thermoset polymers that interacts with platinum electrodes.

HIH-4000 Series

HIH-3610 Series

HIH Series

5. EdgeTech Instruments, Hudson, USA

5.1 RH-CAL Portable Relative Humidity Calibrator from EdgeTech.

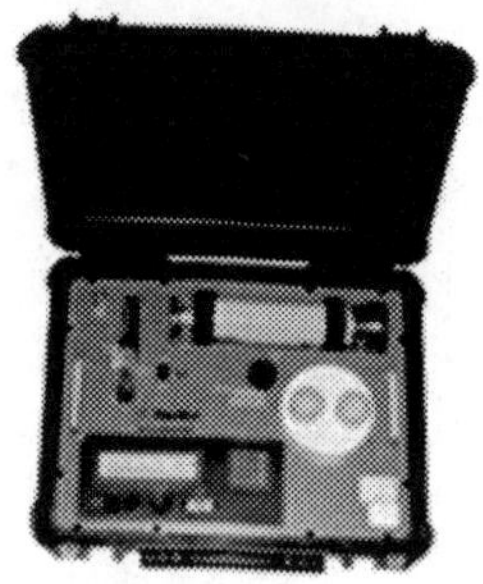

RH-CAL is a microprocessor based, programmable humidity calibration system that is at home in the metrology lab or out in the field performing on site NIST traceable humidity calibrations. The system is claimed to offer the highest accuracy available for both relative humidity and ambient temperature.

Benefits: EdgeTech field proven Optical Chilled Mirror (OCM) technology, completely self-sufficient and portable humidity calibration system, RH Accuracy: ± 0.5%, Range: 5% to 95% and AT [Temperature] - Accuracy: ± 0.2°C, Range: 10°C to 50°C

5.2 Dewmaster – Chilled Mirror Dew Point Hygrometer

Dewmaster is a precision NIST traceable hygrometer with accuracy at an economical price. Dew points can be measured from −75°C to 95°C (chill mirror sensor selection required). Air temperature and pressure sensors are optional. LCD display, Serial and analog outputs, alarms relays. Dew Point lab standard is using proven chill mirror technology.

- Nema 4, panel mount and rack mount configurations available.
- 3 parameter measurement; dew point, temperature, pressure
- 3 line display and analog outputs
- Table top/rack mount
- Nema 4 housing option
- Chilled mirror sensor selection required

5.3 H-STAT Microprocessor-Based Humidistat

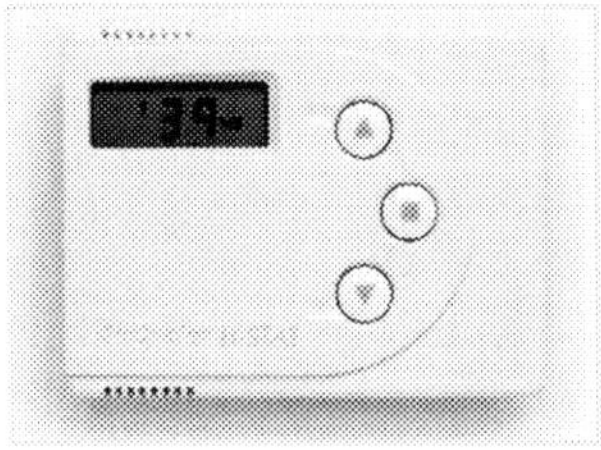

Programmable humidistat

LCD display

User friendly installation with plug-in electronics

Accuracy ± 2%RH from 30 to 80%RH

0-10Vdc output

5.4 Humidity Transmitter H-S

The H-S is a low-voltage microprocessor-based humidity transmitter with a LCD display and voltage output. This transmitter is ideal for monitoring indoor humidity application and HVAC applications where both a display and a voltage output are desired. Options: Temperature sensor to monitor ambient conditions and programmable relay contacts are available.

Features are large LCD showing humidity reading or set point, proven humidity sensor with extraordinary performance in repeatability and durability, user friendly installation/ replacement with plug-in electronic module and base plate.

Measurement range: 10–95% RH, Control Range: 10–95% RH, Accuracy @ 25C: ± 2% RH from 30 to 80%, Repeatability: ± 0.3% RH, Operating Temperature: 5°C to 50°C, Temp effect: 0.05%RH/°F. Power requirement: 24Vdc or 24 Vac, Outputs 0–10 V. Sensor Stability: ±1% @ 50% RH, 5 yr.

5.5 Model 650 Hand held Hygrometer

The Model 650 incorporates maintenance free humidity probe ideal for checking humidity measurements for validation of critical environments. It is rugged, compact and protected by a water-proof plastic housing. The humidity measuring range is 0–100%RH with ±1% accuracy. The probe is equipped with a patented RH sensor. The unit can be upgraded depending on the requirements, which is supported with user friendly Basic Comfort Software, and can be used for programming, data processing and barcoding. The optional barcode pen allows for quick and easy data processing. The modular design allows for upgrades to the instrument as they are developed.

Benefits: Guaranteed two years long stability, humidity sensor is unaffected by water, highest %RH measurement accuracy with a capacitance probe, NIST certification, battery powered and data logging capabilities with small attachable printer.

Description: The precision temperature/humidity measuring instrument includes the basic parameters viz temperature, dew point and %RH and rapid measurement with accurate temperature probes. The measuring instrument extrapolates the final value. Highest precision achieved due to system calibration (measuring instrument + probe). Calibration data are saved in the probe. Calculates all of the physical humidity parameters in the Mollier diagram: %RH, dew point, pressure dew point (tp, tdp), absolute humidity (g/m^3), humidity levels (g/kg pressure-compensated), wet temperature °C and °F, water vapour partial pressure in mbar/hPa, Enthalpy kcal/kg, humidity probes for every application, from low to high humidity levels, guaranteed long-term stability < 1% over 2 years and accuracy up to ±1%.

5.6 Model 650% RH Transmitter

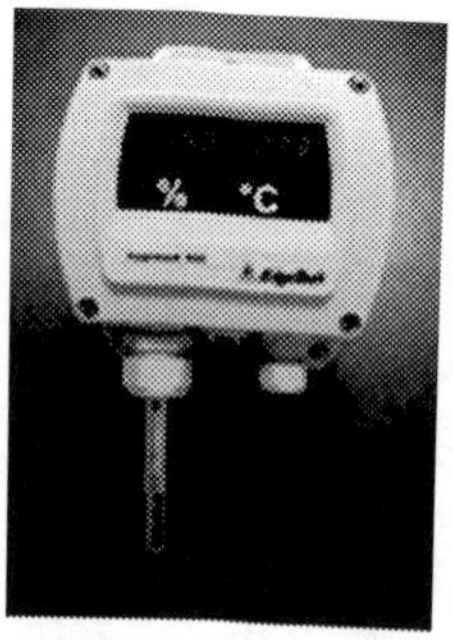

The integrated microprocessor accepts the linearisation and compensation for humidity over the whole range from –40°C to +180°C. This guarantees highest accuracy of up to ± 1% RH. In addition to RH and temperature, dew point can also be calculated. Up to 32 transducers can be analysed on a PC. The optional LED display is easy to read. The Model 650 incorporates ±1% maintenance free humidity probe ideal for checking RH measurements for validation of critical environments. It's rugged, compact and protected by a water-proof plastic housing. The humidity measuring angle is 0–100%RH.

The accuracy of this probe has been tested and proven time and time again in worldwide laboratory tests. The probe is equipped with a patented %RH sensor.

Benefits: Highest precision of ±1% RH, Mollier diagram integrates °C, %RH, td, g/m^3, g/kg., housing with LED display, 2 wire technology, 4 to 20mA for signal and power, variable cable lengths up to 10 m, variable scaling of outputs signals possible and parallel analysis of up to 32 transducers via PC (RS485).

Description: The precision temperature/humidity measuring instrument includes the basic parameters – temperature, dew point and %RH. Rapid measurement with accurate temperature probes. The measuring instrument extrapolates the final value. Highest precision achieved due to system calibration (measuring instrument + probe). Calibration data are saved in the probe. Guaranteed long-term stability < 1% over 2 years. Accuracy achieved up to ±1%. Calculates all of the physical humidity parameters in the Mollier diagram, i.e. relative humidity (%RH) dew point, pressure dew point (tp, tdp) absolute humidity (g/m³), humidity levels (g/kg pressure-compensated).

5.7 Model EC4 – NIST Traceable RH Calibration Chamber

NIST Traceable with accuracy of ±0.5%RH. Large calibration chamber has 4.5Cft internal volume (also available in 20 & 30Cft.), which is extremely stable and maintains uniform environment in wide temperature and RH range. The EdgeTech 4.5 cubic foot calibration chamber combines the quality and durability of the PGC line with the technology of the EdgeTech hygrometer. This chamber can be used for research and production testing, in addition to quality control applications requiring RH and temperature control within very narrow tolerances. The chilled mirror hygrometer provides accuracies of measurement and control. The PGC system furnishes RH and temperature with a chamber uniformity of ±0.5% RH and ±0.3°C. The system, equipped with a single-latched outer door, is easy to use. It also features a clear inner Lexan door and a glove port which can be used for adjusting product and test materials.

Chamber uniformity of ±0.5%RH and ±0.3°C temperature.

High sensor accuracy and reliability; NIST traceable accuracy of ±0.2°C.

Features clear inner Lexan door and glove port which can be used for adjusting product and test materials.

Sensor dew/frost point measurement range up to –40 to +85°C.

Worry-free operation can control RH to ±0.5%.

5.8 Vigilant™ – Multifunction Hygrometer

With dew point (NIST traceable) accuracy of ± 0.2° Servo Lock™, digital control servo, Automatic Balance Control (ABC)., and wide measurement range –75 to +100°C Edge Tech claims of introducing the Next Generation Hygrometer

Highest dew point NIST traceable accuracy of +/-0.2°C
Servo Lock™: digital control servo
Graphical backlit LCD display, Dual alarms
Wide dew point measurement range up to –75 to +100°C
Self-calibrating PRT amplifiers
ABC

The hygrometer is engineered for a multitude of demanding applications and environments. It combines the inherent accuracy and field proven reliability of chilled-mirror technology with the flexibility and speed of digital control within a very compact package. Vigilant is a versatile hygrometer for any process or laboratory environment with its wide –50°C to 90°C measurement range, optional pressure and temperature inputs for calculating any psychrometric variable and with flexible mounting. Using optional accessories, Vigilant can be adapted for bench-top, portable, wall, pipe or rack use. Another practical innovation is the menu-driven graphical backlit LCD display. This hand-free design allows operators to clearly view up to three parameters simultaneously, along with the status of Servo Lock™, ABC, mirror condition, alarms, date and time. No manual scrolling or other distracting selecting is necessary to monitor major indicators. The front panel provides easy access to a fully programmable menu. Servo Lock™, a new digital control scheme improves sensitivity and response characteristics while providing greater control and operational stability. Vigilant's Set Point mirror temperature control allows the user to chill the mirror to a predetermined temperature above dew point and alarm/alert the user if the dew point attains or exceeds this threshold. To provide extended intervals of unattended operation, the Vigilant boasts a new ABC scheme that automatically corrects for the light reducing effects of mirror contaminates. In addition, it provides both analog and digital outputs. 0–5VDC and 4–20 mADC can be assigned to any parameter and are user scalable throughout the entire range. It offers RS232C duplex serial for programming, data output, status indication of control, mirror condition, and optional relays.

5.9 Model 1-C DewTrace – Trace Moisture Analyser

Tracing moisture (0–500 PPMv) measurements of inert gasses with sensitivity below 1 PPMv can be done with fast response and high accuracy. Used in semiconductor industry for monitoring common process gases, gas supplier Industry as a quality control instrument and in electric power industry for checking moisture levels in transformer gases.

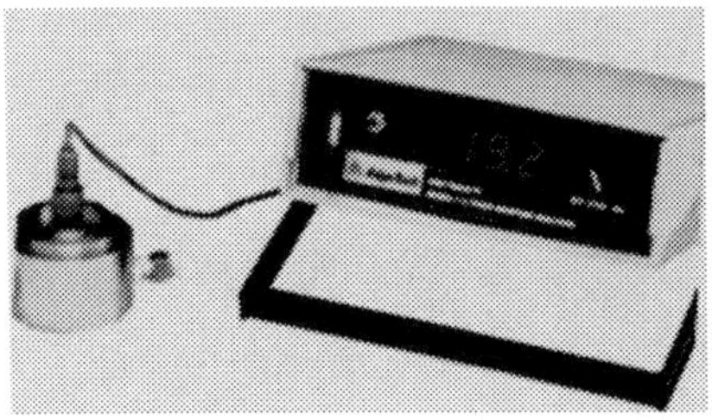

Monitors PPMv moisture levels in relatively clean, dry, inert gases.

Electrochemical sensor with proprietary diffusion membrane.

0.1 to 500 PPMv measurement; sensitivity below 1 PPMv.

Fast response; flow insensitive

The Dew Trace™ moisture analyser utilises an electrochemical (phosphorus pentoxide-P2O5 coated) sensor in combination with a proprietary semi permeable diffusion membrane. This membrane freely passes water vapour molecules while protecting the sensor from larger molecules which might adversely affect performance. The sensing element itself is an interdigitated wafer electrode and is manufactured using a unique procedure. It does not utilise a wire wound component. The cable length between the sensor housing and electronics is of minor importance since the sensor outputs a DC current. A six-foot length is provided. The electronics unit supplies power for the sensor and converts the received analog signal into a moisture value in PPMv which is then digitally displayed on the front panel. The electronics is equipped with a stand-by position on the function switch which allows the cell to remain dry when not sampling, and which also gives indication of proper functioning of the cell. This feature significantly reduces or eliminates the amount of drying time prior to startup. This is also equipped with a Cell Life Indicator which displays the useful life of the sensing element. If the sensing element is replaced, a Calibration Digital Switch, located on the rear panel, matches the new element to the electronics.

Principle of Operation: The electrochemical cell consists of a hygroscopic material, phosphorus pentoxide (P_2O_5), coated on a semiconductor electrode element. Voltages are applied to the electrodes such that adequate field strength is obtained to dissociate absorbed water into hydrogen and oxygen, the current flowing is proportional to the quantity of water dissociated (Faraday's Law of Electrolysis). Thus, this instrument, which measures the dissociation current, makes a fundamental measurement of the moisture present.

Features: This sensor design is easy to install, use, and maintain. The sensing element and membrane are mounted in rugged 316SS housing which can be used in situ with its own mounting flange, or it can be used in a flow stream with the optional 316 SS sample chamber. It is less subject to contamination because of the membrane. There is no need to replace whole sensor when oxide sensors fail. Only the membrane and/or wafer electrode would require replacement which is an easy and inexpensive task.

Measurements are very repeatable. Since measurements are independent of the mass of gas, the inaccuracies and calibration problems associated with most other electrochemical sensors are eliminated.

5.10 Model 137 Aircraft Hygrometer

Rugged, field proven chilled mirror technology with wide measurement and operating range. 2-stage sensor available; The model 137 aircraft hygrometer is a rugged, shock mounted, military qualified instrument for the primary measurement of dew point in flight or in installations with existing flow, such as wind tunnels. It features a miniature remote sensor employing a platinum resistance thermometer, and has 15 feet of interconnecting cable to the control unit. Precision calibration units for the linear analog outputs are included.

Power 115 VAC ±10%, 50 to 1,000 Hz, 50 W nominal. Dew point accuracy to ± 0.5°C with temperature range of −20°C to 60°C. Wide measurement range −60°C to 70°C. 2-stage sensor now available. Weighs less than 13 ½ lbs – including control unit, sensor and cabling.

5.11 Model DPM-99 Medical Gas Dew Point Monitor

Model DPM – 99 Gas Dew Point Monitor exceeds the latest NFPA99 requirements for dew point monitoring of Med Gas, which is easy to install and virtually maintenance free. The NFPA99 has mandated that for Med Gas systems the 'dew point for locally compressed air shall be monitored and alarmed per 4-3.1.9.8 and 4-3.1.1.2(f) to protect from a line pressure dew point rise to 39°F (3.9°C) from a nominal design of 35°F(1.7°C)'. This safeguard was the foundation for the new EdgeTech and AIRVAC Technical Services DPM-99 Med Gas Dew Point Monitor. This specifically designed monitor exceeds the requirements set-forth by the NFPA and was collaborated with AIRVAC Technical Services who are certified by P.IP.E as medical gas inspectors in all 50 US states. The DPM-99 utilises a chilled mirror sensor, one of the most accurate reliable dew point sensing scheme. This primary measurement is directly traceable to N.I.S.T. The DPM-99 also uses a modified Automatic Balance Cycle (ABC) that allows for continuous monitoring of dew point with virtually maintenance-free operation. Dew point is monitored continuously for safety. This consists of sensor, audible and visual alarms, and flow controller contained within an easily mountable rugged aluminium

NEMA-12 enclosure. Only 1 sample line and power connections need to be made to become immediately compliant.

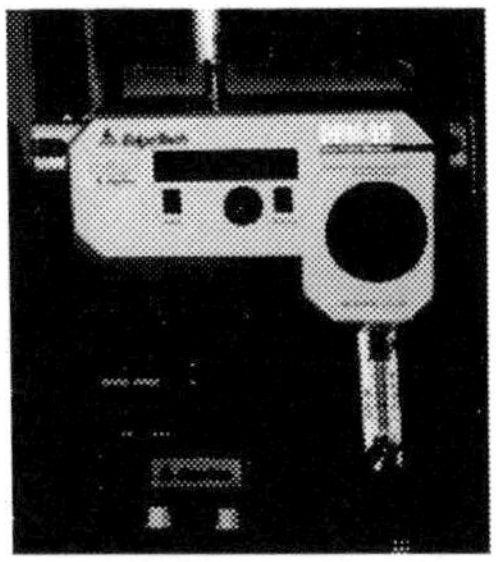

- Performance that exceeds the latest requirements of NFPA99 for Med Gas Dew Point Monitoring
- NIST traceable Dew Point
- Continuous Dew Point Monitoring
- Easy to install, virtually maintenance free
- 16. Model 1500 Portable Dew Point Monitor

Features: Primary measurement technique, NIST traceable, rechargeable batteries, easily transportable in hard protective case, wide measurement range (−75 to 75°C), microprocessor controlled, self-balancing and portable

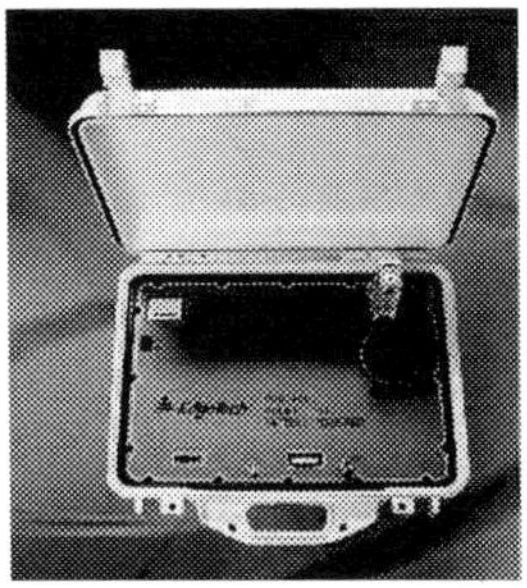

A portable Dew Point hygrometer that integrates the accuracy and reliability of the EdgeTech chilled mirror, with the battery life and ruggedness required for field use. This device incorporates all of the standard features from EdgeTech, such as drift-free NIST traceable accuracy, low maintenance, high durability and ease of use, in a system suited for the demands of field use. This can be used as a field transfer standard, or to spot check process air dryness or Endo gas dew point.

It is a self-contained system with no field assembly required and no need to carry around individual components such as control boxes, sensors, cables, pumps and power cords. Simply attaching the sample line and adjusting the flow (with the integral flow meter) and turning on the power gives the field dew point data with the same chilled mirror accuracy of a lab standard hygrometer. With a nominal battery life of 12 hours, one can collect the data for entire shift or spot check processes for several days.

5.12 Com.Air Complete Dew Point Monitoring System

Com.Air dew point monitoring system designed by EdgeTech offers the highest level of reliability and accuracy for compressed air system dew point measurements, with primary measurement technology fully NIST traceable and utilizing a chilled mirror sensor to deliver consistent drift-free measurements without regular calibrations or sensor replacement. It is easy to install and operate. The rugged aluminium NEMA-12 enclosure houses the sensor, control circuitry, as well as flow control and is wall-mounted to conserve space. Only the sample line and power connection is needed for the system to be operational. It also offers a programmable ABC cycle, which corrects for the light reducing effects of mirror, contaminates and allows for continuous monitoring and virtually maintenance-free operation.

Com.Air offers continuous and virtually maintenance free Dew point monitoring with many standard features:

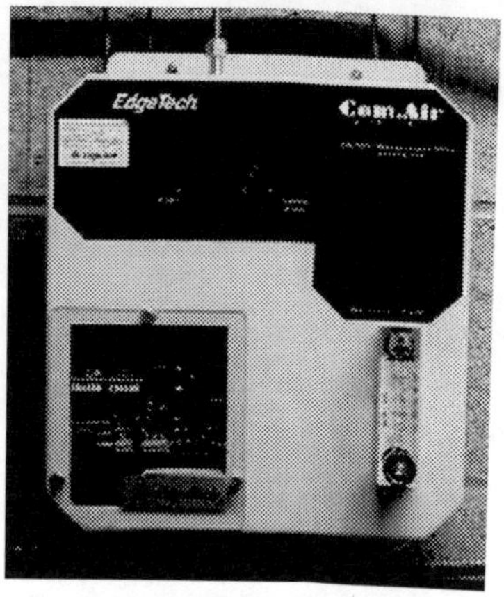

- NIST Traceable
- Virtually maintenance free · Easy to install
- Self-monitoring
- User programmable audible and visual alarms and relay contact
- Programmable ABC
- Widest Available range −103 to 122°F (−75 to 50°C) Dew point monitor for compressed air systems
- Reliable and accurate

5.13 OxyMaster II 16TDP – Oxygen & Dew Point Analyser

- Oxygen and humidity measurement in one unit
- Oxygen measurement range 0–10,000 ppm
- Humidity measurement −0.1–1,000 ppm
- 4–20 mA outputs, RS232 and data logging included
- User selectable measurement range
- Oxygen sensor – a micro fuel cell, P2O5 trace moisture sensor Power 85 – 230 V AC, 24 V DC

☞ *Clear text graphic display*: This Model was made for fast, accurate and economic measurement of oxygen and dew point at trace levels. A simple menu-driven software, touchpad and large LCD-display allow easy and fast start-up of the instrument. The unit measures humidity between −0.1 and 1,000 ppm and oxygen between 0 and 10,000 ppm.

☞ *User-selectable measuring range*: The user can select any range within 0 – 10,000 ppm and 0.1 – 1,000 ppm humidity. Free-programmable concentration alarm and a system alarm give versatility meeting almost any requirement.

- *Data outputs*: For oxygen concentration and dew point 2 x 4-20 mA output is standard. Other outputs are available on request. RS485 is an option.
- *Sensor with long operating life*: The OxyMaster uses a special fuel cell to measure the oxygen concentration. The sensor meets the industrial requirements for accuracy, sensitivity, easy to use and operating life. (To measure the humidity the unit uses a P2O5 probe.)
- *Calibration*: The calibration of the instrument for trace oxygen measurements in gas is done with a calibration gas. The concentration can be chosen freely within the measuring range.
- *Flow-through measuring cell*: The measuring cell is modular and is made of stainless steel. A defective measuring cell can be repaired by replacing the defective part only, rather than the complete unit

5.14 HTPB Series

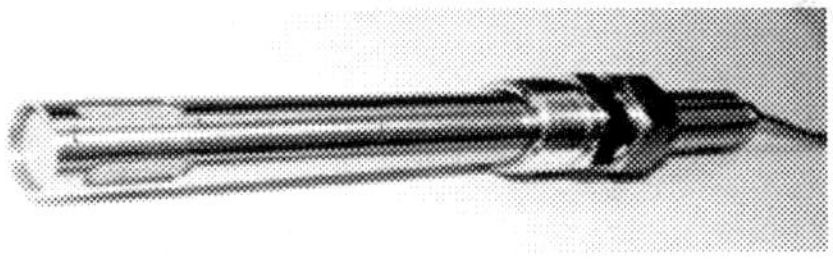

Various models are available which all have in general humidity/ temperature steel probe made of sinter filter, RS 232 and 0–10 V dc outputs, with RH measuring range of 5 to 95% with an accuracy of ±1%.

- **HT75** – Operating temperature −20°C to 75°C, dew point range 30° to −60°C, default outputs: RH, temperature and dew point.
- **HTB75** – Pressure range is (barometric pressure) 300–1,100 mb, operating temperature −20°C to 75°C, default outputs: RH, temperature and pressure.
- **HTP 55** – Pressure range is 0–200 psia, operating temperature −20° to 55°C, default outputs: RH, temperature and pressure.
- **HT115** – Available with remote humidity probe. Operating temperature −20°C to 115°C, default outputs: RH, temperature and dew point-DIS option included.
- **HTB115** – Available with remote (6ft cable) humidity/ high temperature/ barometric pressure stainless steel probe. Pressure range is 300–1,100 mb, operating temperature −20°C to 115°C, default outputs: RH, temperature and pressure, DIS option included.
- **HT115N** – Available with remote humidity/ high temperature stainless steel probe. Operating temperature −20°C to 115°C, default outputs: RH, temperature and dew point
- **Specification:** 0–10 V dc, RS232 outputs, 18–30 V dc power,

Programmable scaling and selectable measurement parameters through RS232 serial port, Wall (W), Remote (R), pipe (P) ¾" thermocouple bore through compression fitting and duct (D) mount configurations-need to specify the mounting configuration at the time of order, Accuracy ±1% nominal, Standard 6 ft. cable length, Stain less steel sinter filter-removable

Options:

- DIS display, 4-20 mA, RS232 and 2 programmable alarm relays, 1 Form A (NO) 3A/250 V ac
- FRSHT75 field replaceable sensors module for RH/T interchangeability accuracy to published specifications.
- FRSHTP55 field replaceable sensor module with RH/T/P interchangeability accuracy to published spec

Sensor Dimensions: Width ¾" OD, Length 8"

5.15 Model WBL-200 Series

WBL200M is a miniature packaged data logger for RH and temperature, housed in a ESD (electro static discharge) enclosure, can transmit up to 70 ft in the line of sight, powers: 3.6 Vdc Lithium coin cell battery, Dimensions: 4.75" (120 mm)L * 1.5"W (38.1 mm) * 0.375" H (9.5 mm).

WBL200T is a data logger for humidity and temperature housed in a aluminium enclosure, can transmit up to 250 ft. in the line of sight, powers: 2 AA lithium batteries, Dimensions: 6.75" L (171.5mm) * 3" W (76mm) * 1" H (25.4mm) WBL200R- (required) is a receiver for both the WBL200M and the WBL200T, power through the USB port on a PC.

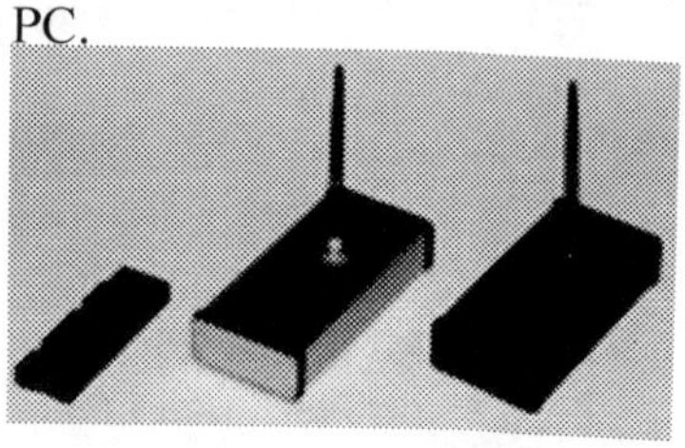

Features: Real time on board data logging capabilities,
Ultralow power consumption.
Up to 3–5-year battery life depends on usage,
Transmission parameters configurable by user,
PC compatible, software included,
Transmit up to 250 feet, 0–100% RH, −40 to 85C° and
FCC and CE approved

Specifications:

Temperature: Sensor: CMOS Range: −55C° to 130C°, *Accuracy*: ±1C° at 25 typical

Humidity: Sensor: Capacitive (polymer), *Range*: 0–100%RH, *Resolution*:

0.3%
Calibrated Accuracy: ±1.5%RH between 0 and10% ±2% between 11 and 55%, ±3%between 56 and100%, *Drift Rate*: 1%/year, *Response Time*: Time constant 3 sec, *On Board Data Logging*: 20,000 data points (Humidity and Temperature), *Real Time Recording*: Transmit data to PC for monitoring and recording, *Reading Intervals*: 1 minute to 60 minutes, *Computer Interface*: RS232, USB, *RF Frequency*: 2.4 GHz. *Coexist with*: WiFi and Blue Tooth, *Typical Battery Life*: WBL200M:1 yr at one sample reading every 15 minutes, WBL200T: 3 year at one sample reading every 15 minutes, *Approvals*: FCC and CE, Optional software for client/server data storage in SQL Server/Oracle

6 Luwa Air Engineering AG Switzerland

6.1 DigiControl 5

Luwa developed the freely programmable digital system DigiControl 5 for regulating, controlling and monitoring of all air handling processes. DigiControl 5 ensures that the air is always ideally conditioned and also allows permanent operation with optimised operating costs.

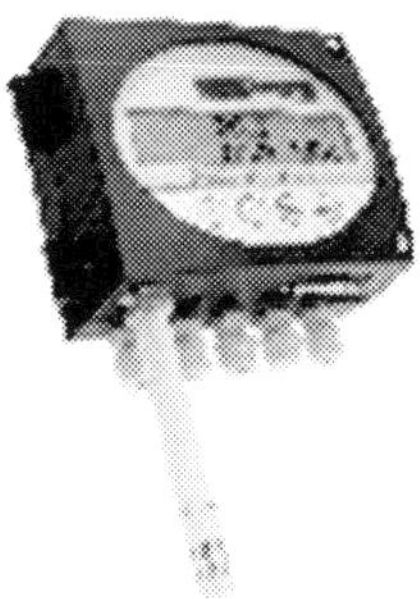

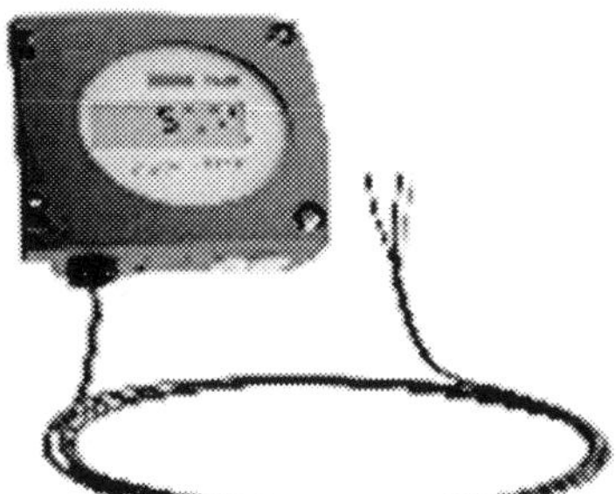

7 Durodyne Canada

7.1 The M-4000 Digital HVAC Analyser

One hand-held, easy to use instrument that can be used to measure: BTU, Air velocity, Temperature and CFM.

The M-4000 measures the humidity, temperature and air velocity with inputted inlet/outlet dimensions which analyses this information to calculate CFM and BTU's. Custom designed solid state circuitry ensures high accuracy.

Batch together all of the functions and features, and the HVAC analyser adds up to a fantastic investment.

The M-4000 helps contractors and engineers make intelligent decisions on balancing and/or maintaining HVAC systems.

- Hand-held, lightweight, simple to use
- Quick response < 5 secDigital-quick, easy readings
- Low Cost!

Applications:

The M-4000 series is a full, five function instrument which measures room conditions and the performance of HVAC equipment – heaters, blowers, furnaces, air conditioners, fume hoods, clean rooms, etc. Contractors, quality assurance staff, operating engineers and building inspectors now have an easy-to-use instrument which will quickly evaluate system design and equipment performance.

They can be used for a wide range of applications, from simple temperature, humidity and air velocity readings to complex multi-outlet computations of BTU's, system balancing or total outlet/output calculations. The HVAC Analyser is extremely easy to use when compared to most air velocity and CFM instruments. The M-4000 is a hand-held instrument and eliminates the need for cumbersome hoods or carrying multiple measurement instruments.

It comes with a high impact plastic carrying case and a Limited One Year Warranty

Specifications	
Instrument case:	High impact plastic
Display:	Large easy to read LCD (4 digit 3/4" h)
Touch Pad:	Large push buttons
Operating Voltage:	9 V DC (alkaline battery 9 V)
Battery Life:	Approx. 8 hours continuous use
Temperature Range:	32°F to 149°F (0°C to 65°C)
Temp. Accuracy:	±2°F
Temp. Response Time:	5 Sec
Humidity Range:	10% to 95%
Humidity Accuracy:	±3%
Humidity Response Time:	10 sec
Air Velocity Range:	1.0 to 49 ft/sec
Air Velocity Accuracy:	±3% or ± 1 digit
Velocity Response Time:	Instantaneous
Weight:	15 Ounces (430 G)
Dimensions:	
Digital Instrument:	8" x 3.5" x 2" (20.3 x 8.9 x 5 cm)
Probe (Sensor):	6.75" x 3" x 2" (17.1 x 7.6 x 5 cm

7.2 M 4000E, M4000ED – Digital HVAC Analyzer

An advanced measurement product with the measurement capabilities of temperature, humidity, air velocity, air volume, capacity; additionally dew point can be measured by Dew Point (M4000ED). The measurement parameters are in English system, i.e. temperature by ° F, Humidity by %RH, air velocity by Ft/sec, width / length by inch, air volume by CFM, power by BTU/H and dew point by ° F

1. Measuring Temperature

Fig. 1

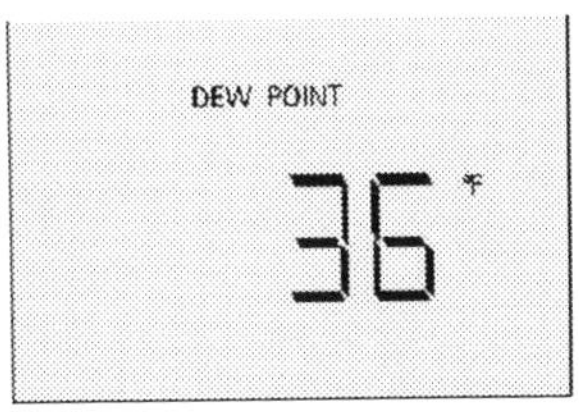

Fig. 2

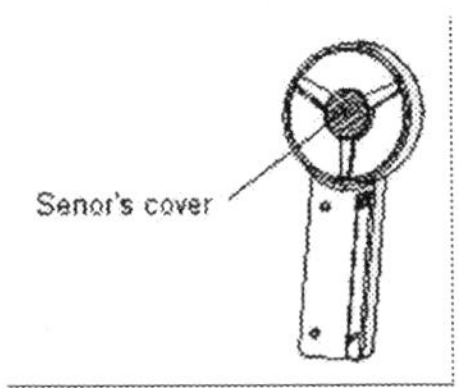

- Press the ON button (on the first time) or MODE button – TEMP will be displayed on LCD (Fig.1).
- Place the probe in front of unit outlet (or any other outlet) so that the sensors cover is facing the outlet.
- The LCD screen will display temperature level in° F. Press D to display Dew Point in ° F. (fig.2)

2. Measuring Humidity Level

1. Press the MODE button – HUMIDITY% will be displayed on LCD. (fig.3)
2. Place the probe in front of unit's outlet (or any other outlet) so that the sensor's cover is facing the outlet.
3. The LCD screen will display relative humidity level in %RH.

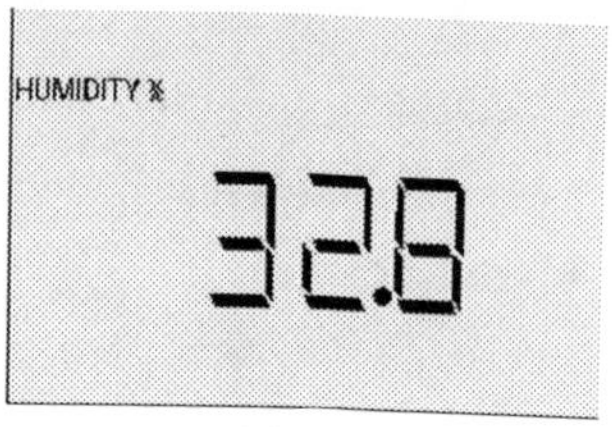

Fig. 3

3. Measuring Air velocity

3.1 Press the MODE button – VELOCITY will be displayed on LCD. (fig.4).

3.2 Place the probe in front of unit's outlet (or any other outlet) so that the sensor's cover is facing the outlet.

3.3 The LCD screen will display Air velocity in feet/second.

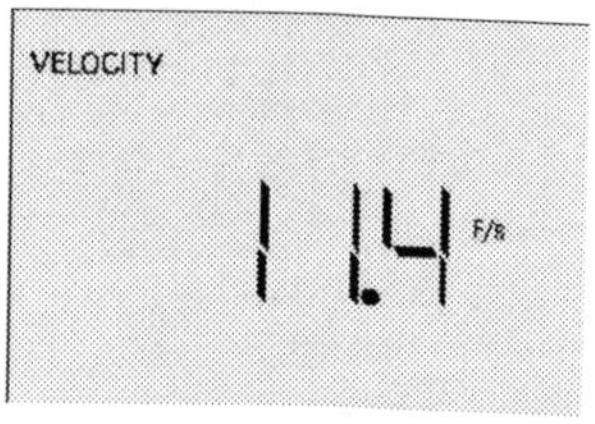

Fig. 4

4. Measuring Air Volume

4.1 Press the MODE button – AIR VOLUME and ENTER WIDTH (flashing) will be displayed on LCD.* (fig 5) The M4000 will request the A\C's outlet width.

4.2 Using Ñ D buttons enter the appropriate measurements in inch.

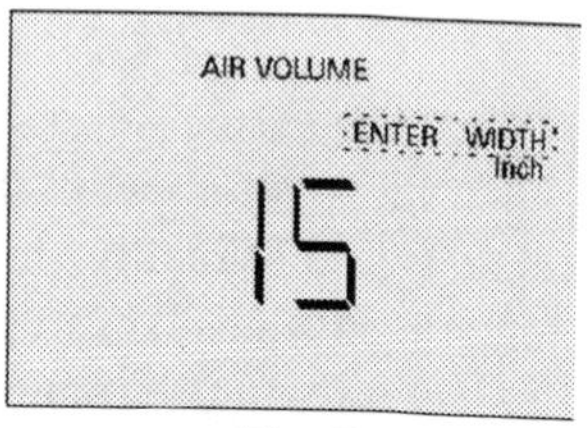

Fig. 5

4.3 Press ENTER to write width to memory. AIR VOLUME and ENTER LENGTH (flashing) will be displayed on LCD.* (fig 6)

4.4 The M4000 will request the A\C's outlet length.

Using Ñ D buttons enter the appropriate measurements in inch.

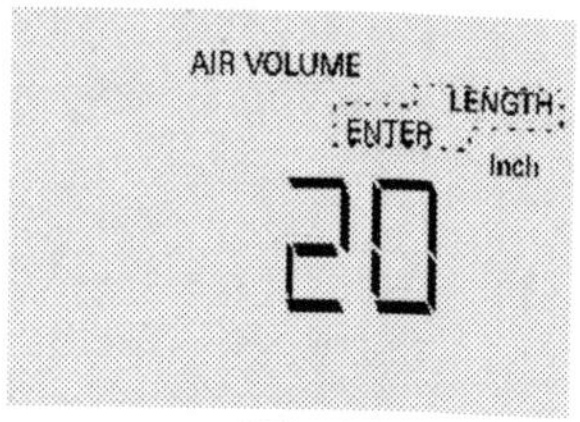

Fig. 6

4.5 Place the probe in front of the unit's outlet so that the sensors cover is facing the outlet.

☞ Press ENTER, the M4000 will start flashing 'WAIT'. 'WAIT' will flash for about 20 sec put the probe on the grill and when 'start average' start to flash then move the probe along the units outlet so as to cover the whole area, aim to get an average reading from all available openings*.

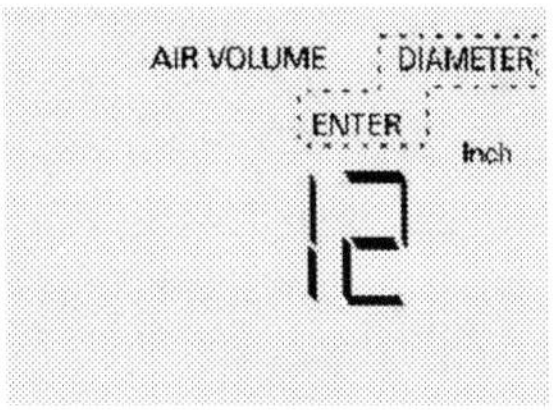

Fig. 7

☞ Try to cover all area in less than 10 sec (fig.8) and repeat again and again for 1 minute ('START AVERAGE' flashing fig.9).

4.6 If you wish to measure round grills – after step 4.1 press ENTER button more than 2 sec (long press) until DIAMETER will be displayed on LCD (fig. 7).

4.7 At the end of one minute the M4000 will beep signaling that measurement time ended. The LCD will display the measured Air Volume in CFM (fig. 10).

4.8 If you wish to redo this action. Or if you wish to measure few openings press ENTER and return steps 4.2–4.7.

4.9 Measure each opening as described up to now and after the last measuring press D and LCD will display the sum of all openings.

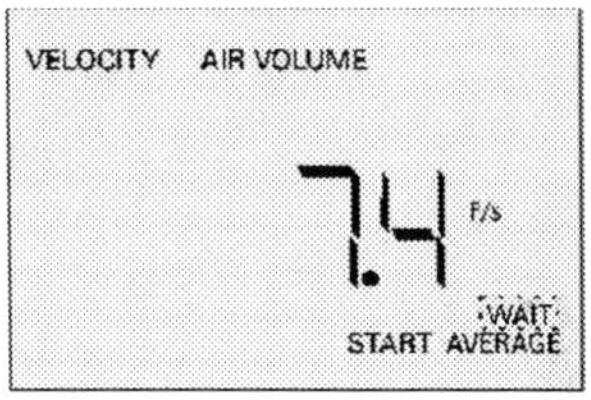

Fig. 8

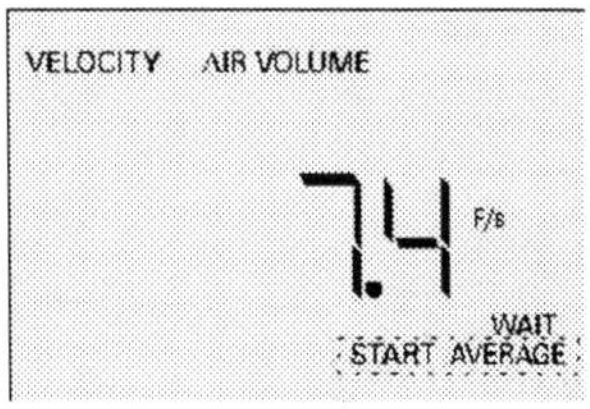

Fig. 9

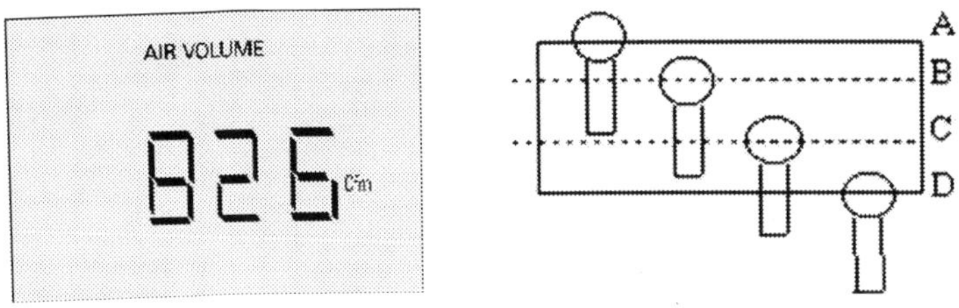

Fig.10

* When moving the probe along the grill's borders (A,D) keep the probe half out of the grill.

5. *Measuring Capacity*

Before you begin to measure switch with MODE to HUM mode. Put the probe in front of the inlet air (or ambient room humidity) and wait until the measure will stable (which promise that the Temp measure stabilised too). Do not put it at output grill – it will take long time to return to normal room temperature and humidity conditions.

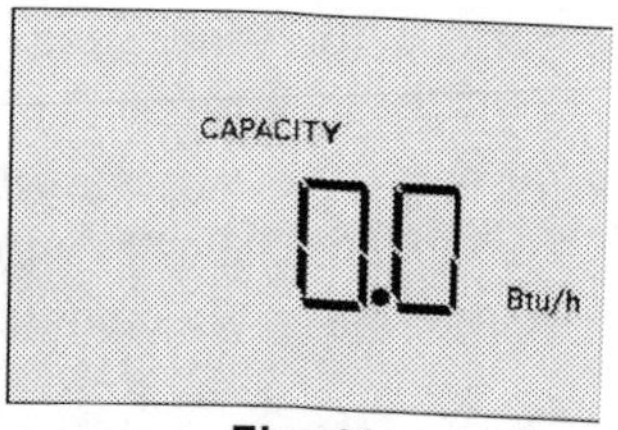

Fig. 11

5.1 Press MODE button – CAPACITY will be displayed on LCD (Fig. 11). (It is recommended to test at HIGH FAN SPEED.)

5.2 Align the probe in front of unit's inlet (or ambient room temperature) and press ENTER.

The LCD will display the HUMIDITY %, CAPACITY, Probe at RETURN AIR and "WAIT" flashing (Fig. 12) The M4000 will start measuring the unit's inlet air parameters while displaying a flashing 'WAIT'. After one minute of measuring, the M4000 will beep and the 'WAIT' will stop flashing a signal that this measurement has ended.

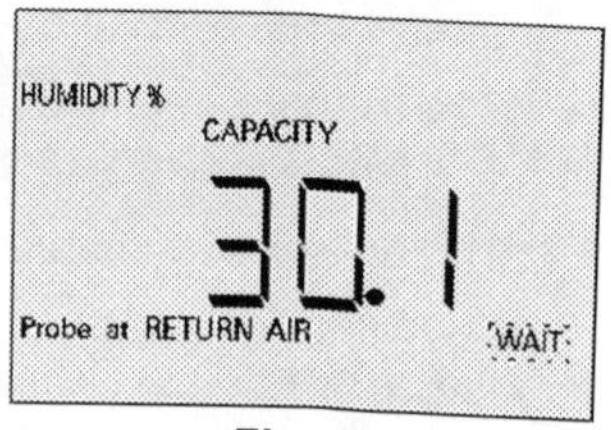

Fig. 12

5.3 ENTER WIDTH flashing will be displayed on LCD.(Fig. 13)

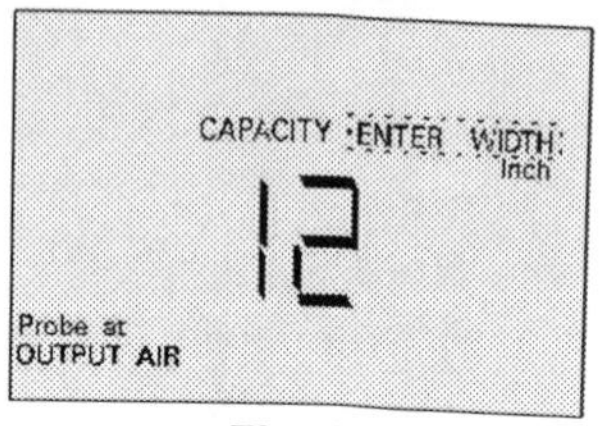

Fig. 13

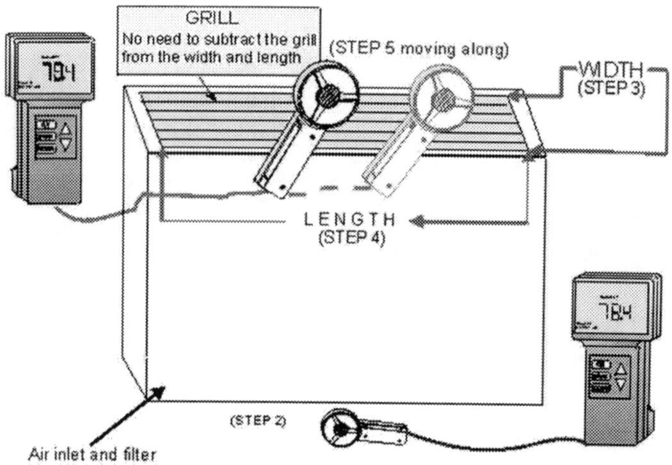

Fig. 14

When measuring return air the sensor box should be in the center (no need to move it during measuring)

5.4 Using the Ñ D buttons enter the outlets width (inch) and press ENTER (fig.14 step 3). ENTER LENGTH flashing will be displayed on LCD.

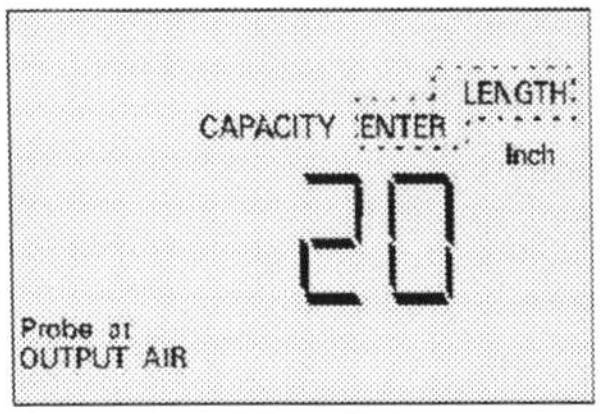

Fig. 15

5.5 The M4000 will request the A\Cs outlet length. Using the Ñ D buttons enter the outlets length (in inch) (fig.14 step 4, Fig. 15)

5.6 If you wish to measure round grills – after step 5.2 press ENTER button more than 2 sec (long press) until DIAMETER will be displayed on LCD. (Fig. 16)

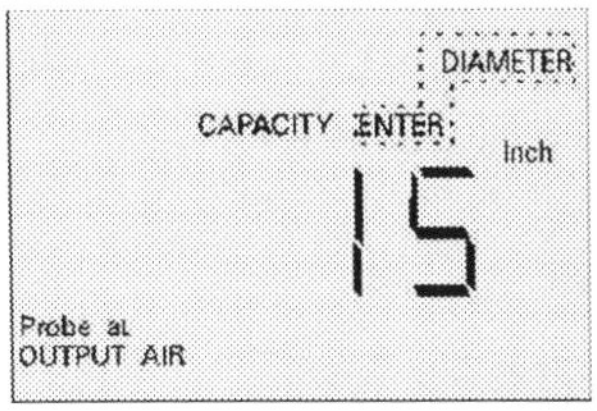

Fig. 16

5.7 Press ENTER "WAIT" will flash (Fig. 17) Place the probe in front of the units outlet. Pay attention that the sensors cover is facing the Grill no need to move it. The M4000 is waiting until sensors will be stable with the new temperature and hum. (about 90 sec) when 'START AVERAGE' will flash (Fig. 17) go to 5.8

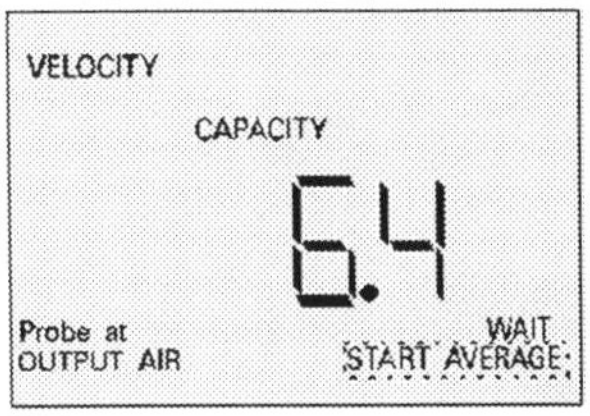

Fig. 17

Important: Read 4.6 (and *) before starting 5.8

5.8 Measure for one minute (Fig. 14 step 5). Make sure you move the probe along the unit's outlet so as to cover the whole area, aim to get an average reading from all available openings. At the end of this minute the M4000 will beep and the LCD screen will display the result in Btu/h (Fig. 18).

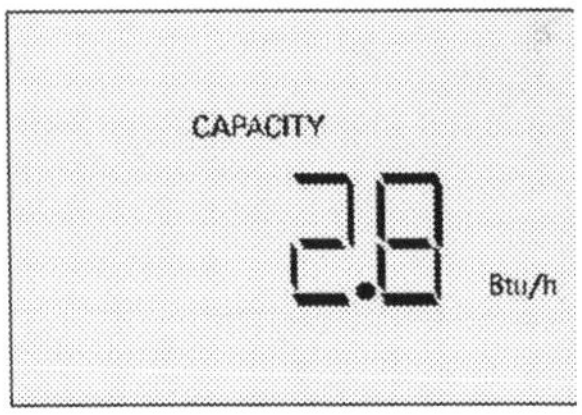

Fig. 18

Redo measuring outlet air: 5.9 In case you wish to redo this action (return air and the grill size are already in memory) press ENTER and repeat Sections 5.4–5.8 above.

Measuring few outlet's opening: 5.10 In case of several outputs openings but the same return air, press ENTER and repeat sections 5.4–5.8 for each one. In this case in section # 5.7 'WAIT' will flash for 20 sec. Each time a measurement ends the LCD will display the Btu/h for that specific outlet, if you wish to know the total Btu/h of all the outlets measured up to now press D.

If you wish to go over the all process again include return air you should wait for about 10 minutes and allow the humidity sensor to return to the humidity room or help it by home ventilator. Then start from section # 5.1.

8 Buyamag Inc. – USA

8.1 Humidity Control Humidistat THC-1

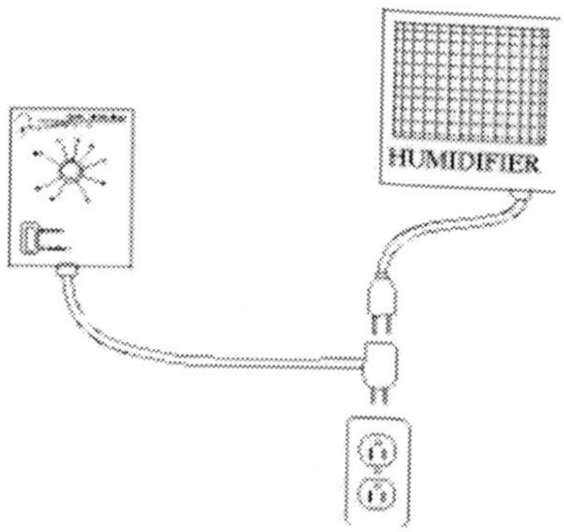

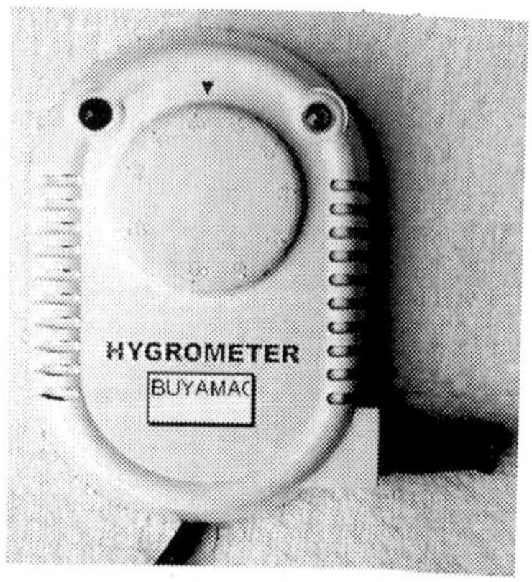

THC-1 Humidity Control is designed to control humidity automatically and accurately. Switch Mode is provided which can be switched between humidify and dehumidify to be the controlled within an enclosed area. The humidity switch controls Humidity Equipment by providing power to it, which can be set and adjust with adjustable setting Knob provided on Control. The Dehumidifying mode will activate fans or other dehumidifying apparatus until level drops approximately 5% and the dehumidifying equipment is disabled until humidity increases again 5%. In the Humidifying mode, this control operates Fogging or other Humidifying equipment by activating switches, motors, valves or pumps. Humidifiers can be precisely controlled according to atmospheric humidity to create ideal conditions for any environment. Sensor allows a 5% (differential) comfort zone between 'on' and 'off' functions. Operating Voltage is 120 V with 15 Amp maximum. This can also be used with many other types of humidity equipment or any other electrically triggered.

8.2 Humidity Control HCH-3

Easy and fast to install and can be connected to regular tap cold water source with water line, tube or hose. Just by plugging the misting fan humidifier power cord into standard three-prong outlet power source, the connection is complete. Air humidity can be automatically controlled by 'THC-1' Humidity Control Humidistat. It can be used for both outdoor and indoor misting fan controls by just hooking it up to any water hose, and the water travels through the hose and is jetted out of 4 misting heads on the misting hub on the front of the fan head. The holes are so small that the water pressure sprays out the water in a fine mist. This unit humidifies and cools large areas for much less expense than an air conditioner and conveniently connects to a standard hose for continuous misting operation cycle. This unique misting fan can reduce the surrounding temperatures by as much as 25 degrees.

Knob and Numbers are provided for desired humidity setting. Red and green lead light indicators provided for RH% maintenance in desired area. Automatic control of mist fog production in desired area with accuracy of ±5% is possible. One humidifier can be integrated with no more than 125V/10A, or 220V/5A for automatic area humidification

Technical Specifications: Humidity Range Control: 30–95% (25°C) with accuracy of **±5%,** *Temperature range*: 0– 60 °C, *No. of Humidifiers Controlled*: One, *Available span operation*: - 2% ~ 10%, *Operation method*: - Electronic (ANALOG), *Relay capacity*: 125 V/10A, 220 V/5A, *Consumption power*: 2.2 W Voltage: 110 V/60 H, 220 V/60 Hz, *Response Method*: High molecular resistant type, *Install Requirements*: 150 cm from the floor (special hole is provided in the back housing), *Weight NP-50*: --- 8Oz., *Size*: 104H x 100 W x 55B (MM) and *Length of Cord*: 1.5 m

9 H. Ikeuchi & Co. Ltd., Japan

9.1 Humidity controller for automatic operation of humidifiers

(Operating Range: 20% RH – 90% RH)

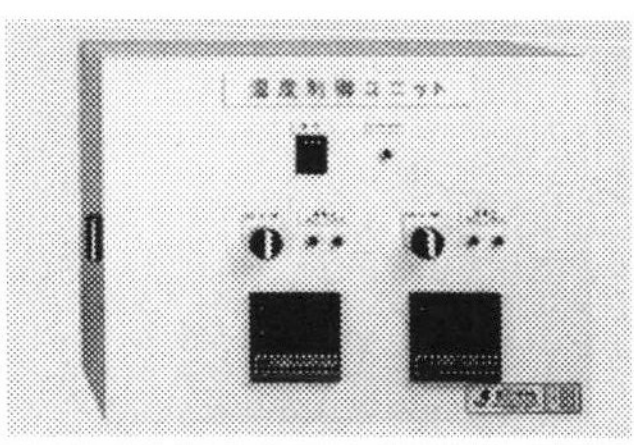

- **Compact Digital Type (RHC-021):** Digital Indicators display present and desired humidity, Compact size to fit any place. Tolerance ± 3%, Supply Voltage 100–240 V
- **Digital Type (RHC-D☆☆☆);** Digital Indicators display present and desired humidity, Multi-channel type available up to 4 channels. Tolerance ± 3%, 'xxx' shows the version of desired specification. (Photo shows the 2 channel type.)
- **Analog Type (RHC-011):** Economical analog type with rotating dial, Compact size to fit any place. Tolerance ± 5%, Supply Voltage 100—240 V

10 Honeywell – USA

10.1 HIH-3610 Series

The HIH-3610 Series humidity sensor is designed for high-volume OEM users. Direct input to a controller or other device is made possible by this sensor's linear voltage output. With a typical current draw of only 200 A, the HIH-3610 Series is ideally suited for low drain, battery operated systems. Tight sensor interchangeability reduces or eliminates OEM production calibration costs.

Individual sensor calibration data is available. The HIH-3610 Series delivers instrumentation-quality RH sensing performance in a low cost, solderable SIP (Single In-line Package). Available in two lead spacing configurations, the RH sensor is a laser trimmed thermoset polymer capacitive sensing element with on-chip integrated signal conditioning. The sensing element's multilayer construction provides excellent resistance to application hazards such as wetting, dust, dirt, oils, and common environmental chemicals.

10.2 Humidity Sensor HIH-3610 Series

- *Parameter Condition: (Relative Humidity): RH Accuracy*(1) ±2% RH, 0–100% RH non-condensing, 25°C, *V supply*=5 V DC, *Interchangeability* ±5% RH, 0–60% RH; ±8% @ 90% RH typical, *Linearity* ±0.5% RH typical, *Hysteresis* ±1.2% RH span maximum, *Repeatability* ±0.5% RH, Response Time, 1/e 15 sec in slowly moving air at 25°C and Stability ±1% RH typical at 50% RH in 5 years.
- *Power Requirements: Voltage Supply*: 4 VDC to 5.8 VDC, sensor calibrated at 5 VDC,

- *Current Supply*: 200 A at 5 VDC, *Voltage Output*: V supply = 5 VDC
- *Drive Limits: V out* = V supply (0.0062 (Sensor RH) + 0.16), typical @ 25°C (Data printout option provides a similar, but sensor specific, equation at 25°C.), 0.8 VDC to 3.9 VDC output @ 25°C typical, *Push/pull symmetric*; 50 A typical, 20 A minimum, 100 A maximum, *Turn-on* 0.1 sec
- *Temperature Compensation*: *Effect @ 0% RH*: True RH = (Sensor RH)/(1.093-0.0012T), T in °F., *Effect @ 100% RH*: True RH = (Sensor RH)/(1.0546-0.00216T), T in °C ±0.007 %RH/°C (negligible)., −0.22% RH/°C (<1% RH effect typical in occupied space systems above 15 °C (59 °F))
- *Humidity Range*: *Operating*: 0 to 100% RH, non-condensing., *Storage*: 0 to 90% RH, non-condensing
- *Temperature Range*: *Operating*: −40°C to 85°C (−40°F to 185°F)., *Storage*: −51°C to 125°C (−60 °F to 257°F)
- *Package*: Two numbers Three pin, solderable SIP in molded thermoset plastic housing with thermoplastic cover. Handling Static sensitive diode protected to 15 kV maximum
- Notes:
- Extended exposure to 90% RH causes a reversible shift of 3% RH.
- This sensor is light sensitive. For best results, shield the sensor from bright light.

11 Jackson Systems. Indianapolis, USA

11.1 White-Rodgers Programmable 4H/2C Thermostat with Humidity Control

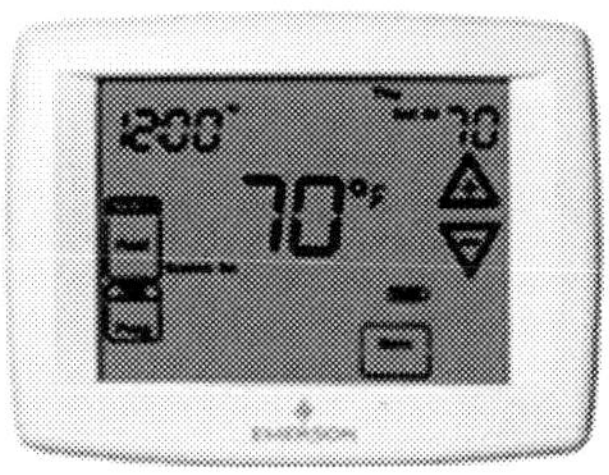

The Big Blue™ Universal Humidity Thermostat is designed to deliver a new level of comfort. It features the largest display on a touchscreen operated thermostat, as well as simple and logical installer setup and programming. Features like variable keypad security to prevent tampering, and temporary and vacation overrides, add convenience.

These thermostats offer power stealing assist for maximum battery life, remote sensing, as well as the exclusive Cool Savings™ feature providing the ultimate in comfort control and energy savings. The ±1° temperature accuracy will keep your customers comfortable and protect their HVAC system. Emerson® Blue is the result of smart, innovative thinking

11.2 Humidistats

Provides automatic control of humidification equipment for optimum humidity control

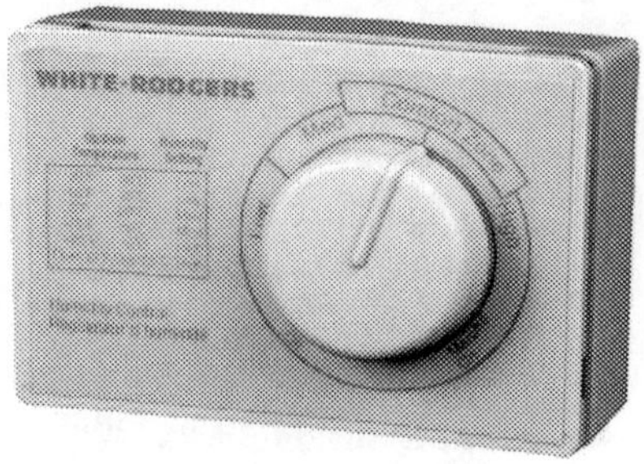

FEATURES

- Choice of horizontal or vertical mounting.
- Line voltage or low voltage.
- Snap-Action switch.
- Mounts on 2” x 3” single gang box.
- Pigtail leads.

Dimensions: Horizontal 23/4”H x 4 /2” W x 27/ 6”D., Vertical 4 /2”H x 23/4” W x 27/ 6”D

2271-100

Model Number	Description:	Switch Action	Range	Differential	Contact Ratings
2271-100	Humidistat (Horizontal)	Open on Rise	10% to 60% RH	5% RH	120 VAC, .5A, Pilot Duty, 30 VAC, 60 VA

11.3 Dehumidistats

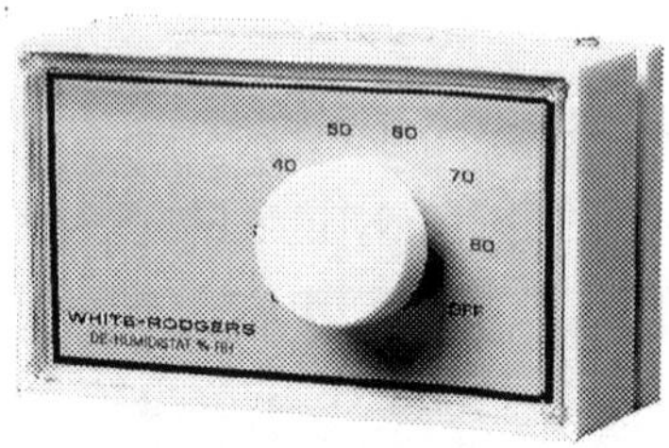

Provides automatic control of dehumidification equipment or operates damper/exhaust fans to control humidity levels

FEATURES: White color case. Choice of horizontal or vertical styling, Line voltage or low voltage, Snap-Action switch, Mounts on 2” x 3” single gang box and Pigtail leads.

Model Number	Description	Switch Action	Range	Differential	Contact Ratings
2273W-21	Dehumidistat (Horizontal)	Close on Rise	30% to 80% RH	5% RH	120 VAC, 1.0A Pilot Duty 30 VAC, 60 VA
2274W-21	Dehumidistat (Vertical)	Close on Rise	30% to 80% RH	5% RH	120 VAC, 1.0A Pilot Duty 30 VAC, 60 VA

Dimensions: Horizontal 23/4”H x 4 /2” W x 27/ 6”D., Vertical 4 /2”H x 23/4” W x 27/ 6”D

11.4 Relative Humidity Transmitters both Analog and Digital

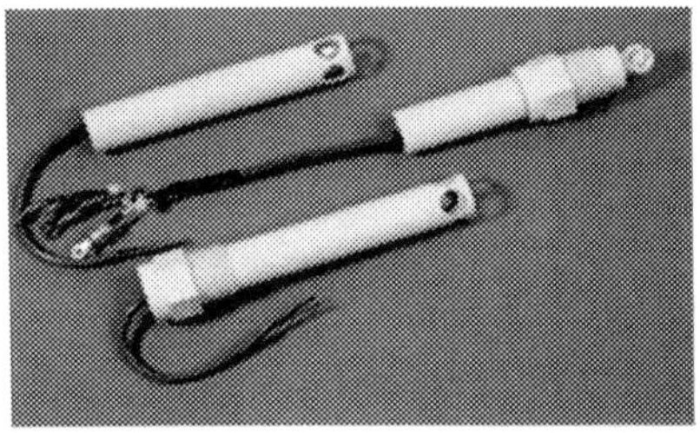

Features: Rugged epoxy encapsulation
Voltage, current, or digital output
Optional dust filter
Individually point calibrated
Combination humidity, temperature transmitters available
Threaded or wall box mount
High quality at low cost

General Specifications: *Power:* 4–20 mA loop (2 wire), 7– 24 VDC (3 wire)., *Ambient operating range*: 0 to 70 C., *Storage Temperature:* −25°C to 85°C., *Temperature coefficient* 0.1% RH/C., *Linearity* ±2% RH., *Operating range (non-condensing)* 5–98% *Time constant (moving air) 20 sec* – 2 minutes

Output Ranges: 4–20 ma, 0–1 V, 0–5 V, 0–10 V, Digital RS 232. Special ranges and custom calibration available

12 Omega Engineering Inc. USA

12.1 iTHX-2 – Temperature + Humidity Industrial iServer Micro Server

View Temperature + Humidity with a Web Browser: The NEWPORT iTHX transmitter lets monitoring and recording temperature, relative humidity and dew point over an Ethernet network or the Internet with no special software except a Web browser. It serves Active Web Pages to display real time readings, display charts of temperature and humidity, or log data in standard data formats for use in a spreadsheet or data acquisition program such as Excel or Visual Basic.

- Virtual Chart Recorder
- Web Server
- Temperature
- Relative Humidity
- Dew Point
- Accurate Readings
- Email Alarms
- Data Logging
- No Special Software Required

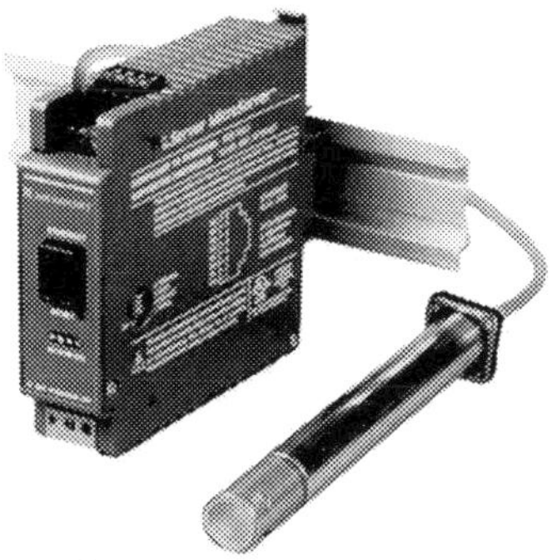

Display and Chart Two Channels: The iTHX transmitters come complete with a temperature and humidity probe for measurement of a single location. With the addition of a second probe, the iTHX transmitter can measure and display humidity, temperature and dew point in a second location up to 20 feet away. The transmitter can display and chart absolute measurements in both locations, or a differential measurement between the two locations. The second probe requires no change to the basic iTHX transmitter hardware, firmware or software. A second probe can be added at the time of purchase or in the future. NEWPORT offers a choice of industrial probes in 2" and 5" lengths and a wand style for ambient indoor applications.

Probe Options: Select *nothing (leave field blank)* for Standard 5" Industrial Probe and 20' Cable with Stripped Wire Leads, −2 for Industrial 2" Probe and 3' Cable with Stripped Wire Leads (substitution for 5" Probe)

Replaces THDX & KTX models!

12.2 Temperature/Relative Humidity Chart Recorder CTH200 Series

The CTH200 thermo-hygrograph is an economical, self-contained recorder. The user can set the chart speed at one revolution for every 1 day, 8 days or 32 days. Simultaneous temperature and humidity readings are plotted on the easy-to-read, 165 mm (6.5"), circular chart.

Designed for wall-mounted operation in a vertical, upright position, the CTH200 comes with easily replaceable charts and recording pens. The CTH200 uses the same pens and chart paper as the CTH100 Series. Simultaneous view of both temperature and RH. Recordings are possible. Neat appearance and compact size; good for various applications. Easy, convenient chart replacement. Variable recording times: 1 day, 8 days or 32 days. Measurements in °F or °C

12.3 Thermohygrograph – Single Height

- Humidity section: Range: 0 to 100%, accuracy: ± 3% between 20% and 80%, sensing element: specially treated human hair.
- Recording section: Either battery operated quartz clock with simple mechanism for changing rotation from daily to weekly or Clockwork clock – Daily or weekly rotation Recording is effected by easily changed fibre tipped pen.

☞ Temperature section: Range either 0 to 50°C or −5 to 55°C, accuracy : ±0.5°C, sensing element: bi-metallic coil (corrosion protected)

General Description: Mechanism machined & fabricated from high quality brass, polished and lacquered. The unit is supplied with fabricated, vented sheet steel cover finished in white enamel with convenient carrying handle.

Fairmount Single Height Thermohygrograph

This compact single height instrument neatly combines both temperature and relative humidity records on one standard height chart thereby allowing installation inside a standard size Stevenson Screen if required.

12.4 Thermohygrograph (Instrument #4)

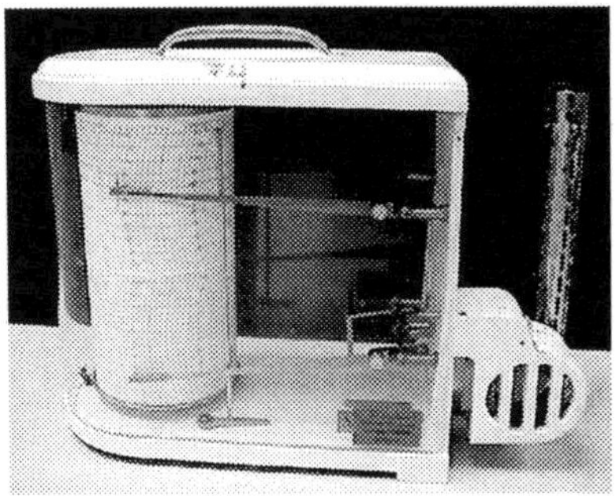

The thermo-hygrograph records RH% and temperature continuously for a given time period. A chart marked off into hours and days is fastened to the rotating drum. The drum is regulated by a timing device so that the pen tips move horizontally one hour on the chart for each hour that actually passes. The top pen responds vertically to changes in temperature. The bottom pen responds to changes in humidity. Therefore, temperature is plotted on the top graph and RH% on the bottom graph of the chart. This was done continuously for one week as shown on the sample chart.

To read the two graphs on the chart:

1. The small numbers across the top indicate time. For example, 12=noon, 18=6:00pm, 24=12:00 midnight.
2. Find the desired time and day. Follow the orange curved line down to its intersection with the %RH or temperature curve, and then follow the orange horizontal line to read the corresponding %RH or temperature.

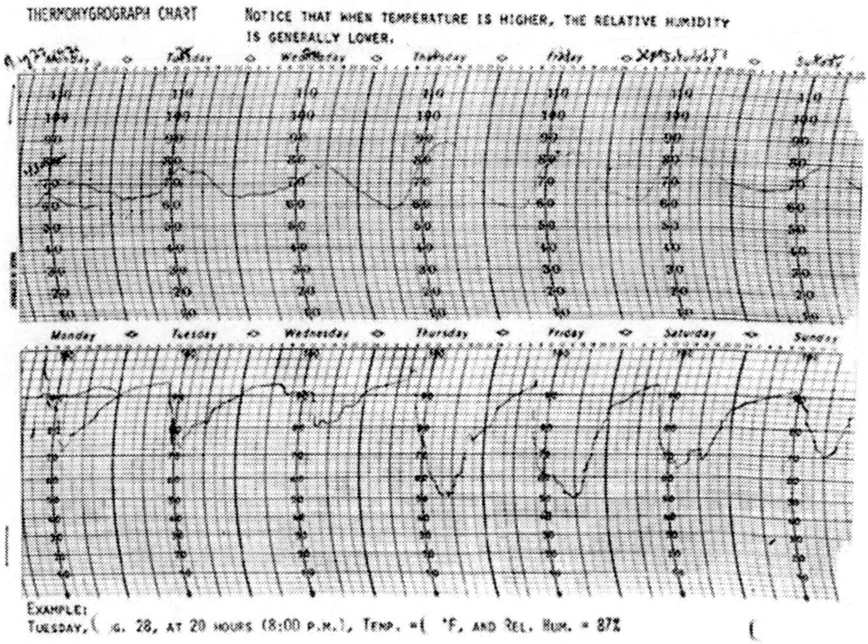

12.5 Sling Psychrometer (Instrument #5)

Using the Sling Psychrometer we can determine RH%, dew point temperature and current temperature.

Principles: The Sling Psychrometer determines RH% by measuring the evaporation of water into the surrounding air. Two thermometers are placed in flowing air, one thermometer bulb being covered by a wet wick. Evaporation from the wetted bulb causes its temperature to be depressed relative to that of the dry bulb. In dry air, evaporation will be rapid and the depression will be greater, giving a low RH% reading. In humid air, there will be little evaporation and the depression will be small, giving a high RH% reading. If the air is completely saturated, no evaporation can take place, so both the wet and dry bulb thermometers will give the same reading. This equates to 100% RH. The RH% can be read off the slide rule calculator integrated into this model of the Sling Psychrometer. An independent 'slide rule' calculator and Standard Humidity Tables can also be used.

Operation: The Sling Psychrometer consists of three main parts: 1) the inner frame with wet and dry bulb thermometers, 2) the wet and dry bulb scale on the reverse side of the inner frame and 3) the outer case with the relative humidity scale.

Taking a Reading: Open the instrument by withdrawing the inner frame from the case. Thoroughly wet the wick using the dispensing water bottle provided, or immersing it in water for about 30 sec. Be sure that both the exposed wick and that portion coiled in the wick container are thoroughly saturated. (Hot water must not be used as it may damage the thermometers.) The wick will remain moist for several hours. Always check that the wick is wetted before taking any readings and ensure no moisture remains on the dry bulb. To take a reading, set the Psychrometer at right angles, and, holding the case, rotate the frame for 30–60 sec at between 2 and 3 revolutions per second. Stop revolving the instrument and note the wet and dry bulb temperatures. Wet-bulb temperature should be read first and as quickly as possible for highest accuracy. Delay in reading may cause error. Close the instrument and use the slide rule calculator to determine the %RH.

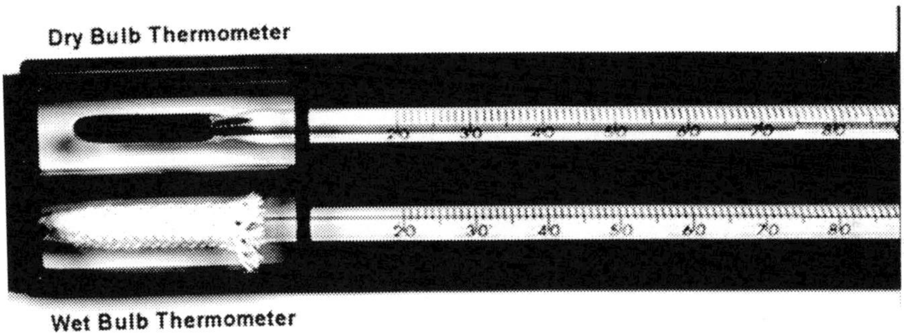

Using the Slide Rule Calculator: Slide the inner case in or out until the wet- and dry-bulb temperatures you recorded are in line with each other (similar to a slide rule). The calculator has two scales: the upper scale should be used for dry bulb temperatures below 70°F and the lower scale should be used for dry-bulb temperatures above 70°F. Near 70°C either scale can be used.

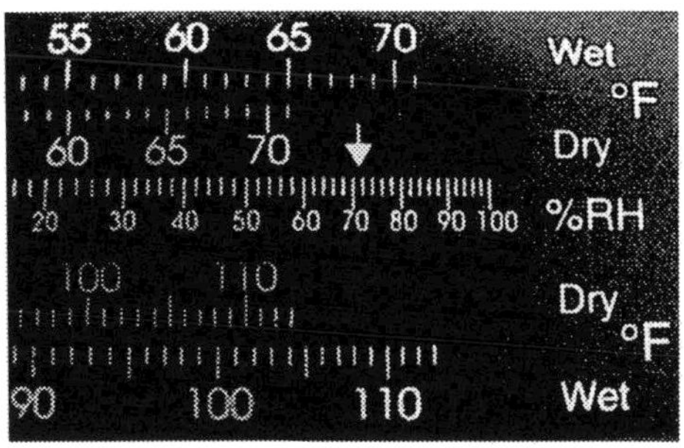

If the wet bulb reading is 62°F and the dry bulb reading is 68°F, the RH% shown by the white arrow is 69–70%.

13 Dickson USA

13.1 Temperature and Humidity Monitoring Instruments

Pharmaceutical, life science and bio-tech companies have long been familiar with Dickson's THDX and KTX model chart recorders. Now, Dickson has released two new designs (TH8 and KT8) which incorporate value, performance and accuracy; all things expected from the instrumentation for a lab or plant. The new TH8 line of 8-inch Temperature/Humidity Chart Recorders and the new KT8 line of K-Thermocouple Remote Sensing 8-inch Chart Recorders have a number of added features designed to make them newly practical in pharmaceutical applications – from larger charts for readability, to smaller footprint designs, to higher quality metal components, and more.

There are three different models of Dickson TH8 8" Temperature/Humidity Chart Recorders and five different models of KT8 K-Thermocouple Remote Sensing Chart Recorders, each optimised for different humidity and/or temperature ranges. Users are able to interchange 24-hour, 7-day or 31-day charts as desired, opt to display temperature or humidity, and specify preferred traceable calibration methods.

Benefits of TH8 and KT8

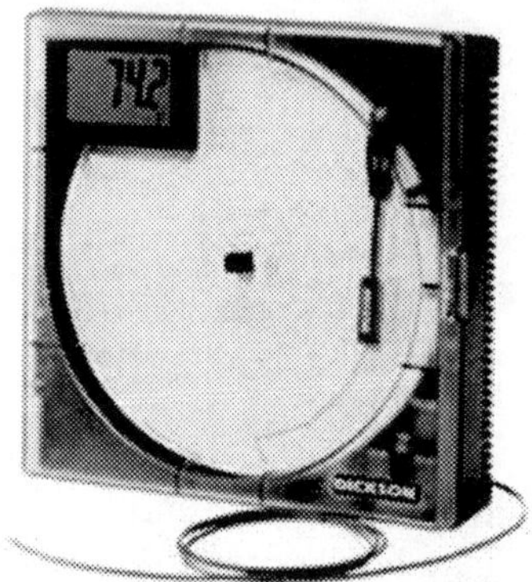

KT8

- ☞ Large 8" charts providing 84% more readable area than 6-inch designs with large digital displays viewable from 20-foot distances. Remote probes ideal for tight locations, clean rooms or incubators.
- ☞ Audio and visual alarms
- ☞ Compact and durable enclosure that is 41% smaller than previous models
- ☞ Dew point recording and display
- ☞ Flip-up pen arm for easy chart and pen changes
- ☞ Standardised and easily replaced k-thermocouple remote probes (in the KT8 models)

This new chart recorder replaces the older Dickson THDX and KTX models with the same functionality but now these new TH8 and KT8 recorders have even more features.

14 Canalair Service S.R.L Italy

14.1 The Supervision System

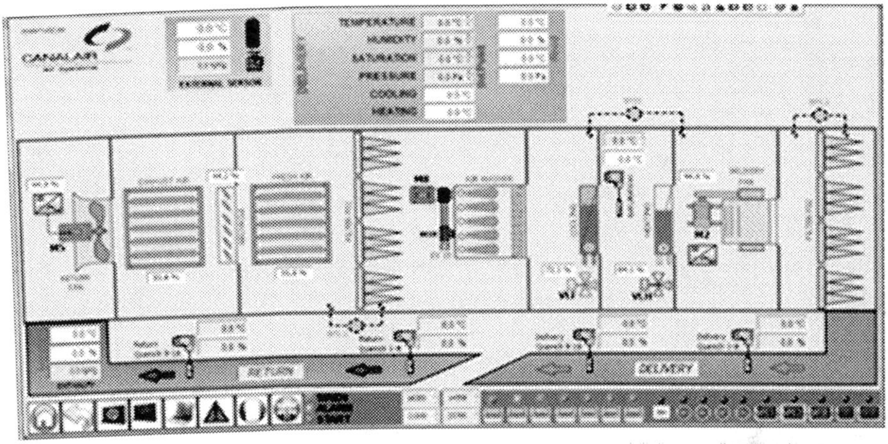

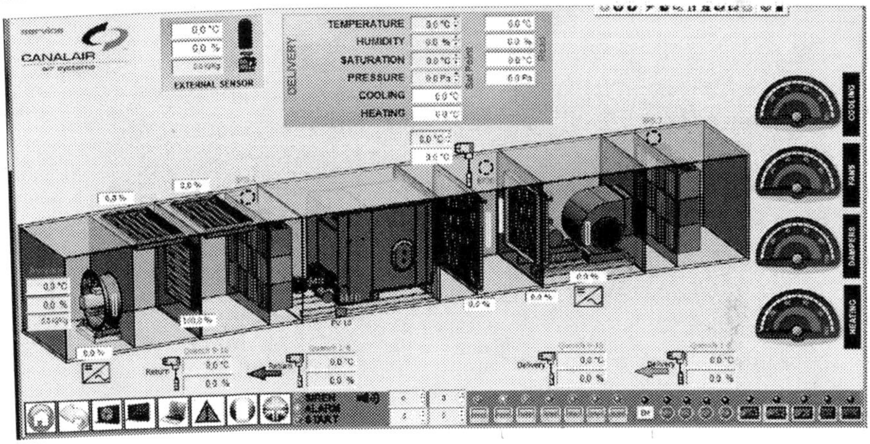

The supervision system by Canalair Service Srl ensures the high precision in managing the events related to the air treatment and conditioning plants functionality. The operator has available in real time all the needed information related to the plant situation, having the opportunity to intervene quickly, in order to minimise the production stops and consequently improve the plant efficiency. The graphics supplied by our system are immediate and intuitive and represent also and historical archive of the plant activities and functionality.

15 Carel Industries, Padova, Italy

15.1 Air-Handling Unit Control Solutions

- CAREL: Leaders for more than 30 years in control and humidification systems for HVAC/R applications. Innovation, experience and technological development applied to air-handling units (AHUs)
- Air-handling unit applications feature a very wide variety of system types, with differing degrees of specificity. CAREL can supply solutions for all installations, whatever their complexity: modular controllers, a wide range of humidifiers and accessories to ensure increased integration, up to complete system supervision so as to ensure total reliability and maximum energy saving. The system approach is combined with flexibility, both in relation to hardware, i.e. architecture and components, and software, starting with the standard application dedicated to AHUs, completely customisable using 1tool.

15.1.1 pCO Sistema+:

- Energy saving controllers, accessories and management algorithms designed to achieve maximum energy saving Fieldbus System solution based on serial integration of devices: optimised management of a higher quantity of information connectivity programmable controllers with plug-in options for communication with the most commonly-used BMS systems. The advantages of an integrated solution for air-handling unit control, based on programmable controllers and a complete range of accessories.
- Configurability: New I/O chip technology means up to 10 channels can be configured as digital/ analogue inputs or analogue outputs, so as to connect all types of probes and smart actuators for AHUs (inverters, dampers, valves, humidifiers, etc.).
- Programmability: The 1tool development environment allows complete software customisation: control logic, display of parameters and serial communication. pCO5+ can be programmed directly from a PC, using a 'plug & play' key or via USB (on models where available).
- User interfaces: A complete range of programmable displays for all performance, price and appearance requirements. The high-tech and attractive touchscreen terminal can be used to create intuitive interfaces with complete information, ideal for integrating complex units such as AHUs.

- Technology: A new microprocessor guarantees high program execution speed, effectively controlling even faster transients. Trends in different variables or events (e.g. alarms) can be saved for long periods.
- Communication: pCO Sistema+ offers up to 7 communication ports to interface via the most commonly-used protocols. All pCO Sistema+ components can be connected to pLAN or BMS networks, for remote system management. Fieldbus can be used to control smart actuators, i.e. devices featuring serial communication.
- Backward compatibility: The new pCO5+ controller guarantees backward compatibility with the entire range of pCO3 and pCO5 controllers, both regarding the software and hardware, thus protecting and guaranteeing the investments and know-how of CAREL customers.

15.1.2 Integrated Solutions for Energy Saving

- Increasingly advanced and complete solutions to meet the demands of manufacturers and installers. Maximum performance and compatibility.
- Temperature and humidity probes: CAREL offers a complete range of temperature and humidity sensors for different uses: probes for residential or industrial environments, duct-mounted and immersion, all featuring special care in the design and the possibility of RS485 communication via CAREL and Modbus protocols.
- Air quality probes: VOC, CO_2, CO_2+VOC: These analyse air quality and are ideal for ventilation and air-handling systems in residential and commercial environments. Main functions: air quality measurement; analysis of contamination by polluting gases; sensitivity threshold settings and efficient management of room ventilation.
- Air or gas flow switch: These measure absence or insufficient flow rate in ducting, activating a switch.
- Differential pressure transducer: Differential pressure transducers use a sensor to provide a voltage or current signal, calibrated and compensated for different temperatures. These are ideal for measuring low pressure values in air-conditioning systems, rooms, laboratories and cleanrooms (non-corrosive atmospheres)
- Differential pressure switch: Ideal for control and safety applications, signalling when fans stop or filters are blocked.
- Inverter: Inverters represent the maximum in efficiency, as they modulate unit operation and consequently optimise energy consumption. The wide operating range, easy installation and high level of protection against electromagnetic disturbance mean these devices can be easily integrated into the system.

15.1.3 Evaporative Cooling: Energy Saving and Lower System Capacity

- In an AHU, in certain circumstances, the supply air can be cooled by humidifying it adiabatically (Direct Evaporative Cooling); the same is also true for the exhaust air, which can be cooled without limits in terms of humidity. This in turn lowers the temperature of the incoming outside air, via a heat exchanger, reducing power consumption and maximum cooling capacity.
- Electrical panels: CAREL can supply complete electrical panels for AHU control, integrating the power and control electronics, using high-quality components and developing solutions in compliance with system requirements and standards in force.
- Frost protection thermostat: This protects heat exchangers (evaporator coils) by managing electric heaters, making sure the temperature never falls below the preset safety value.
- Evaporative cooling: Air can be effectively cooled by exploiting the evaporation of atomised water: the change in state absorbs energy from the air, which is cooled. Evaporation of 100 kg/h of water absorbs 69 kW of heat, for power consumption of less than 1 kW.
- Adiabatic humidifiers: Adiabatic humidifiers exploit the property of atomised water to provide significant advantages in terms of accuracy and efficiency in temperature and humidity control. Precise and continuous modulation of humidification capacity and very low energy consumption make Humifog and optimist ideal solutions for application in air-handling units, in both cooling and heating operating modes.

15.1.4 Standard AHU Control Application

- The standard application guarantees maximum adaptability to every type of air-handling unit.
- This software solution incorporates all of CAREL's know how in the management of air-handling units. The type of AHU to be controlled is not based on selecting pre-configured units, but rather the actual devices installed on the AHU (ON/OFF or modulating devices: dampers, coils, humidifiers, fans, etc.) and then setting their parameters. The position of each input and output can be chosen, then making changes on site during installation. Configuration is made simple by an extremely intuitive and customisable user interface, set out into menus divided by macro-area: only the required parameters relating to the functions that have been enabled are shown on the screens.

- Full connectivity
  - Up to 6 serial probes over RS485 Modbus, Integration of fan inverters via serial, BMS options for interfacing via the most common supervision protocols, Integrated MP Bus® solution for controlling Belimo serial actuators.
- Management of main AHU functions
  - Various models of supply and return air fans, ON OFF and modulating; control of outside air, mixing and exhaust dampers; control of cooling and heating operation using PID algorithms with coil management; free cooling/free heating based on temperature and enthalpy; management of heat recovery units; cascade control of different devices; humidity control with dehumidification and isothermal and adiabatic humidification; alarm management with log; management of up to 4 time bands per day, plus management of days of the week and special days; 4 generic control loops available for additional functions.
- Custom versions: Using 1tool, the development environment for CAREL programmable controllers, customers have a library of standard modules available that can be fully configured to optimise control of the main system components or integrate different devices.

15.1.5 *Solutions for Connectivity and Communication*

- pCO Sistema: CAREL's experience in supervision and integration. Information management as the basis for energy saving.
- Up to 7 communication ports available with the most commonly used protocols in HVAC/R applications, for complete system management: pLAN, BMS1, BMS2, FieldBus1, FieldBus2, USB host, USB device.
- pLAN: One multi-master RS485 communication port to connect terminals and manage distributed intelligence
- BMS: One RS485 BMS port, as standard, also available in the optically isolated version, configurable with CAREL and Modbus® RTU protocols. A vast range of optional cards can also be installed during commissioning, to interface with the most commonlynused communication standards: Modbus® (RTU & TCP/IP), BACnet™ (MS/TP & IP), SNMP, LonWorks®, Konnex® and Johnson METASYS®.
- Fieldbus: One RS485 FB port, as standard, also available in the optically isolated version, configurable with CAREL and Modbus® RTU protocols. This can be used to manage smart actuators such as inverters, EC fans and variable flow pumps, without the cost of an additional card.

15.1.6 Solutions for System Monitoring and Supervision

Instruments for centralised remote system management. Supervision for efficiency

Completing the solutions for AHUs, a global monitoring and supervision solutions, ensures synergic management of the entire installation. Easy to use, complete control over the systems managed, sophisticated configuration for alarm notification and analysis tools are all features that make CAREL supervisors a winning solution in terms of reliability, security of information and energy saving.

- tERA: The service ERA A cutting-edge proposal for information management and optimizing the performance of complete systems. Reduced maintenance costs and improved customer service, thanks to remote analysis. The ready-to-use solution for your service center: security and flexibility in providing your customers service... with just a few simple clicks!
- Global monitoring: CAREL can propose the best possible system to suit specific needs. The embedded plug & play solution and tailor-made software reduce installation and configuration times. The data are saved in a specific database, guaranteeing integrity and reliability of the information managed.
- Plant Watch PRO: Compact embedded solution for small systems with max. 30 devices.
- Plant Visor PRO: Embedded solution for medium/large installations up to max. 300 devices.

Appendix – 8

Dehumidifiers

1 Bry-Air (Asia) Pvt. Ltd. Gurgaon - India

1.1 Desiccant Dehumidifier

This is compact custom designed environmental control equipment. It is claimed to solve problem of any size relating to humidity by the Bry-Air Dehumidifiers. They offer the simple and economical solution to humidity control which can maintain RH as low as 1% or even lower at a constant level, regardless of ambient condition during production, processing, storage and packaging.

FFB Dehumidifier

Largest capacity range in compacts – from 170 to 2000 cmh (100 to 1177 cfm). The new series Bry-Air FFB, 170 through 2000 cmh (100 to 1177 cfm) desiccant dehumidifier line is designed to integrate 'small footprint and finish' of a commercial unit with 'the ruggedness' of an industrial dehumidifier.

Range : 170 to 2000 cmh (100 to 1177 cfm) CE **Certified - FFB (Fluted Flat Bed) Dehumidifiers**

The BRY-AIR® FFB Compact Dehumidifier is CNC fabricated with powder coated finish and incorporates a high performance fluted metal silicate desiccant synthesized rotor which allows for:

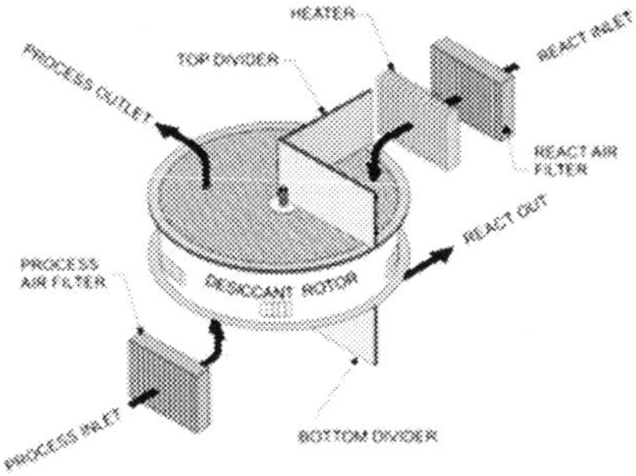

- Reduced foot print.
- Increased performance.
- Lowest volume/weight per CMH.
- Innumerable mounting options.
- Light weight and can be installed above the false ceiling eliminating the need for floor space.

This removes moisture through a process of continuous 'physical adsorption'. The moisture is adsorbed in the dehumidification sector by the fluted, desiccant synthesised rotor and is exhausted in the reactivation sector by a stream of hot air in a counter flow.

Model	Capacity
FFB 170	170 cmh
FFB 300	300 cmh
FFB 600	600 cmh
FFB 1000	1000 cmh
FFB 1500	1500 cmh
FFB 2000	2000 cmh

Innumerable mounting options: Floor, ceiling, wall, table and trolley. The dehumidifier can be mounted on top of false ceiling also.

FVB Dehumidifiers

The BRY-AIR FVB (Fluted Vertical Bed) is the latest in desiccant dehumidifier technology. It incorporates a custom-made fluted metal silicate desiccant synthesisedrotor which ensures highest moisture removal at lowest energy cost. It has a small footprint, thus allowing for space saving, and is available in two ranges FVB senior and FVB junior with capacities ranging from 1500 cmh (880 cfm) to 7600 cmh (4473 cfm).

FVB JUNIOR		FVB SENIOR	
Model	Capacity	Model	Capacity
FVB 1500	1500 cmh	FVB 5000	5000 cmh
FVB 2100	2100 cmh	FVB 6200	6200 cmh
FVB 3000	3000 cmh	FVB 7600	7600 cmh
FVB 4000	4000 cmh		

Features

- Lowest volume/weight per cmh
- Capacity: 1500 cmh (880 cfm) to 7600 cmh (4473 cfm)
- Self-contained unit, CNC manufactured with powder coated finish
- Can be packaged with pre/after cooling, bypass filter elements
- Highest moisture removal efficiency at lowest energy cost
- Standard inbuilt heat recovery
- Easy to install andmaintain
- Can maintain RH as low as 1% or even lower, regardless of ambient conditions
- With high performance, metal silicate fluted desiccant synthesisedrotor
- A totally self-contained unit
- Can be packaged with pre/after cooling, bypass ducting andmixing chambers, volume controls, filter elements, etc. for complete environment control

How does it work?

Range : 1500 cmh (880 cfm) to 7600 cmh (4473 cfm)New FVB (Fluted Vertical bed) Dehumidifiers

The moisture-laden air enters through the process inlet and passes through the desiccant rotor, where the moisture is absorbed by the desiccant. The dry, dehumidified air is delivered to conditioned area through the process outlet.

The rotor then rotates to the reactivation sector and the moisture is removed from the desiccant rotor, by blowing hot air through the reactivation outlet to the atmosphere. The hot desiccant rotor is cooled by a small portion of the process air before rotating back to the process sector. Thus, the adsorption process is continuous.

CNC fabricated with powder coated finish, the Bry-Air FVB can be packaged with

* Pre/after cooling
* by pass ducting and
* mixing chambers

Volume controls, filter elements, etc., are provided for complete environment control in a single package for any application.

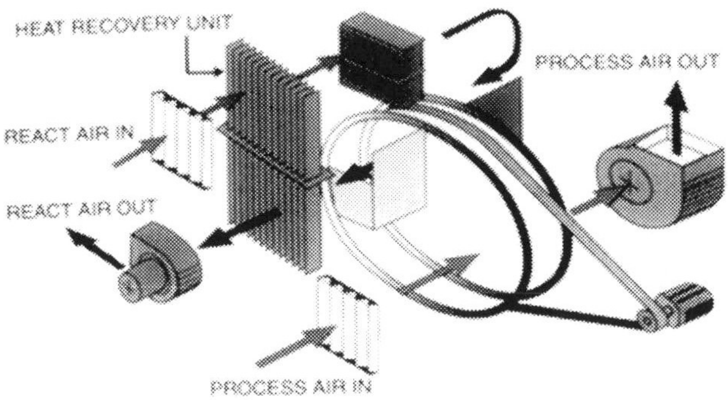

The EcoDry Dehumidifying Rotors are the result of many years of in-depth study and research on countless permutations and combinations of substrates, desiccants and chemical processes. Every parameter has been considered to maximiseperformance, durability and applicability.

Advantages

- Adsorbent, non-toxic, non-flammable, fully water washable.
- Insitu synthesised metal silicate desiccant on an inert inorganic fibre substrate.
- Active desiccant 80% of the media weight, so as to ensure high performance and minimal heat carry-over.
- Rotor is non-flammable. The net organics in the honeycomb media less than 2%.
- Special edge hardened media surface to ensure a smooth surface and long life of both, media and the seal contacting it.
- Rotor perimeter flange extend media and seal life.
- Rotor incorporates robust internal structure with perimeter flange for industrial quality, durability and easy serviceability.

FnP is specially designed for Food and Pharma industry

- Extra hygiene
- Fire resistance and safety
- Weatherproof options
- Efficient controls
- 'Zero' leakage &high structural strength
- User friendly features
- With EcoDryinside
- Bry-Smart controls

FnP – Dehumidification System

Bry-Air Dehumidifiers are available as standard units as well as packaged with pre-cooling, pre-heating, after-cooling, after-heating, heat recovery, etc. for the most cost-efficient environment control in a variety of industrial applications.

Packaged Systems...
Customized Solutions

1.2 Dehumidifier Digital Controller (DDC)

- Bry-Smart Dehumidifier Digital Control
- Smart controller for smart age
- Complete equipment diagnosis in a minute

Online support to customers

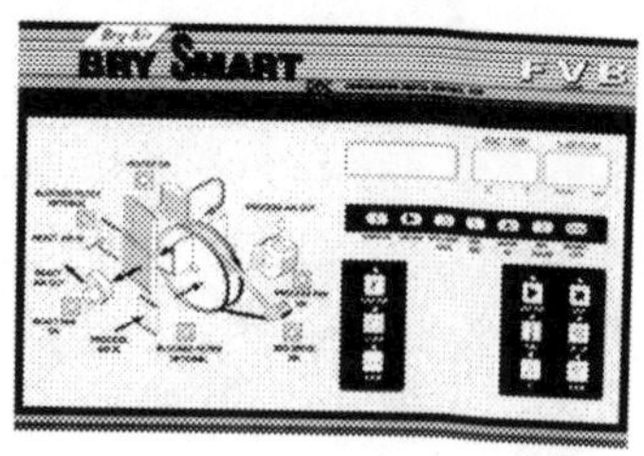

Appendix – 9

Fume and dust control, air ventilation hose and ducts

1. Dunham Rubber and Belting Greenwood USA

M/s Dunham Rubber and Beltings have developed varieties of hoses for different purposes. Here some of their products are discussed.

1.1 General Purpose Hose and Ducting

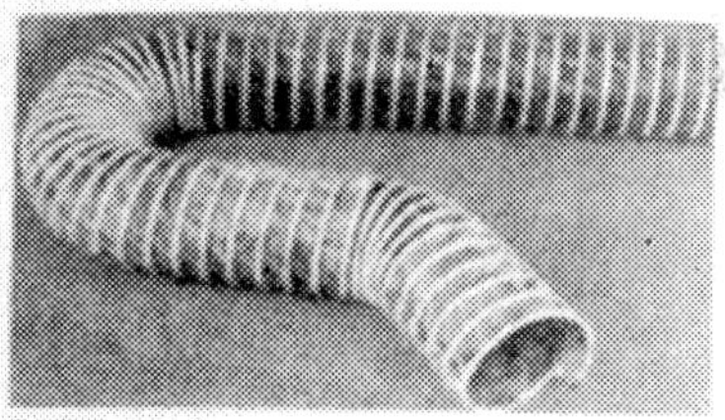

General purpose duct and hose consist of products made from PVC materials and double or single ply cotton and neoprene materials used as material handling and blower hose for light duty conveying of dust, woodchip and grass clippings. PA-EX series hose for higher temperature applications and TD-S series hose where the hose is regularly dragged around from place to place. R series hose is used for air, fume, dust and light duty material handling applications.

1.2 Suction, Transport and Material Handling Hose

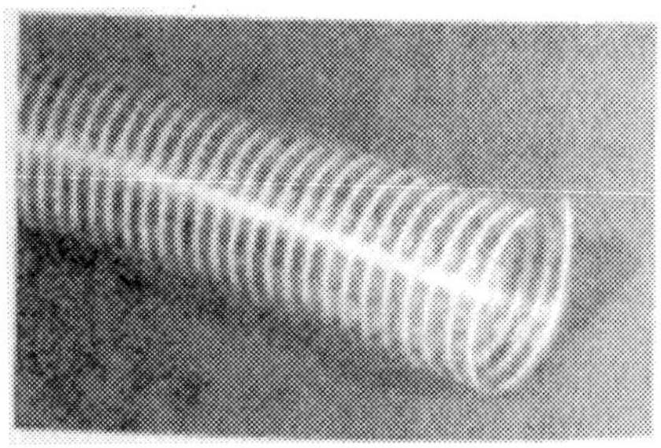

Plastic hoses and duct are used in a tremendous range of industrial applications; from dust collection, fume control, pellets and scrap transfer, leaf and woodchip collection, insulation blowing, etc. The Master-Pur series of hose including the durable Master-Pur Step FG and Master TPR L are a versatile multiple hose for a wide range of processes and application are available in this section. Additionally, series ARH, UFD-AP, CVD-AP and 333 Insulation Blowing hoses are also available and described below.

- ☞ UFD has superior abrasion resistance, flex fatigue resistance, allows users to locate blockages and handles higher pressure and vacuums than standard ducting. Materials used in UFD-Clear are FDA acceptable, available with cuffed end finishes.
- ☞ CVD-AP is economical and has smooth interior to minimise friction loss, crush resistance. Helix provides added durability where the hose is being dragged. It is light weight, durable and has great bending characteristics. It is used for sawdust collection, ventilation, fume removal, light duty material handling and leaf collection.
- ☞ RFH-W.045 has good abrasion resistance and hence ideal for applications where dragging is involved. The wear strip helps prevent premature wear-through, which is molecularly bonded to the hose wall. The wear strip will not delaminate.
- ☞ TBH-W wear strip provides greater wear resistance in applications where the hose may be dragged. It has excellent tear, abrasion resistance and moisture and drag resistance. The wear strip is molecularly bonded to the hose and hence do not delaminate. Ideal for dust collection, ventilation and leaf and lawn collection applications.
- ☞ UFD-AP has smooth interior, lightweight and highly flexible. It can be a replacement for thicker-walled EDPM hoses. The clarity of the hose allows the user to check blockages. The helix provides added durability in applications where hose is being dragged. It is good for dust collection, leaf and woodchip collection, for transferring stones, pellets and scrap.
- ☞ UFD.045 has superior abrasion resistance. The colour of the hose allows for the user to locate blockages. It is good for leaf and lawn collection.
- ☞ UFD.060 has superior abrasion resistance. The colour of the hose allows for the user to locate blockages. It is good for leaf and lawn collection.
- ☞ Following tables gives the comparisons

Type	Temperature range	Construction	Standard Colours	Standard Lengths	Size Range
UFD	−65 to 200°F	Thermoplastic polyurethane reinforced with a wire helix	Clear or black	25 in., 50 in.	2 in – 18 in.
CVD-AP	10–160°F	Clear polyvinylchloride reinforced with a rigid polyvinylchloride helix	Clear with black helix	25 in., 50 in.	4 in – 8 in

Contd.

Contd.

Type	Temperature range	Construction	Standard Colours	Standard Lengths	Size Range
RFH-W.045	−60 to 275°F (intermittent to 300°F)	Heavy-walled thermoplastic rubber reinforced with a wire helix and an external polypropylene wear strip.	Black with orange wear strip	10 in (seasonal)	8 in, 12 in.
TBH-W	−40 to 225°F	100% polypropylene with a wire helix and a polypropylene wear strip.	Black with a green wear strip.	25in., 50in.	2 – 12 in
UFD-AP	−65 to 200°F	All thermoplastic polyurethane hose reinforced with an ABS helix.	Clear with a safety yellow helix.	25in., 50in.	1.5 – 8in
UFD.045	−65 to 200°F	Heavy-walled thermoplastic polyurethane reinforced with a wire helix.	Clear or translucent blue.	25in., 50in.	4 – 8in.
UFD.060	−65 to 200°F	Heavy-walled thermoplastic polyurethane reinforced with a wire helix.	Translucent blue.	25in., 50in.	4-18in.

1.3 Suction and Blower Hoses for Air Conditioning, Ventilation and Welding Fumes

Items within this section include utility blower hose and large volume air conditioning/ ventilation hoses. Additionally, hoses for suction plants, tent heating. Weld fume extraction and low pressure applications are also available within this section. Many varieties of material construction and end finishes are available including sewn ends, funnel cuff, belt loop and pull tab, enclosed belted cuff and more.

1.4 Medium and High-Temperature Hose

Dunham rubber offers a wide range of hoses and ducting capable of handling medium to high temperature applications. This includes vehicle and engine construction applications, chemical plants, exhaust fume equipment, granule dryers, aggressive media extraction, suction plants, paint spray extraction, aircraft and defense industry applications. Master-clip CAR for diesel engine fume exhaustion, U-lok 1500 for service up to 1500°F for furnace construction.

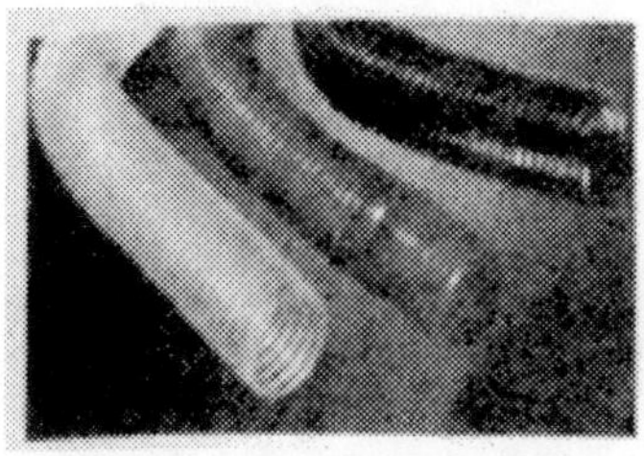

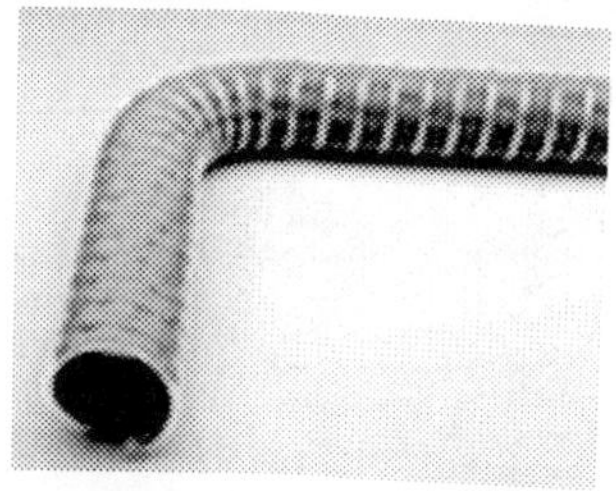

1.5 Gas and Fume Exhaust Hoses

These systems draw harmful fumes and gasses out of the working environment, promoting worker safety. Other products in this section are designed for use in the ship building industry, furnace construction applications, infrared dryer plants, iron and steel works, weld fume extraction and bulk handling technology applications.

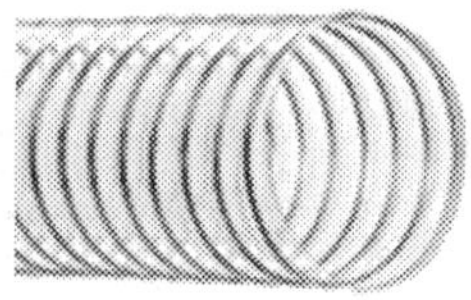
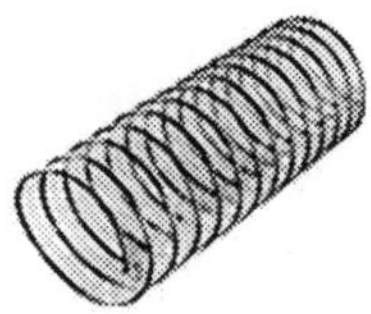

Dunham rubber offers a wide range of products for gas and fume removal and extraction. Options are available to cover a wide range of applications in a variety of operating conditions. For example, the Carflex line of products is ideal for extraction of engine exhaust gases with temperatures that range up to 570°F. This product is ideal for overhead hose reel applications in service garage environments or in slotted channels above floor or under floor suction extraction operations.

1.6 Super Vac-U-Flex

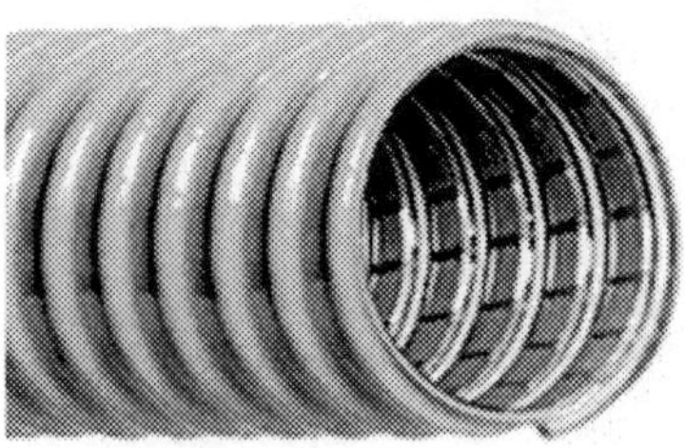

It is made of high strength fibre-reinforced PVC hose cover bonded to coated spring steel wire helix. It is grey in colour and can withstand temperatures from −20°F to 150°F. Its main features are Cuffed ends, Fume removal, Moisture resistant, Oil resistant and UV/ozone resistant. The applications include Air, Chemical fumes, Dust, Exhaust and Ventilation, General purpose, Material handling, Moisture/hydraulic, Oil mist, Special purpose, Vacuum and suction.

2 Hi-Tech Duravent USA

2.1 RFH Series Hoses and Duct

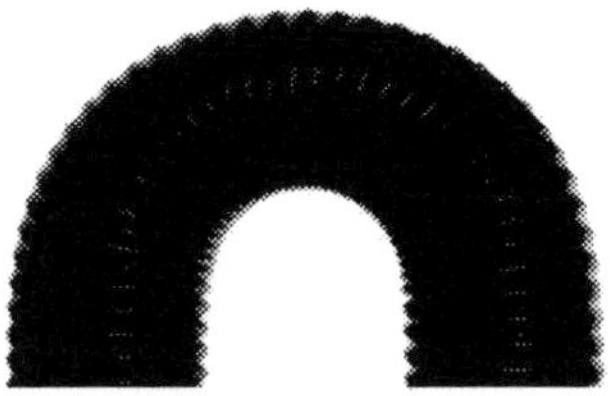

Hi-Tech Duravent RFH series hoses offer wider temperature range, have superior chemical resistance and better abrasion resistance. They are versatile and have wear strip capabilities offering better UV, moisture and weathering resistance. They have outstanding flex fatigue resistance, air tight and will not set to shape of box when packed. They are available in different series like RFH, RFH-plus, RFH-white, RFH-W and RFH.045. They are used for chemical fume removal, fume control and cooling, dust control and ventilation and lint control.

(a) Commercial Vacuum and Carpet Cleaning Hoses

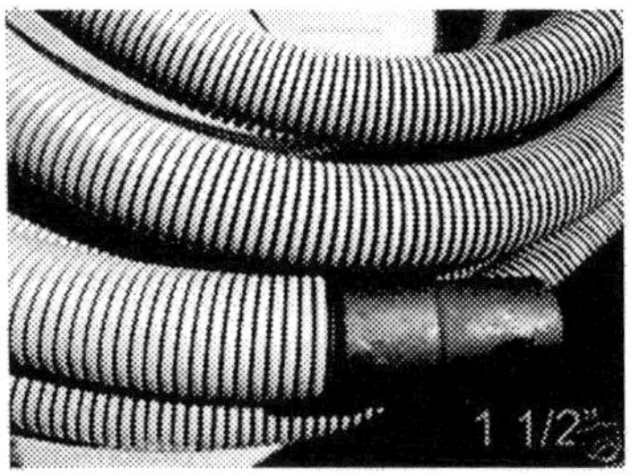

Commercial hoses are used for industrial and commercial vacuum cleaners, ventilating or cooling industrial machinery, bilge pump hose, car wash systems, carpet cleaning and more. The hoses of this category are moisture resistant, UV/Ozone resistant, oil resistant, chemical resistant, drag resistant and fume resistant.

(b) Dusting Clamps and Accessories

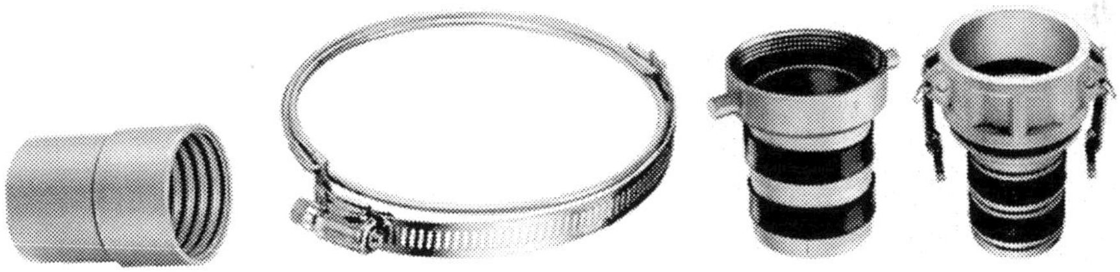

Varieties of ducting bridge clamps and connectors are available with most of the hose manufacturers and suppliers. They are available in different sizes and shapes depending on the type and size of hose.

3 Guangzhou PaiPu Ventilation Equipment Co ltd China

3.1 PVC Flexible Air Duct

The PVC flexible air duct are made from fire retardant materials and steel wire helix and have maximum durability. These air ducts are used with ventilators for fume removal, bringing fresh air to manholes, tanks and other enclosures. They are tough weather resistant with dual differential flow and low friction loss for moving air and fumes.

4 Excel Airtechnique India

4.1 Fume Extraction Systems

Fume and smoke produced inside a factory is a contaminant. A control device and an extraction device is needed to control the buildup of concentration of fume and/or smoke to provide a good working environment. The fume extraction system by Excel Airtechnique involves well-designed hoods a scrubber and has optimum exhaust capacity.

5 Dundas Jafine Ontario Canada

5.1 Flexible Aluminium Ducting – Class 0 Air Duct

Dundas Jafine – Class 0 Air duct is ideal for use in general purpose exhaust applications. Totally, noncombustible and can be bent into any configuration. It can operate at high temperature of up to 435°F.

6 Dunne Cleaning Specialists Inc. Denmark

6.1 Duct Cleaning

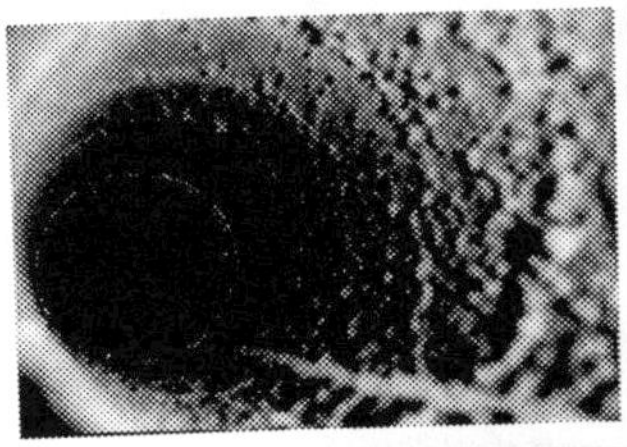
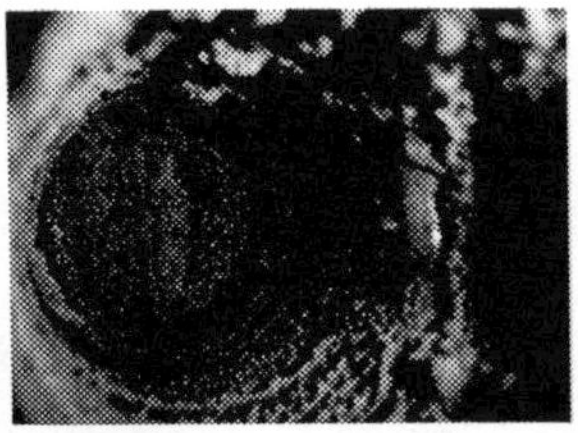
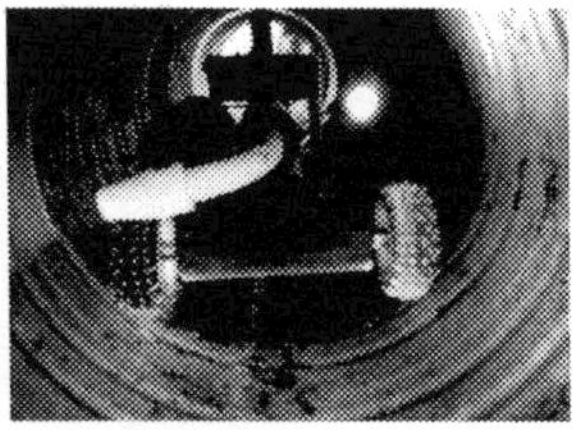

Danduct Clean is the world leading supplier of duct cleaning equipment and are supplying air duct cleaning equipment all over the world. The Danduct clean ductclean equipment are developed exactly to ensure good and healthy ventilation hygiene and thereby ensure the buildings indoor air quality (IAQ) being as high as possible.

The most effective way to clean air ducts and ventilation systems is to employ Source Removal Methods of cleaning. This requires professional duct cleaning equipment such as the Danduct Clean DC4 brush cleaning machine and duct cleaning robots like the multipurpose robot. The dirt and debris can then travel down the ducts to the ventilator, which removes it from the system.

6.2 Micro Inspector

New Innovation. The new remote controlled Danduct Clean inspection camera for professional inspection of HVAC systems. The compact 4-wheel camera system has two cameras, a colour camera on the front and a rear black and white camera. The low weight and compact size enables the new Micro Inspector to go into small ducts also.

Digital Viewing

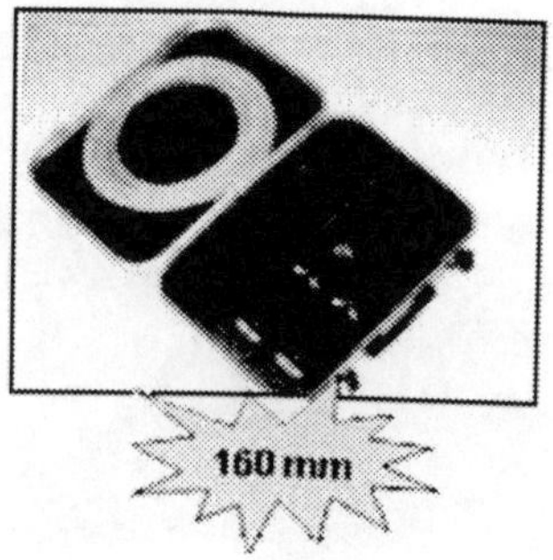

Everything can be viewed on the small LCD colour monitor; or use the additional GrabBee program to view and record everything on your laptop. Making images and video clips have never been easier!

The Micro Inspector is easily controlled from the joystick, and the powerful LED lights provide you a perfect view of the inside of the duct work. Chassis with two cameras and LED light.

- Control box with joystick
- 30 metre cable
- LCD TV/monitor
- Transport box
- Micro inspector dimensions:
- Length: 165 mm
- Width: 150 mm
- Height: 80 mm
- Weight: 1.5 kg

6.3 TecCam

With a small investment, you will have a great marketing; not only to show your client the existing condition of the ductwork and components, but perhaps even more to see the result of your cleaning and to show your customer that you did an excellent job.

TecCam is an easy to use lightweight manual inspection tool that can reach up to 12 m inside the ductwork;, it weighs less than 2gkg. The black and white camera head with infrared light is positioned at a flexible shaft that can pass numerous 90° bends. It fits through 42 mm openings and can maneuver through most turning valves. The infrared light source provides exceptional illumination with reduced glare on the duct surface.

There is no need for separate push rods, the single shaft design integrates the power and imaging cables within the flexible shaft. It is a very resistant tool, with a tough outer jacket, that helps protect the wires from damage. Also it has great adaptability and can be attached to most cleaning tools including duct brushes, whips and skippers.

The TecCam is delivered with a power supply and connection to a monitor. The image from the camera can be viewed on any monitor with an AV-connection.

For easy transport and to make it easy to carry, TecCam is delivered with a holder. Packing the equipment have never been easier, there are no large carrying cases. At the same time, the TecCam is well protected in the holder during transport.

The Danduct Clean TecCam is a part of the extensive range of inspection and cleaning equipment that we offer to our customers.

Technical data for the TecCam:

Length of shaft:	12 m
Camera:	Black & White, Pal or Ntsc
Light:	Infrared
Power supply:	110 V or 220 V
Weight:	1.7 kg

6.4 Ventilators

Danduct Clean supplies strong ventilators that can easily be moved from room to room – even stairs are no obstacle.

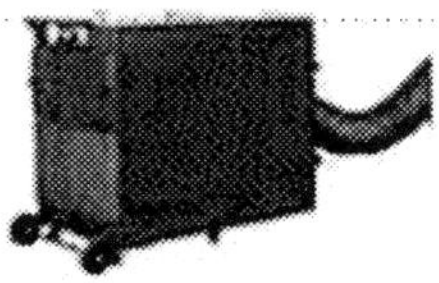

Features

VT4000-VT4001

2 models with airflow 4500 m^3/h

Easy to clean and transport

With the new Danduct Clean VT4000 ventilator unit you have the perfect all-in-one solution to extract and filtrate all dust and debris from the heating, ventilation and air conditioning systems.

The mobile unit is equipped with a standard 16-ml HEPA filter, which allows you to use it even in strict surroundings like a hospital. For easy cleaning and to minimise downtime, the filter unit is equipped with a rotor clean system, which will also extend the lifetime of the filter significantly. For extra filtration, a HEPA filter cassette can be installed.

For easy transport, it can easily be taken apart and put in the back of a truck. The VT4000 is a very powerful ventilator with airflow of 4.500 mi/h. For easy operation, it has a control panel with vacuum meter, am meter and on/off switches.

VT4001 vacuum and filter unit with a build-in and replaceable EU6 filter and a HEPA filter cassette. Modular build-up and low weight.

VT1500, VT3000 and VT10000

- 3 models with airflow varying from 2000 to 10000 m^3/h
- 2 ventilators can be combined resulting in an airflow of up to 20000 m^3/h
- Easy to move
- Can be connected to a 16 m^2 HEPA filter

Choosing the right size of ventilator is a paramount issue in creating a sufficient negative airflow in the ducts. Test has been showing that 10 meter pr. second / 33 feet pr. second will transport the dust and most other particles to the filter.

The range of ventilators provides airflow from 2.000 to 10.000 m^3/h and by connecting two ventilators together they can move as much as 20.000 m^3/h.

The ventilators can be connected to a filter unit with 16 m^2 HEPA filter. The HEPA filter is classified between H13 and H14 with a filtration of 99.97% of all particles down to 0.3 micron. The HEPA filter is mostly used in hospitals or other sensitive areas.

Technical Specifications

VT4000-VT4001:

	VT4000	VT4001
Air flow:	4500 m3/h	4500 m3/h
Motor		
- KW:	2.2	2.2
- Amp:	13.4	13.4
- Volts:	230	230
Filter:	Replaceable 16m2 Hepa filter	Replaceable EU6 filter + Hepa filter cassette
Filter efficiency:	Hepa = 99,97% of all particles over 0.3 microns.	EU6 = 65% of all particles over 0.5 microns. Hepa = 99,97% of all particles over 0.3 microns
Vacuum unit		
- L x H x W:	850 x 1150 x 750 cm	850 x 1150 x 750 cm
- Weight:	118 kg	118 kg
Filter unit		
- L x H x W:	900 x 1610 x 750 cm	600 x 1150 x 750 cm
- Weight:	122 kg	52 kg
Hose size:	250 mm	250 mm
Noise:	68 DB	68 DB

VT1500, VT3000 and VT10000:

Type	Power	Amp	HP	Airflow
VT 1500	240 V / 50 Hz	7.9	1,5	2000 m3/h
	110 V / 60 Hz	14.0	1,5	2000 m3/h
VT 3000	240 V / 50 Hz	12.5	2,5	4500 m3/h
	400 V / 60 Hz	5.3	3,0	4500 m3/h
VT 10000	400 V / 50 Hz	17.5	10,0	10000 m3/h

7 Dunne Cleaning Specialists – Broadview – IL, USA

7.1 Duct Cleaning

A vacuum hose (minimum of 10 inch in diameter) is attached separately to the supply and then to the return duct with two separate hooks to clean the entire duct system. If only 1 hook is used, half the system will not be cleaned and in a short time the dirty side will contaminate the cleaned side.

Brushes and electric blowers are very ineffective in duct cleaning although they are very inexpensive. Running a brush down the duct from the grills makes an awful mess when pulled back into the room. A blower directed into the duct will only remove debris. To complete clean the duct work, a power whip must be fed through the vents and the trunk lines in the basement.

A power whip with a high pressure air sweep ball will fit into virtually every inch of duct work and blast lose and get airborne the caked on contaminants and debris in both the supply and return ducts; which once airborne are carried out by the large industrial 25HP vacuum located outside the building. It removes as much debris as possible from the entire supply and return ductwork to collection bag.

8 Climavent UK

8.1 Fume Extraction Systems

Climavent dust and fume extraction solutions ensure a safe and comfortable working environment in industrial processes around the world. For example, optimum dust control is achieved by supplying dust extraction at source, utilising extraction arms, down draught benches, dust control booths and fixed extraction hoods. The exhaust for these systems is provided by Climavent's extensive range of dust extraction units that include wet dust exhaust units, reverse jet filters, auto shaker units and mobile extraction units. Powerful suction is guaranteed because all dust and fume extraction units use superior quality filtration systems, centrifugal fans and bespoke ductwork systems.

Fume extraction units are engineered using the same innovative approach to effectively handle the obnoxious fumes and odours generated in specific processes. Our product range also includes mobile dust extraction units and portable fume extraction units.

Climavent Fume Extraction Systems have been built to handle the obnoxious fumes and odours in a range of industrial, commercial and educational environments. The Units are available in a variety of sizes and finishes, and each unit can be customized to suit specific applications for use. Fume extractors can be wall or stanchion mounted, or with portable fume extraction fans and self-supporting extraction arms. Depending on the application Fume Extraction accessories, carbon filtration units and pipe work can be manufactured in Polypropylene, UPVC, stainless steel and powder coated mild steel. Units can be built to occupy single positions in a laboratory or to offer multi point extraction along production lines.

Appendix – 10

Filters and dust collectors

Varieties of filters are available for filtering air fed to air washer plant and the air being let out from the machines such as Static Flat Screens, Static Zigzag Screens and Rotary Screens. There are systems for continuously cleaning the machines like travelling cleaners. Following are some of the commercially available filters. The descriptions given are as claimed by the manufacturers.

1. Static Flat Screen Filters

1.1 AeroTech Equipments & Projects (P) Ltd Gurgaon, India

1.1.1 Fine Filters

Efficiency claimed	99.9%
Initial pressure drop	6.5 mm WG at aerated flow in clean condition
Normal rating	5 micron
Media	Synthetic nonwoven polyester, nylon or polypropylene fabrics.
Construction	The filter housing type and flanged type housing thickness as per specific requirements.
Testing	Tested as per BS 6540 (1985)
Servicing	Filter cleaning recommended when the pressure drop across the filter is double that of initial pressure drop. These filters are washable and can be washed with any household detergent.
Application	AeroTech Micrometric Fine Filters are best suited for computer centers, ventilation system, telephone, modern paint shops, nuclear power plants, textile mills and miscellaneous industrial ventilation.

1.1.1 Pre-Filter

Efficiency Claimed	90%
Initial Pressure drop	2.5 mm WG at aerated flow in clean condition
Normal rating	20 micron and 40 micron
Media	Synthetic nonwoven + HDP Mesh

Contd.

Contd.

Pressure Drop	Initial pressure drop ~2. 5mm wg and recommended final pressure drop is 5–10 mm wg
Operating Conditions	Filter can withstand temperature up to 80°C and relative humidity of 100% at 0 condensation. Filter is washable or can be cleaned by applying air pressure and reused.
Application	Enviro tech pre-filters are most suited for: air-handling units (AHUS), air conditioning systems, power plants, chemical and fertiliser complexes, green houses, automatic drain holes, paint shops, textile mills, cement factories, laminar air flow units, biosafety hoods, fan filter units (FFU), gas turbine industry, compressor, diesel engines, nuclear power station, CD-ROM production industry, picture tube production industry, semi-conductor industry, aeronautical industry and miscellaneous

1.5. Yan Zhang Shandong Aobo Environmental Protection Technology Co., Ltd, Qingdao, China

1.5.1 Activated Carbon Nets Plate Air Filter

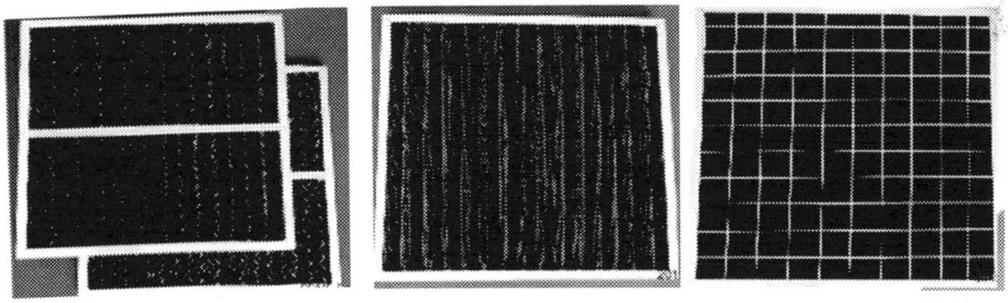

Activated carbon filter plate may be an aluminium frame, galvanised frame, stainless steel frame and so on, the thickness can be produced according to customer requirements, using high-quality active filter carbon, carbon fibre felt. The frame area can be extended with pleated filters or with flatbed. It has a large dust holding capacity and low initial resistance.

1.6. Eddie Wu Kunshan Laidel Filtration Equipment Co., Ltd. Shanghai. China

1.6.1 Activated Carbon V-bank Filter with High Air Flow

Features

- ☞ Using carbon particles and ultrafine synthetic fibres composite material, effectively removing harmful substances in the air.
- ☞ Larger filter area, reliable performance with high adsorption capacity and removal efficiency.

☞ Light weight, easy installation and maintenance.

☞ Frame is made of ABS, easy to dispose in accordance with environmental protection requirements.

Application: Central air conditioning and ventilation systems, to eliminate light concentrations of organic pollutants, sulfur dioxide, nitrogen dioxide, with the function of removing of dust and gas contamination at the same time.

1.7 Luwa Air Engineers AG, Switzerland

1.7.1 MultiCell Filter MCV by Luwa

The Multi Cell Filter MCV with its modular and space saving concept is suitable for machine exhaust air cleaning of blowroom and carding machines as well as in air conditioning plants for the dust and fly filtration of the room return air. It is a highly compact filter of modular construction with an automatic filter cleaning robot. Its compact design reduces the floor space requirements by 1/3rd compared to the other automatic fine filter. It is particularly suited for cleaning the contaminated return air containing lint and fine dust and thanks to its strong suction system it can even handle the sticky weaving heat.

V-shape filter cells and suction robot.

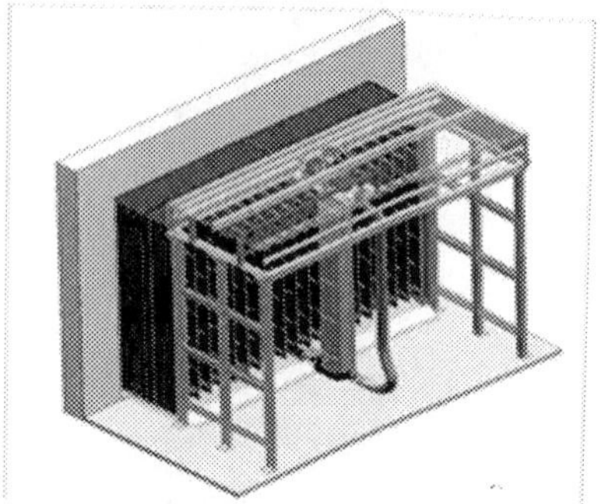

MultiCell filter MCV as filter wall in air conditioning Plant

Category

- Filtration: Air-Dust class
- Cleaning: Automatic – Integrated with Dust Collector or Cyclone
- Capacity: Up to 200,000 m^3/h

MultiCell filter MCV installed in a Compact Filter Unit CFU for Blowroom.

Contd.

Type	Flange or box type
Frame	Tested as per BS 6540 (1985)
Testing Method	As per BS-2831

1.1c Metallic Viscous Filters/Wire Mesh

Efficiency claimed	65%
Initial pressure drop	1 mm WG at rated flow in clean condition
Media	Wire mesh
Type	Box type (Sliding fitting)
Frame	GI / aluminium
Servicing	Metallic frames are washable
Application	AeroTech Micrometric Fine Filters are best suited for computer centers, ventilation system, telephone, modern paint shops, nuclear power plants, textile mills and miscellaneous industrial ventilation.

Fine Filters

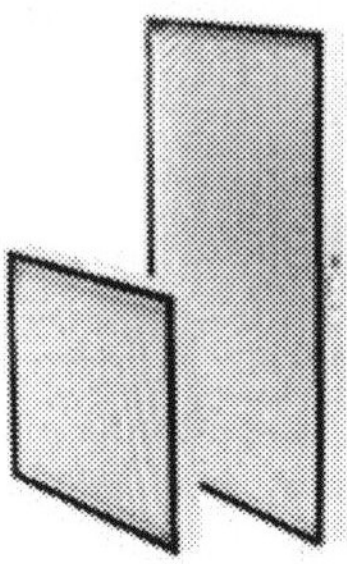

Pre-filters

Metallic Viscous Filters

1.2 Nisarga Filters, Bangalore, India

1.2.1 Fine Air Filters

These are exclusively designed to filter 5 micron particles with efficiency of 95% and above. Ergonomically designed these filters are widely used as an intermediate filter to the HEPA filter and pre- filter. Our Fine Filters are widely used in various industries for air conditioning, ventilation and air handling units.

Size Standard	Capacity CFM	Grade	Average Arrestance
610X610X100 mm	588 to 1,000 cfm	F5	95%
610X610X150 mm	1,000 to 2,000 cfm	F5	95%
610X610X305 mm	2,000 to 2,470 cfm	F5	95%

1.2b Pre-Filter

These are one of most favoured selections of clients owing to its robust construction, compact design and durable performance. Our pre-filters are widely used to remove dust and pollution and to provide fresh air. These are extensively used in kitchen booth, cinema halls in automobile industries

Size Standard	Capacity CFM	Grade	Average Arrestance
510 X 610X50 mm.	580 to 1,000	G 2 to G 4	G 2 – 60% to 80%
S10X61OX 150 mm	1,000 to 2,000	G 2 to G 4	G 3 – 80% to 85%
610X610X305 mm	2000 to 2470	G 2 to G 4	G4->90%

1.3 Louisville, KY USA

1.3.1 Robust Pleated Air Filter – The Perfect Pleat Ultra

The Perfect Pleat Ultra from AAF international is designed for air filtration applications across general industry and has a pleated design that has enabled the manufacturer to eliminate the traditional metal support grid without sacrificing structural resilience. The pleat provide extra surface for filtration, and also make the filter strong and robust.

1.4 Enviro Tech Industrial Products, Delhi India

1.4.1 Pre-Filter Pleated Type

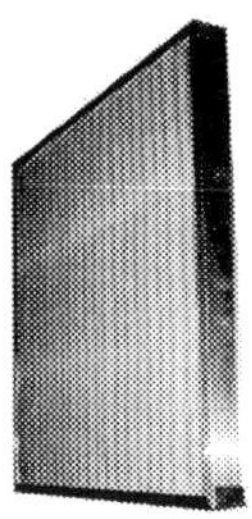

Pre-filter is used for coarse pre-filtration, provision against insects, textile fibres, pre-filtration in HVAC and coarse applications. Our pre-filters are available in the range of 15 to 20 micron (G3, G4 Class) and 10 micron (F5, F6)

Features :

- Low initial pressure drop
- Light in weight and easy to handle
- Cleanable / washable high cost / performance ratio
- High cost / performance ratio
- Uniform pleating ensure better airflow

Applications :

- HVAC air-conditioning systems.
- Industrial ventilation systems.
- Coarse filtration and pre-filtration

Specifications:

Type	Flange / Box Type
Frame	Galvanized steel / CRCA powder coated / aluminium anodised / SS 304
Separator	Aluminium
Filter Media	HDPE media / non-woven synthetic media
Gasket	Neoprene rubber gasket
BS EN 779 Class	G3, G4
Filter Efficiency/Arrestance	G3, G4 50–80% Arrestance F5 80–90% Arrestance
Euro vent Class	EU 3, EU 4

1.4.2 *Micro Vee Fine Filters*

These have been specially designed to meet the requirement of clean air in general ventilation and air conditioning system in hospitals, synthetic fibre industries and compressor applications.

Construction: The heart of the filters is a reinforced washable synthetic nonwoven filter media/fire retardant glass fibre media pleated in a format along with the support of wire/ HDPE netting. The mediums are carefully selected to offer the best possible performance to suit the specific need. In order to obtain uniformity of dust loading the filter pleats are kept apart by means of aluminium separators.

Media: The media is a special synthetic non-woven bounded fibre designed to give high efficiency and long service life. It is pressed in the form of a fabric and given special treatment which prevents fungus growth. An alternative media in superior felt is also available, which gives long run trouble free services, and economical in usage.

Features:

- Efficiency: 99.9% down to 5 microns
- Pressure drop: 6.5 mm of wg., at rated flow

- Frame: Galvanized steel sheet/MS/ aluminium/ stainless steels sheet
- Finish: Painted/ powder coated
- Filter media: Synthetic non-woven filter media supported with HDPE mesh and GI/aluminium Expanded mesh
- Media: Pleated
- Sealant: Available
- Type: Box type / Flange type

Specifications of Standard Models:

Model No.	Nominal Dimension Millimeters Inches		Rated Capacity @ 6.5 WG M3/ Hr CFM		App. Wt. (Kg.)
RE - 6001	254x254x85	10x10x3.25	170	100	5
RE - 6002	508x254x85	20x10x3.25	330	195	8
RE - 6003	508x254x152	20x10x6	560	330	10
RE - 6004	508x381x152	20x15x6	890	525	15
RE - 6005	508x381x152	20x20x6	1220	710	18
RE - 6006	610x610x150	24x24x6	1,670	1,050	25
RE - 6007	610x610x305	24x24x12	3,000	1,800	50

1.4.3 *Micro Vee Filter*

The Micro Vee filters are designed to arrest finer particles up to 5 microns. It is normally used as a secondary filter after pre-filters to have prolonged service life. This range is used in any air handling application.

Features:

- Easy to clean
- Easy installation
- Precisely design

1.4.4 *Fine Filter*

Fine Filter using the advanced technology in accordance with the set industry standards. Owing to its long service life and easy installation features, this filter is widely demanded among clients. Furthermore, the filter is strictly tested by skilled quality controllers on various quality parameters.

Features: Long service life, Easy to install, Low maintenance Specifications, High and

Medium Efficiency, Extended Surface Supported Pleat Air Filters, Media Pack Design, Extended-Surface Rigid Air Filter with Synthetic Media, Ducted, Disposable Ceiling Filter Modules for Clean room Applications, Ducted, Disposable Ceiling Filter Modules for the Most Stringent Requirements, Sturdy Economical Disposable Panel Filters, Lightweight Washable Filters for Average Dust Loading Conditions.

1.4.5 Air-Handling Unit Filter

A wide range of air-handling unit filter, offered in varied sizes and specifications as per the application requirements of the clients. The filter is manufactured using superior quality raw material and employing contemporary technology. In addition, the quality examiner conducts a series of tests on this range in order to eradicate any kind of defect in it.

Features:

High dust holding capacity

- Low maintenance
- Durability

Specifications:

Efficiency	Standard HDPE pleated type filters have an efficiency of 90% for 10 micron size particles. Pre-filters with an efficiency rating of 80% to 90% to capture 10 micron to 25 micron size particles are also available on order.
Frame Type	Available in flange or box type
Frame Material	Standard model is constructed in galvanised steel. Other models with aluminium anodised/stainless steel frame with or without powder coating are also available on order.
Filter Media	HDPE and non-woven synthetic.
Sealant	PU-based epoxy
Support	Large frame size models are made wobble resistant using GI/aluminium/SS support
Face Guard	No face guard with standard model is supplied. On order/request filter can be protected with expanded mesh on both sides
Nominal Depth	Depth size is available from 1. 0 inches to 6 inches (pre HDPE pleated and rod type)

Contd.

1.7.2 Fine Dust Filter FPP

Luwa FP-P Fine Dust Filters remove air contamination such as fine dust, smoke, vapour, soot, pollen and bacteria and are therefore ideally suitable as final filters or as pre-filters for HEPA- or ULPA-filters in air conditioning installations. They are suitable for all standard filter applications, especially with those requiring increased service life, safety and versatility. Luwa FP Filters are available in 6 efficiencies1), 4 nominal sizes, 2 depths2) and 2 models (NT/HT).

Application parameters

Continuous operating temperature: 70°C (on request, versions for 120°C available).

Pressure drop:

- Recommended final pressure drop: 450 Pa
- Max. final pressure drop (endurance strength): 800 Pa
- Bursting press. (new filter): 1000 Pa

Admissible relative humidity: 100%

Fine dust filter FP-P

1.7.3 Unitary System

Luwa's compact air conditioning unit is an extended development to Luwa's Compact Filter Unit. The combination of filter plant with the air conditioning unit will allegedly be the ideal solution for space constraint. The unit is made in a stable self-supported structure, either floor or ceiling mounted. Since the unit is equipped with the TexFog technology, no water tank is needed and therefore expensive civil structure can be avoided. Luwa's Automatic Panel Filter is a modular system for dust and fibre laden extract air, particularly suitable for spinning and weaving plants. With the high filter efficiency, it makes a simple task to reach Occupational and Safety Health Administration standards for the return or exhaust air. The filter cells are automatically cleaned by means of suction nozzles, mounted on a robot that moves in and out of the cells and laterally from cell to cell. The filter is of modular design and can be extended from 5 to 18 cells. Its advantages include 20% increased filter area with redesigned filter panels, active floor cleaning due to new floor suction nozzle, simplified mechanical shifting and suction device, increased cleaning capacity with variable robot speed and variable suction nozzles even with high load of dust and fibres.

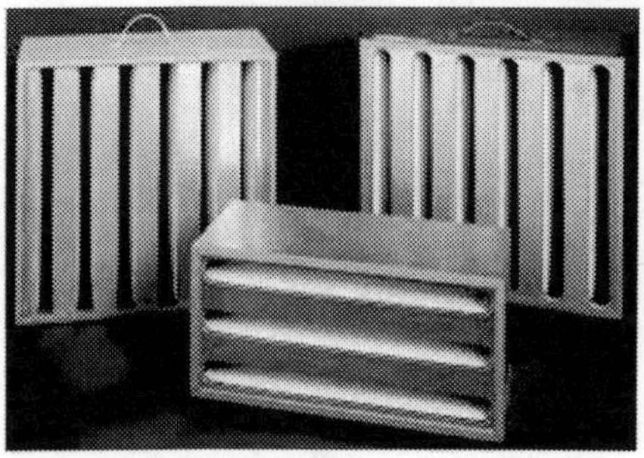

Luwa N Ultra Filter

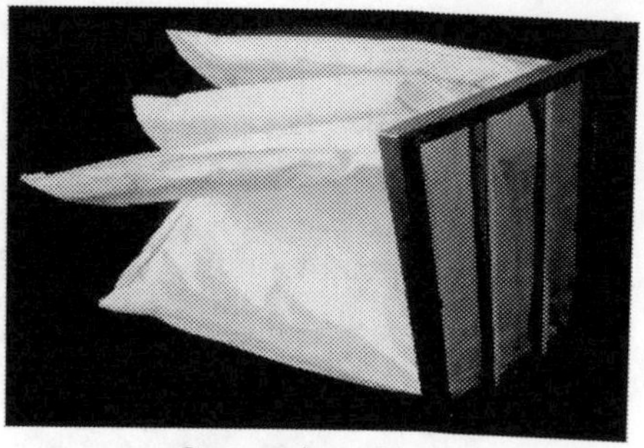

Coarse bag filters

1.7d Static Water Filter – SWF

Water strainers and screen are used in water circuits with low contamination and short pipe routing to protect the equipment from clocking. The water is passing through a screen which need to be cleaned manually. Depending on the expected contamination the filters are installed either inline or in open inside the water collection tank.

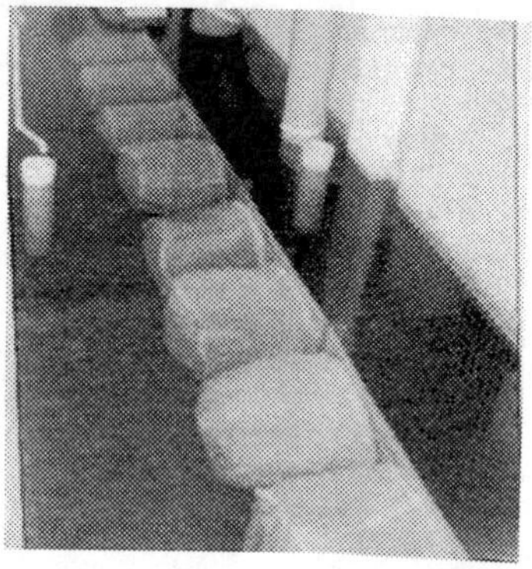

Category:

Filtration: Water – Sludge

Cleaning; Manual

Capacity: 20m^3/h per cell

2 Rotary Filters

2.1 Keystone Ahmedabad India

2.1.1 Rotary Air Filters by Keystone

This equipment is specially designed for handling high volume of air laden with fibre and dust returning along with the air from the department or from waste recovery units. The fluff, fibre and dust arrested on the filtering

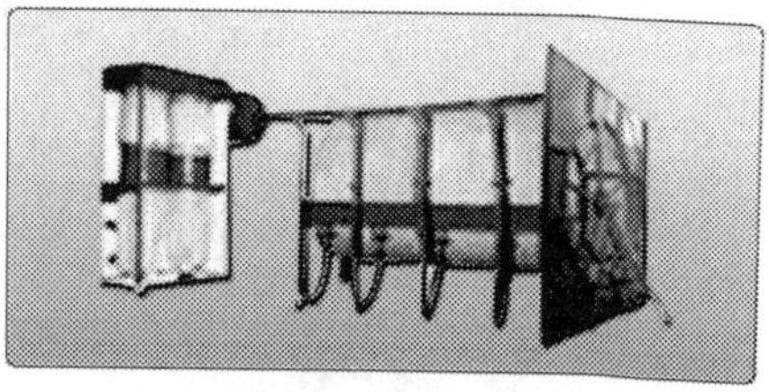

media are sucked out by travelling suction nozzles, moving along the outer surface of the drum and are deposited in collection bags for easy disposal.

The clean air is then either recirculated or exhausted in to the atmosphere. Continuous filtering and cleaning enable a uniform flow of return air at constant pressure. A variety of filter media like polyurethane foam and nonwoven material are available for varied applications.

2.1.1 Rotary Screen filter

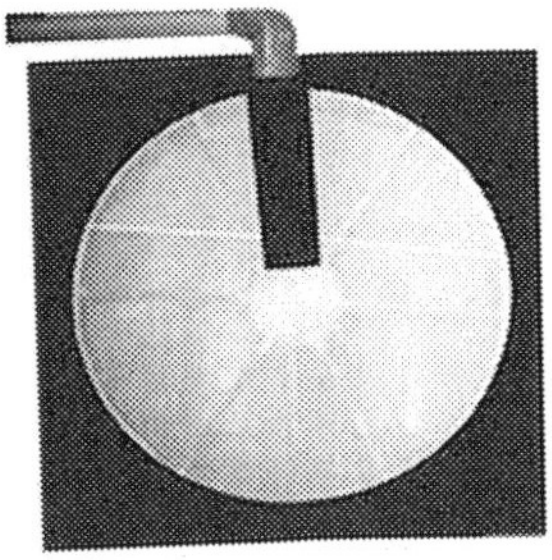

Rotary Screen filter is used for fibre and waste separation mainly to clean air laden with high concentration of dust and fibres. Its high filtering property and separation process with low dimension makes it practically versatile. The dust, fluff and fibres arrested on the filtering media are sucked out by travelling suction nozzles, moving along the outer surface of the drum and are deposited in collection bags for easy disposal. The clean air is then either recirculated or exhausted in the atmosphere. Continuous filtering and cleaning enable uniform flow of return air at constant pressure. A variety of filter media like polyurethane foam and nonwoven material are available for varied applications.

2.1.3 Rotary Water Filter

Keystone make Rotary Water Filter is for automatic and continuous cleaning of water in Air Washer Tank for Humidification Plant. This equipment is an automatic cleaning device with rotating perforated stainless steel drum mounted on SS Frame. It has SS Screen filter drum, high pressure spray nozzles with drive mechanism with geared motor and dust separated with high-pressure water spray nozzles assembly. It is much suitable for automatic cleaning of Air Washer water tank in Air Conditioning and Humidification Plant, and for Chilled Water Spray Humidification Plant.

2.2 Shri Jai Shakti Corporation, Ahmedabad, India

2.2.4 Drum Filter

The design of the Drum filter completely departs from most previous filter systems. The filter drum is stationary and can be bolted directly to a wall opening. The incoming air flows from the inside to the outside, leaving the drum through the whole filter surface. This means that the air inside the filter chamber is clean. Rotating and changing suction nozzles on the inside of the drum continuously vacuum off any dust and waste from the filter media. Only very little air is required to clean the filter medium as the suction nozzles are very small.

The nozzles are fluidic optimized and touch the filter media, which guarantees high and efficient cleaning. The drum filter combination TFC features an additional coarse particle filter in form of a prefilter disc that is installed at the air intake side of the drum filter. Coarse particles will adhere on the rotating disc while the fine dust passes through it into the drum. A stationary suction nozzle cleans the prefilter disc. The suction nozzles inside the filter drum and the pre-filter disc can be driven by the same motor!

The drum filters are perfectly suitable for energy efficient filtration of large volumes of air with fibrous dust such as wood, polystyrene, insulation material, natural material, etc.

Advantages

- Clean filter chamber since air flows from the inside to the outside
- Drive is easily accessible on the clean gas side
- Regenerative filter unit
- Continuous cleaning, therefore no pressure fluctuations within the system
- Precise adaptation to the total air volume due to the modular design
- No dust build-up between prefilter and filter drum
- Energy efficient operation due to low pressure loss
- Sturdy construction
- Energy efficient

Application in various industrial sectors includes textile industry, nonwoven industry, woodworking industry, paper and cellulose industry, tobacco industry, fibreglass industry, hygienic industry and automotive industry.

2.3 Fine Air Systems, Ahmedabad, India

2.3.1 *Rotary Air Filter*

A fully automatic rotary filter for a fine filtration of the micronic dust particles and recirculation of the department air coming out of an exhaust plant. They work at very low filtration velocity. A special filter cloth is used to achieve a fine filtration. With automatic waste cleaning nozzle arrangement over the filter cloth with a dust collector, it is possible to achieve a steady exhaust rate from the department without fluctuation in RH, development of hot spots and infiltration of air from one department to another department.

2.4 Luwa Air Engineers AG. Switzerland

2.4.1 *Fine Air Filter – RAF*

The rotating fine air filter is used wherever coarse dust and fibre are in low concentrations and needs to be extracted from room or machine exhaust air. It is characterised by the way of its constructions having lengthy drum wrapped with nonwoven filter fabric rotating on its axis and the air to be filtered flows through it inwards from outside. Along the length it has got vacuum nozzles travelling back and forth almost touching the filter surface and removes the accumulated layer of fibres and dust.

Category

Filtration: Air-Dust Class

Cleaning: Automatic – Integrated with Dust Collector or Cyclone

Capacity: Up to 200000 m^3/h

2.4.2 *Rotary Air Filter LDF*

The Luwa Rotary Air Filter LDF is a fine dust filter for exhaust air with a high concentration of dust and fibres. Due to the automatic cleaning of the filter drum with a suction device, maintenance is easy.

- Airflow rate maximum 250,000 m^3/h
- Dust concentration 0–80 mg/m^3
- Static pressure 250–400 Pa
- Diameter 2,000, 2,500 or 3,000 mm

- Length 1,700–5,100 mm (6,800 mm)

Requires installation inside prefabricated housing or inside a concrete structure.

2.4.3 RPF – Automatic Pre-filtration

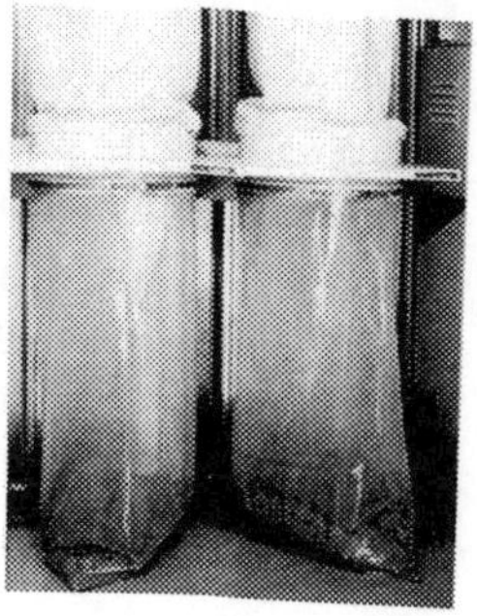

The Luwa rotary pre-filter unit is used as a part of air engineering plants to provide preliminary filtration of air steams heavily charged with waste, especially in automatic waste disposal systems. Once all the waste is separated from the good fibres, it is necessary to segregate the reusable and non-reusable waste as well. Non-reusable waste is separated out with the automatic rotary pre-filter unit, which can then be disposed of correctly so as not to harm the environment and surroundings in any way. The air stream is conducted axially to the slowly rotating pre-filtration disc via feed pipes through the back panel or via a connecting box. The rotational rate, mesh size and flow velocity, which are selected in accordance with the type of application, cause a build-up of a layer of waste which has an additional effect of retaining fine dust. The waste deposited on the filter screen is removed by means of a fixed, radially mounted suction nozzle which feeds a disposal station.

Category

Filtration: Air-Dust class

Cleaning: Automatic – Integrated with Fibre Separator or Waste Separator

Capacity: Up to 150,000m^3/h

Advantages: claimed are low cost, individual adaptation to extremely varying operational conditions, small space requirement and easy integration, low power consumption, multiple filter classifications available, increasing separation efficiency with simultaneous reduction of the load of the final filtration stage, simple to maintain and operate, semi or fully automated waste disposal.

Technical Data – Dimensions and Installation

Width	1,520 / 2,128 mm
Height	1,520 / 2,128 mm
Length	912mm
Rated Air Flow	12,000–75,000M^3/Hr
Nominal Pressure Loss	200–400Pa
Drive Power Consumption	0.17 kW
Stripper Fan Power Consumption	4–6kW

2.4.4 *Rotary Water Filter – RWF*

In order to protect the pump and the damping equipments, fibres and dust particles are separated from the water circuit by the filter fabric. The flow through the filter fabric leaves the dirt on the fabric so that it is transported by the rotary motion of the filter drum to the top. At the correct position, the dirt is washed out from inside and flushed to the sludge collecting channel.

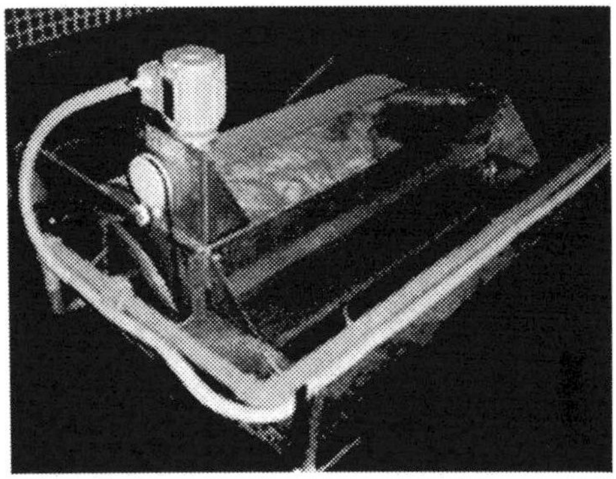

Category:

Filter: Water – Sludge

Cleaning: Automatic

Capacity; Up to 120 m^3/h

Automatic Water Filter WDF

Stainless steel rotating cylinder with nylon mesh filter, spray backwash and motor drive.

Low power consumption

Reliable operation

Easy maintenance

2.5 Excel Airtechnique, India

2.5.1 *Rotary Air Filter*

A fully automatic rotary filter is preferred when a fine filtration of the micro-dust particles and recirculation of the department air coming out of an exhaust plant is required. Normal filters work at very low filtration velocity. In rotary air filter developed by Excel Airtechnique, a special filter cloth is used to achieve fine filtration. With automatic waste cleaning nozzle arrangement over the filter cloth with a dust collector, it is possible to achieve a steady exhaust rate from the department without fluctuation in RH%, development of hot spots and infiltration of air from one department to another department.

2.6 Manstock Engineering, Pakistan

2.6.1 Rotary Air Filter

The Manstock Rotary Air Filter is designed to remove fibres generated through processing of natural and man-made materials. The filter is centrally placed for receiving fibre/waste laden contaminated air, the fibre/waste is deposited on a rotating drum, and is continuously removed from the drum surface via oscillating suction nozzles. The clean air passes through the filter, and is ready for re-circulation into the department or to be exhausted into the atmosphere.

Manstock's Rotary Air Filter provides long and trouble-free operation in industrial conditions. It has a solid and rigid construction, and many special design features are included. The metal cage construction offers higher free area. The drum surface is covered with high quality filter medium. For different applications, different types of filter media are used. The filter medium is removable and washable, and it can be easily re-mounted. The filters are shipped in knocked down condition and allow easy installation at mill sites.

2.6.1 Rotary Water Filter

Manufactured using stainless steel sheet and stainless steel mesh, the Manstock Rotary Water Filter has a sturdy construction, is rust-resistant, and is most suitable for long and trouble free operation in industrial conditions. Contaminated water with fibres/waste passes through the wire mesh of the Rotary Water Filter. The fibres adhere to the surface and are continuously removed from the filter and collected in a tray. The clean water is continuously returned to the water reservoir for re-use. The fractional horsepower motor with gear drive ensures uninterrupted operation at low power consumption.

2.7 Dev Engineers, Ahmedabad, India

2.7.1 *Rotary Water Filter*

Water is filtered continuously in this device, unlike others where the water is returned to a central location. The products are manufactured using high-quality raw material that is procured from authentic vendors of the market. It has the features of technical strength, capability, high performance, easy installation, low maintenance and corrosion resistance.

2.8 Manvi Textile Air Engineers Pvt Ltd, Mumbai, India

2.8.1 *Rotary Air Filter*

Rotary air filter shall have rotating drum made of steel grid mesh with suitable mounting frames, driven by a geared motor at low rpm. It is designed to provide large filtration area. It is handling high volume of air laden with fibre and dust returning along with return air from the department. Drum shall be covered with a suitable filter medium as per the application, pressure drop and dust level. The fluff, fibre and dust shall be arrested on the surface of the filter media covering the rotating drum surface, and it will be sucked by suction nozzle arrangement with to and fro motion connected to flexible suction hose. This arrangement shall continuously clean the fluff and dust arrested on filter media and collect in separate collection bags. Suction fan with collection unit shall be house in separate room adjoining the filter room.

Low pressure drop is maintained due to high pressure cyclone fan, which is designed to work effectively to remove not only the fluff collected on the surface of filter media but also the dust arrested in pores of filter media. This lead to uniform flow of return air at constant high pressure to remove continuously fluff and dust from department so as to maintain clean department.

2.9 Jiangsu Jingya Environment Technology Co. Ltd. China

2.9.1 *JYWIII Type Extrinsic Type Filter*

It is mainly used for the dust air filtration in the textile industry and return air filtration of air conditioner system. It can filter and collect the tiny and

miscellaneous dust from air. Each section has only one suction nozzle with difficult stoppage and less installed power of dust collecting ventilator. It is driven by reciprocating mechanisms of link chain by suction nozzle, and hence very low fault rate. The draught fan pressure difference control the operation in intervals. The equipment consumes less energy during operation.

- Filter area: 10 m^2 to 48 m^2
- Disposed air volume: Dedusting system 8,500 m^3/h to 115,000 m^3/h
- Return air filtration: 60,000m^3/h to 280,000 m^3/h
- Installed power: 0.55 kw

2.9.2 JYS-II Series Internal Flushing Water Filter

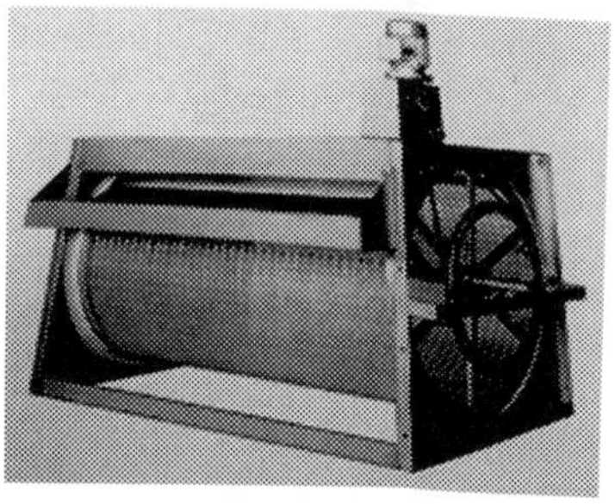

It is used for the filtration of spraying water and chilled water in air conditioning system in cotton, hemp, fur, chemical fibre, weaving and other industries. As it has excellent purifying and filtration capacity it also used for wastewater containing miscellaneous dust and suspending particles. With fashion structure, reliable transmission and low fault rate, it has reliable operation. The cone-shaped plastic rubber sealing ring is applied to seal the filtration cage and framework, therefore, the sealing effect is good. The major materials are manufactured by applying stainless steel, nylon and other anti-corrosive materials. Hence the whole machine has excellent anti-corrosive performance.

The dirty stuff on the wire mesh will be automatically cleaned under the dual function between brush rollers and spraying device, therefore, the filtration effect is good. Filtered water volume ranges between 100 t/h and 400 t/h. Installed power ranges from 0.25 kW to 0.37 kW.

2.10 Airmaaster Technolozies India (P) Ltd, Coimbatore, India

2.10.1 Blow Room & Cards Miracles with Air

Consists of Primary filter, Secondary Filter, Transport Fans, Compactors / Waste Collection Bags and Main Centrifugal Fan

Primary Filter

Primary filter has a disc with flange either stationary or rotary covered with a coarse filter mesh of nylon or stainless steel to separate heavy waste particles & fibres from the waste generated by spinning machines sucked / thrown by the Main Centrifugal fan. The filtered waste is continuously sucked by a nozzle which is either stationary or rotating.

Secondary Filter

The types of Secondary filters are: Rotary Drum Filter (RDF), Stationary Drum Filter (SDF) and Multi-drum Filter (MDF)

Rotary Drum Filter

Rotary drum filter has rotating drum fabricated of punched sheets and the punched sheet is covered by filter media to extract the fine dust which has escaped through the primary filter. The dust deposited on the filter media is collected by a set of traversing nozzles.

Stationary Drum Filter

The stationary drum filter has round flange on suitable frame work and can be bolted directly to a wall opening / sheet metal enclosure. The incoming air flows from the inside to the outside through the filter media. This means that the air inside the filter room is clean. A set of rotating and traversing suction nozzles on the inside of the filter media drum continuously sucks the dust and short fibres from the filter media.

Multi Drum Filter

The multi drum filter has a set of drums arranged concentrically in suitable frame work. The filler media is fixed between the adjacent drums and so on, so that the filter media takes a cylindrical shape. A set of rotating and traversing suction nozzles clean the filter media continuously by sucking the dust and short fibres from the filter media.

Transport Fans

The transport fans are high speed centrifugal fans with straight bladed impellers to collect the waste deposited on primary and secondary filter media surfaces. The size of the transport fan depends on the filtration surface area.

Compactors / Waste Collection Bags

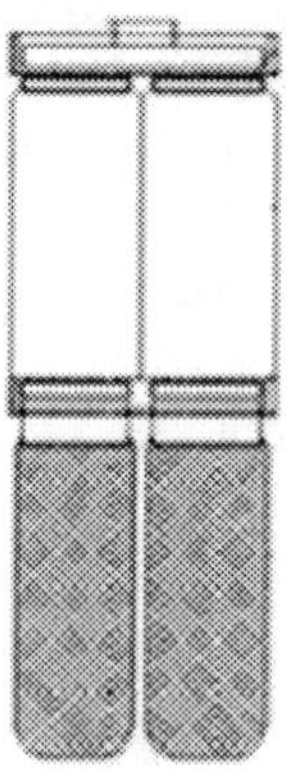

The compactors have cyclone formation unit to form a cyclone and extract the waste sucked by the transport fans and deliver the same to delivery chamber where a rotating screw is fitted so that the waste is delivered in a compact form for ease of handling and transportation. In case of synthetics and blends if the customer intends to collect the waste in loose form for reuse waste collection bags are used in place of compactors.

Main Centrifugal Fan

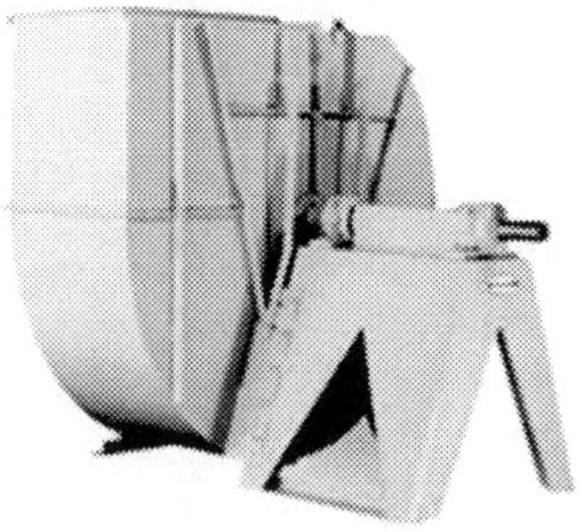

The main centrifugal fans are made of backward curved blades for energy efficiency, with single leaf / aero-foil leaves depending upon the quantity of air to be handled. The fans are provided with single inlet in case of waste collection system with sheet metal enclosure and double inlet in case of waste collection system with civil construction type.

2.11 Coimbatore Air Control Systems Private Limited, Coimbatore, India

2.11a *Automatic Rotary Drum Filter*

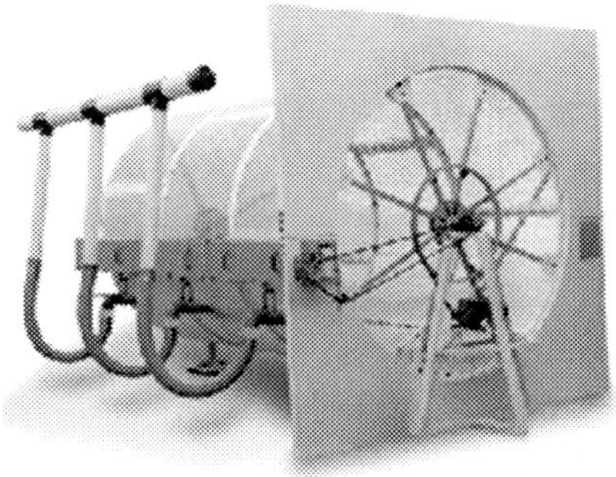

A clean environment contributes to better performance and higher productivity, especially in textile mills. Heavy dust and fibre fly can have a direct impact on the quality of the yarn. Coimbatore Air Control Systems, in answer to this need has introduced Rotary Drum Filter with high filtering efficiency and low maintenance.

This filter is made of perforated MS sheets with study MS supporting arrangements. The drive for the drum filter is through 0.5 HP gear motor, and the suction nozzle is connected to 5 HP high pressure suction fan. The dust removed from the filter is collected in the cyclone compactor / collection bag.

The linear movement of the nozzle synchronised with the rotation of the drum of uniform cleaning and the nozzle is specially designed for optimum suction pressure.

The synthetic type of filter media is fixed on the surface of the rotary drum to clean the cotton fibre / dust from the department exhaust trench with high filtering efficiency and less pressure drop

3 Dust Collectors

3.1 Luwa Air Engineers AG. Switzerland

3.1.1 Dust Separator & Screw Compactor

Dust Separator – Luwa Cyclone: Space-saving central final filtering for all types of waste coming from air filtering plants. Using the vortex and gravity forces a cyclone separates heavy particles from the air flow without a filter media. The air is entering tangentially in the cylindrical housing and flows in a helical pattern. Heavier particle will not be able to rotate at the same velocity as the air and are striking at the outside wall. Once separated at the wall, the waste material falls down wherein usually a screw compactor is present to further make it more compact for disposal. The separating of the dust is made according to the cyclone principle with downstream fine filtration of the conveying air. The thoroughly filtered air can be blown into the room or led back into a pre-filter. The filter cartridges charged with dust are automatically blown off by the impact of the compressed air blast at timed intervals.

- Category: Air-Dust Class
- Cleaning: Manual changing of bags – can be integrated with a Luwa briquette press.
- Capacity: 500 to 10,000 m^3/h

Screw Compactor: The Screw Compactor can be applied downstream of any dust-and fibre-separating group, thus conveying the waste automatically to a disposal station, in a continuous and space-saving way. Arriving waste is seized by the warm shaft of the Screw Compactor and pushed through a tube. The compacting is controlled by a pneumatic cylinder which restricts the tube exit by clamping jaws.

Luwa cyclone dust separator

Screw Compactor

Advantages

Dust Separator

- Automatic filter cleaning.
- High filtering efficiency
- Low space requirement, room height <3 m.
- Free standing
- Low consumption of compressed air, since duration and intervals of the air blasts are adjustable

Screw Compactor

- Automatic discharge of waste
- Smaller waste volume
- No dust emission, because the dust is contained in the waste
- No necessity to replace bags
- Conveyance in ascending slope up to 30° is possible

3.1.2 Fibre Separator

FSB & FSC

The Luwa Fibre Separators FSB and FSC are designed for filtering the waste out of conveying air flow and to discharge it in the room, in a compact form and without dust emission. This is a combination of a screen condenser and a screw compactor and used to separate textile waste, such as fibres of fibres mixed with shell particles or dust from a conveying air flow. The waste material collected is accumulated at the perforated metal conus, from where it is conveyed downwards by a worm shaft. Thanks to the conical form of perforated metal conus and to the resistance of the spring plate diaphragm, the material gets condensed. The fibres separated are disposed in easily compactable forms in bags, waste or compacting chambers.

Features: Smaller waste quantities can be pressed directly into a waste bag / bin, for disposal. Larger quantities can be discharged into a silo or into the filling hopper of a baling press.

Category

Filtration: Air – Fibres

Cleaning: Automatic discharge to a silo – Can be integrated with the Luwa baling press

Capacity: 2,000 to 8,000 m3/h

Advantages

- Disposal without dust emission.
- Integrated control permits autonomous operation

- The reduced construction height allows installation in rooms with low head room
- Continuous or pressure-controlled operation
- Intermittent feed is possible

Technical Data

Type	FSB	FSC
Air flow rate (m^3/hr)	1,200–4,000	4,000–8,000
Waste throughput (kg/hr)	max. 150	max. 250
Rated output of motor (kW)	0.3	0.37
Compacting factor	approx. 2:1, depending on type of material	

3.1.3 Dust Collector – DC

Dust collectors are the end stage filters which are collecting the concentrated dust waste removed from an automatic filter. The air which is conveyed by means of a conveying fan enters the dust box on the DC and escapes through the vertical filter hoses. The waste carried by the air is held back on the inner surface of the filter hoses or drops down into the dust bags as the air velocity drops.

Category

- Filtration: Air-Dust class
- Cleaning: Manual changing of bags – can be integrated with a Luwa briquette press
- Capacity: 500 to 2000 m^3/h

3.1.4 Universal Waste Separator

In Luwa universal waste/air separator for feeding presses and recycling machines, waste can be discharged in a compressed form, by means of a downstream screw compactor. For feeding recycling machines, the waste is continuously sucked off. For feeding processes or silos, the waste separator is installed directly above the filling hopper.

The separation of the waste is achieved by rotating screen drum. The waste is deposited on the screen drum, and a counter rotating bucket wheel strips the accumulated waste off from the screen, conveying it downwards at the same time. The differential pressure control allows constant operational pressure and deposit thickness on the screen, thus ensuring also the separation of dust.

Category

- Filtration: Air-Fibres
- Cleaning; Automatic discharge to a silo – can be integrated with the Luwa Baling Press
- Capacity: 4,000 to 12,000 m^3/h
- Advantages are reliable cleaning of the screen drum by the driven bucket wheel and dust separation through the fibrous deposit. Optimally harmonised with other Luwa components for disposal of waste.

Technical Data

Type	**WSA**	**WSB**
Air flow rate (m^3/Hr)	3,000–8,000	7,000–16,000
Negative Pressure (Pa)	4,000	4,000
Waste throughput (kg/Hr)	450	900
Rated output of motor (kW)	0.25	0.37
Operational Differential Pressure* (Pa)	500–1400	500–1800

*according to type of waste

3.1.5 Waste Separator WS

The separation of waste is achieved by a rotation screen drum. A bucket wheel removes the accumulated waste off the screen, so it is easy to maintain. The differential pressure control allows for constant operational pressure and the deposit thickness on the screen.

Airflow rate 3,000 to 16,000 m^3/h

Waste throughput 50 to 4,000 kg/h

Maximum negative pressure 4,000 Pa

3.1.6 Fibre Separator FS

This is a waste separator for mainly cotton. It reduces the waste volume. The disposal is without dust emission.

Airflow rate 1,200 to 8,000 m^3/h

Waste throughput Max 250 kg/h

Compacting factor Approx.2 (depends on type of waste)

3.1.7 Dust Separator DS

This is a space saving, central and final filtering stage for all types of dust coming from air filtering plants. The filtering efficiency is very high and due to automatic filter cleaning the maintenance is very easy.

Airflow rate maximum 1,800 m^3/h

Waste throughput max 30 kKg/h

Dust contamination in exhaust air <0.2 mg/m^3

A screw compressor can be used for the economic compacting of the separated dust.

Compacting factor approx. 2–3 (depends on the type of waste)

3.1.8 Exhaust Cleaning

In a first step, the exhaust is getting washed. Secondly the air passes an ionisation unit where particles get electrically charged. Subsequently water is injected and the charging pollutants cling to the water droplets which are separated at the mist eliminator. In the last stage air is passing through high voltage precipitator, where the smallest dust particles are collected at the discharge plates.

Category

Filtration: Exhaust air – oil, gasses

Cleaning: Automatic

Capacity: 5,000 to 10,000 m^3/h

3.2 Jet Tools North America La Vergne USA

3.2.1 Vortex Dust Collector

DC-1100VX-CK Dust Collector, 1.5HP 1PH 115/230V, 2-Micron Canister Kit

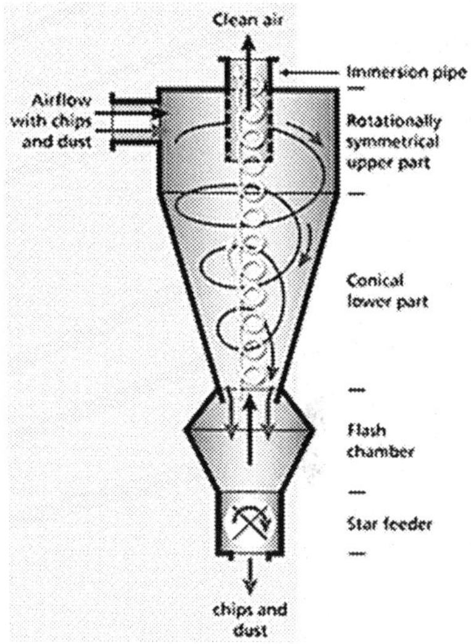

The JET DC-1100VX features the exclusive Vortex Cone which improves chip separation to prevent clogging of the filter, and increases packing efficiency of the collector bag. The high air velocity (CFM) stands up to any competitive specifications, and the industrial controls are designed for years of durable use. Collection bags have a snap ring for fast installation and removal. Casters offer maneuverability within the shop.

- Vortex Cone improves chip separation and collector bag packing efficiency
- Eliminates premature filter clogging for sustained performance
- Single-stage design for economical and quiet operation
- Industrial controls designed for years of trouble-free use
- Quick-connect collection bags with elastic band for fast, easy installation and removal

3.3 Spin Air Systems Coimbatore Pvt Ltd, India

3.3.1 Automatic Waste Evacuation Systems

The compactor in this unit is vertically operable within the plane of the wall for trash compaction purposes. Beneath the compactor, a trash receiving container is provided, which is horizontally slide able from a loading position.

Automatic Waste Removal System along with Fibre Separator Compactors are available

3.4 Zakład Urządzeń Techniki Powietrza Sp. z o.o. Poland

3.4.1 Aerotech Dust Collectors

Cyclones: The cyclone has tangential inlet which gives the gas flow rotating moving whereby the dust particles are driven towards the envelop surface where they are screwed down. Aero Tech cyclones are supplied in MS sheet / Stainless steel/MS with FRP lined construction. They are available in DECA and DECH type depending upon the collection efficiency

Aerotech Wet scrubber: Aerotech – Dews wet collector is a low pressure, high efficiency collector of coarse dust which settles in water rapidly. The efficiency is as high as 95% for particle size up to 5 micron. It is without water circulating pump.

Bag filters: Aero tech bag filter contains a number of hose made of textile material. The dust laden air/gas is passed through the walls of the hoses from inside to outside or outside to inside. The dust is deposited as a layer on the inside/outside of the hoses. The dust is dislodged to the dust hopper by shaking or reversing high pressure jet of air.

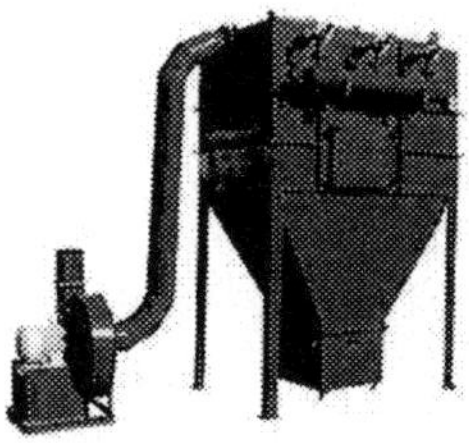

3.5 Sai Aqua fresh new Delhi, India

3.5.1 Dust Master Filters

Dust master filters supplied by Sai Aquafresh are one among the number of air pollution control devices supplied. It catches even the minute dust particles and ensures clean air.

3.6 China Hangzhou Huahong Machinery Co Ltd, China

3.6.1 Regenerated Cotton Short Fibre Unit MTR 288

The unit is specially designed to handle short staple cotton and dust chimney velvet. It does not damage fibre and its daily production capacity is around 3,000 kg for 8 hour shift. It contains a large capacity blower treating more than 400 kg waste material and producing 110–150 kg velvet per hour.

Large capacity blower

Dust Chimney

3.6.1 Second-Class Dust Collector Model 4-600

This device can separate tiny impurity from air discharged by blower. It is an ideal dust collector in embossing, stripping, oil-press and recovering for cotton-processing. Main features are little resistance, low material consumption and small area. This device accords with national standard of China in dust density.

Technical parameters

1. Size 800 mm dia and 3,200 mm height
2. Handling capacity 2,950 m^3/H
3. Dust density lesser than 120 mg/m^3

3.7 Jiangsu Jingya Environment Technology Co. Ltd. China

3.7.1 JYFO type Honeycomb Type Dust Collecting Unit

JYFO honeycomb type dust removal unit is a high efficient and energy saving dedusting equipment with 24 patents. It is widely applied in air conditioner dedusting system in cotton, fur, hemp, chemical fibre, paper making and other textile industries. This can filter and collect fibres and dusts from air.

Features

- High dust removal efficiency. The dust concentration in the air is less than 0.85 mg/m^3 after filtration.
- Small dedusting resistance. The total filtration resistance is less than 250 Pa.
- Small installed power and low energy consumption.
- Compact structure and small floor area.
- High degree of automation: The overall machine is controlled by PCL programme. It can be equipped with touch-screen operation interface, remote monitoring device and infra-red fire alarm apparatus.

- The machine is completely sealed and the machine room environment is clean.
- It can adopt the fibre dust treatment with different features.
- The filter bag can be replaced easily. It is installed in the structure of frameless cylindrical cage which is convenient to operate.
- Long service life of filter bag. The dust collection mode is advanced with less friction between suction port and filter bag, and difficult hardening filter material.

3.8 AESA Air Engineering Private Limited – France, India, Singapore and China

3.8.1 Textile Dust and Waste Removal

Pre-filter Type TVM

Volume: 15,000–130,000 m^3/h – Filtering resistance: 500 Pa max

3.8.a.2 Fibre Compactor

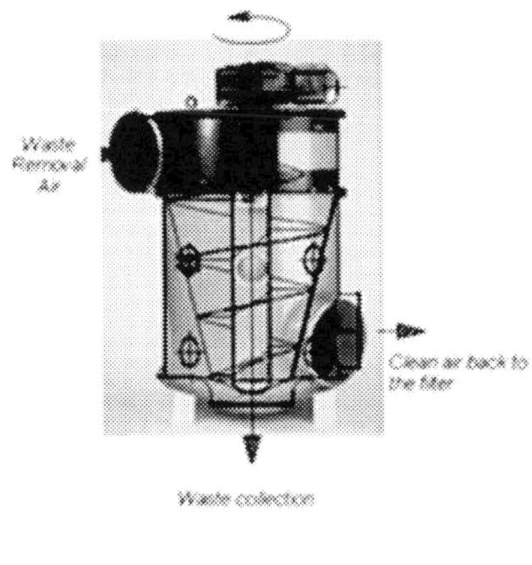

Type FKA

Airflow: up to max 8,000 m^3/h

Separated waste volume: up to 400 kg/h

Type FKC

Airflow: up to max 3,500 m^3/h

Separated waste volume: up to 200 kg/h

3.8.2 Drum Filter type TFB

Waste collector

1–5 sections

Air volume: 10,000–200,000 m^3/h

Air Pressure drop: 200–500 Pa

Room Air

Air volume: 46,000 –270,000 m3/h

Filter medium: AN 300

Cyclone – Screw Compactor for Textiles

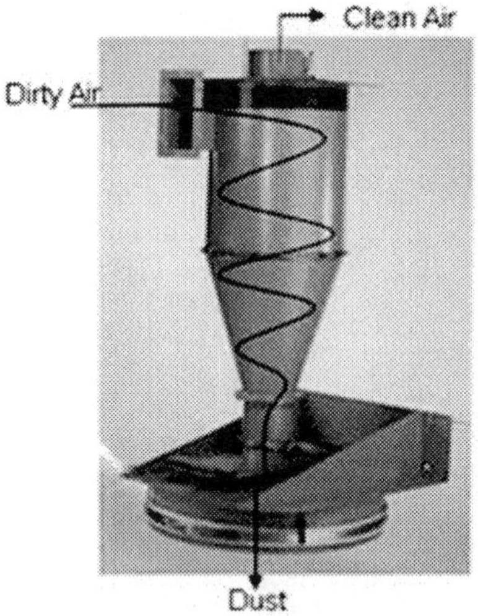

Cyclone Type ZSA
(For fine dust)

Air volume: 300–1,000 m^3/h

Dust Quantity: up to 40 kg/h

Power Screw type CPS 200

(For Central Waste collection)

Maximum volumetric capacity: 0.2 m3/h

3.8.3 Central Dust Systems for Textiles

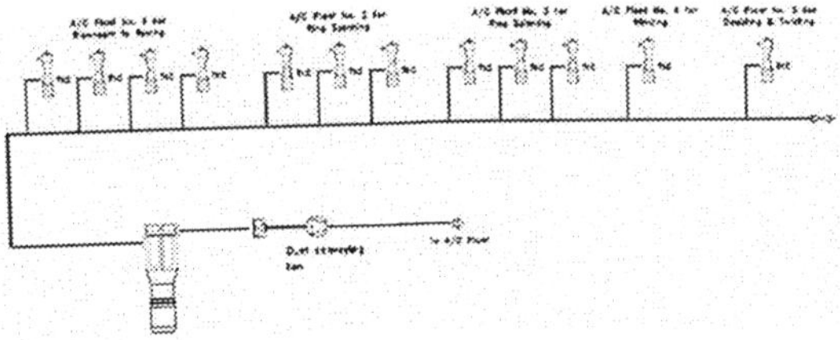

Pneumatic Damper type TKC 200

2 pneumatic cylinder with 2/5-way solenoid valve

A flap, retracted in open position avoid fibre accumulation

High efficiency dust separator type Jetfilter

Bale Press Systems

Horizontal Presses for continuous and automatic pressing of waste and fibres

Silos with Discharge Device AVA for storage of material

Automatic separation into different qualities for different purposes

Pressure 15–50 tons

3.8.4 Briquetting Machine

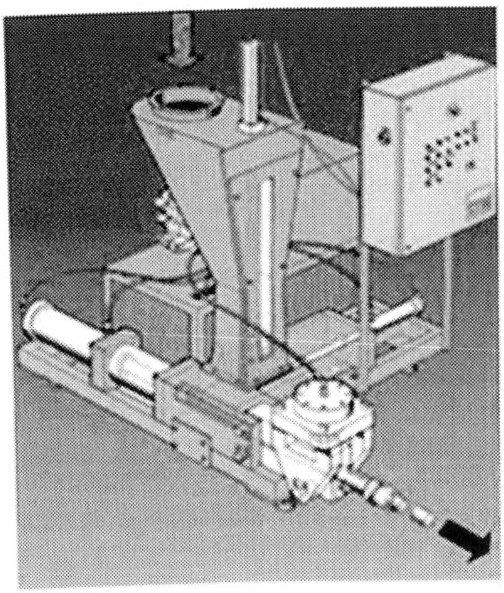

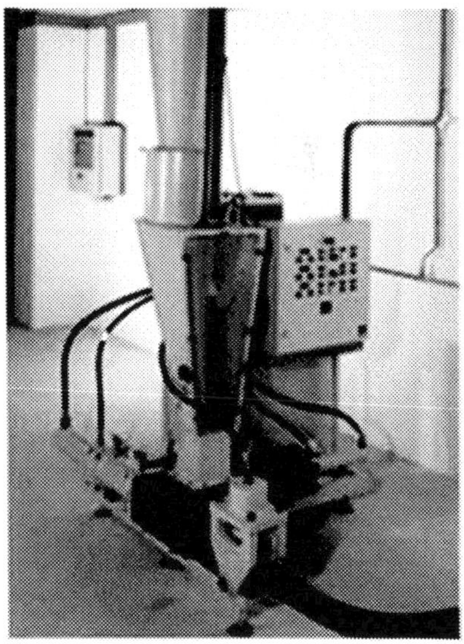

For the production of briquettes from dust and short fibres, Hydraulic drive and automatic control – Volume: up to 150 kg/h

3.8.5 Central Vacuum Cleaning System for Textile Mills

Vac 'airs waste collector

External body in heavy steel, pointed with access door, designed to resist at extreme high suction pressure

security filtration cartridges in bottom

3.9 Coimbatore Air Control Systems, Coimbatore, India

3.9.1 Self-Cleaning Disc Filter

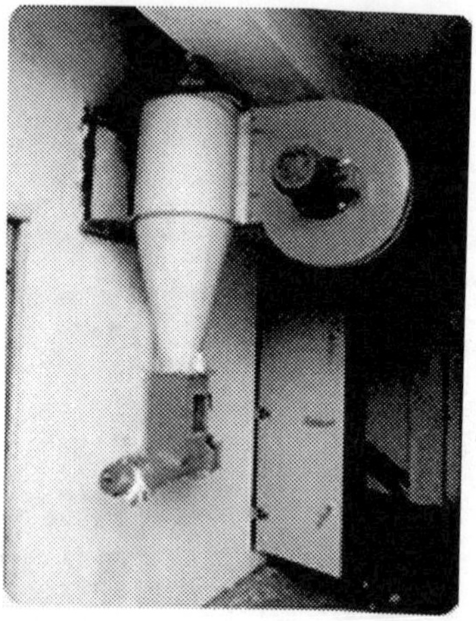

This filter is made of MS square tube with stainless steel filter media with sturdy MS supporting arrangements. The drive for the suction nozzles is through 0.5 HP gear motor and the suction nozzle is connected to 3 HP high pressure suction fan. The dust removed from the filter is collected in the cyclone compactor / collection bag.

4 Travelling Cleaners

4.1 Luwa Air Engineers AG Switzerland

4.1.1 Travelling Cleaner SpinTravClean (STC) for spinning mills

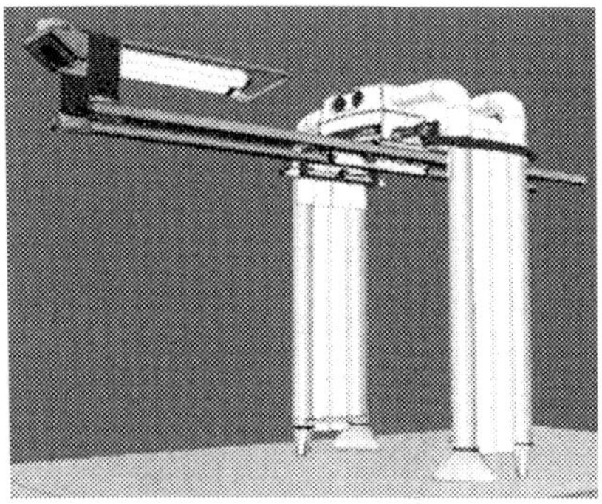

The SpinTravClean (STC) combines innovative technology and heavy duty design. Providing clean machine working zones, the STC improves the product quality, the working conditions and the overall efficiency due to less broken ends and reduced number of faults. Applications are primarily in spinning, winding and twisting.

Different drive propulsion systems are available for the respective application conditions:

- Belt drive – the choice for smaller and standard applications, simple and reliable.
- Direct drive – the choice for longer tracks and sophisticated tasks. These systems come with PLC controller and frequency inverter; it gets the power from conductors or on request via conductor chain.
- Combining best-in-class engineering design with modular architecture, application knowledge, and manufacturing excellence, Luwa supplies travelling cleaners that maximise productivity at competitive prices.

Performance features

- Sturdy construction with cantilevers.
- Mechanical auto-reverse when hoses touch obstacle.
- Automatic waste collection system.
- Stable thanks to twin-rail track.
- Easily customizable configuration.
- Advantages are proven reliability, lowest and easy maintenance and removable filter for easy cleaning.

4.1.1 LTC (LoomTravClean)

Luwa Loom TravClean (LTC) is powerful travelling cleaner for today's weaving applications. The LTC ensures a clean environment, improves the working conditions and the overall efficiency due to less downtime of your

production machines, less broken ends and a reduced number of faults. Applications are primarily in weaving.

The LTC comes with unique features as oscillating nozzles, separated air circuits and free programmable controls (PLC). This ensures highest cleaning efficiency and a significant quality improvement to your fabric - for any application from silk to jute.

Quick, accurate and reliable mounting is provided by an efficient design. Flexibility and connectivity is implemented into every product to enable not only fast installation but also rapid on-site customisation for specific applications.

The revolutionary oscillating nozzles efficiently clean large loom areas and can cover multiple loom types in a row. Safety aspects are covered through a collision avoidance system with a customizable reaction, thus also avoiding wear and tear. Special rails for a wide track ensure safe and smooth travelling, allowing huge spans up to 8 m without need of additional supports. For process zone air-conditioning, the Luwa LoomSphere is foreseen to be integrated into the track.

Performance features

- Twin-rail track for unique stability for straight track or round track
- Sturdy construction with cantilevers
- Oscillating nozzles for multiple direction cleaning
- Collision avoidance sensors
- Automatic waste discharge and disposal
- Design and selected materials ensure highest productivity and reliability with lowest cost of operation (e.g. sleek construction for avoiding fibre clogging). It is stable due to use of twin-rail track.

4.2 Industrial Air Solutions, Inc. Raleigh, NC 27675 USA

4.2.1 Power Vac Electric for indoor debris removal

The Electric Power Vac commercial vacuum sweeper manufactured by IAS Industrial vacuum is suited for quick cleanup of litter and debris found on floor. This can be used on variety of surfaces such as concrete, tile, carpet and more. It can sweep 40,000 sq ft per hour. Maintenance is easy as it has no moving parts. Its unique engineering provides for the collection of dissimilar litter. It allows objects outside the width of the machine also to be vacuumed. The quick-tilt feature offers a Snap-On Dolly assembly to allow the introduction of large objects into the air stream. The 11 cubic feet capacity polyester felt bag has a zippered screened vent for superior dust control.

4.3 Steinemann AG | Wilerstrasse, Switzerland

4.3.1 Steinemann Vacuum System

It is a centralised vacuum system. Powerful vacuum accelerates manual cleaning of looms. This system increases production efficiency, relieves the workers. The waste is collected and a centralised place and can be further processed.

- Production efficiency increases by 1–2%
- Volume capacity blower 1,140–2,600 m^3/h
- Max. set vacuum 280–350 mbar
- Simultaneously used hoses 1–4
- Average power consumption 6.5–22 kW

4.4 Jiangsu Jingya Environment Technology Co Ltd, China

4.4.1 JYQJ Overhead Travelling Cleaner

JYQJ travelling cleaner is used in ring spinning of spun yarns, roving, spooling, assembly winding, weaving and other places in textile mills. It enhances yarn and fabric quality and reduces labour force. The patented flexible air regulation technology helps the clearer to realise the optimal application effects. The high-performance draught fan has the high efficiency and low energy consumption. The excellently processed railing is equipped with safety devices to avoid derailment. High-quality tangential belt and rational driving technology results in long life of tangential belts. It is available with multiple specifications depending on the end-user requirements. It has less maintenance and hence more economical.

4.4.2 JYQJ-BJ Intelligent Loom Overhead Travelling Cleaner

Features

- The flexible straight line, turning-circle and other types of railway configuration can satisfy and adapt different demands.
- Patented technology applied with features of single-side aspirant, down blowing and ash discharge cleaning.
- Catheter-type (closed) side wires are applied for power supply with features of safety and reliability.
- With the humanised design, photo-electricity induction, intelligent suspension or returning, it can avoid the interference in people and object during operation.

- The blowing and suction function configuration by high-efficient and diversified multi-tubes can enlarge the effective cleaning range and make the cleaning result more rational.
- The automatic ash discharge device or centralised collection device after overboard ash discharge can be optional.
- It is simple to maintain.

5. Ionizers

5.1 Wein Products Inc. Los Angelis, USA

5.1.1 High Density Negative Ioniser for Air Supply VI 2500

The Wein high density negative ionizer takes where nature stops by electronically generating a powerful stream of negative ions. The spiral emitter on the VI-2500 negative ionizer produces 70 trillion negative ions per second in intermittent pulses which ‘wash’ the air at a rate of nearly 100 feet per minute. This removes particles as small as .01 Microns. It emits no detectable ozone so don’t expect the ozone smell that ozone generators create.

- Ion output: 450 trillion ions/sec cm^2 at 2.5 cm
- Ion Monitor: Nneon discharge pulse monitor
- Coverage: 800 sq. ft. (40’ x 20’) room
- Shown to help treat Seasonal Affective Disorder Depression in some people.
- Power use: Less than 6 watts.
- Particle effectiveness: As small as .01 micron
- Purified airflow output: Greater than 100 feet/minute
- Ozone output: No detectable ozone (well below OSHA standard)
- Negative Ionizer Approvals: Listed UL adapter Measured with Miniram monitor type PDM-2 (Serial No.2028). GCA Corp. Environmental Instruments Technology Division. Measured with Dwyer Air Flow Vanometer, Model 50.

Negative Ions are molecules that have gained an extra electron. These are beneficial to the environment because they help attract dust, cigarette smoke, pet dander, pollen, mold spores, viruses and bacteria; thus, helping eliminate them from the air you breathe. Have you ever noticed how fresh air smells after a rainfall or near a waterfall? If so, then you have experienced ionization in its grandest form. In addition, negative ionization has been shown to elevate mood and reduce stress. Scientists at Columbia University showed that negative ions can also combat depression and Seasonal Affective Disorder (SAD).

References and further reading

1. Technifax – Air washer plants in Textile Industry.
2. S. G. Wakade – Psychrometrics – 31st All India Textile Conference Sholapur 1974.
3. Wikipedia Encyclopedia.
4. Literature supplied by Luwa India pvt Ltd, Bangalore.
5. Literature supplied by Manvi Textile Air Engineers, Mumbai.
6. Activated carbon add nets plate air filter – YAN ZHANG Shandong Aobo Environmental Protection Technology Co., Ltd, Qingdao, China.
7. Activated carbon v-bank filter with high air flow – Eddie Wu Kunshan Laidel Filtration Equipment Co., Ltd. Shanghai. China.
8. Dust Separator & Screw Compactor – http://www.luwa.co.uk/pf/dust%20sep.html
9. Fibre Separator – http://www.luwa.co.uk/pf/fibre%20sep.html
10. Haijiang integrated auto membrane filter press/energy saving filter press – http://haijiang-filterpress.en.alibaba.com
11. Fine Air Systems Ahmedabad – http://www.fineairsystems.com/rotary.html
12. Rotary Water Filter by Man stock Engineering – http://www.manstock-engg.com/air_handing.html
13. RPF Rotary Pre Filter – www.luwa.co.uk/pf/rpf.htm
14. Vortex Dust Collector – http://www.jettools.com/us/en/view-series/vortex-cone/Vortex
15. Waste Separator – http://www.luwa.co.uk/pf/waste%20sep.html
16. Eurovent 4/21 Mann_Hummel Vokes air filter – http://www.vokesair.com/
17. Dry Ventilation – http://www.fineairsystems.com/dry.html
18. Paul A. Funk* and Kevin D. Baker, Dust Cyclone Technology – A Literature Review, Engineering and Ginning, *The Journal of Cotton Science* 17:40–51 (2013) 40, http://journal.cotton.org, © The Cotton Foundation 2013.
19. Troubleshooting – Zero Water – https://www.zerowater.com/troubleshooting.aspx
20. Samarth Air Washer Plants and Components.
21. Keystone Air systems.
22. Dri-Steem – Outdoor Enclosures for Humidifiers Archive news Story – Aug 22, 2003.

23. Joan Crowe – Proper design of vapor retarders requires close attention to temperature – AJA.
24. Brade Engineering – Hot water condensate units.
25. Cotton Mill Dust may be carrier as well as culprit in respiratory ills – News from Harward Medical, Dental and Public Health Schools, March 21, 2003.
26. CQ Gaobiao Tech Ltd – Dehumidifiers.
27. Draft-Air India Ltd – Creating the right atmosphere always.
28. Carl Ian Graham PE – High performance HVAC – Whole building Design guide.
29. Motion Tracking – Event Sentry.
30. Engineering tool Box.
31. The Old farmer's Almanac Webcam – Yankee Publishing Inc.
32. Air Engineering and Refrigeration History – national Academy of Engineering, USA.
33. Air Conditioning: Improving the way of life – Consumer information – Koldfax News.
34. Yahoo Answer.com.
35. WordNet.com.
36. Beverly T. Lynds – About Temperature.
37. Berryhill, Robert A – Air Washer Piping – United States Patent 3965690.
38. Air Conditioning Calculations, Codes and Standards – HVAC&R.
39. American Moistening Company.
40. Hashem, Feisal – Centrifugal Humidifier with saw tooth ridged impingement surface – United States Patent 6601778.
41. Sensorland.com – Online information center for sensing and measurement.
42. Controlling Moisture by Mechanical Design in Highly Efficiency Homes – Toolbase Services July 2004.
43. Titan products – Measurement devices for control systems – England.
44. Hiross Flexible Space System – Italy.
45. Roger L Horward – Breaking tradition – A better way of Controlling Economizers – November 1, 2006.
46. Hi-tech Duravent Hoses.
47. Jargon Buster by Rapid Climate Control.
48. Loom and Room Conditioning system – US patent Nov 23, 2004.
49. Ronald Yaeger and Gary D. Wolf – Method and Apparatus for conditioning un-recycled ambient air – US Patent 25 Apr 1995.
50. Method and Apparatus for temperature and humidity control within a chamber – US Patent 4183224.

51. Arima M, Hara EH., Katzberg JD – A fuzzy logic and rough sets controller for HVAC systems – WESCANEX95, Communications, Power and Computing. Conference Proceedings. IEEE Volume 1, Issue 15–16 May 1995, Pages 133–138.

52. Jet Spray atomizing by J. S. Humidifiers.

53. G. Srikanth, Sangeetha Nangia and Atul Mittal – Energy efficient FRP Axial Flow Fans – Science Tech Entrepreneur Vol 8 No 5 Sept –Oct 2000 The Hindu.

54. Humidification – Controlling the air we breathe – The Armstrong Humidification Handbook.

55. Glenn Elert – Refrigirators – Physics Hypertext book.

56. The Design Standards – Section 15880 San Diego State University Division 15 – Mechanical.

57. B.P Ager and J.A Tickner – The control of Microbiological Hazards associated with Air conditioning and Ventilation Systems.

58. Dr H. R. Sheikh – Social Compliance in Pakistan's textile industry – Textile Institute of Pakistan.

59. Existing HVAC systems analysis by Spencer Engineers Inc.

60. Literature supplied by Excel Airtechnique.

61. William G. Suggs – Supply air grill condensation elimination method and apparatus – US Patent 19 Aug 1997.

62. Veriteq – The trouble with humidity – Hidden challenges of humidity calibration.

63. Ventilation principles (COPE) by National Research Council, Canada.

64. Keecha Trikomen President ASHRAE Thailand Chapter 2001-2001 – Humidity control for tropical climate.

65. Understanding HVAC System Design issues by the Air conditioning contractors of America (ACCA).

66. K. R. Chandran and P. Muthukumaraswamy – SITRA Energy Audit – Implementation Strategy in Textile Mills.

67. Air Duct System Design by Dynacomp. Inc.

68. Air Duct Cleaning by Air Tech Environmental System.

69. Moriguchi, Tetsumo Yamaoka, Akira – Cooling device condenser and air conditioning system – United States Patent 6718790 dated 13 Apr 2004.

70. S. Ashok – Energy Savings and Audits in Textile Industry – ashoksvs@eth.net.

71. Steve Spielmann – New Technology enhances cooling tower maintenance.

72. PLC technology and its use in highway traffic system applications – Autooonet Automated Online.

73. Ashoke Sarkar – UVC lamps can keep AHU clean – Air Conditioning and Refrigeration Journal Oct-Dec 2005.

74. Shankar Sapaliga – Duct Cleaning – Putting all the pieces together – Air Conditioning and Refrigeration Journal Oct-Dec 2005.

75. Yashwant Jhaveri – Chemical cleaning of finned coils – Air Conditioning and Refrigeration Journal Oct-Dec 2005.
76. M. El-Morsi, S.S. Klein Fellow ASHRAE and D.T. Reindl – Air Washers – A new look at a vintage technology – Air Conditioning and Refrigeration Journal Jan-Mar 2005.
77. M.M. Roy – Humidification for Textile Mills – Air Conditioning and Refrigeration Journal Jan-Mar 2005.
78. S. Sankaran – Air Conditioning for Synthetic Fibre Plants – Air Conditioning and Refrigeration Journal Jan-Mar 2005.
79. Isaac Ray PhD, Croll Reynolds – Air pollution control in the textile industry: A technology for the 21st Century – Air Pollution Control Systems, Westfield N.J.
80. Central Pollution Control Board, India – Note on Textile Industry.
81. Natural Gas Technologies by energy solutions Center.
82. Bryce Dvorak – Energy efficient Humidification.
83. Water Quality Answers by Water Quality Association, USA.
84. S.P. Patel – Humidification in Textile Mills – textile Engineering tablet II, The Textile Association India Education System.
85. How and Air Door/Air Curtain works by TMI International.
86. Clean room – Industry Design Thumb rule by Aeromech Equipments Pvt Ltd.
87. UV Air Cleaners & Ultraviolet Water Purifiers for Health by American Air and Water Inc., UV Disinfection Systems.
88. J. Edward Smith and Ralph B. Whisnant – Evaluation of a Teflon based Ultraviolet light system on the disinfection of water in a Textile Air Washer – Pollution Prevention Challenge Grant from the North Carolina department of Natural Resources and Community Development.
89. High efficiency, low power air filtration system using a charged liquid – Techwatch, Sri Lanka – Aug 2003.
90. Specifying Humidification Systems for mission critical environments – Liebert Engineering White Paper.
91. Frequency of Air changes per hour – A key consideration in selecting air purification systems – A report by Airistar Technologies LLC. Frequently asked questions on humidification – by PLASTON.mht.
92. The Armstrong Humidification handbook HB – 501.
93. A Source book for green and Sustainable building
94. B. Purushothama – Moisture relations in Textiles – Thesis submitted to Bangalore University 1973
95. B. Purushothama – A study on Moisture gain in cone winding – Journal of Textile Association – Jan 1978.
96. B. Purushothama – Fundamentals of Textile Mill management – The Textile Association India, Miraj – 1980.

97. B. Purushothama – A Practical guide to Quality Management in Spinning – Published by Woodhead Publishers India 2011.

98. B. Purushothama – Principles of Textile Testing and Statistics – The Textile Association (India) 2011.

99. B. Purushothama – Process control in Spinning – The Textile Association (India) 2011.

100. B. Purushothama – Hand book on Cotton Spinning Industry – Woodhead Publishing India – 2015.

101. Indian Patent Database Search Online.

102. Creel Humidification in Ring frames by SITRA (Patent No 174964 dated 2 Jul 1990).

103. A control system for humidification plant in textile industry by SITRA – Patent No 648/CHE/2005 dated 27 May 2005.

104. Humidifier for local control of relative humidity of textile process – ATIRA – Patent No 176935.

105. Rakesh Pramodbhai Shah, Nilesh Prafulchandra Varia and Devanand Prafulchandra Varia – Cotton Humidification and Conditioning System (Patent Application No 28/MUM/2007A dated 26 Jan 2007.

106. Geiger Stephan, Leu Karl and Subuis Robert – Air Humidifier and Evaporation Mat – International Patent WO/2007/090313 dated 16 Aug 2007, Application No PCT/CH2007/000064.

107. Helmut Stueble – Process for cooling and conditioning air – US Patent No 283026 dated 5 Mar 1996.

108. Bertoldo, Franco – An apparatus and a method for humidifying a continuous textile material – European Patent EP1428923 – 06/16/2004.

109. Ruben D Restrepo MD RRT FAARC and Brian K Walsh RRT-NPS FAARC – AARC Clinical Practice Guideline Humidification During Invasive and Noninvasive Mechanical Ventilation: 2012.

110. Olli Seppänen and Jarek Kurnitski. Moisture control and ventilation, WHO Guidelines for Indoor Air Quality: Dampness and Mould. http://www.ncbi.nlm.nih.gov/books/NBK143947/.

111. Lessons by IIT Kharagpur NPTEL online – Refrigeration and Air conditioning – Mechanical Engineering. http://nptel.ac.in/courses/Webcourse-contents/IIT%20Kharagpur/Ref%20and%20Air%20Cond/New_index1.html.

112. Jason Derrick How Ultrasonic Humidifiers Work – Principle of Operation – http://blog.stulz-ats.com/how-ultrasonic-humidifiers-work.

113. The WBDG Sustainable Committee – Enhance Indoor Environmental Quality (IEQ) – https://www.wbdg.org/design/ieq.php.

114. Martin Holladay, GBA Advisor – Vapor Retarders and Vapor Barriers – Answers to persistent questions about vapor diffusion – http://www.greenbuildingadvisor.com/blogs/dept/musings/vapor-retarders-and-vapor-barriers.

115. Don Prowler, FAIA – Donald Prowler & Associates Whole Building Design-Revised and updated by Stephanie Vierra, Assoc. AIA, LEED AP BD+C – Vierra Design & Education Services, LLC – Last updated: 03-22-2012 – https://www.wbdg.org/.

116. Watershed Protection Plan Development Guidebook E:\APPENDIX B.doc B-1 Northeast Georgia Regional Development Center Appendix B Description of Commonly Considered Water Quality Constituents. https://epd.georgia.gov/sites/epd.georgia.gov/files/related_files/site_page/devwtrplan_b.pdf

117. Hitachi Infrastructure Systems (Asia) Pte.Ltd. Principles of Electrostatic precipitator and factors affecting performance. http://www.hitachi-infra.com.sg/ services/energy/dustcollection/principle/dustcollection.html.

118. Climavent Fume Extraction Systems - http://www.climavent.co.uk/fume-extraction-systems/.

119. Vaisala Veriteq - Data Loggers - https://www.cik-solutions.com/en/catalog/monitoring-datalogger/veriteq/hardware/.

120. Multi-Wing Group – Denmark - X-large diameter fans - http://www.multi-wing.com/Products/BrowseByDiameter/Xlarge.

121. HVLS Fans: Four Important Factors to Consider Prior to Making a Purchase Decision Large-Diameter Fans Meet Manufacturer's Cooling Needs, http://www.impomag.com/article/2002/04/large-diameter-fans-meet-manufacturers-cooling-needs.

122. Wikipedia, the free encyclopedia High-volume low-speed fan - https://en.wikipedia.org/wiki/High-volume_low-speed_fan.

123. Humidification in Textile Mill - http://www.cottonyarnmarket.net/OASMTP/HUMIDIFICATION%20IN%20TEXTILE%20MILL.pdf.

124. Steve Hale, Technical & Business Development manager, Condair plc - The importance of humidity control in textile processing - http://www.condair.co.in/knowledge-hub/the-importance-of-humidity-control-in-textile-processing.

Printed in the United States
by Baker & Taylor Publisher Services